D1217848

# TREATISE ON ANALYTICAL CHEMISTRY

*A comprehensive account in three parts*

*PART I*

THEORY AND PRACTICE

*PART II*

ANALYTICAL CHEMISTRY OF
INORGANIC AND ORGANIC COMPOUNDS

*PART III*

ANALYTICAL CHEMISTRY IN INDUSTRY

# TREATISE ON ANALYTICAL CHEMISTRY

Edited by I. M. KOLTHOFF
*School of Chemistry, University of Minnesota*

and PHILIP J. ELVING
*Department of Chemistry, University of Michigan*

## PART II

## ANALYTICAL CHEMISTRY OF INORGANIC AND ORGANIC COMPOUNDS

### VOLUME 16

*Functional Groups*

INTERSCIENCE ® PUBLICATION

*JOHN WILEY & SONS, New York–Chichester–Brisbane–Toronto*

Copyright © 1980 by John Wiley & Sons, Inc.

All rights reserved. Published simultaneously in Canada.

Reproduction or translation of any part of this work
beyond that permitted by Sections 107 or 108 of the
1976 United States Copyright Act without the permission
of the copyright owner is unlawful. Requests for
permission or further information should be addressed to
the Permissions Department, John Wiley & Sons, Inc.

**Library of Congress Cataloging in Publication Data** (Revised)

Kolthoff, Isaak Maurits, 1894–
    Treatise on analytical chemistry.

    "An Interscience publication."
    Includes bibliographies and index.
    CONTENTS: pt. 1.  Theory and practice.--pt. 2. Analy-
tical chemistry of inorganic and organic compounds.
    1.  Chemistry, Analytic.  I.  Elving, Philip Juliber,
1913-    joint author.  II.  Title.
QU75.K64       543       78-1707
ISBN 0-471-05857-2

Printed in the United States of America

10 9 8 7 6 5 4 3 2 1

43
81t
t.2
/.16

# TREATISE ON ANALYTICAL CHEMISTRY

## PART II

## ANALYTICAL CHEMISTRY OF INORGANIC AND ORGANIC COMPOUNDS

## VOLUME 16

With the cooperation of EDWARD W. D. HUFF-MAN, *Huffman Laboratories, Inc.* and JOHN MITCHELL, JR., *E. I. du Pont de Nemours & Co,* as section advisors.

## AUTHOR OF VOLUME 16

R. F. MURACA

## Acknowledgement

In view of the comprehensive nature of the Treatise, the editors have felt it desirable to consult with experts in specialized fields of analytical chemistry. For the sections of Part II dealing with "Organic Analysis," they have been fortunate in securing the cooperation of Dr. Edward W. D. Huffman and of Mr. John Mitchell, Jr. The competence of the authors of the individual chapters in the sections on organic analysis, combined with the Messrs. Huffman's and Mitchell's broad knowledge of the area, has resulted in a succinct and critical treatment of elemental and functional organic analysis as well as of certain areas basic to organic analysis. The constructive help of the authors and of the Messrs. Huffman and Mitchell in the preparation of Volume 11 and the following volumes is acknowledged with gratitude.

**R. F. Muraca**
*College of Notre Dame, Belmont,
California*

# PART II. ANALYTICAL CHEMISTRY OF INORGANIC AND ORGANIC COMPOUNDS

## CONTENTS VOLUME 16

### SECTION B-2.  Organic Analysis II.  Functional Groups

**Azoxy Group.**  By *Ralph F. Muraca*

# Nitro and Nitroso Groups. By *Ralph F. Muraca*

# Nitrate and Nitrite Ester Groups.  By *Ralph F. Muraca*

## Nitrile, Isocyanide, Cyanamide, and Carbodiimide Groups. By *Ralph F. Muraca*

# TREATISE ON ANALYTICAL CHEMISTRY

*A comprehensive account in three parts*

*PART I*

THEORY AND PRACTICE

*PART II*

ANALYTICAL CHEMISTRY OF
INORGANIC AND ORGANIC COMPOUNDS

*PART III*

ANALYTICAL CHEMISTRY IN INDUSTRY

# AZOXY GROUP

By R. F. Muraca, *College of Notre Dame,*
*Belmont, California*

**Contents**

# I. INTRODUCTION

The azoxy group:

$$-N{=}N-$$
$$\downarrow$$
$$O$$

may be viewed as a nitrogen analog of the nitro group:

$$O$$
$$\uparrow$$
$$-N{=}O$$

The azoxy group is readily formed by oxidation of aryl azo compounds and thus is usually encountered with aromatic radicals attached to it. Only a few aliphatic azoxy compounds are known, but they have rapidly become of great importance; for example, the first knowledge of an $\alpha,\beta$-unsaturated azoxy compound was gained recently with the isolation of the natural products macrozamin and elaiomycin. It was also the first instance in which hydrogen was found on the carbon adjacent to the azoxy group.

The sulfur analog of the azoxy group is the azothio group:

$$S$$
$$\uparrow$$
$$-N{=}N-$$

which readily loses sulfur to form the azo group.

The structure of the azoxy group has only recently been established with certainty; in early years, the two nitrogen atoms and the oxygen atom were assumed to form a three-membered ring (Kekulé). The

oxygen atom in the azoxy group has an $N$-oxide configuration which
may be represented as

$$-N{=}N- \quad \text{or} \quad -\overset{+}{N}{=}N-$$
$$\downarrow \qquad\qquad\quad | $$
$$O \qquad\qquad\quad O^-$$

In addition to the *cis-trans* isomerism made possible by the double bond
in the azo group, the disposition of the oxygen atom on the nitrogen of
the azo group gives rise to two additional isomers. Thus the *trans* isomer
of a monosubstituted azoxybenzene can exist in two forms:

$$\underset{\text{"Alpha"}}{H_5C_6\overset{N=N}{\underset{O}{\diagup}}\diagup^{C_6H_4X}} \qquad\qquad \underset{\text{"Beta"}}{H_5C_6\overset{N=N}{\underset{O}{}}\diagup^{C_6H_4X}}$$

Both isomers can be reduced to the same azo compound.

## II. SYNTHESIS AND OCCURRENCE

There are two major synthetic routes for the formation of arylazoxy
compounds, one based on the alkaline reduction of aryl nitro compounds, and the other on the oxidation of aryl azo compounds with
strong reagents. Azoxy compounds also may be formed whenever
resistive nitrogen functional group compounds are oxidized; for example, $p$-nitroazoxybenzene is formed by the action of monopersulfuric
acid on $p$-nitroaniline (17). Sensitive compounds such as 4-dimethyl-
aminoazobenzene, however, form an azoxyamino-oxide (97).

Symmetrical aromatic azoxy compounds are usually prepared by
refluxing aromatic nitro compounds with methanol and sodium or
potassium hydroxides (129):

$$4\,C_6H_5NO_2 + 3\,CH_3OH + 3\,NaOH \rightarrow 2\,C_6H_5{-}N{\overset{\overset{\textstyle O}{\uparrow}}{=}}N{-}C_6H_5 + 3\,HCOONa + 6\,H_2O$$

If the reduction takes place at low temperatures, aldehydes are
formed instead of carboxylic acids:

$$2\,C_6H_5NO_2 + 3\,CH_3OH \xrightarrow{\text{NaOH}} C_6H_5N{\overset{\overset{\textstyle O}{\uparrow}}{=}}N{-}C_6H_5 + 3\,CH_2O + 3\,H_2O$$

The formation of azoxybenzene actually takes place by an aldol-type
condensation of nitroso compounds and hydroxylamines formed as
intermediates in the reduction of nitrobenzene:

$$C_6H_5NO + C_6H_5NHOH \rightarrow C_6H_5 \overset{\overset{\text{O}}{\uparrow}}{-N{=}N} -C_6H_5 + H_2O$$

Unsymmetrical aryl azoxy compounds that cannot be obtained by oxidation of the corresponding azo compounds (e.g., 4-dimethyl amino-azobenzene) may be prepared by condensation of aryl hydroxylamines and aryl nitroso compounds in the presence of alkali. However, when a *p*-substituted phenylhydroxylamine is condensed with nitrosobenzene, a mixture of unsubstituted azoxybenzene and *p,p'*-disubstituted azoxybenzene is formed because the following equilibrium is established rapidly (94):

$$ArNO + Ar'NHOH \rightleftharpoons Ar'NO + ArNHOH$$

More importantly, the condensation offers a route for the formation of the alpha and beta isomers of *trans-p*-substituted phenylazoxybenzene (3); the isomers, formed in low yield, are separable by absorption on alumina.

Symmetrical azoxy compounds can also be made electrochemically; conversion of dinitro-, iodoxynitro-, and iodosonitrobenzenes to azoxy compounds can be accomplished by cathodic reduction as a paste (containing carbon for increasing conductivity) with a zinc anode in an electrolyte of magnesium bromide at a pH of about 7.0 (74). In general, electroreduction of nitro compounds in alkaline media results in the formation of azoxy compounds either as by-products or as the major products, but *o*- and *p*-nitrophenols, nitroanilines, *N*-nitroaryl compounds, *N,N*-disubstituted anilines, and certain hindered nitro compounds are exceptions. Nickel cathodes in alkaline solutions are preferred because their low hydrogen overvoltages prevent reduction of azoxy compounds. Moreover, azoxy compounds are insoluble and precipitate out of the zone of reaction; if the azoxy compound is kept in solution however, azo compounds form at high current densities and hydrazo compounds at lower densities. An excellent summary of electrolytic reductions has been prepared by Popp and Schultz (99).

The reduction of nitro compounds for conversion to azoxy compounds may also be effected by zinc dust in alcoholic sodium hydroxide (24), ferrous sulfate and sodium hydroxide (2), lithium aluminum hydride in ether (93), alkaline sodium stannite (45), glucose in alkaline medium (95,52), alkaline sodium arsenite (23), alcoholic solutions of sodium or potassium alkoxides (79), sodium 2-hydroxyethoxide in ethylene glycol (114), or phosphine (29). Many nitro compounds can be reduced to azoxy derivatives in acidic solutions, but most preparations require an alkaline medium. Azoxybenzene can be produced on an industrial scale by heating nitrobenzene with alkali and molasses in a petroleum solvent (55).

Symmetrical as well as unsymmetrical azoxy compounds can often be made by oxidation of azo compounds (see Section III. B. 2 of "Azo Group," Vol. 15) with 30% hydrogen peroxide in glacial acetic acid; other oxidizing agents are also effective, for example, peracetic acid (61) and peroxybenzoic acid (11,12). Hydroxylamines and hydrazones (56) are easily oxidized to azoxy compounds; for example, phenylhydroxylamine is converted to azoxybenzene (15) when heated in air on a steam bath, and $\alpha$-naphthylhydroxylamine under the same conditions forms $\alpha$-azoxynaphthalene (121).

Unsymmetrical azoxy compounds may be prepared by oxidation of indazole oxides (18,19), by condensation of a nitroso compound with an hydroxylamine (94), by selective substitution on an aromatic azoxy compound, or by the reaction of Grignard reagents and alkylated organonitrosohydroxylamine tosylates (54,110).

Azoxybenzene and $\alpha$-4-nitroazoxybenzene are formed (along with nitrodiphenylamines) by the photolysis of nitrosobenzene (101). Azoxy compounds are formed as by-products when aromatic nitro compounds are reduced with stannous chloride (47). An azoxy compound, 2,2'-dicarboxy-3,3',5,5'-tetranitroazoxybenzene ("white compound"), is formed by an oxidation reaction in the continuous production of trinitrotoluene (39,71).

The bisazoxy aromatic compounds and the trisazoxybenzenes are usually prepared by oxidation of the corresponding azo compounds. Mixed azo-azoxy compounds have also been prepared; for example, azobisazoxybenzene:

$$C_6H_5-N=N-C_6H_4-N=N-C_6H_4-N=N-C_6H_5$$

obtained by sodium arsenite reduction of $p$-nitroazoxybenzene (23).

Cyclic azoxy compounds are prepared by special methods; six- and seven-membered rings are usually involved, for example:

*o,o'*-Azoxybiphenyl

Ethyl-2,2'-azoxydiphenylmethane-4,4'-dicarboxylic acid

A few azothio compounds have been prepared by reaction of aryl compounds with sodium in boiling toluene (84):

$$C_6H_5N:S:O \longrightarrow C_6H_5\overset{\overset{S}{\uparrow}}{N}=N-C_6H_5$$

The azoxy compounds are usually found as the *trans* isomers; only a few of the geometric *cis* isomers have been synthesized (31,34,124). The formation of azoxy compounds by oxidation of the corresponding azo derivatives with peroxybenzoic acid has been studied extensively (9–11), and it has been shown that *trans*-azobenzene yields *trans*-azoxybenzene, and *cis*-azobenzene yields the corresponding azoxy isomer when the reaction is performed in the dark to prevent isomerization. Peculiarly, reduction of either isomer of azoxybenzene yields *trans*-azobenzene; apparently, during reduction the azoic nitrogen bond becomes single and permits free rotation. On the other hand, aryl *trans*-azoxy compounds may be isomerized to *cis* isomers by action of sodium methoxide, although the reverse isomerization does not take place (28).

Azoxy compounds, particularly the symmetrical aryl azoxy compounds, are used as acaricides (118) and insecticides (119); polymers have been prepared which contain arylene groups connected to one another by azoxy groups (27). Azoxybenzene has been found to increase the breakdown potential of transformer oils (7). Selective chelating agents have been prepared by introducing the azoxy group into a multidentate chelate (40,41).

The activity of some simple azoxy compounds on microorganisms that destroy rubber has been found to be negligible (77), but some azoxy compounds seem to possess antiseptic qualities.

With few exceptions, azoxy compounds are products of synthesis, and only *cycasin* (76) and *macrozamin*, the toxic β-primeverosides of *Macrozamia* spp. (80–82), and the antibiotic *elaiomycin* (109) have been isolated from natural products. *Hygroscopin A* is presumed to contain an azoxy group (91).

### III. PROPERTIES

Nearly all azoxy and azothio compounds are colored pale yellow through red; the azoxy compounds have such well-defined and separated melting points that they are often used to characterize and identify azo compounds. However, some of the azoxy compounds behave as anisotropic liquids at temperatures just above their melting points (liquid crystals). For example, 4,4′-azoxyanisole gives a turbid liquid at its melting point (118°C) which clears at 138°C; the reverse process occurs on cooling (111). Most azoxy compounds are insoluble in water.

## A. PHYSICAL PROPERTIES

The physical properties of representative azoxy compounds are given in Table I. The melting points of the *cis* and *trans* isomers of the azoxy compounds are considerably different, but their colors (to the eye) are quite similar. The alpha and beta isomers usually have different melting points and are similarly colored. As expected, all isomers show differences in their absorption spectra.

The *cis* aliphatic azoxy compounds usually have higher melting points and higher dipole moments than the *trans* aliphatic azoxy compounds (81); the higher melting points suggest dimerization:

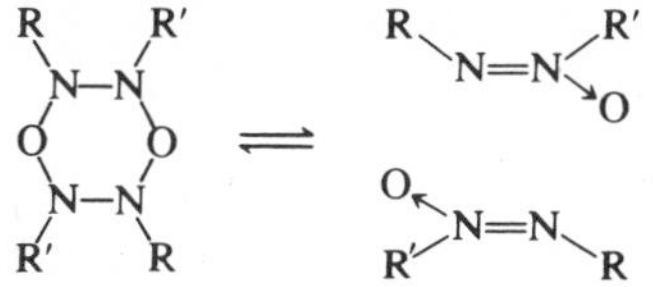

Trisazoxybenzene can have four positional isomers, $\alpha-\alpha-\alpha$, $\alpha-\alpha-\beta$, $\beta-\alpha-\beta$, and $\beta-\alpha-\alpha$; for example, the $\beta-\alpha-\alpha$ isomer is represented by the following formula:

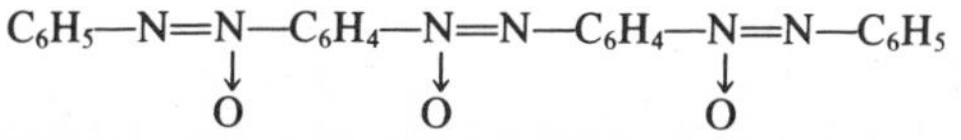

$$C_6H_5-N=N-C_6H_4-N=N-C_6H_4-N=N-C_6H_5$$
$$\downarrow \qquad \qquad \downarrow \qquad \qquad \downarrow$$
$$O \qquad \qquad O \qquad \qquad O$$

The total number of isomers is much larger, owing to the possibility that each of the substituent groups may be in the *cis* or *trans* configuration. As expected, each of the compounds should have a different melting point; the $\alpha-\alpha-\beta$ compound does not react with bromine, the $\alpha-\alpha-\alpha$ and the $\beta-\alpha-\beta$ compounds yield monobromo-substituted products, and the $\beta-\alpha-\alpha$ yields a dibromo product.

There is a possibility of the widespread existence of crystalline polymorphs of azoxy compounds; a crystalline transformation paralleling that of monoclinic-to-rhombic sulfur has been observed in azoxybenzene in the range of 0–30°C (123). Liquid-crystal formation also occurs in azoxybenzene and the azoxytoluenes (67), as well as in the azoxyphenetoles and the azoxyanisoles. Nematic liquid-crystal compositions containing azoxycompounds are used in the electro-optical display devices of small computers and watches.

Exposure of aqueous methanolic solutions of azoxybenzene to daylight causes a gradual rearrangement to 2-hydroxyazobenzene; the rate

## TABLE I
### Physical Properties of Representative Azoxy Compounds

| Compound | Molecular weight | Color and form | Melting point | Boiling point | Solubility | Ref.[a] [b] |
|---|---|---|---|---|---|---|
| Azoxybenzene (cis) | 198.23 | — | 87 | — | Alc., ether, lig. | B16[2], 313 |
| Azoxybenzene (trans) | 198.23 | Light yellow rhomb. | 36 | d. | ether, conc. sulf. | B12[1], 1038 |
| 1,1'-Azoxynaphthalene | 298.35 | Yellow or red orthorhomb. plate (alc.) | 127 | — | — | B12[1], 1039 |
| 2,2'-Azoxynaphthalene | 298.35 | Yellow or red needles (alc., gl. acetic) | 164 | — | Benz., chl. | B16, 635 |
| 2,2'-Dimethoxyazoxybenzene | 258.27 | Yellow prism (methanol) | 81 | — | Benz., alc., ether, chl., acetone | B16[2], 326 |
| 4,4'-Dimethoxyazoxybenzene | 258.27 | Yellow monocl. | 118.5–118.6 | — | Water, benz. | B16[2], 335 |
| 2,2'-Azoxybenzenedicarboxylic acid | 286.25 | Pale yellow tricl. prism or leaf | 254 | — | Alc., acetone, pyridine | B16[2], 336 |
| 4,4'-Azoxybenzenedicarboxylic acid | 286.25 | Yellow amor. powder | 240 d. | — | Pyr. | B16[2], 315 |
| 4-Bromoazoxybenzene | 277.13 | Deep yellow plate | 93.5–94.5 | — | Alc., benz., methanol | 81 |
| ω-Azoxy-p-chlorotoluene | 259.72 | Needles | 103 | — | Alc. | 82 |
| 3-Azoxy-p-menthane | 322.54 | — | 82 | — | Alc. | 82 |
| Azoxycyclohexane | 210.32 | — | 22–23 | — | Alc. | 81 |
| Azoxymethane | 74.08 | Colorless liq. | — | 98 | Alc. | 81 |
| 1-Azoxypropane | 154.21 | Colorless liq. | — | 67[20] | Alc. | 81 |

[a]The number following the letter B indicates the volume of the main series of Beilstein, *Handbuch der organischen Chemie*; the superscript refers to the first, second, etc., supplement; and the following number is the page in the supplement.

[b]*Handbook of Chemistry and Physics*, 45th ed., Chemical Rubber Co., 1964–1965.

of conversion is proportional to the intensity of illumination (32). Since many azoxy compounds in solution undergo rearrangements and isomerization in light, it is necessary to use freshly prepared solutions for analytical work and to protect them at all times from light.

The results of studies of the photolysis of azoxymethane by Gowenlock (60) show that production of nitrogen is a primary process:

$$CH_3—N=N—CH_3 \rightarrow \cdot CH_3 + N_2 + OCH_3$$
$$\downarrow$$
$$O$$

presumably taking place simultaneously with a reaction leading to the formation of nitrous oxide:

$$CH_3—N=N—CH_3 \rightarrow \cdot CH_3 + N_2O + \cdot CH_3$$
$$\downarrow$$
$$O$$

and the pyrolytic reactions yielding dimethyl ether:

$$CH_3—N=N—CH_3 \rightarrow CH_3OCH_3 + N_2$$
$$\downarrow$$
$$O$$

The free methyl radical also reacts with azoxymethane to form methane:

$$CH_3 + CH_3—N=N—CH_3 \rightarrow CH_4 + \cdot CH_2(N=NO)CH_3$$
$$\downarrow$$
$$O$$

## B. CHEMICAL PROPERTIES

The stability of the azoxy group toward a variety of chemical reagents is one of its most characteristic features. The azoxy group is not attacked by dilute acids or by sodium hydroxide solution. As a rule, the group resists oxidizing agents and often shows considerable stability toward many reducing agents. The only chemical reactions of the azoxy group which are of analytical import are reduction, oxidation, and attack by concentrated sulfuric acid.

### 1. Reduction

The reduction of the azoxy group usually proceeds in stepwise fashion:

Azoxy → azo → hydrazo → amine

Since the azoxy and azo compounds are usually highly colored, the disappearance of color signals reduction to the hydrazo group; reduction to the amine group can be accomplished with vigorous agents. However, the rates of reduction of the various functional groups are largely controlled by the nature of the substituents attached to the groups, and since it is difficult to predict quantitatively the extent of reduction afforded by a given reagent, there are no general reductimetric *analytical* procedures; any method selected for use must be thoroughly checked with pure samples of the compounds that will be encountered in subsequent analyses. There are, however, suitable *preparative* procedures for reducing azoxy compounds to hydrazo compounds and to amines.

The mechanism and the rate of reduction of azoxybenzene and some of its derivatives by titanium(III) chloride in aqueous alcohol were studied by Stephen and Hinshelwood (108). When azoxybenzene is reduced in acidic media, about 20% is converted directly to aniline without passing through the azobenzene stage, and the remainder is converted to a mixture of benzidine and aniline by a stepwise reduction with azobenzene and hydrazobenzene as intermediaries. The initial reduction of azoxybenzene is first order with respect to both azoxybenzene and titanous chloride (like azobenzene), and the predominant reducing species appears to be $Ti(OH)_2^+$; however, the entire reduction process is an example of the second-order consecutive reactions for which kinetic expressions have been found. The influence of substituent groups on the rate of reduction of the azoxybenzene derivatives appears to be considerably smaller than on the reductions of nitro and azo compounds. Reduction of azoxy compounds by titanium(III) in citrate media was studied by Belcher et al. (20) for possible use as an analytical method, but the mode of reduction of azoxybenzene was found uncertain, for no reduction could be obtained in acidic media; in contrast, azoxybenzene is quantitatively reduced by chromous chloride in hydrochloric acid at room temperature (25).

Quantitative reduction of the azoxy group can be effected by hydrogen iodide at elevated temperatures and pressures in accordance with a procedure outlined by Aldrovandi and De Lorenzi (1); the liberated iodine is determined by titration with standardized thiosulfate solution. Reduction with hydrogen bromide in acetic acid at room temperature may yield a variety of products, including hydrazo derivatives (75).

Azoxybenzene and some of its derivatives are reduced to amino compounds by alkali-metal sulfides ($Na_2S_2$) in water, alcohol, and ethylene glycol monoethyl ether solutions; disproportionation of intermediate hydrazo compounds appears to take place (64,65).

Mild reducing agents convert the azoxy group to an azo group; however, the reduction is difficult to control, and often the azoxy group is converted to the hydrazo group or to two amino groups. Ordinarily, azoxy compounds can be converted to azo compounds by zinc dust in aqueous or alcoholic alkali (see Section V.B.1.c), low molecular weight azoxy compounds can be reduced to azo compounds by distillation from dry iron powder (125), but reduction of azoxy compounds by zinc dust in alkali may also form hydrazo compounds; in these instances, the reduction is allowed to proceed until the reaction mixture becomes colorless or nearly so. However, the nature of the substituents on the aryl nuclei of aryl azoxybenzenes and the reactivity of the products largely determine the route of the reduction; for example, zinc dust in acetic acid reduces 2,2'-dimethyl-5,5'-dichloroazoxybenzene to the corresponding hydrazo derivative, but reduction with zinc dust in alkali gives a low yield because the hydrazo compound is decomposed by alkali (106). The following equations (106) are intended to emphasize the complexity of the interaction of azoxy compounds with reducing systems; thus, as indicated:

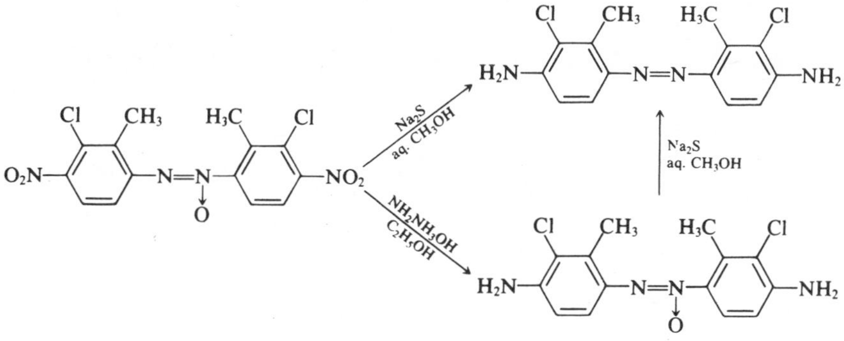

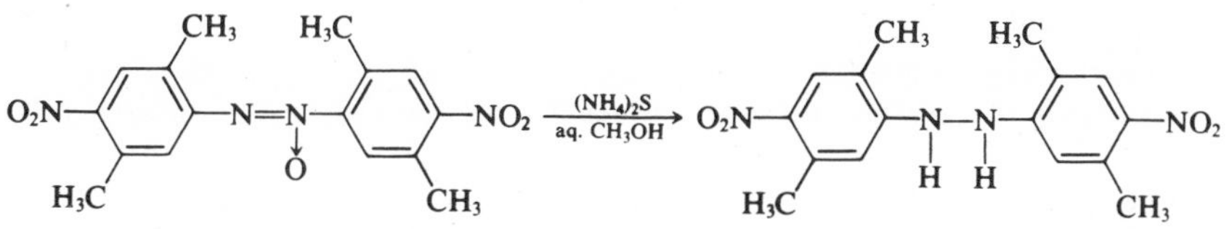

the reduction of 2,2'-dimethyl-3,3'-dichloro-4,4'-dinitroazoxybenzene is influenced by the position of the two electron-donating chlorine substituents *ortho* to the two electron-sink nitro groups. The attack on the azo group by ammonium sulfide (instead of the more vulnerable nitro groups) may possibly be the result of the positive inductive effect of the methyl groups; apparently, the electrometric shift of electrons from the azo group to the nitro groups is prevented, and the electron density on the azo group is kept high enough to promote attack by ammonium sulfide (73).

Hydrazine hydrate and Raney nickel catalyst appear to be able to convert azoxybenzene to azobenzene and subsequently to hydrazobenzene (69); since nitro groups are reduced to amine groups by this combination of reagents (51), and hydrazobenzene can be reduced to aniline, the azoxy group can be cleaved by prolonged reduction (14). The reducing action of pentacyanocobaltate(II) in aqueous solutions also seems to be quite specific with respect to the type of functional group reduced, but it is also sensitive to substituents on the aryl nuclei; nevertheless, it may prove to be quite a useful reagent since reductions are carried out at room temperature and at a hydrogen pressure of 1 atm. The utilization of hydrogen is easily followed in an apparatus such as that described by Gould and Drake (59a), and the extent and rate of reaction can thus be readily measured. The catalyst composition found effective has a cyanide/cobalt ratio of 5:1 at a cobalt concentration of 0.15 $M$ (78). Hydrogen with nickel catalysts in organic solvents reduces azoxy compounds to hydrazo compounds (117). Platinic oxide in acidic methanol is an effective catalyst (81).

Vigorous reducing agents such as stannous chloride or tin and hydrochloric acid ultimately cleave the azoxy group to two amines which can be separated and characterized to afford identification of the original azoxy compound. However, even these apparently straightforward reductions cannot be applied in every instance; for example, since azoxybenzene is reduced to aniline, but $p,p'$-dichloroazoxybenzene is reduced to $p,p'$-dichlorobenzidine, it is evident that the benzidine rearrangement which occurs readily in acidic media often controls the route of the reduction. Other reducing agents also exhibit peculiarities; for example, the reduction of $m$-azoxydisulfonic acid with sodium amalgam or ammonium sulfide gives rise to the corresponding azo compound, but sulfur dioxide is without action.

Cyclic azoxy compounds are reduced to azo and hydrazo compounds, and the cyclic starting material can be regenerated by hydrogen peroxide.

Azothiobenzenes readily lose sulfur when warmed with zinc dust; the corresponding azo compounds are formed. Sulfur is also lost when solutions of azothio compounds are passed over alumina (84).

Studies of the electrolytic reduction of the azoxy group have largely focused on the behavior of the simplest symmetrical aryl azoxy compound, azoxybenzene. For example, Chaung et al. (33) found that the voltammetric behavior of the systems azoxybenzene-azobenzene-hydrazobenzene at a graphite electrode in 50% ethanolic solutions indicated that the azobenzene-hydrazobenzene system is less reversible at a graphite electrode than at a mercury electrode. As indicated before, reductions in strongly acidic electrolytes favor the benzidine rearrangement. Thus, Lob (86) reduced azoxybenzene in alcohol and concentrated sulfuric acid to benzidine with a mercury cathode, but Elbs

(43) obtained *p*-aminophenol in acidic solutions. Elbs and Schwarz (44) obtained the corresponding hydrazo compound by electrolytic reduction of an alcoholic solution of *p,p'*-diamino-*o*-azoxytoluene in sodium acetate. Bergman and James (21) found that azoxy compounds in acidic solutions are easily reduced electrolytically to azo, hydrazo, or amino compounds and, accordingly, that azoxybenzene formed benzidine as well as azobenzene, hydrazobenzene, and aniline.

The polarographic behavior of an equimolar mixture of azoxybenzene and hydrazobenzene was investigated by Costa (35) and Wawzonek and Fredrickson (123) in buffered aqueous methanolic solutions over a pH range of 2.0–11.1.

Azoxybenzene is reduced largely to azobenzene by anodically generated, lower-valent aluminum (116) or unipositive magnesium (89,127).

## 2. Oxidation

The azoxy group is extremely resistant to oxidation, but the nature of the substituents attached to the group may degrade its resistance; hydroxyaryl substituents are particularly effective in rendering possible facile oxidation of the azoxy group. Thus, while azoxybenzene withstands the attack of warm 2% potassium permanganate in alkaline solution, monosubstituted hydroxyazobenzene is oxidized to the isodiazotate, $C_6H_5N{=}N{-}OH$, with concurrent formation of nitrosobenzene and nitrobenzene. Furthermore, the alpha isomer is immediately reactive, whereas the beta isomer reacts slowly [24 hr (5)]. Under similar conditions, however, *p,p'*-azoxyphenol is completely degraded to oxalic acid.

Many substituted azoxy compounds, like azo compounds, can be degraded completely by hot nitric acid, especially if the oxidation is performed in sealed tubes, but azoxybenzene is so resistant that Petriew (98) was able to prepare it by oxidation of azobenzene with chromic acid in acetic acid solution in a closed tube at 150–250°C. [Oxidation by chromic acid is the basis for several quantitative methods for nitrogen-functional group compounds (8,37,72,92).] Cold nitric acid of specific gravity 1.45–1.54 nitrates the aryl nuclei of the less sensitive aromatic azoxy compounds; azoxybenzene may form trinitroazoxybenzene. Differences between the alpha and beta forms are observed; for example, acid of 1.45 specific gravity nitrates only the beta form of *p*-mononitroazoxybenzene, the proximity of the oxygen atom of the unsymmetrical azoxy group acting as a shield for the benzene ring of the alpha form (4):

Alpha:   $C_6H_5$—N=N—$C_6H_4NO_2$ $\xrightarrow[d.=1.45]{HNO_3}$ no reaction
                        $\downarrow$
                        O

Beta:   $C_6H_5$—N=N—$C_6H_4NO_2$ $\xrightarrow[d.=1.45]{HNO_3}$ $O_2NC_6H_4$—N=N—$C_6H_4(NO_2)$
                       $\downarrow$                                          $\downarrow$
                       O                                            O

Sensitive azoxy compounds, such as the hydroxyaryl azoxy compounds, can be nitrated by nitrous acid; here, again, the aryl nucleus is protected by the proximity of the oxygen atom of the azoxy group (6):

$HOC_6H_4$—N=N—$C_6H_5$ → $HOC_6H_3$—N=N—$C_6H_5$
          $\downarrow$                      |        $\downarrow$
          O                     $NO_2$      O

$HOC_6H_4$—N=N—$C_6H_5$ → no action
          $\downarrow$
          O

Azobenzene and azoxybenzene resist oxidation by periodic acid (115).

Electrochemical oxidation of azobenzene also produces azoxybenzene, but side reactions lead to a mixture of products such as $p,p'$-azodiphenol, $p$-phenylazophenol, and 4,4′-di($p$-hydroxyphenylazo)biphenyl (46).

### 3. Action of Acids

The azoxy group in aryl compounds is weakly basic; it resists attack by dilute or concentrated hydrochloric acid as well as dilute nitric or sulfuric acid. Hahn and Jaffé (62,63) have determined $pK_a$ values for the conjugate acids of a series of azoxybenzenes; as expected, electron-releasing substituents increase, and electron-withdrawing substituents decrease, the base strength of the azoxy group. Since acidity affects the spectral absorption of azoxybenzenes, colorimetric procedures should be designed to afford rigidly reproducible hydrogen ion concentrations.

Azoxybenzenes, when treated with concentrated sulfuric acid, are converted to hydroxyazobenzenes by the Wallach rearrangement (122); for example, azoxybenzene largely forms $p$-hydroxyazobenzene:

$C_6H_5$—N=N—$C_6H_5$ → $C_6H_5$—N=N—$C_6H_4OH$
          $\downarrow$
          O

In general, when an aromatic azoxy compound is added to cold concentrated sulfuric acid, and the mixture warmed to about 95°C, there is a rapid darkening of color, eventually approaching deep red or brown, owing to the intermediate formation of an addition complex of quinoid structure (58). Azo compounds also form red solutions in concentrated sulfuric acid, but the colorations usually do not deepen on heating. In some instances, chlorosulfonic acid also may be used to effect the Wallach rearrangement.

The Wallach rearrangement has been extensively studied and is often the basis of preparative methods for hydroxyaryl azo compounds (95,96,100,107); however, the reaction mechanism has not been clearly elucidated. Experiments with azoxy compounds containing $N^{15}$, $C^{14}$, and $O^{18}$ (102–105) suggest that isomerization proceeds through a symmetrically constituted intermediate with a cyclic structure having the oxygen atom between the nitrogen atoms; on the other hand, it is postulated that the reaction takes place by means of ionic intermediates which are formed through nucleophilic attack by $OSO_3H^-$ or $OSO_2Cl^-$, followed by hydrolysis (57,107):

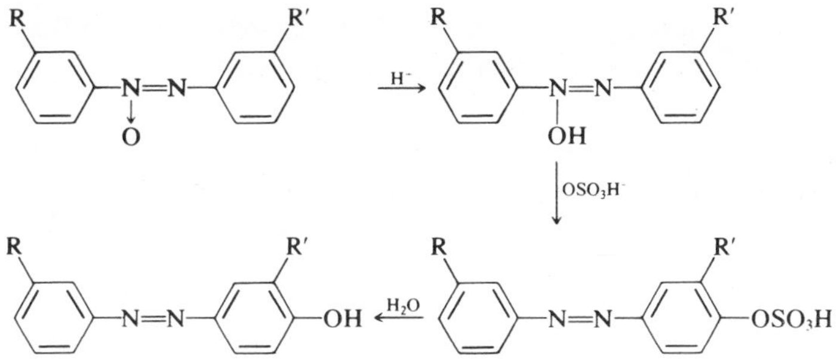

The Wallach rearrangement does not yield exclusively *para*-substituted hydroxyazobenzenes; Bamberger (16) first showed that a small amount of *o*-hydroxyazobenzene was formed as well as the *para* compound; Lachman (79) found that about 2% of the *ortho* isomer can be obtained. When the *para* positions of an azoxybenzene are occupied, the oxygen of the azoxy group is removed; thus 4,4'-dichloroazoxybenzene yields 4,4'-dichloroazobenzene (62,66). If only one of the *para* positions of an azoxybenzene is occupied, the oxygen atom will migrate to the open position in the other ring. The alpha and beta isomers of mono-*para*-substituted benzenes yield the same azobenzene; this behavior seems to confirm that the reaction proceeds via a mechanism involving the cyclic structure noted in the preceding paragraph (62).

The initial, intermediate, and final colorations observable when azoxy compounds are treated with concentrated sulfuric acid can be used as a means of identification; the behavior is so characteristic that it can serve as a generic test for azoxy aromatic compounds. (See Section V.B.1.e.).

In sharp contrast to the aryl azoxy compounds, the secondary alkyl azoxy compounds are readily degraded by acids (as well as alkalies) (81):

$$\begin{array}{c} CH_3 \\ \diagdown \\ CH_3 \end{array}\!\!CH\!-\!N\!=\!N\!-\!CH\!\!\begin{array}{c} CH_3 \\ \diagup \\ CH_3 \end{array} \xrightarrow{\ H^+\ } 2\,(CH_3)_2C\!\!=\!\!O + NH_2\!-\!NH_2$$

The primary alkyl azoxy compounds are also cleaved, for example:

$$CH_3CH_2\!-\!N\!=\!N\!-\!CH_2CH_3 \xrightarrow{\ H^+\ } CH_3CH_2COOH + CH_3CH_2NHNH_2$$

The azothiobenzenes are decomposed to anilines by action of hydrochloric acid; hydrogen bromide also cleaves them, but the aniline derivatives may be brominated (84). Concentrated sulfuric acid forms sulfonated products instead of products of the Wallach transformation; concentrated nitric acid yields nitrated products.

## C. GENERAL REFERENCES

There are comparatively few general references on azoxy compounds; most compendia are reviews on the structural problems of the azoxy group.

Zollinger, Heinrich, *Azo and Diazo Chemistry, Aliphatic and Aromatic Compounds*, Interscience, New York, 1961, translated by H. E. Nurston. Contains a brief section on azoxy compounds.
Taylor, T. W. J., and Wilson, Baker, *Sidgwick's Organic Chemistry of Nitrogen*, Oxford Press, London, 1942. A comprehensive survey of nitrogen compounds, including a brief section on azoxy compounds.
Bigelow, H. E., *Chem. Rev.*, **9**, 117 (1931). An excellent monograph describing the evolution of views on the structure of azoxy compounds and a summary of synthetic methods; outdated.
Urbanski, Jerzy, *Wiad. Chem.*, **13**, 125 (1959). A more recent review on the evidence for the structure of the azoxy group. [See *Chem. Abstr.*, **53**, 15943 (1959)].
Millar, I. T., and H. D. Springall, *A Shorter Sidgwick's Organic Chemistry of Nitrogen*, Clarendon Press, Oxford, 1969. A shortened version of the revised editions of "Sidgwick" (see above); includes many new references and modern concepts.

## IV. TOXICITY AND INDUSTRIAL HYGIENE

The carcinogenic activity of azoxy compounds has not been investigated thoroughly because there are very few uses for them in industry. Some azoxy compounds are used as parasiticide vapors alone or in combination with insecticides such as Lindane, Aldrin, or DDT. Studies of the toxic hazards arising from the use of such vapors have been performed; no deleterious effects are observed on exposure of birds or of small mammals, such as dogs, guinea pigs, and mice, at azoxybenzene levels of the order of 0.4 gram/(day)(1000 ft$^3$) (13). On the other hand, it has been found that azoxyethane (subcutaneous injection) induces local sarcoma and adenocarcinoma in the intestinal tract (38). Thus it appears that the azoxy compounds have about the same level of carcinogenicity as the azo compounds; accordingly, they should be handled in the same fashion (see Section IV of "Azo Group," Vol. 15.). Undoubtedly, many of the tests described in the section just cited can be readily modified to make them applicable to azoxy compounds.

## V. QUALITATIVE DETERMINATIONS

### A. PREPARATION OF SAMPLE

Nearly any procedure may be used for concentrating the sample or for dividing it into various fractions for simplicity of analysis because azoxy compounds are usually quite stable. Since azoxy compounds may be toxic, all samples should be handled with caution. Also, analytical procedures involving the reduction of azoxy compounds must be done carefully owing to the extreme toxicity of the resulting amines.

### B. DETECTION OF AZOXY GROUP

A general reaction for the azoxy group is not known; the sequence of colors formed when a sample is placed in concentrated sulfuric acid and then heated may be taken as fairly conclusive evidence that the sample contains an aromatic azoxy compound. A combination of chemical tests must be used for other azoxy compounds, and often physical methods must be invoked (e.g., infrared spectrophotometry).

### 1. Chemical Methods

Chemical methods for detecting the azoxy group are largely dependent on its transformation to other functional groups.

#### a. RESISTANCE TO OXIDATION

The azoxy group resists attack by oxidizing agents such as hydrogen peroxide and glacial acetic acid, peracetic acid, and peroxybenzoic acid; hence compounds containing the azoxy group usually can be recovered unchanged from such oxidizing media. The presence of other oxidizable functional groups must be taken into consideration.

### b. CLEAVAGE TO AMINES

The formation of amines by reductive cleavage may provide a means of detecting the presence of an azoxy group, particularly after demonstration of resistance to oxidation (see above). Reductive cleavage to amines is best performed with tin and hydrochloric acid, and the resulting amines may be detected in accordance with established procedures.

### c. REDUCTION TO AZO GROUP

Controlled reduction of the azoxy group in many compounds will convert it to an azo group which can be detected by the methods described in Section V.B. of "Azo Group," Vol. 15. For reduction to the azo group, azoxy compounds may be treated with mild reducing agents such as zinc dust and alkali. A typical procedure is given below (125):

Add 10 grams of zinc dust in 2- or 3-gram portions to about 2 grams of the powdered azoxy compound in 10 ml of 25 wt.% NaOH solution to which sufficient alcohol has been added to dissolve the sample. The mixture becomes warm because of reaction of the zinc with the base. (It may be necessary to heat the mixture on a water bath for about 15 min under reflux.) Flood the reaction mixture with about 100 ml of cold water to throw out the azo compound (with excess zinc); filter off the solids and extract the azo compound with petroleum ether or other solvent. Recrystallize the azo compound and perform tests to verify the presence of the azo group.

Reduction of the azoxy group by the above procedure (as well as others) can lead to complete reduction to hydrazo group compounds or even amines. Generally, the reduction is controlled so that the mixture is still highly colored after sufficient zinc has been added; in difficult cases, it may be necessary to try variations of the reducing procedure given above. Often it is not possible to prevent nearly exclusive formation of hydrazo group compounds; in these instances, it is best to let the reaction proceed to the hydrazo state and then to convert the hydrazo compound to an azo compound by dissolving it in alcohol and oxidizing it with a mild oxidant or even with a stream of air (53). One of the color reactions of the azo group compounds (see Table IV of "Azo Group," Vol. 15) may be found useful for verifying conversion of the sample to an azo compound.

Azoxy compounds that are volatile can be reduced to azo compounds by distillation from iron powder:

Dry about 15 grams of iron powder on a water bath; transfer to a distillation flask, and mix well with about 5 grams of the azoxy compound. Distil the azo compound slowly from the mixture. Wash the azo compound with dilute (about 10%) hydrochloric acid, and crystallize it from a suitable solvent (e.g., petroleum ether) or redistil it.

The azo group compounds formed by reduction of azoxy compounds can be reoxidized to the original azoxy compounds (assuming that other functional groups are not irreversibly affected by the reduction); thus formation and regeneration of the azoxy group can serve as a method of detection.

### d. REDUCTION TO HYDRAZO GROUP

In many instances, reduction of the azoxy group in certain compounds goes beyond the azo group stage, and hydrazo group compounds are formed. Reduction with zinc and acetic acid, zinc in alkali, or ammonium sulfide, or catalytically with hydrazine hydrate and Raney nickel, or with pentacyanocobaltate(II) see Section III.B.1) may be used to convert azoxy compounds to hydrazo compounds. In acidic media, the hydrazo compounds tend to undergo the benzidine or semidine rearrangement. In any event, the resulting compounds, whether they be hydrazo group compounds or benzidines, can be identified, and their presence taken to indicate that the azoxy group was present in the original substance. Hydrazo group compounds can be reoxidized to azo group compounds and thence to azoxy group compounds, and the capacity to be alternately reduced and reoxidized to the original form may be regarded as evidence of the presence of the azoxy group in compounds which do not have other functional groups that are irreversibly reduced or oxidized. A typical procedure for reducing azoxy group compounds to hydrazo group compounds follows (53):

Dissolve about 10 grams of the azoxy compound in the least amount of alcohol, and add 35 ml of 40% sodium hydroxide solution. Reflux the mixture on a steam bath with an excess of zinc dust. Usually the mixture becomes red in a short time; continue the refluxing for several hours until the mixture becomes nearly colorless. Allow the mixture to cool, and then precipitate the hydrazo group compound with water. If the precipitate appears colored, dissolve it in alcohol containing a few drops of acetic acid, add some zinc dust, heat to decolorize, and reprecipitate the colorless hydrazo compound by addition of water; dry in a vacuum desiccator.

### e. COLORS IN SULFURIC ACID

Aromatic azoxy compounds undergo the Wallach rearrangement when treated with concentrated sulfuric acid (see Section III.B.3). When the compound is first placed in the concentrated acid at room temperature, a light-colored solution is formed; when the temperature of the mixture is quickly raised to 95°C, the color generally increases in intensity, to red or red-brown and then fades somewhat as the conversion to hydroxyazo compounds becomes complete. Less than a minute is required for the reaction to subside. Polynitroazoxy compounds require longer heating and temperatures higher than 95°C, but the darkening cannot be mistaken for charring since the solution rapidly loses color on further heating. Azo compounds also form colors when dissolved in concentrated sulfuric acid, but their solutions seldom change color when heated. The Wallach rearrangement has been recommended as a generic test for aromatic azoxy compounds (53).

## 2. Physical Methods

Infrared spectrophotometry provides a definitive method for detecting the presence of the azoxy group, and it is often used to confirm chemical evidence. Absorption in the ultraviolet region is insufficiently defined to permit positive detection of the azoxy group, but in many instances it may be used advantageously to support evidence gained from absorption in the infrared regions and from chemical tests.

### a. INFRARED SPECTROPHOTOMETRY

The oxygen atom in the azoxy group makes the azo bond unsymmetrical and thus gives rise to definite absorption bands in the infrared region. In aromatic azoxy compounds, asymmetric vibrations arising largely from $N$=$N$ stretch occur at 1480–1450 cm$^{-1}$ (6.76–6.90 $\mu$) and symmetric vibrations, mainly from $N \rightarrow O$, occur at 1335–1315 cm$^{-1}$ (7.49–7.60 $\mu$) (88,126). The asymmetric band of hexafluoroazoxymethane is shifted to a shorter wavelength (6.37 $\mu$), and the symmetric band to a longer wavelength (doublet at 7.80 and 7.94 $\mu$) (70).

The aliphatic azoxy compounds are considered to have absorption bands at 1530–1495 cm$^{-1}$ (6.54–6.69 $\mu$), characteristic of the asymmetrical $N$=$N$ stretch (82), at 1342–1285 cm$^{-1}$ (7.45–7.78 $\mu$) for the symmetrical $N$=$N$ stretch, and at 1310–1250 cm$^{-1}$ (7.63 to 8.00 $\mu$) for —N—O stretching (81).

Mixed aryl-aromatic azoxy compounds generally show $N$=$N$ absorption at 1475 to 1465 cm$^{-1}$ (6.78–6.83 $\mu$) and N—O absorption at 1300 to 1280 cm$^{-1}$ (7.69–7.81 $\mu$) (128).

The azothio group also exhibits two absorption bands; the band at 1465–1445 cm$^{-1}$ (6.82–6.92 $\mu$) represents chiefly $N$=$N$ stretch, and the band at 1071–1058 cm$^{-1}$ (9.34–9.45 $\mu$) is mainly $N \rightarrow S$ stretch (48).

b. Ultraviolet Sectrophotometry

Aliphatic azoxy group compounds show two prominent absorption bands in the ultraviolet region (81,82). A strong band in the vicinity of 220 nm is attributable to an electronic transition ($E$-band) involving the —N=N— system. A weak band in the vicinity of 270 nm can be attributed to a radical transition ($R$-band) of the azoxy group. Since the $R$-band in azomethane occurs at 350 nm and is very weak, it is evident that coordination of an oxygen atom has induced a hypsochromic shift and an increase in absorption characteristic of the effect of substitution on $R$-bands (30).

The disubstituted azoxybenzenes usually show absorption in three regions, approximately at 230, 260, and 320 nm, corresponding to the $E_1$-, $E_2$-, and $K$-bands, respectively. The $R$-band absorption of the azoxy compounds, which usually occurs in the visible region in the spectra of the azobenzenes, is superposed nearly indistinguishably on the 320-nm $K$-bands, and serves to distinguish many aromatic azoxy compounds from the corresponding azo compounds. The band at about 230 nm also appears in the spectra of phenylnitrones and the azobenzenes and sometimes is seen as two separate peaks; the double peaks are not found in the spectra of the azo compounds (59). Other features of the uv spectra of the aromatic azoxy compounds are discussed in Section V.C.2.b.

## C. IDENTIFICATION OF AZOXY COMPOUNDS

Azoxy compounds are usually identified by conversion to easily purified and easily prepared derivatives. The derivatives are prepared by reactions involving the azoxy group itself or the substituents on the group. The physical methods most often used to identify the aromatic and many of the aliphatic azoxy compounds are infrared spectrophotometry and ultraviolet spectrophotometry, largely because it is a simple matter to compare the spectrum of an unknown with a catalog of fiducial spectra. Nuclear magnetic resonance spectra are particularly useful for determining the structures of aliphatic azoxy compounds, and it is anticipated that in the near future nmr data will be more plentiful.

The aromatic azoxy compounds can form complexes with metal ions and the ammonium ion or their salts. The identification of metals or the ammonium ion is a simple matter with the analytical procedures described in Part II, Section A, of this *Treatise* or in other texts on qualitative analytical chemistry. Knowledge of the presence and identity of metals in the azoxy group compound to be identified is especially desirable to avoid misinterpretation of color tests and the formation of unwanted precipitates. The presence of inorganic matter can be detected by igniting a sample at a low temperature, but separate tests must be used to detect ammonium ion.

### 1. Chemical Methods

There is no well-developed method which can be used to identify any class of azoxy compound; thus procedures for identification usually involve reductive cleavage and subsequent identification of the amines or partial reduction to azo or hydrazo group compounds followed by identification of the product.

### a. REDUCTIVE CLEAVAGE TO AMINES

As was indicated in Section III.B.1, the azoxy group can be cleaved to amines by a strong reducing agent such as tin and hydrochloric acid. Symmetrical aliphatic or aromatic azoxy compounds form only one amine (provided that the azoxy group substituent is not cleaved into additional amines); the unsymmetrical azoxy compounds form mixtures of amines which usually must be separated before identification.

A typical procedure for reductive cleavage of the azoxy group consists of refluxing the azoxy compound with tin and hydrochloric acid:

Place about 1 gram of the azoxy compound and 2 grams of granulated tin in a small flask. Connect the flask to a reflux condenser, and pour in about 10 ml of water or alcohol; follow with small portions of 10% hydrochloric acid, vigorously shaking the flask and its contents after each addition. Use a total of about 20 ml of the acid. Warm the mixture on a steam bath to complete the reaction; transfer the hot contents of the flask to a small amount of water (10–15 ml), and add enough 40% sodium hydroxide to precipitate and then dissolve tin hydroxide. Extract the clear solution with ethyl ether or petroleum ether, dry the ether with a desiccant, and remove by careful evaporation.

Depending on the nature of the azoxy compound, other reductive methods may also be used, for example, those given in Section V.C.1.c of "Azo Group," Vol. 15. Furthermore, since the greater number of aromatic azoxy compounds can be considered to be derivatives of the corresponding azo compounds, the methods outlined in the section just cited may be used for identifying the amines resulting from reductive cleavage.

Hydrazo compounds derived from azoxy compounds (see Section V.B.I.d) can be readily reduced to amines by sodium hydrosulfite (procedure given in Section V.C.1.b(4) of "Azo Group," Vol. 15) in neutral or alkaline solution, or by tin or stannous chloride in hydrochloric acid.

## b. Reductive Cleavage by Hydrogenation

Cleavage of the azoxy group by hydrogenation is preferable to cleavage by other reduction methods because the resulting mixture of amines contains a minimum of extraneous products and subsequent separations of amines are simplified. Catalytic hydrogenation under pressure in special equipment is usually simply performed. However, identification of azoxy compounds in the laboratory can be accomplished simply and more efficiently by catalytic hydrogenations performed at atmospheric pressure in apparatus which provides nearly continuous indication of the rate or amount of hydrogen taken up by a sample (see Section V.C.1.c of "Azo Group," Vol. 15). For suitable catalytic systems, see Section III.B.1.

## c. Reduction to Azo Group Compounds

Many aromatic azoxy compounds can be partially reduced to the corresponding azo compounds. Representative procedures for such reductions were given in Section V.B.1.c, but care must be taken to modify the conditions of reduction or even to use entirely different methods of reduction to ensure conversion to azo group compounds (colored) rather than to hydrazo group compounds (essentially colorless). Many of the aromatic azo compounds have sharp boiling or melting points; they can also be identified by their infrared or ultraviolet spectra and occasionally by their X-ray diffraction patterns (see Section V.C.2 of "Azo Group," Vol. 15).

## d. Reduction to Hydrazo Group Compounds

A great many azoxy compounds, particularly the aromatic ones, are readily reduced with zinc dust and alkali to the corresponding hydrazo group compounds by the methods already described in Section V.B.1.d. The hydrazo group compounds or their acetyl or benzoyl derivatives can be identified by their melting points. In certain instances, the hydrazo group compounds can be converted to identifiable benzidine derivatives by boiling in acids, provided that the number of products formed by the rearrangement is not too great.

Simple hydrazo compounds can be rearranged in the following:

Shake about 2 grams of the hydrazo compound with about 50 ml of a molar solution of hydrochloric acid at room temperature until the substance dissolves completely (about 0.5 hr). Heat the solution to about 50°C, and if necessary add water to keep all salts in solution; filter the warm solution. Add an excess of 3 *M* sodium hydroxide solution to the cold filtrate, filter off the solid, wash with water, and recrystallize from water or alcohol. If no material separates, bring the solution to neutrality and extract with a suitable solvent, such as diethyl ether or petroleum ether.

### e.  COLOR REACTIONS

The color changes accompanying the Wallach rearrangement often are sufficiently characteristic to permit identification of a limited number of the aromatic azoxy compounds. This method of identification is especially useful for confirmation of the results obtained from other methods, and provides one of the simplest ways for differentiating isomers. The following procedure is applicable to a large number of aromatic azoxy compounds (58):

Put about 0.5 mg of the substance in a test tube, and add 1 ml of 98% sulfuric acid at room temperature. Mix well, and note the color of the solution. Immerse the tube in a water bath held at 95°C. Within 3–5 sec, note the color of the solution. Quickly return the tube to the water bath, and note the color within 30–45 sec later. If the color does not intensify within a few minutes, remove the tube from the bath and gradually heat the mixture to boiling; note whether the color deepens.

The test can be performed on a very small scale; a tiny crystal of the sample or even the residue obtained from evaporation of a drop of a solution of the sample is treated on a watch glass or in the bottom of a porcelain crucible with a drop of sulfuric acid, and the temperature brought to 95°C.

The colors observed are compared with the listings in Table II or with the results of parallel tests made with known aromatic azoxy structures. Some aromatic azo N-oxides, such as benzocinnoline oxide, do not undergo the Wallach rearrangement and thus will not give characteristic changes of color when heated in concentrated sulfuric acid. Aromatic compounds containing reactive iodine groups (e.g., *p*-iodophenol and *o*-iodoaniline) show color darkenings which might be mistaken for the colorations characteristic of the aromatic azoxy compounds; these changes, however, occur very slowly in comparison to those observable with the azoxy compounds (58). The azo compounds also form colored solutions in concentrated sulfuric acid, but the colors are only slightly altered when heated.

### f.  WALLACH REARRANGEMENT

Isomerization of azoxybenzenes to the corresponding *p*-hydroxy-azobenzenes takes place easily in concentrated sulfuric acid; the resul-

TABLE II

Colors Formed by Azoxybenzenes in Sulfuric Acid (58)

| Compound | Color | | | |
|---|---|---|---|---|
| | Initial | 3 sec at 95°C | 30 sec at 95°C | Boiled |
| Azoxybenzene | Lemon yellow | Very deep red | Very deep red | — |
| 3,3'-Difluoro- | Lemon yellow | Orange | Deep orange-red | — |
| 3,3'-Dichloro- | Lemon yellow | Orange-red | Deep orange-red | — |
| 3,3'-Dibromo- | Lemon yellow | Orange-red | Bright red | — |
| 4,4'-Dichloro- | Yellow | Orange-red | Bright blood red | — |
| 3,3'-Diiodo- | Yellow | Dark red | Intense, dark red-brown | — |
| β-4-Bromo- | Yellow | Deep red-brown | Intense, dark red-brown | — |
| 3,3'-Dinitro- | Yellow | Yellow | Yellow | Red-brown to very light brown |
| 3,3',5,5'-Tetranitro- | Pale yellow | Pale yellow | Light yellow-brown | Red-brown to very pale brown |
| β-4-Nitro- | Pale yellow | Orange-red | Deep blood red | — |
| α-4-Nitro- | Greenish yellow | Greenish yellow | Orange-red | |
| 2,2'-Dimethoxy- | Yellow-brown | Olive green to deep green to very dark royal blue | Deep bluish purple | — |
| 4,4'-Dimethoxy- | Yellow-brown | Deep red-brown | Deep chocolate brown | — |
| α-4-Bromo- | Yellow-orange | Deep red-brown | Intense, dark red-brown | — |
| 4,4'-Dimethyl- | Orange | Intense, dark blood-red | Deep red-brown | — |
| 4,4'-Diphenyl- | Deep red | Deep Tyrian purple | Purplish brown | — |

25

ting azobenzenes can be readily identified by their ultraviolet, visible, and infrared absorption spectra or by the other methods noted in Section V.C of "Azo Group," Vol. 15. The procedure given below for effecting isomerization of aromatic azoxy compounds has broad applicability, but it may need to be modified to permit efficient recovery of certain hydroxyazobenzenes:

Dissolve about 0.005 mol of the azoxy compound in 20 ml of sulfuric acid (sp. gr., 1.84), and allow to stand at room temperature for about 1 hr; some compounds may require longer times and temperatures of the order of 40–50°C. Pour the mixture into water to precipitate the azo compound; it may be necessary to add salt to force sulfonates out of solution. The azo compound may be purified by dissolution in dilute alkali and reprecipitation with acid or salt, or it may be converted to a more tractable derivative.

Chlorosulfonic acid at − 8°C can occasionally be used in place of sulfuric acid (87). Another procedure is given in Section V.C.1.e of "Azo Group," Vol. 15.

### g. Nitro and Bromo Derivatives of Aromatic Azoxy Compounds

Only the beta isomer of p-nitroazoxybenzene is nitrated by acid of 1.45 specific gravity; similarly, only the beta isomer is brominated by bromine, and in both instances the substitution occurs in the *para* position (See Section III.B.2). Thus nitration and bromination afford relatively simple procedures for distinguishing alpha and beta isomers of the aromatic azoxy compounds that have open *para* positions.

Both rings of many of the aromatic azoxy compounds can be nitrated with a mixture of fuming nitric acid in phosphoric acid (120) or fuming nitric acid in glacial acetic acid (106).

### 2. Physical Methods

Identification of the azoxy compounds by physical methods is straightforward, but because of a marked scarcity of data it has been necessary to employ the usual degradative and derivative chemical methods for clarifying structural configurations. At present, absorption spectrophotometry is the principal physical method used for identifying azoxy compounds, but recent developments indicate that nuclear magnetic resonance and even mass spectroscopy will displace spectrophotometry for the elucidation of structures. Combinations of chromatographic separations and infrared and ultraviolet spectrophotometry are particularly useful. Edwards et al. (42) have developed a standard chromatographic-spectrophotometric technique for separation and identification of aromatic nitrogen compounds, including some aromatic azoxy compounds. Mixtures of compounds in benzene are adsorbed on silicic acid and developed with 19:1 benzene-acetone; azoxy compounds are easily separated from azo compounds and can be confirmed by their absorption maxima in the uv-visible spectrum.

## a. Infrared Spectrophotometry

The infrared spectra of the azoxy compounds are remarkably similar to those of the corresponding azo compounds; the differences which are apparent are largely due to the $N \rightarrow O$ frequencies, as described in Section V.B.2.a. Because of the similarity of spectra (see Fig. 1), the azoxy compounds can often be identified by reference to the spectra of azo compounds (see Section V.B.2.a of "Azo Group," Vol. 15).

## b. Ultraviolet Spectrophotometry

The salient features of the ultraviolet absorption spectra of the azoxy compounds were summarized in Section V.B.2.b, where it was indicated that the two absorption bands exhibited by aliphatic azoxy compounds and the three absorption bands of disubstituted azoxybenzenes are sufficiently characteristic to support infrared evidence of the presence of the azoxy group. The positions of the absorption maxima in the uv spectra of typical aromatic azoxy compounds are given in Table III.

Gore and Wheeler (59) found that the $E_1$-band at about 2300 Å in the azoxybenzenes is not appreciably shifted by substitution; changes in absorption intensity are observable, however. The band appears in spectra of the phenylnitrones as well as in spectra of the azobenzenes, and is attributed to electronic transitions in benzene rings; the band is considered to be the $E$-band of benzene at 2050 Å ($\epsilon = 6300$) displaced by substitution of the azoxy group. On occasions, the $E_1$-band of the azoxybenzenes appears as two peaks, perhaps because of unequal polarization induced by the proximity of the oxygen atom to one ring (alpha and beta isomers). Obviously, double maxima are not observed in aromatic azo compounds.

The second ($E_2$) band observed at about 2500 Å in the spectra of azoxybenzene and some of its derivatives seems to originate in transitions involving one of the benzene rings and the azoxy group. The $E_2$-band is often submerged by the much stronger $E_1$-band of many disubstituted compounds. In general, E-bands in the spectra of aromatic azoxy compounds are indicators of the $\pi - \pi^*$ transitions of the aromatic nucleus.

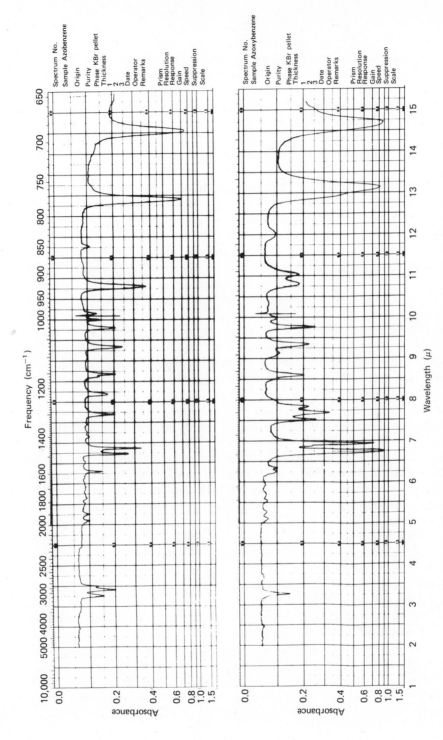

Fig. 1. Infrared spectra of KBr pellets containing azobenzene (top) and azoxybenzene (bottom).

# TABLE III

Position of Maxima and Corresponding Log ε Values in Absorption Spectra of Some Aryl Azoxy Compounds in Ethanol (wavelengths in nanometers; log ε values in parentheses)

| Compound | $E_1$-Band | $E_2$-Band | $K$-Band | $R$-Band | Ref. |
|---|---|---|---|---|---|
| Azoxybenzene | 231 (3.93) | 261 (3.87) | 323 (4.16) | | 9 |
| α-2-Phenylazoxynaphthalene | 216 (4.51) | 276 (4.28), 285 (4.23) | 340 (4.30) | | 9 |
| β-2-Phenylazoxynaphthalene | 214 (4.59), 241 (4.15) | 259 (4.24), 291 (4.08) | 331 (4.30) | | 9 |
| 2,2'-Azoxynaphthalene | 216 (4.71), 263 (4.41) | 277 (4.37), 289 (4.24) | 346 (4.41) | | 9 |
| β-1-Phenylazoxynaphthalene | 220 (4.54), 269 (4.12) | | 378 (4.06) | | 9 |
| β-1,2'-Azoxynaphthalene | 216 (4.76), 254 (4.25) | 262 (4.29), 291 (4.04) | 386 (4.17) | | 9 |
| 1,1'-Azoxynaphthalene | 218 (4.93), 262 (4.11) | 288 (3.91) | 364 (4.06) | | 9 |
| o,o'-Azoxytoluene | 235 (3.96) | | 311 (3.94) | | 9 |
| 4,4'-Dinitro-α-azoxybenzene | 227 (3.88) | 281 (3.75) | 342 (4.14) | | 59 |
| 4,4'-Dinitro-β-azoxybenzene | | | 337 | | 36 |
| 4,4'-Dimethoxy-α-azoxybenzene | 242 (4.06) | | 356 (4.41) | | 59 |
| 4,4'-Dimethoxy-β-azoxybenzene | | | 340 | | 36 |
| 2,2'-Dimethoxyazoxybenzene | 227 (3.59) | | 304 (3.41), 330 (3.37), 342 (3.37) | | 59 |
| 2,2'-Dichloroazoxybenzene | 223 (4.04) | | 304 (3.88) | | 59 |
| 4,4'-Dichloroazoxybenzene | 223 (3.96) | | 330 (4.27) | | 59 |
| $C_6H_5NO=NSO_2C_6H_4CH_3(p)$ | 236 (3.96) | | 290 (4.1) | 500 (2.0) | 49 |
| $p\text{-}(C_2H_5)NC_6H_4NO=NSO_2C_6H_4CH_3(p)$ | | 247 (4.0) | 465 (4.5) | | 49 |

The *K*-band of the 4- and the 4,4′-substituted azoxybenzenes at about 3200 Å undergoes pronounced bathochromic shifts in the sequence usually observed in other spectra, that is, $I < F < Me < Cl < Br < NO_2 < OMe < Ph$. The effect of iodine substitution is abnormal, especially since the order of mesomeric effects is $F < Me < Cl < Br < I < OMe < Ph$. K-band absorptions represent $\pi - \pi^*$ transitions of the chromophoric group.

The uv spectra of a few alpha and beta isomers of the azoxybenzenes have been studied (see Table III). Figure 2 shows the characteristic differences usually observed in the spectra of the alpha and beta isomers of a *trans*-azoxybenzene and the corresponding azo compound. The

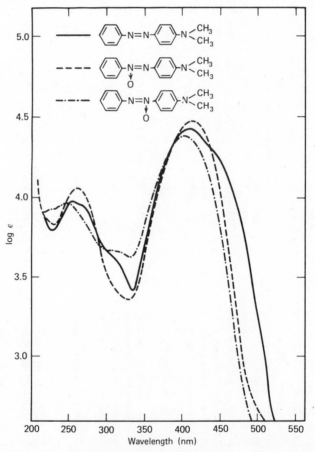

Fig. 2.  Absorption spectra of 4-dimethylaminoazobenzene and the alpha and beta isomers of 4-dimethylamino-azoxybenzene. [From Pentimalli (97).]

$R$-bands are characteristic of conjugated system $n - \pi^*$ transitions of the azoxy group and the aromatic nucleus.

Substitutions in *meta* positions have a slight effect on the $K$-band of aromatic azoxy compounds; however, since the nitro group is a powerful -$M$ substituent, it will reduce electron availability in the aromatic rings when it is in the *meta* position. As a consequence, new bands at about 3150 Å appear in the spectra of the 3,3'-dinitroazoxybenzenes, and the intensity of the entire $K$-band is decreased. *Ortho* substitutions lead to definite hypsochromic shifts and a decrease in the intensity of the $K$-band. Steric hindrances to coplanarity play an important part in the shifts and intensities of the $K$-band; large groups (e.g., iodo-) often cause splitting of the band. Singh et al. (107) have made comparisons of the spectra of a number of similarly substituted azo and azoxy compounds; their data are valuable for identifying azoxy compounds and for differentiating azoxy from azo compounds.

The uv spectra of disubstituted *cis*-azoxybenzenes show smaller bathochromic shifts of the $K$-bands (also lower intensity) than are observed for the corresponding *trans* derivatives (90).

The uv spectra of the azothiobenzenes (83) are similar to those of the azoxy compounds except for the bathochromic shifts, which are the result of the greater chromophoric effect of the azothio group. *Para*-substituted compounds show absorption maxima near 2300 Å (log $\epsilon$ = 4.1) and at 4250–4500 Å (log $\epsilon$ = 4.2), except for the bathochromic shifts observed in compounds such as $(CH_3)_2NC_6H_4N(S){=}NC_6H_5$ and $C_6H_5{-}C_6H_4N(S){=}NC_6H_5$. *Ortho* compounds have absorption maxima near 2600 Å (log $\epsilon$ = 3.9) and 3600–3700 Å (log $\epsilon$ = 3.8).

c. Nuclear Magnetic Resonance

Nuclear magnetic resonance spectra of *cis*- and *trans-p,p'*-dimethylazobenzene show a single methyl resonance, but in the spectra of the corresponding azoxy derivatives there are two methyl peaks separated by about 7 Hz because the alkyl groups exist in different magnetic environments (124). A similar separation was observed by Webb and Jaffé in the spectrum of carefully purified *cis-p*-methylazoxybenzene at 0°C, and since the methyl resonances at 112.6 and 105.5 Hz downfield of tetramethylsilane corresponded exactly to the two methyl peaks produced by *p,p'*-dimethylazoxybenzene, two *cis* isomers were indicated. Furthermore, as the solutions warmed to room temperature, two additional resonance peaks appeared at 124.4 and 118.3 Hz, indicating the formation of the "alpha"- and "beta"-*trans* isomers. In time, the higher field signals disappeared and the lower ones increased in intensity. In analogy with the observed methyl resonances of the alpha- and beta-*trans* isomers, the resonance at 112.6 Hz was assigned to the alpha-*cis* isomer and that at 105.5 Hz to the beta-*cis* isomer.

The chemical shifts observed in the nmr spectra of the azoxybenzenes are consistent with the assumption that the methyl resonance of *trans-p*-methylazobenzene will be affected by the degree of perturbation of the system induced by further substitution, oxidation, or change of configuration. The downfield chemical shifts observable in unsymmetrical aliphatic azoxy compounds reflect a reduction of electron density at both nitrogens and permit determination and confirmation of structures (50,110); however, the low solubility of *cis* compounds in organic solvents limits the general application of nuclear magnetic resonance. The azoxy structure of *cycasin* was confirmed by proton magnetic resonance (76).

### d.  PARACHOR

The parachor values of a series of azoxy compounds near their melting points were obtained by Bigelow and Keirstead (22) and compared with values computed from constants given by Sugden (112) and with the azoxy group value of 66.7 reported by Sugden et al. (113). In most instances, the calculated parachor values were in good agreement with those obtained from measured physical constants. There were some discrepancies; in a few instances, chromatographic separations performed on the melts indicated that the discrepancies were attributable to the presence of decomposition products. On the other hand, the experimentally determined parachor value for the compound believed to be *o*-nitro-*m,m'*-dichloroazoxybenzene was sufficiently different from the computed value (578.0 calc. vs. 555.3) to throw doubt on the assigned value (no decomposition products were detected). In another instance, the measured parachor value indicated that the compound could not contain the azoxy group. The studies of Bigelow and Keirstead (22) suggest that the parachor equivalent for the azoxy group in the simpler azoxybenzenes is 59.2.

### e.  MASS SPECTROMETRY

Very little work has been published on the mass spectrometry of azoxy compounds. The high-resolution mass spectrum of the major peaks obtained from azoxybenzene (see Fig. 3) may be compared with the mass spectrum of azobenzene (see Fig. 11 of "Azo Group," Vol. 15). Accurate mass measurements were made ($\pm 0.0005$ amu)

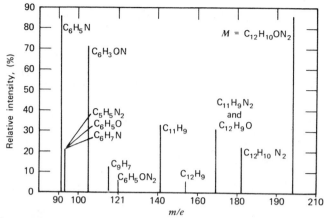

Fig. 3. Major peaks in high-resolution mass spectrum of azoxybenzene.

at the upper end of the spectrum. As shown in Fig. 3, the major peaks at the upper end of the spectrum are the molecular-ion peak ($M$), $M - C_6H_5NO$ ($m/e$ 91), $M - C_6H_7N$ ($m/e$ 105), $M - CHON_2$ ($m/e$ 141), and $M - O$ ($m/e$ 182). The strong bonding of the oxygen atom in the structure is indicated by the intensity of the molecular-ion peak and the fragment at $m/e$ 105; the oxygen atom also appears in less intense peaks, in the interesting skeletal rearrangement resulting in the formation of a fragment at $m/e$ 169 ($C_6H_5$—O—$C_6H_4$), and in the fragment at $m/e$ 93, $C_6H_5O$.

The mass spectra of seven aromatic azoxy compounds have been studied by Bowie et al. (26), and mechanisms for several rearrangement processes based on oxygen migration are described; several of the rearrangements are indicated in Fig. 3, for example, $M^+ - N_2O - H$ to form a diaryl ion ($m/e$ 153); $M^+ - CO - N_2 - H$ ($m/e$ 141); $M^+ - C_6H_5 \cdot NO \cdot (m/e$ 91).

The mass spectrum of azobenzene is similar, for example, in respect to the rearrangement of $M - N_2$, the scission of a phenyl group, and the rearrangement that permits scission of the N-N double bond. Dubov and Khokhlova (38a) have found the mass spectrum of hexafluoroazoxy-methane to be similar to that of hexafluoroazomethane with respect to the absence of a molecular ion, the maximum peak ($CF_3^+$), and the presence of an $N_2^+$ peak resulting from the elimination of nitrogen; the lesser intensity of the $N_2^+$ peak in the azoxy compound spectrum is attributed to the necessity for rupture of the $N^+$-$O^-$ coordination bond. A predominant splitting of the N-N double bond in hexafluoroazoxy-methane (in comparison to the azo analog) is in agreement with the results of thermal decomposition studies at 250–300°C.

## VI. QUANTITATIVE DETERMINATIONS

There has been little incentive to develop quantitative methods for the azoxy group and azoxy compounds because they are of very minor commercial importance. The azoxy compounds that are produced for scientific studies or for investigation of potential commercial use are usually analyzed by the Dumas method; a minor number are analyzed by Kjeldahl procedures.

Reductimetric procedures employing titanium(III) solutions have been tried with azoxybenzene and found to be not rigidly quantitative; it is unfortunate that this azoxy compound has been chosen for establishing the merits of titanium(III) and other reductimetric procedures, for azoxybenzene (like azobenzene) is peculiarly resistive to attack by the reducing agents useful for analytical work. As was indicated in earlier sections, the relative ease of reduction of the azoxy (and the azo) group is controlled in large measure by the substituents on the aromatic nuclei; in the near future reductimetric procedures will undoubtedly be found applicable to the quantitative determination of large numbers of highly substituted aromatic azoxy compounds.

### A. TITRATION WITH REDUCING AGENTS

Investigations of the use of titanium(III) solutions as titrants for the azoxy group have in general indicated that azoxybenzene is not reduced quantitatively. Hulle (69a) recommends determination of azoxy groups by titanous titrations. Stephen and Hinshelwood (108) have studied the kinetics of the reduction in acidic solutions and have found that azoxybenzene is only partly reduced to aniline. A submicromethod, in which titanium(III) is used in a citrate medium, yielded fairly acceptable results with azoxybenzene when reaction times of the order of 20 min were employed (20). Only 2 equiv. of reductants are consumed within 3 min; 4 equiv., corresponding to the formation of benzidine, are consumed in 20 min. For complete reduction of the azoxybenzene to aniline, 6 equiv. are required:

Azo:  $2 H^+ + 2 e + R-N(O)=NR \rightarrow R-N=N-R + H_2O$

Benzidine:  $4 H^+ + 4 e + R-N(O)=NR \rightarrow R-NHNH-R + H_2O$
$$\downarrow$$
$$NH_2-R-R-NH_2$$

Aniline:  $6 H^+ + 6 e + R-N(O)=NR \rightarrow 2 RNH_2 + H_2O$

The titration of 4,4′-dimethoxyazoxybenzene by titanium(III) in citrate medium proceeds smoothly and nearly quantitatively; 4 equiv. of reductant are consumed and the amount is independent of time. The essential steps in the procedure found satisfactory for the azoxy compounds are as follows (20):

A sample of from 50 to 80 $\mu$g is dissolved in 0.2–0.3 ml of ethanol and then mixed with 0.4 ml of water and 0.3 ml of 20% aqueous potassium citrate. The resulting solution is deoxygenated by a stream of nitrogen in an apparatus which permits titration in an inert atmosphere (such as nitrogen). A measured excess of titanium(III) sulfate (250 $\mu$l) is added, and the whole allowed to remain from 5 to 20 min (according to the nature of the sample). Then 0.2 ml of 12 N sulfuric acid solution is added, followed by 100 $\mu$l of iron(III) ammonium sulfate solution and 0.04 ml of a 10% aqueous solution of ammonium thiocyanate. After 5 min, the excess iron(III) is titrated with titanium(III).

A general method for determining azoxy compounds titrimetrically with chromium(II) chloride has been described by Bottei (25). Azoxybenzene (0.8–0.9 mequiv) was dissolved in 1:1 sulfuric acid and deoxygenated; after 15 min, a measured excess of 0.1 N chromium(II) chloride was added and the mixture allowed to stand at least 3 min; the excess of chromium(II) was back-titrated with ferric alum (see Section VII of "Azo Group," Vol. 15). The determinations were quantitative. Attempts at direct titration with potentiometric end-point detection 60°C were abandoned because the reductions were too slow near the end points. Azoxybenzene consumed 4 equiv. of reductant/mol; azobenzene consumed only 2 equiv., but other azo compounds required 4 equiv. (formation of hydrazo compounds or benzidines) and gave an average error of 0.27%.

## B. POLAROGRAPHIC METHODS

There have been no reports of the systematic application of polarography for the quantitative determination of compounds containing the azoxy group. However, the studies of Wawzonek and Fredrickson (123) have established that azoxybenzene is smoothly reduced at the dropping mercury electrode when well-buffered solutions of 50% isopropanol and water are used to maintain the compound in solution.

The polarographic reduction of azoxybenzene at pH 2–12 involves the irreversible transfer of four electrons to form hydrazobenzene; the product may rearrange to benzidine. In general, the half-wave potential is a linear function between pH 2 and about 7; at values between 7 and 11, the half-wave potential is still a linear function, but it has much less slope than at the lower pH values (123). Costa (35) found that the alpha and beta isomers of $p$-hydroxyazoxybenzene gave single polarographic steps at $-0.78$ and $-0.80$ V, respectively, at pH 7. Four electrons are involved in the irreversible reduction which forms an unstable hydrazo derivative, $p$-$HOC_6H_4NHNHC_6H_5$; the hydrazo compound reacts readily with unreduced $p$-hydroxyazoxybenzene to form the stable but more easily reduced $p$-hydroxyazobenzene. The $p$-hydroxyazobenzene is reduced irreversibly at $-0.6$ V with the transfer of four electrons to aniline and $p$-hydroxyaniline. Hence the polarograms of the isomers of $p$-hydroxyazoxybenzene show several waves. The progressive additions of electrons from azoxy to azo to hydrazo groups, however, are not resolved as polarographic waves.

Costa (35) also studied the polarographic behavior of the *alpha* and *beta* isomers of $p$-methoxyazoxybenzene. Characteristic of the aromatic azoxy compounds, the half-wave potentials were single steps at increasing potentials ($-0.4$ to $-1.1$ V) with increasing pH. The *beta* isomer, $p$-$CH_3OC_6H_4N(O)$=$NC_6H_5$, was reduced at a slightly more negative potential.

Polarographic procedures intended for quantitative determinations of the azoxy group require rigid control of pH; 0.1 M acetate buffers in 50% alcoholic solutions are recommended. Furthermore, any polarographic procedure developed for a specific azoxy compound should be checked at various levels of concentration of sample in order to determine the effect of the interaction of reduced products with unreduced sample. Dilute solutions of azoxy compounds must be protected from light. A procedure that has been developed specifically for analysis of samples from a given origin should be checked frequently by inspecting the entire polarogram to ensure that the isomer content of the samples has not changed and that inclusions of new sorts of impurities are not influencing the polarographic behavior of the azoxy group. For example, Holleck and Holleck (68) have found that carboxymethylcellulose and triphenylphosphite shift the half-wave potential toward more negative values, while gelatin shifts it toward more positive values; reduction occurs as a surface process in the absence of inhibitors and as a space process when inhibitors are present. Gelatin promotes proton transfer in the adsorbed layer.

## C. KJELDAHL METHODS

The procedures given in Sections VII.B and VI.A.3 of "Azo Group," Vol. 15, for the determination of the nitrogen contents of azo group compounds are also applicable to azoxy group compounds. Section III.B.1 may be consulted for assistance in the selection of appropriate reducing agents for the azoxy group.

## D. IODOMETRIC METHOD

The procedure given by Aldrovandi and De Lorenzi (1) for reduction of the azoxy group by hydriodic acid is reported to be quantitative. The method involves heating the sample in a sealed tube with hydriodic acid; the liberated iodine is determined by titration with standard thiosulfate solution. The procedure is particularly attractive for analysis of the rather volatile aliphatic azoxy compounds.

## VII. LABORATORY PROCEDURES

The azoxy group in many compounds, especially the highly substituted aromatic types, can often be determined by titration with titanous chloride by methods such as those described in Section VII of "Azo Group," Vol. 15. However, quantitative reduction of azoxy groups can be accomplished more readily by titration with chromous ion in strongly acidic solutions; the following procedure is an adaptation of the method of Bottei (25).

### A. CHROMOUS CHLORIDE REDUCTION

The azoxy group is reduced only to the hydrazo stage:

$$-N{=}N- + 4\,H^+ + 4\,e \rightarrow -N{-}N- + H_2O$$
$$\underset{O}{\big|} \qquad\qquad \underset{H\;\;H}{\big|\;\;\big|}$$

Inasmuch as chromous solutions are sensitive to oxidation, titrations must be performed under oxygen-free conditions; also, it should be noted that other functional groups may also be reduced.

**Reagents**

*Standard chromous chloride,* 0.1 N. Half-fill the storage flask of the apparatus described below with amalgamated, mossy reagent-grade zinc, and then fill to capacity with a solution of 0.1 M chromic chloride (in 0.1 M HCl). The amalgamation is performed with a dilute solution of mercury chloride (in HCl) in sufficient quantity to provide 2% mercury. Allow the reduction to proceed overnight. Store the solution in the storage flask under hydrogen which has been freed of oxygen by bubbling through a solution of chromous chloride in normal sulfuric acid that is in contact with amalgamated zinc. Standardize the chromous solution by titration of 25-ml aliquots (mixed with 25 ml of 12 N HCl) of an aqueous 0.1000 N solution of cupric sulfate pentahydrate (84a).

*Ferric alum solution,* 0.1 N. Dissolve 26.6 grams of $NH_4Fe(SO_4)_2 \cdot 12\,H_2O$ in water containing 28 ml of concentrated sulfuric acid, and dilute to 1 liter.

**Apparatus**

A suitable titration cell can be made from a small 200- or 300-ml three-neck flask. One neck bears a stopper with a small hole through which is passed the tip of the buret; a two-hole stopper in the center neck permits insertion of a gas delivery tube (loosely, so that it can be raised or lowered) and loosely supports a small glass tube that has a platinum wire sealed to its end; the third neck is outfitted with a stopper that supports a salt bridge or, preferably, a saturated calomel electrode-bridge combination of the type ordinarily used with pH meters. Agitation of the titration medium is best accomplished with a magnetic stirring bar.

Preparation, storage, and dispensing of standard chromous solutions can readily be effected in the apparatus shown in Fig. 9 of "Diazonium Group," Vol. 15, and described in Section VII.C of "Diazonium Group"; alternatively, the apparatus of Lingane and Pecsok (84a) may be used.

**Procedure**

Dissolve about 1 mequiv. of the azoxy group compound in about 25 ml of 1:1 sulfuric acid solution contained in a titration flask; set the flask in place under the buret which contains chromous titrant, and pass oxygen-free inert gas through the solution for about 15 min to purge the flask and its contents of oxygen.

Dispense an accurate volume of about 20 ml of the chromous titrant into the sample solution and mix for 3 min (or more) by passing a slow stream of inert gas through the liquid in the flask. Back-titrate the excess of chromous ion with standardized ferric alum solution; since platinum is a catalyst for the decomposition of chromous ion by hydrogen ion, lower the platinum electrode into the solution just before back titration. Determine the equivalence point potentiometrically by any convenient method; the end-point breaks are usually quite pronounced (e.g., 150–500 mV).

## REFERENCES

1. Aldrovandi, R., and F. De Lorenzi, *Ann. Chim. (Rome)*, **42**, 298 (1952); through *Chem. Abstr.*, **46**, 10051 (1952).

2. Alway, F. S., and W. O. Bonner, *J. Am. Chem. Soc.*, **27**, 1107 (1905).

3. Anderson, W., *J. Chem. Soc.*, **1952**, 1722.

4. Angeli, A., *Gazz. Chim. Ital.*, **46**(ii), 82 (1916).

5. Angeli, A., *Gazz. Chim. Ital.*, **51**(i), 35 (1921).

6. Angeli, A., D. Bigiavi, and G. Carrara, *Atti Accad. Lincei*, **31**(v), 439 (1922); through *Chem. Abstr.*, **17**, 1447 (1923).

7. Angerer, L., *Nature*, **199**, 62 (1963).

8. Arshid, F. M., et al., *J. Soc. Dyers Colour.*, **69**, 11 (1953).

9. Badger, G. M., R. G. Buttery, and G. E. Lewis, *J. Chem. Soc.*, **1953**, 2143, 2156.

10. Badger, G. M., and G. E. Lewis, *J. Chem. Soc.*, **1953**, 2147.

11. Badger, G. M., and G. E. Lewis, *J. Chem. Soc.*, **1953**, 2151.

12. Badger, G. M., and G. E. Lewis, *Nature*, **167**, 403 (1951).

13. Baker, A. H., G. F. H. Whitney, and A. N. Worden, *Lab. Pract.*, **8**, 3–10, 26 (1959); through *Chem. Abstr.*, **54**, 21621 (1960).

14. Balcon, D., and A. Furst, *J. Am. Chem. Soc.*, **75**, 4334 (1953).

15. Bamberger, E., *Berichte*, **33**, 113 (1900).

16. Bamberger, E., *Berichte*, **33**, 3192 (1900).

17. Bamberger, E., and R. Hubner, *Berichte*, **36**, 3803 (1903).

18. Behr, L. C., *J. Am. Chem. Soc.*, **76**, 3672 (1954).

19. Behr, L. C., E. G. Alley, and O. Levand, *J. Org. Chem.*, **27**, 65 (1962).

20. Belcher, R., Y. A. Gawargious, and A. M. G. MacDonald, *J. Chem. Soc.*, Suppl. 1, **1964**, 5698.

21. Bergman, I., and J. C. James, *Trans. Faraday Soc.*, **50**, 60 (1954).

22. Bigelow, H. E., and K. F. Keirstead, *Can. J. Chem. Res.*, **24B**, 232 (1946).

23. Bigelow, H. E., D. McIntosh, and W. M. McNevin, *Trans. Roy. Soc. Can.*, Sect. III, **23**, 119 (1929).

24. Bigelow, H. E., and D. B. Robinson, *Organic Syntheses*, Vol. 22, John Wiley, New York, 1942, p. 28.

25. Bottei, R., *Anal. Chim. Acta*, **30**, 6 (1964).

26. Bowie, J. H., G. E. Lewis, and R. G. Cooks, *Chem. Commun.*, **1967**, 284.

27. British Pat. 907, 105 (Oct. 3, 1962), P. F. Holt, A. N. Hughes, and R. S. W. Braithwaite; through *Chem. Abstr.*, **58**, 589 (1963).

28. Brough, J. N., B. Lythgoe, and P. Waterhouse, *J. Chem. Soc.*, **1954**, 4069.

29. Buckler, S. A., et al., *J. Org. Chem.*, **27**, 794 (1962).

30. Burawoy, A., *J. Chem. Soc.*, **1939**, 1177.

31. Calderbank, K. E., and R. J. W. LeFevre, *J. Chem. Soc.*, **1948**, 1949.

32. Campbell, N., A. Henderson, and D. Taylor, *Mikrochem. Ver. Mikrochim. Acta*, **38**, 376 (1951).

33. Chuang, L., L. Fried, and P. J. Elving, *U.S. AEC COO-1148-91* (1964): through *Chem. Abstr.*, **63**, 7904 (1965).

34. Cook, A. H., and D. G. Jones, *J. Chem. Soc.*, **1939**, 1309.

35. Costa, G., *Ann. Triest. Univ. Trieste*, Sect. 2, **2223**, 115 (1953); through *Chem. Abstr.*, **48**, 4331 (1954).

36. Costa, G., *Gazz. Chim. Ital.*, **85**, 548 (1955); through *Chem. Abstr.*, **49**, 14483 (1955).

37. Desai, N. F., and C. H. Giles, *J. Soc. Dyers Colour.*, **65**, 639 (1949).

38. Druckrey, H., et al., *Z. Krebforsch.*, **67**(1), 31 (1965); through *Chem. Abstr.*, **63**, 12098 (1965).

38a. Dubov, S. S., and A. M. Khokhlova, *Zh. Obshch. Khim.*, **34**, 586 (1964).

39. Dunstan, I., H. G. Adolph, and M. J. Kamlet, *Tetrahedron*, **20**, Suppl. 1, 431 (1964).

40. Dziomko, V. M., and K. A. Dunaevskaya, *Acta Chim. Acad. Sci. Hung.*, **32**, 223 (1962); through *Chem. Abstr.*, **58**, 5028 (1963).

41. Dziomko, V. M., and K. A. Dunaevskaya, *Zh. Obshch. Khim.*, **31**, 68 (1961); through *Chem. Abstr.*, **55**, 23393 (1961).

42. Edwards, W. R., O. S. Pascual, and C. W. Tate, *Anal. Chem.*, **28**, 1045 (1956).

43. Elbs, K., *Z. Elektrochem.*, **2**, 472 (1896).

44. Elbs, K., and B. Schwarz, *J. Prakt. Chem.*, **71**, 562 (1901).

45. Evans, I., and H. S. Fry, *J. Am. Chem. Soc.*, **26**, 1161 (1904).

46. Fichter, F., and W. Jaeck, *Helv. Chim. Acta*, **4**, 1000 (1921).

47. Flurscheim, B., and T. Simon, *J. Chem. Soc.*, **93**, 1463 (1909); through *Chem. Abstr.*, **3**, 171 (1909).

48. Foffani, A., et al., *Tetrahedron Lett.*, No. 11, 21 (1959).

49. Freeman, H. C., et al., *J. Chem. Soc.*, **1952**, 3384.

50. Freeman, J., *J. Org. Chem.*, **28**, 2508 (1963).

51. Furst, A., and R. E. Moore, *J. Am. Chem. Soc.*, **79**, 5492 (1957).

52. Galbraith, J. W., E. F. Degering, and E. F. Hitch, *J. Am. Chem. Soc.*, **73**, 1323 (1951).

53. Gaudry, R., and K. F. Kierstead, *Can. J. Res.*, **27B**, 897 (1949).

54. George, M. V., R. W. Kierstead, and G. F. Wright, *Can. J. Chem.*, **37**, 679 (1959).

55. German Pat. 228,722 (May 1, 1908), Chemikalienwerk, Griesheim; through *Chem. Abstr.*, **5**, 592 2156 (1911).

56. Gillis, B. T., and K. F. Schimmel, *J. Org. Chem.*, **27**, 413 (1962).

57. Gore, P. H., *Chem. Ind.*, **1959**, 191.

58. Gore, P. H., and G. K. Hughes, *Anal. Chim. Acta*, **5**, 357 (1951).

59. Gore, P. H., and O. H. Wheeler, *J. Am. Chem. Soc.*, **78**, 2160 (1956).

59a. Gould, C. W., and H. J. Drake, *Anal. Chem.*, **23**, 1157 (1951).

60. Gowenlock, B. G., *Can. J. Chem.*, **42**, 1936 (1964).

61. Greenspan, F. P., *Ind. Eng. Chem.*, **39**, 847 (1947).

62. Hahn, C-S, and H. H. Jaffé, *J. Am. Chem. Soc.*, **84**, 946 (1962).

63. Hahn, C-S, and H. H. Jaffé, *J. Am. Chem. Soc.*, **84**, 949 (1962).

64. Hashimoto, S., and J. Sunamoto, *Kagaku To Kogyo*, **37**(1), 45 (1963).

65. Hashimoto, S., and J. Sunamoto, *Kagaku To Kogyo*, **37**(5), 230 (1963); through *Chem. Abstr.*, **60**, 437 (1964).

66. Heumann, K., *Berichte*, **5**, 913 (1872).

67. Hodkin, A., and D. Taylor, *J. Chem. Soc.*, **1955**, 489.

68. Holleck, L., and G. Holleck, *Z. Naturforsch.*, **19b**, 162 (1964); through *Chem. Abstr.*, **60**, 14126 (1964).

69. Hornsby, S., and W. L. Peacock, *Chem. Ind.*, **1958**, 858.

69a. Hulle, E. V., in Houben–Weyl–Müller, *Methoden der Organischen Chemie*, Vol. 2, 4th ed., Theime, Stuttgart, 1953, p. 705.

70. Jander, J., and R. N. Hazeldine, *J. Chem. Soc.*, **1954**, 919.

71. Joshi, S. A., and W. P. Patwardhan, *Curr. Sci.*, **22**, 239 (1953).

72. Jurecek, M., V. Novak, and P. Kozak, *Talanta*, **9**, 72 (1962).

73. Khalifa, M., *J. Chem. Soc.*, **1960**, 1854.

74. King, M. V., *J. Org. Chem.*, **26**, 3323 (1961).

75. Kobayashi, M., *J. Chem. Soc. Jap.*, **74**, 968 (1953); through *Chem. Abstr.*, **49**, 2989 (1955).

76. Korsch, B. H., and N. V. Riggs, *Tetrahedron Lett.*, No. 10, 523 (1964).

77. Kost, A. N., I. T. Nette, and N. V. Pomortsera, *Vestn. Moskov. Univ. Ser. Mat., Mekh. Astron., Fiz. Khim.*, **14**(3), 213 (1959); through *Chem. Abstr.*, **54**, 13708 (1960).

78. Kwiatek, J., I. L. Mador, and J. K. Seyler, *J. Am. Chem. Soc.*, **84**, 304 (1962).

79. Lachman, A., *J. Am. Chem. Soc.*, **24**, 1178 (1902).

80. Langley, B. W., B. Lythgoe, and L. S. Rayner, *J. Chem. Soc.*, **1951**, 2303.

81. Langley, B. W., B. Lythgoe, and L. S. Rayner, *J. Chem. Soc.*, **1952**, 4191.

82. Langley, B. W., B. Lythgoe, and N. V. Riggs, *J. Chem. Soc.*, **1951**, 2309.

83. Leandri, G., and A. Mangini, *Bull Sci. Fac. Chim. Ind. Bologna*, **15**, 51 (1957); through *Chem. Abstr.*, **51**, 14417 (1957).

84. Leandri, G., and P. Rebora, *Gazz. Chim. Ital.*, **87**, 503 (1957).

84a. Lingane, J. J., and Pecsok, R. L., *Anal. Chem.*, **20**, 425 (1948).

85. Linten, E. P., C. H. Holder, and H. E. Bigelow, *Can. J. Res.*, **B19**, 132 (1941).

86. Lob, W., *Berichte*, **33**, 2333 (1900).

87. Lukashevich, V. O., and T. N. Kurdyumova, *Zh. Obshch. Khim.*, **18**, 1963 (1948); through *Chem. Abstr.*, **43**, 3800 (1949).

88. Maier, W., and G. Englert, *Z. Elektrochem.*, **62**, 1020 (1958).

89. McEwen, W. E., et al., *J. Am. Chem. Soc.*, **78**, 4587 (1956).

90. Muller, E., and E. Hory, *Z. Phys. Chem. Soc.*, **A162**, 281 (1932).

90a. Muraca, R. F., unpublished data.

91. Nakazawa, K., *J. Antibiot. (Jap.)*, **I**, 329 (1956).

92. Novak, V., *Collect. Czech. Chem. Commun.*, **28**, 3443 (1963).

93. Nystrom, R. F., and W. G. Brown, *J. Am. Chem. Soc.*, **70**, 3738 (1948).

94. Ogata, Y., M. Tsuchida, and Y. Takagi, *J. Am. Chem. Soc.*, **79**, 3397 (1957).

95. Opolinick, N., *Ind. Eng. Chem.*, **27**, 1045 (1935).

96. Parsons, I., and J. C. Bailar, *J. Am. Chem. Soc.*, **58**, 268 (1936).

97. Pentimalli, L., *Tetrahedron*, **5**, 27 (1959).

98. Petriew, H., *Berichte*, **6**, 556 (1873).

99. Popp, F. D., and H. P. Schultz, *Chem. Rev.*, **62**, 19 (1962).

100. Schultz, G., *Berichte*, **17**, 464 (1884).

101. Shamma, M., J. K. Whitesell, and P. L. Warner, Jr., *Tetrahedron Lett.*, No. 43, 3169 (1965).

102. Shemyakin, M. M., et al., *Dokl. Akad. Nauk SSSR*, **135**, 346 (1960); through *Chem. Abstr.*, **55**, 11337 (1961).

103. Shemyakin, M. M., et al., *Izv. Akad. Nauk. SSSR, Ser. Khim.*, **1963**(7), 1339; through *Chem. Abstr.*, **59**, 12619 (1963).

104. Shemyakin, M. M., V. J. Maimind, and B. Vaichunaite, *Chem. Ind.*, **1958**, 755.

105. Shemyakin, M. M., V. J. Maimind, and B. Vaichunaite, *Izv. Akad. Nauk. SSSR, Otdel. Khim. Nauk.*, **1960**, 566; through *Chem. Abstr.*, **54**, 24474 (1960).

106. Singh, J., et al., *Can. J. Chem.*, **40**, 1921 (1962).

107. Singh, J., et al., *Can. J. Chem.*, **41**, 499 (1963).

108. Stephen, M. J., and C. Hinshelwood, *J. Chem. Soc.*, **1955**, 1393.

109. Stevens, C. L., et al., *J. Am. Chem. Soc.*, **80**, 6088 (1958).

110. Stevens, T. E., *J. Org. Chem.*, **29**, 311 (1964).

111. Stewart, G. W., *J. Chem. Phys.*, **4**, 231 (1936).

112. Sugden, S., *The Parachor and Valency*, Geo. Rutledge, London, 1930.

113. Sugden, S., J. B. Reed, and H. Wilkins, *J. Chem. Soc.*, **127**, 1525 (1925).

114. Tadros, W., M. S. Ishak, and E. Bassili, *J. Chem. Soc.*, **1959**, 627.

115. Tanabe, H., *J. Pharm. Soc. Jap.*, **76**, 1023 (1956); through *Chem. Abstract*, **51**, 2598 (1957).

116. Tsai, T. T., W. E. McEwen, and J. Kleinberg, *J. Org. Chem.*, **25**, 1186 (1960).

117. U.S. Pat. 1,589,936 (June 22, 1926). O. W. Brown and C. O. Henke; through *Chem. Abstr.*, **20**, 3016 (1926).

118. U.S. Pat. 2,987,436 (June 6, 1961), R. J. Geary; through *Chem. Abstr.*, **55**, 21468 (1961).

119. U.S. Pat. 3,052,667 (Sept. 4, 1962), W. E. Hanford, J. W. Copenhaver, and H. R. Davis; through *Chem Abstr.*, **58**, 2405 (1963).

120. Urbanski, T., and J. Urbanski, *Bull. Acad. Polon. Sci. Cl. (III)*, **6**, 305 (1958); through *Chem. Abstr.*, **52**, 19994 (1958).

121. Wacker, L., *Annalen*, **317**, 375 (1901).

122. Wallach, O., and L. Belli, *Berichte*, **13**, 525 (1880).

123. Wawzonek, S., and J. D. Fredrickson, *J. Am. Chem. Soc.*, **77**, 3988 (1955).

124. Webb, D. L., and H. H. Jaffé, *J. Am. Chem. Soc.*, **86**, 2419 (1964).

125. Wild, F., *Characterisation of Organic Compounds*, Cambridge Press, London, 1960.

126. Witkop, B., and H. M. Kissman, *J. Am. Chem. Soc.*, **75**, 1975 (1953).

127. Yang, J. Y., W. E. McEwen, and J. Kleinberg, *J. Am. Chem. Soc.*, **80**, 4300 (1958).

128. Zawalski, R. C., and P. Kovacic, *J. Org. Chem.*, **44**, 2130, (1979).

129. Zinin, N., *J. Prakt. Chem.* **36**, 93 (1881).

# NITRO AND NITROSO GROUPS

BY R. F. MURACA, *College of Notre Dame, Belmont, California*

**Contents**

# I. INTRODUCTION

Compounds that contain the nitro group are of great commercial importance, for they are widely employed as intermediates in the manufacture of other chemical materials and are often used per se as explosives. As a consequence, considerable attention has been given to the development of general analytical procedures for the nitro group and to the search for methods by which specific nitro compounds can be determined. In contrast, the nitroso compounds are of only minor industrial importance; fortunately, procedures for their analysis are very similar to those used to determine the nitro group.

Nitro compounds are classified by the atom to which the $—NO_2$ group is attached:

1. *C*-nitro compounds, in which the nitro group is attached directly to a carbon atom (the true nitro compounds), that is, $\equiv C—NO_2$.

2. *N*-nitro compounds, or the nitramines, $\equiv N—NO_2$ (see Table I).

3. *O*-nitro compounds, which are the esters of nitric acid, $\equiv C—O—NO_2$.

Nitrosation processes give rise to compounds containing the —NO group, and these are also conveniently classified by the atom to which the functional group is attached:

1. *C*-nitroso compounds, containing the $\equiv$C—N$=$O group.
2. *N*-nitroso compounds, containing the $=$N—N$=$O group (nitrosamines).
3. *O*-nitroso compounds (the nitrous acid esters), containing the $\equiv$C—O—N$=$O group.

Compounds with a nitro and a nitroso group on the same carbon are called pseudonitroles.

The *C*-nitro and the *C*-nitroso compounds can be further subdivided in accordance with the number of hydrogen atoms on the carbon, following the nomenclature of the alcohols; thus compounds with the

TABLE I
Types of Nitramines

| Type | Example | |
|---|---|---|
| Primary nitramine | Ethylnitramine | $C_2H_5NHNO_2$ |
| Primary isonitramine | Ethylisonitramine (tautomer of ethylnitramine) | $C_2H_5N=NOH$ with O above |
| Primary nitramide | Nitrourethane | $C_2H_5-O-\overset{O}{\overset{\|}{C}}-NHNO_2$ |
| Secondary nitramine | Diethylnitramine | $C_2H_5\overset{NO_2}{\overset{\|}{N}}-C_2H_5$ |
| Secondary isonitramine | Methyl methylisonitramine | $CH_3O-N=N-CH_3$ with O above |
| Secondary nitramide | Methylnitrourea | $CH_3-\overset{O_2N}{\overset{\|}{N}}-\overset{O}{\overset{\|}{C}}-NH_2$ |
| Nitrimine | 2-Nitriminobutane | $C_2H_5-\overset{CH_3}{\overset{\|}{C}}=NNO_2$ |
| Polynitrimine | Cyclonite (RDX) (Hexahydro-1,3,5-trinitro-1,3,5-triazine) | (structure of RDX) |

group $-CH_2-NO_2$ are primary C-nitro compounds, and those that have the nitro or nitroso group attached directly to an aryl nucleus are obviously tertiary derivatives.

Compounds containing the O-nitro and the O-nitroso group are esters, and analytical procedures for their determination are different from those for the analysis of other nitro and nitroso compounds; therefore they will be discussed under "Nitrate and Nitrite Ester Groups."

## II. SYNTHESIS AND OCCURRENCE

There are many general methods for nitrating and nitrosating compounds, and a number of procedures have been developed to produce specific compounds in large quantities; only the principal reactions of general applicability can be noted here. As a rule, the methods employed for attaching nitro and nitroso groups involve vigorous attack on a starting material by nitric acid, nitrous acid, or a number of active nitrogen-containing reagents; a variety of isomers are usually formed along with some degradation products, and the result of an analysis for a given principal nitro or nitroso compound must always be considered tentative until the effect of adventitious material has been evaluated. It is often helpful for the analytical chemist to be aware of the synthetic route used to prepare a sample submitted for analysis. Some theoretical aspects of nitrosations by use of nitrous acid have been reviewed by Turney and Wright (696).

### A. C-NITROPARAFFINS

A limited number of C-nitroparaffins have been produced since 1940 on an industrial scale by the direct nitration of alkanes with nitric acid at 400–450°C. A variety of mononitro substitution products are formed, even when a pure alkane is treated with nitric acid; for example, propane yields 1-nitropropane and 2-nitropropane, as well as nitroethane and nitromethane through scission of carbon bonds. Reaction mixtures are usually separated by fractional distillation (13,701,702), and it follows that a commercial sample may often contain a group of related compounds with one in preponderance. The vapor-phase nitration of alkanes by nitric acid or nitrogen dioxide has made available nitroparaffins in sufficient quantities to serve as chemical intermediates and as solvents, and serious consideration has been given to the use of nitromethane as a monopropellant for rocketry and as an automotive fuel (455). Nitrocyclohexane and 2,2-dinitropropane are also prepared by liquid-phase nitration.

The laboratory synthesis of $C$-nitroparaffins by the classical method of Victor Meyer (492) involves the reaction of silver nitrite with an alkyl iodide in ether solution. Although nitromethane is readily and nearly quantitatively produced from methyl iodide, the higher alkyl halides (more than four carbons) produce isomeric $O$-nitroso compounds as well. Variations of the Victor Meyer reaction utilize alkyl halides and sodium nitrite in dimethylformamide or dimethyl sulfoxide (379); primary as well as secondary nitroparaffins can be produced but $O$-nitrisation (nitrite esterification) is a competing reaction and the tertiary alkyl halides yield scarcely any of the desired product.

A variety of compounds can be converted to nitromethane; hence these compounds can be detected by the sensitive and specific colorimetric test for nitromethane described in Sections III.C.5.a and V.B.1.a(2). Nitromethane can be obtained as follows (185):

1. $XCH_2COONa + NaNO_2 \rightarrow CH_3NO_2 + NaX + NaHCO_3$
2. $(CH_3)_2SO_4 + NaNO_2 \rightarrow 2CH_3NO_2 + Na_2SO_4$
3. $Na(CH_3)SO_4 + NaNO_2 \rightarrow CH_3NO_2 + Na_2SO_4$
4. $CH_3I + NaNO_2 \rightarrow CH_3NO_2 + NaI$

Primary and secondary $C$-nitroparaffins can also be synthesized by oxidation of oximes with trifluoroperoxyacetic acid in a solvent (e.g., acetonitrile):

$$R_2C{=}NOH \rightarrow R_2CHNO_2$$

A basic material is used to neutralize the trifluoroacetic acid formed in the reaction (166,270).

Tertiary $C$-nitroparaffins are readily synthesized by oxidation of tertiary carbinamines with potassium permanganate:

$$(CH_3)_3C{-}NH_2 \rightarrow (CH_3)_3C{-}NO_2$$

Paraffins containing a tertiary hydrogen are easily nitrated to $C$-nitro compounds by nitric acid.

Kolbe's method for preparation of $C$-nitroparaffins (368) is based on the distillation of a mixture of an aqueous solution of alkali-metal nitrite and an $\alpha$-halogen-substituted fatty acid (729):

$$CH_2ClCOONa \rightarrow CH_2(NO_2)COONa \rightarrow CH_3NO_2 + NaHCO_3$$

When the halogen atom is attached to a tertiary carbon atom, a small yield of a *pseudonitrole* is obtained in place of the expected $C$-nitroparaffin:

$$(CH_3)_2CBrCOOH \xrightarrow{\text{NaNO}_2} (CH_3)_2C \Big\langle \begin{matrix} NO \\ NO_2 \end{matrix}$$

The pseudonitroles can also be obtained by treating oximes with nitrogen tetroxide or by the action of nitrous acid on secondary nitroalkanes (see Section III.C.3.a).

Aliphatic polynitro compounds (C-nitro) are used as explosives; each compound is produced by special methods of which only a few are examples of general procedures for the preparation of C-nitro compounds. For example, 2,2-dinitropropane is made by oxidizing a pseudonitrole; the compound is more powerful an explosive than TNT. A comprehensive review of the chemistry of aliphatic polynitro compounds and their derivatives has been prepared by Noble et al. (520). Tetranitromethane is prepared commercially by treating acetic anhydride with fuming nitric acid at room temperature (105); its use as a rocket propellant has been evaluated.

The addition of nitrogen tetroxide to olefinic double bonds and the action of concentrated nitric acid on ketones are examples of general reactions for producing polynitro compounds, but there are many competing side reactions; thus the action of nitric acid on acetone also produces methylnitrolic acid, $CH_3COCNO_2(NOH)$, and the addition of nitrogen tetroxide to olefinic compounds yields various amounts of mixed vic-C-nitro and O-nitroso compounds as well as the desired vic-di-C-nitro products.

The nitrolic acids, or the oximes of nitroaldehydes, $RC(NO_2)$=NOH, are weakly acidic compounds which are pale yellow or white explosives when in the solid state, but give blood-red anions in solution; they are usually encountered in analytical methods as the products of reaction of nitrous acid with a primary nitroalkane.

C-Nitroolefins can be prepared by treating C-nitroalkanols with dehydrating agents such as $KHSO_4$, or phthalic anhydride (85). Nitroethylene polymerizes to a white powder which is a weak explosive but is too unstable for general use. Compounds containing the nitroethylene radical, C=C—$NO_2$, are used as fungicides or insecticides (704); they may be synthesized by removal of acetic acid from the acetate of a nitroalcohol (264). Compounds such as the ω-nitrostyrenes are produced by alkaline condensation of nitroparaffins with benzaldehyde in the presence of amines (263).

C-Nitroalcohols are used as chemical intermediates, as hardening agents for protein adhesives, and in the manufacture of resins. They are synthesized commercially by condensation of a nitroparaffin and an aldehyde in the presence of a basic catalyst (see Section III.B.5).

Chloropicrin, $Cl_3CNO_2$, is used as a soil disinfectant and a tear gas; it may be synthesized by action of chlorine or a hypochlorite on picric acid or its salts (733), or by the reaction of hypochlorites with nitromethane (705).

## B. *N*-NITROPARAFFINS

*N*-Nitroalkyl compounds are usually called nitramines (or, sometimes, nitroamines) and may be primary, R—NH—NO$_2$, or secondary, R$_2$=N—NO$_2$. The structures reflect a relationship to nitroamine itself, NH$_2$—NO$_2$; in turn, nitroamine may be regarded as the amide of nitric acid. With a primary nitramine, an isomerism seems to be possible with an isonitramine:

$$-N\overset{\displaystyle NO}{\underset{\displaystyle OH}{\Big\langle}} \quad \text{or} \quad -\overset{+}{N}\overset{\displaystyle NOH}{\underset{\displaystyle O^-}{\Big\langle}}$$

but there is no evidence that nitrosohydroxylamines are interconvertible with nitramines; thus the analytical chemistry of nitrosohydroxylamines will not be considered in this section.

The primary aliphatic nitramines are best prepared by direct nitration of an alkyl amine with acetone cyanohydrin nitrate in basic solution (165):

$$C_2H_5NH_2 + (CH_3)_2C(ONO_2)CN \xrightarrow{NaOH} C_2H_5NHNO_2 + (CH_3C(OH)CN$$

Alternative synthetic routes involve nitration of an acylated amine (such as a urethane) to form a nitramide (or a nitromethane) which can be hydrolyzed to yield the nitramine:

$$CH_3NHCOOC_2H_5 \xrightarrow{HNO_3} CH_3N(NO_2)COOC_2H_5 \xrightarrow{H_2O} CH_3NHNO_2$$

and nitration of an *N,N*-dichloroamine with subsequent reduction of the intermediate chloronitramine (644):

$$RNCl_2 \rightarrow RNClNO_2 \rightarrow RNHNO_2$$

Nitroguanidine,

$$\overset{\displaystyle H_2N}{\underset{\displaystyle HN}{\Big\rangle}}C-NHNO_2$$

is a representative *N*-nitro aliphatic compound that is produced commercially. A continuous process for its synthesis (480) involves feeding sulfuric acid and guanidine nitrate (2.35:1) into a circulating sulfuric acid solution of guanidine nitrate at 35°F; part of the reaction mixture is continuously withdrawn and diluted to precipitate the nitro compound. Nitroguanidine and its derivatives are used in flashless military propellants. The explosive RDX (cyclonite), hexahydro-1,3,5-trinitro-1,3,5-triazine, is also produced in large quantities; see Table I.

The secondary nitramines may be prepared by oxidation of the corresponding nitrosamines by hydrogen peroxide in trifluoroacetic acid (163), by action of nitrogen tetroxide on amines at −80°C (726), or by action of acetone cyanohydrin nitrate on unbranched secondary alkyl amines (165). Branched-chain secondary amines are converted to the corresponding nitramines by nitrogen pentoxide at $-30°C$ (167).

The nitramido group is simply an R—N—$NO_2$ group in which the nitrogen and the alkyl residue are derived from an amide function. For example, the nitration of ethylsulfonmethylamide forms a secondary nitramide:

$$C_2H_5SO_2NHCH_3 \xrightarrow{HNO_3} C_2H_5SO_2N \begin{matrix} CH_3 \\ NO_2 \end{matrix}$$

Nitrourea, $H_2N$—CO—$NHNO_2$, is an example of a primary nitramide.

Nitrimine compounds are formed by action of dinitrogen tetroxide or nitrous acid on oximes, and their structure is considered to be as follows:

$$R_2C{=}N{-}NO_2$$

Many nitrimines are unstable in aqueous media, but cyclonite (see Table I) is quite stable and is an important explosive.

An $N$-nitroaldimine may be formed by condensing nitroamine with an aldehyde such as furfuraldehyde:

$$C_4H_3OCH{=}N{-}NO_2$$

## C. C-NITROAROMATICS

The aromatic nitro compounds constitute the most important members of the tertiary $C$-nitro compounds. Nitration takes place readily in the liquid phase; thus they are more easily prepared than the aliphatic $C$-nitro compounds. Vast quantities of aromatic nitro compounds are used as dyes, explosives, medicinals, and chemical intermediates. Although the aromatic $C$-nitro compounds containing only nitro groups are useful (e.g., as explosives), as a general rule the majority of these compounds contain other functional groups, for example, $C$-nitroazo dyes, 2,4-dinitrophenol [used as a weed killer (690)], and a wide variety of phosphate esters such as parathion ($O,O$-diethyl-$O$-$p$-nitrophenyl-phosphorothioate) that are employed as agricultural insecticides (534); nitroalkylazoaryl compounds may be useful as diesel fuel ignition improvers (706).

Aromatic $C$-nitro compounds are almost always prepared by direct nitration of aromatic nuclei; the most common nitrating agent is a mixture of nitric and sulfuric acids, but alkali-metal nitrates and nitrogen tetroxide are also used, together with sulfuric acid. Although aromatic nuclei are readily nitrated, if the aromatic ring already contains a nitro group, further nitration is difficult unless elevated temperatures and highly concentrated nitric and sulfuric acids are used; when a second nitro group is attached, it is generally in the *meta* position. A third group is difficult to attach to an aromatic ring by direct nitration, and resort is made to other synthetic routes; for example, activating groups are first attached to the dinitro aromatic, or the nitration is performed on a substituted aromatic. Thus 1,3,5-trinitrobenzene can be prepared by decarboxylation of 2,4,6-trinitrobenzoic acid:

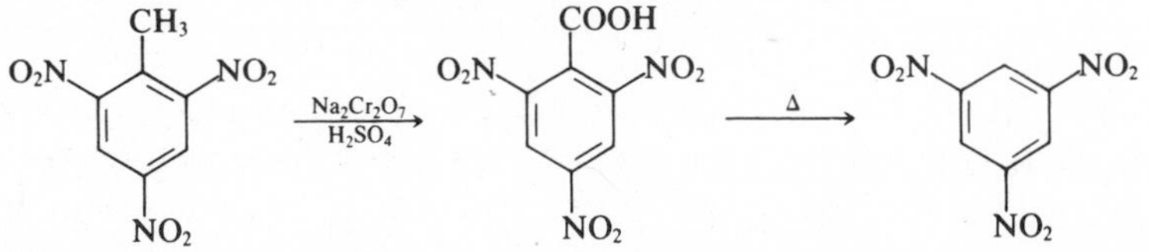

The presence of substituents on an aryl ring determines the position of the entering nitro group, and, as expected, *ortho*, *meta*, and *para* isomers are formed chiefly because of the relative electron-attracting or electron-repelling qualities of the substituents, although other factors may also exert directive influences.

Aromatic *C*-nitro compounds are prepared by special methods when direct introduction of a nitro group is not possible by using nitric acid, alkali-metal nitrates, nitrogen tetroxide, or catalyzed (mercury) mixtures of nitric and sulfuric acid. Sensitive functional groups, like —$NH_2$, are often converted temporarily to less sensitive groups (e.g., —NH—$COCH_3$) before direct nitration.

Alkyl *O*-nitro compounds in sulfuric acid (e.g., ethyl nitrate) may be used to prepare aromatic *C*-nitro compounds. In alkaline media, the alkyl nitrates are valuable reagents for attacking active methylene groups; for example, phenylnitromethane is readily produced by reaction of benzyl cyanide with ethyl nitrate in the presence of sodium ethoxide.

Mixtures of acetic acid or acetic anhydride with nitric acid are often employed to form nitro derivatives of reactive substances, Acetyl nitrate, $CH_3COONO_2$, which is probably present in these mixtures, is also a useful nitrator; mononitroaryl compounds are not nitrated further by this reagent.

Aromatic *C*-nitro compounds are produced on a commercial basis by oxidation of *C*-nitroso or *C*-amino compounds with Caro's acid (mono-persulfuric acid, $H_2SO_5$), peracetic acid (241), or trifluoroperacetic acid (163). The diazonium group (from an aryl *C*-amino compound) may be replaced with the —$NO_2$ group by reaction with sodium nitrite in the presence of cuprous salts (Sandmeyer reaction).

### D. N-Nitroaromatics

The aromatic nitramines are also examples of $N$-nitro compounds, but they are produced by methods somewhat different from those used for their aliphatic counterparts. For example, primary aromatic nitramines are conveniently prepared by the oxidation of an aromatic diazotate with ferricyanides, or by treating an aryl diazonium perbromide with alkali.

The primary nitramines have the grouping C—NH—NO$_2$, and the secondary nitramines have the grouping

$$\begin{array}{c} C \\ \diagdown \\ \diagup \quad N-NO_2; \\ C \end{array}$$

however, the N—NO$_2$ group can also exist in nitrimines, which have the grouping

$$\begin{array}{c} R \\ \diagdown \\ \diagup \quad C{=}N-NO_2. \\ R \end{array}$$

Secondary aromatic nitramines are prepared by direct nitration of secondary amines; a characteristic example of the application of this method is the formation of 'tetryl' by nitration of dimethylaniline:

The nitrimines are prepared by reaction of oximes with nitrous acid:

$$R_2C{=}NOH + HNO_2 \rightarrow R_2C{=}N-NO_2 + H_2O$$

However, most nitrimines are prepared by specific reactions.

### E. C-NITROSOPARAFFINS

The primary, RCH$_2$NO, and secondary, R$_2$CHNO, nitrosoparaffins are somewhat difficult to prepare, and for some time it was assumed that they were not capable of more than transient existence owing to rapid isomerization to the oxime (234):

$$R_2CHNO \rightleftharpoons R_2CNOH$$

However, more recent work has demonstrated that they can be isolated and that they are relatively stable. Ordinarily, the nitrosoparaffins exist as dimers, which are colorless; the monomers are blue or green.

Tertiary C-nitrosoparaffins are formed in disappointingly small quantities by oxidation of the corresponding tertiary amines with Caro's acid, and scarcely at all from the primary and secondary amines. Poor yields are also obtained when nitroparaffins are reduced to hydroxylamino compounds and subsequently oxidized:

$$RNO_2 \rightarrow RNHOH \rightarrow RNO$$

Thus a number of C-nitrosoparaffins can be prepared by partial reduction of the nitro group with zinc dust and ammonium chloride (or zinc dust and boiling ethanol) and subsequent oxidation with aqueous ferric chloride.

Emery and Neilands (162) have demonstrated that N-substituted hydroxylamines (and hydroxamates) can be oxidized by periodate to cis-nitrosoparaffins; apparently, an intermediate is formed (as with periodate oxidation of glycols) which forces the alkyl radicals into a cis configuration and thus prevents tautomerization to oximes. Trifluoronitrosomethane must be prepared by special methods (311).

Emmons (164) has described a method for preparing primary, secondary, and tertiary C-nitroso compounds; it involves oxidation of an amine or a ketimine, $R-N=CR_2$, with neutralized peracetic acid in methylene dichloride. C-Nitrosoparaffins can be formed by pyrolytic or photochemical decomposition of tert-alkyl nitrites (235) and by combination of free radicals (265). Gowenlock and Trotman (235) have described the preparation of $C_1$ to $C_5$ nitrosoparaffins by pyrolysis of alkyl nitrites in a flow system; the cis and trans isomers were isolated or detected as monomers and dimers.

C-Nitrosoketones are formed, in part, as products of reaction of alkyl nitrites on ketones that have a tertiary hydrogen; the chief product is an oximino compound for example, $CH_3COCH=NOH$. C-Nitroso-haloparaffins are prepared by the action of halogens on C-oximes:

$$R-C=NOH + X_2 \rightarrow RCX-NO + HX$$

In turn, the halogen compound (a mesohalogenonitroso paraffin) can be converted to a pseudonitrole by silver nitrite;

$$R_2C{\overset{\displaystyle X}{\underset{\displaystyle NO}{\diagdown}}} + AgNO_2 \rightarrow R_2C{\overset{\displaystyle NO_2}{\underset{\displaystyle NO}{\diagdown}}} + AgX$$

Pseudonitroles (nitronitrosoalkanes) are the chief product of reaction of nitrous acid with secondary nitroalkanes. The pseudonitroles are bimolecular and colorless when crystalline, but they exhibit the characteristic deep blue color of a C-nitroso monomer when molten or dissolved. They are insoluble in aqueous acids or alkalies (see Section III.C.3.a). Other methods for producing C-nitrosoparaffins are summarized by Gowenlock (234); it is of interest to note that partial reduction of nitro compounds seldom forms the corresponding nitroso products in good yield.

### F. N-NITROSOPARAFFINS

The primary aliphatic nitrosamines (nitrosoamines) are essentially unknown; the secondary nitrosamines are prepared by reaction of a secondary aliphatic amine with nitrous acid, nitrosyl chloride, or nitrogen tetroxide (726):

$$H_3C \diagdown NH + HONO \rightarrow H_3C \diagdown NNO$$
$$H_3C \diagup \qquad\qquad H_3C \diagup$$

N-Nitrosodimethylamine

Although preparative methods for N-nitrosoparaffins predominantly involve the action of nitrous acid (acidic solutions) on secondary amines, compounds such as N-nitrosodimethylamine (dimethylnitrosamine) often can be produced in surprising amounts in neutral solutions containing carbohydrates under oxidizing conditions, and (especially) when an aldehyde such as formaldehyde is present. Formal studies of the effect of pH on the formation of secondary nitrosamines are needed to elucidate the mechanism of nitrosamine formation.

The N-nitrosamides (nitrosoamides) have the general formula R'CON(NO)R and are usually encountered as derivatives of N-nitrosoacetanilide, for example, $CH_3CON(NO)C_6H_5$. Accordingly, they can be produced by action of nitrous acid on an anilide in acetic acid, or by reaction of an acid anhydride with a diazotate:

$$C_6H_5N{=}NONa + (CH_3CO)_2O \rightarrow CH_3C \diagup\!\!\!\!\!\!^O \diagdown NC_6H_5$$
$$\underset{N=O}{|}$$

### G. C-NITROSOAROMATICS

The direct oxidation of aryl amines to C-nitroso compounds is possible in many instances because of the apparent stability of these compounds; thus oxidation of aniline and many of its derivatives by permonosulfuric acid (Caro's acid) (305) and by perbenzoic or peracetic acid (134) often leads to satisfactory yields of aromatic C-nitroso compounds.

Tertiary aromatic amines readily form C-nitroso compounds when treated with nitrous acid; the —NO group is introduced into the aromatic ring in the *para* position:

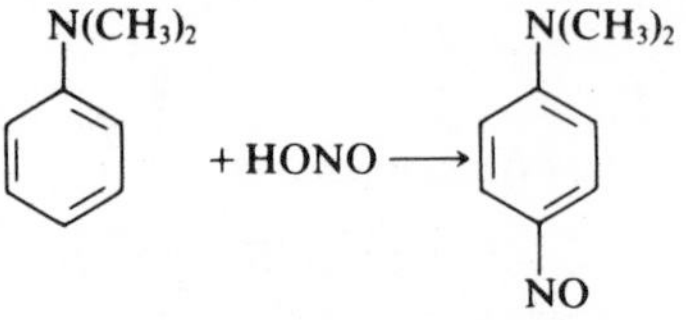

p-Nitroso-N,N-dimethylaniline

The reaction of nitrous acid with phenol yields p-nitrosophenol; the resulting compound is tautomeric with quinone monoxime:

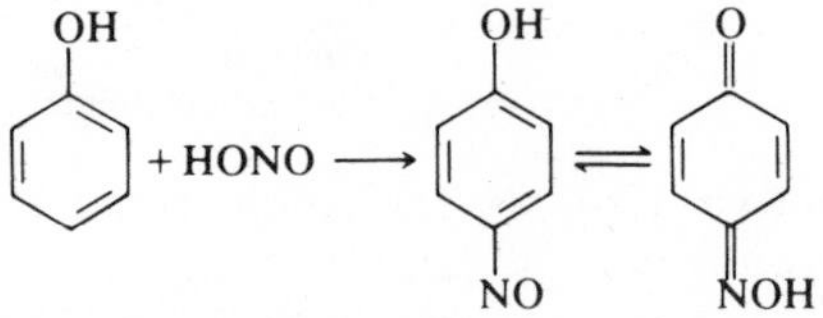

As demonstrated by studies of nuclear magnetic resonance spectra and supporting evidence from ultraviolet and infrared data, p-nitrosophenol and several of its methyl derivatives exist predominantly in the benzoquinone monoxime form in organic solvents; in aqueous solutions, the corresponding potassium salts have the negative charge largely on the oxime oxygen (524). For example, a solution of p-nitrosophenol contains about 85% of p-benzoquinone monoxime; the isomers can be isolated in two crystalline forms, one pale green and the other colorless. Alkaline solutions of p-nitrosophenol are reddish green. Substituted phenols can be nitrosated directly; dihydric phenols form dinitroso compounds.

Most aromatic C-nitroso compounds are synthesized by oxidation of N-aryl hydroxylamines, which, in turn, are prepared by partial reduction of aromatic C-nitro compounds:

$$ArNO_2 \rightarrow ArNHOH \rightarrow ArNO$$

Typically, reduction of C-nitro compounds is accomplished by zinc and ammonium chloride, or zinc dust and boiling ethanol. Aqueous dichromate is frequently used to oxidize N-aryl hydroxylamines (307), but reagents such as $FeCl_3$, $H_2O_2$, and hypochlorite are employed in special instances. Nitrobenzene can also be converted to nitrosobenzene by electrolysis in special cells; however, the method is not generally applicable, for the electrolytic conversion of aromatic *C*-nitro compounds to *C*-nitroso derivatives cannot be controlled readily, and a variety of products is formed.

Small yields of nitrosobenzene are formed by reduction when nitrobenzene vapor is passed over hot iron (582); *C*-nitroso compounds in poor yield result when aromatic *C*-nitro compounds are heated with iron powder (582).

Hexanitrosobenzene is obtained by heating trinitrotriazidobenzene; it is a stable, nonhygroscopic compound that is more sensitive to impact than is tetryl (694). Dinitrodinitrosobenzene is a powerful explosive.

## H. *N*-NITROSOAROMATICS

The aromatic secondary nitrosamines constitute the majority of known aryl nitrosamines; they are readily prepared by action of nitrous acid on aryl secondary amines, for example:

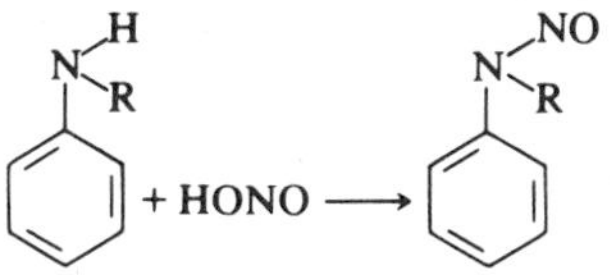

*N*-Nitroso-*N*-alkylaniline

The analytical reagent, cupferron (ammonium salt of *N*-hydroxy-*N*-nitrosobenzenamine), is prepared by action of an alkyl nitrite and dry ammonia on phenylhydroxylamine (469):

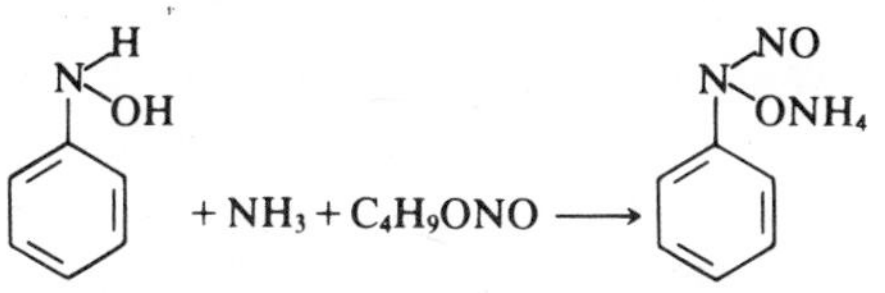

Cupferron

The nitrosoamines corresponding to primary aromatic amines may be viewed as tautomers of aromatic isodiazo hydroxides (see also "Diazonium Group," Vol. 15):

$$C_6H_5NHNO \rightleftharpoons C_6H_5N=NOH$$

Aryl nitrosamides are very reactive substances prepared from aryl amides by direct nitrosation:

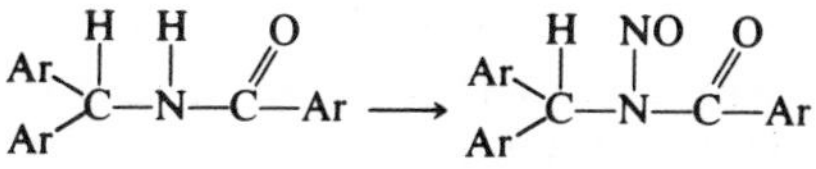

It is believed that their reactivity stems from tautomerism with corresponding diazo compounds, for example, an $N$-nitrosoacetylamine with the diazoacetate:

$$Ar—N(NO)COCH_3 \rightleftharpoons Ar—N{=}N—OCOCH_3$$

## I. OCCURRENCE

The nitro compounds find widespread use in industrial syntheses of dyes, explosives, and drugs, as well as in theoretical studies; a partial listing of their applications would be pointless. It is of interest, however, to note the unexpected presence of a variety of nitro compounds in nature, especially since nitro compounds are quite poisonous.

$\beta$-Nitropropionic acid was the first aliphatic nitro compound to be found in nature. It was identified as a hydrolytic product of hiptagin, a glucoside present in the bark of the tree *Hiptage mandoblata* and in the berries of the karaka tree (102), as well as in the fungal mold *Aspergillus flavus* (93). It has been shown to be the toxic component of the *Indigofera* species used for forage and ground cover (506). A naturally occurring aromatic nitro compound, chloramphenicol [D-threo-$N$-(1,1'-dihydroxy-1-$p$-nitrophenylisopropyl)dichloroacetamide],    was isolated from *Streptomyces venezuela* (157,572), and a synthetic product is produced commercially as the antibiotic chloromycetin. Nitrofurfural is used as a bacteriostat, and $N,N$-dimethyl-$p$-nitrosoaniline has germicidal activity. One of the products of the growth of *Aspergillus wentii* is 1-amino-2-nitrochloropentane carboxylic acid; it affects the structure of higher plants.

Violet roots and some tropical plants contain $\beta$-nitropropionic acid, and the antibiotic rufomycine is a peptide which contains a nitrotyrosine structural unit. A derivative of nitrophenanthrene appears to be responsible for the healing qualities of *Aristolochia clematitis*, and it is reported that the odor of certain plants is due to nitro compounds. The nitro derivative of an imidazole (Flagyl) is efficient in the treatment of *Trichomona vaginalis*. Nitro compounds are widely used as drugs in poultry feeds (coccidiostats).

Nitroso compounds are encountered much less frequently than nitro compounds. Practically all of the industrial utilization of the nitroso compounds is represented by the nitroso dyes Naphthol Green (C.I. 10020) and the Gambines (nitrosonaphthols); quinone oximes are used as vulcanizing agents, and $p$-nitrosodimethylaniline is chiefly a dye intermediate but also serves as a vulcanization accelerator.

Aromatic nitroso compounds appear to be intermediates in the bacterial degradation processes of compounds such as $p$-nitrobenzoic acid and aromatic primary amines. Ferroverdin, a green pigment that is produced by species of *Streptomyces*, includes an aryl nitroso structure (25a).

### III. PROPERTIES

The nitro group is a hybrid of equivalent resonance structures of which two characteristic examples are as follows:

$$-\overset{+}{N}\diagdown_{O^-}^{O} \rightleftharpoons -\overset{+}{N}\diagup_{O}^{O^-}$$

The hybrid structure has a full positive charge on the nitrogen, and, depending upon the substituent group attached to the nitrogen, the dipole moments range between 3.5 and 4.0 D. The high dipole moments indicate that there is an axis of symmetry through the nitrogen atom and that the negative charge is uniformly distributed between the two oxygen atoms. Hence a very probable structure is

$$-\overset{+\delta}{N}\diagup^{O^{-\delta/2}}_{\diagdown O^{-\delta/2}}$$

The symmetrical structure is also supported by X-ray studies. The polar character of the nitro group is evident because its presence leads to a lower volatility in compounds than is found when a ketone group is present; for example, nitromethane boils at 101°, whereas acetone boils at 56°C. On the other hand, solubilities of nitro compounds are surprisingly low; 1 part of nitromethane is soluble only in 10 parts of water, but acetone and water are mutually soluble.

The structure of the nitroso group also reflects a strong tendency for partitioning of charges on both nitrogen and oxygen atoms, but since resonance structures cannot occur as with the —$NO_2$ group, the electronic structure of the nitroso group is modified in accordance with the nature of the atom or radical attached to it. In general, nitroso group structures involving small formal charges on the atoms are preferred, and in some nitroso compounds there is a greater tendency for the disposition of two quartets in the space around an atom than for the formation of a two-electron bond (429).

Tertiary aliphatic and aromatic C-nitroso compounds are quite stable; they are usually monomeric and have a blue or green color in the gas phase or in dilute solution, but they often may be isolated as yellow to colorless solids or liquids. The color changes appear to be the result of dimerization; thus dimerization of nitrosobenzene occurs readily, and the presence of N-N bonds in the dimers has been indicated by X-ray diffraction studies (723):

$$C_6H_5N{=}O \rightleftharpoons \underset{O}{\overset{C_6H_5}{>}}\!N{-}\!N\!<\!\overset{O^-}{\underset{C_6H_5}{}} \rightleftharpoons \underset{^-O}{\overset{C_6H_5}{>}}\!N{=}N\!<\!\overset{O}{\underset{C_6H_5}{}}$$

*trans* dimer

The tendency for dimerization is often influenced by structure; for example, crystal structure analyses of *p*-bromonitrosobenzene and 2,4,6-tribromonitrosobenzene prove that these colorless compounds are dimeric, and data obtained by Webster (723) for the normally green crystals of *p*-iodonitrosobenzene show that the solid is monomeric. In contrast, *o*-iodonitrosobenzene is dimeric owing to steric hindrance.

A comprehensive review of the structure of nitroso group compounds has been prepared by Wagniere (719). It is important to note that *C*-nitroso compounds dimerize whereas *N*-nitroso compounds do not; thus *N*-nitroso compounds do not show color changes in going from one *n*-mer form to another (234). A large number of the simple-structured primary and secondary *C*-nitroso compounds cannot be isolated because they rapidly isomerize to oximes:

$$R_2CHNO \rightleftharpoons R_2C{=}NOH$$

The dissociation of dimeric primary and secondary *C*-nitrosoparaffins is often complicated by the irreversible rearrangement of the monomeric form (blue color) to the isomeric oxime. On the other hand, monomeric nitroso compounds condensed as solids at about $-100°$ produce the dimer instead of the oxime when melted at about $-80°$ (234).

As would be anticipated, *C*-nitroso dimers can exist as *cis* or *trans* forms; the *trans* dimer is nearly always formed because it represents a more stable configuration:

$$\underset{O}{\overset{R}{>}}\!N{=}N\!<\!\overset{O}{\underset{R}{}}$$

*cis*-Nitrosoalkane dimers can be formed, in general, by the pyrolysis of alkyl nitrites ($O{-}NO_2$); the products are collected on the walls of a liquid-oxygen-cooled vessel (235). The monomeric nitroso compound condensed on a cooled surface is oriented in a regular manner, with the oxygen atoms of the nitroso group attached to the cooled surface; as the surface is warmed, the oriented monomer reacts to form a white *cis* dimer even though this is less stable than the *trans* dimer. *Trans* dimers are formed by all other methods.

## A. PHYSICAL PROPERTIES

The physical properties of a short series of nitro compounds are given in Table II; it is hoped that the list of compounds in the table will suggest to the analytical chemist the wide variation of properties that can be encountered. C-Nitroparaffins can be purified by the method of Kornblum et al. (381). Introduction of the nitro group into the generic structure of a hydrocarbon produces an expectedly large rise in boiling point and an increase in density; for example, most nitro compounds are heavier than water.

The mononitroparaffins are usually colorless, pleasant-smelling, high-boiling substances that are virtually insoluble in water. The nitroparaffins of low molecular weight are good solvents for polar compounds, especially when used in combination with alcohols. The high molecular weight mononitroparaffins are crystalline, and all polynitroparaffins are waxlike solids at room temperature. The mononitroparaffins are considered stable; however, nitromethane can be detonated with a blasting cap, and the higher mononitro compounds may explode if heated in closed containers. In contrast, most polynitroparaffins are not very stable, and it is best to regard them as treacherous explosives; there exists a relationship between the constitutions of organic compounds and their characteristics as explosives (441). [see also Section V.A of "Nitrate and Nitrite Ester Groups.")

Aromatic nitro compounds are usually high-boiling, yellow to red solids, which are insoluble in water unless solubilizing groups are present; a few aromatic nitro compounds are oily liquids with characteristic odors (e.g., nitrobenzene), but most are sharply crystalline. The mononitro aromatics are generally quite stable, but the polynitro compounds are shock-sensitive and explosive. Aromatic nitro hydrocarbons are good solvents for most organic compounds and a surprising number of inorganic salts; however, the oxidizing nature of the nitro group may be a disadvantage, especially with sensitive compounds.

Many simple nitrosoparaffins and tertiary and aromatic C-nitroso compounds are colorless, yellow, or orange solid or liquid dimers which revert to monomers in dilute solution, in the gas phase, or at elevated temperatures; the monomers of the aliphatic C-nitroso compounds are usually blue, and those of the aromatic are green. In fact, color can in many instances be used to distinguish C-nitroso from C-nitro compounds.

Dimerization and isomerization of primary and secondary C-nitroso compounds are reactions which are not completely understood; these reactions may actually compete:

TABLE II

Physical Properties of Typical Nitro Compounds

| Compound | Molecular weight | Color and form | Melting[b] point | Boiling[b] point | Solubility[b] | Ref.[a] |
|---|---|---|---|---|---|---|
| *C*-Nitroparaffins | | | | | | |
| Nitromethane | 61.04 | Liq. | −28.5 | 100.8 | Alc., ether | B1[2], 40 |
| Trinitromethane | 151.04 | Liq. | 14–15 | 45–7[22] | Water, alc. | B1[3], 116 |
| Tetranitromethane | 196.03 | — | 13 | 126 | Alc., ether | B1[2], 47 |
| Chloropicrin, Cl₃CNO₂ | 164.38 | Colorless liq. | −64 | 112 | Alc., acet. | B1[3], 113 |
| 2-Nitropropane | 89.09 | Liq. | −93 | 120 | Alc., ether | B1[2], 79 |
| 2-Nitro-1-butanol | 119.12 | — | −47 | 105[10] | Alc., ether | B1, 370 |
| 2-Ethyl-2-nitro-1,3-propanediol | 149.05 | Needles | 57–58 | d. | Water, alc. | B1, 483 |
| *N*-Nitroparaffins | | | | | | |
| *N*-Nitrodimethylamine | 90.08 | Needles | 58 | 187 | Water, alc. | B4[2], 342 |
| Nitroguanidine | 104.07 | Needles | 246–247 | — | Alk. | B3[2], 100 |
| Cyclonite (RDX), hexahydro-1,3,5-trinitro-1,3,5-triazine | 222.12 | Orthorhomb. | 203.5 | — | Acet. | B26[2], 5 |
| *C*-Nitroaromatics | | | | | | |
| Nitrobenzene | 123.11 | Liq. | 5.7 | 210.8 | Alc., ether | B5[2], 171 |
| 2-Nitrotoluene | 137.14 | Yellow | −2.9 | 220.4 | Alc., ether | B5[2], 243 |
| 2,4,6-Trinitrotoluene | 227.13 | Orthorhomb. | 82 | 240 exp. | Alc. | B5[2], 268 |
| 2,4′-Dinitrodiphenylamine | 259.23 | Red needles | 222–223 | — | Chl. | B12[2], 337 |
| *N*-Nitroaromatics | | | | | | |
| *N*-Nitroaniline | 138.12 | Leaves | 46–47 | exp. | Water, alc. | B16[2], 343 |
| *N*-Nitro-*N*-methyl-2,4,6-trinitroaniline (tetryl) | 287.15 | Yellow prisms | 129 | 187 exp. | Sl. sol. alc., ether | B12, 770 |

[a] The number following the letter "B" indicates the volume of the main series of Beilstein, *Handbuch der Organischen Chemie*; the superscript refers to the first, second, etc., supplement; and the following number is the page in the supplement.

[b] Abbreviations: Alc. = alcohol; acet. = acetone; d = decomposition; exp. = explosion; superscripts are pressures in mm of Hg.

$$R_2CHNO \rightleftharpoons (R_2CHNO)_2$$
$$R_2CHNO \rightleftharpoons R_2C\!=\!NOH$$

The dimeric $C$-nitroso compounds can exhibit *cis-trans* isomerism (see above). Conversion of a *trans-C*-nitrosoparaffin to the *cis* isomer can be brought about by irradiation with ultraviolet; the reverse process occurs readily in nonpolar solvents without irradiation (235).

With few exceptions, the nitramines are colorless. Although many nitramines decompose explosively when heated, and some of the more prominent explosives are nitramines, the N—NO$_2$ group is inherently quite stable (contrasted with the diazonium group). The primary nitramines, RNHNO$_2$, give an acidic reaction in water and, when treated with dilute alkali, form salts (e.g., RNNaNO$_2$) which can react with alkyl halides to furnish neutral secondary nitramines:

$$RNNaNO_2 + CH_3I \rightarrow R(CH_3)NNO_2 + NaI$$

Primary $N$-nitroso compounds are very frequently colorless and do not show the blue colorations characteristic of monomeric $C$-nitroso compounds even though there is a tendency for dimerization. They are definitely acidic compounds but are decomposed by acids. Secondary aliphatic $N$-nitroso compounds (nitrosamines) are soluble in water and are essentially neutral or very weakly basic (268). Mixed aromatic and aliphatic nitrosoamines are weakly basic and insoluble in water, but can form salts such as the hydrochlorides. All nitrosamines are decomposed by acids.

Physical properties of typical nitroso compounds are given in Table III. It is of interest to note that the boiling points of dialkylnitrosamines are higher than would be expected from a consideration of their molecular weights and by comparison with alkyl nitrites (268).

Nitro and nitroso compounds can undergo a variety of complex reactions when exposed to light, but it is difficult to make generalizations because the products of photoinduced reactions are determined in large measure by the structures connected to the nitro and nitroso groups. For example, certain compounds may dissociate into free alkyl and aryl radicals; others may suffer cleavage of the N—O bond, and nitro compounds frequently rearrange to form O—NO compounds. A large number of compounds undergo *cis-trans* photoisomerization or photodimerization. In fact, photolysis of nitro and nitroso compounds provides a synthetic route for production of compounds with other nitrogen-containing functional groups. For the purposes of analytical chemistry, it is necessary to recognize that many nitro and nitroso

TABLE III
Physical Properties of Typical Nitroso Compounds

| Compound | Molecular weight | Color and form | Melting[b] point | Boiling[b] point | Solubility[b] | Ref.[a] |
|---|---|---|---|---|---|---|
| *C*-Nitrosoparaffins | | | | | | |
| Nitrosomethane | 45.036 | Colorless needles (dimer) | 120–122 | — | EtOH | B1[3], 105 |
| 2-Nitroso-2-methylpropane | 87.116 | Colorless prisms (dimer) | 76 | d. | EtOH | B1[1], 129 |
| 2-Nitroso-2-methylbutane | 101.143 | White needles, prisms (dimer) | 50 | — | EtOH | B1[1], 139 |
| 1-Chloro-1-nitrosoethane | 93.508 | Colorless blades (dimer) | 64–65 | — | Insol. $H_2O$ | B1[2], 69 |
| 1-Chloro-1-nitrosobutane | 121.56 | Colorless (dimer) | 41–42 | 92 d. | Insol. $H_2O$ | B1[2], 86 |
| 2-Chloro-2-nitrosopentane | 135.588 | Blue liq. | — | 32[23] | Insol. $H_2O$ | B1[2], 99 |
| Pseudonitroles | | | | | | |
| 2-Nitroso-2-nitropropane | 118.076 | Colorless, monoclin. | 76 d. | — | EtOH, $CHCl_3$ | B1[3], 260 |
| 2-Nitroso-2-nitrobutane | 132.103 | White prisms | 58 | — | EtOH | B1[1], 124 |
| 3-Nitroso-3-nitro-2-methyl-butane | 146.130 | Blue oil | — | 60 d. | EtOH | B1[1], 141 |
| 3-Nitroso-3-nitropentane | 146.130 | White tablets | 63–66 | — | EtOH | B1[1], 133 |
| 2-Nitroso-2-nitropentane | 146.130 | Deep blue oil | — | 59 d. | EtOH | B1[1], 133 |

| | | | | | | |
|---|---|---|---|---|---|---|
| Nitrolic acids | | | | | | |
| Methylnitrolic acid | 104.049 | Colorless needles | 64–68 d. | — | $H_2O$, EtOH, alk. | B1[1], 92 |
| Ethylnitrolic acid | 118.076 | Colorless rhomb. | 88 d. | — | $H_2O$, EtOH, alk. | B1[1], 189 |
| Propylnitrolic acid | 132.103 | Colorless prisms | 66 | — | $H_2O$, EtOH, alk. | B1[1], 247 |
| N-Nitrosoparaffins | | | | | | |
| N-Nitrosodimethylamine | 74.077 | Yellow oil | — | $150^{7.55}$ | EtOH | B4[2], 585 |
| N-Nitrosodiethylamine | 102.131 | Liq. | — | $61–63^{12}$ | EtOH | B4[2], 617 |
| N-Nitrosodipropylamine | 130.184 | Liq. | — | $59–61^{1.5}$ | EtOH | B4[2], 628 |
| N-Nitrosodiisopropylamine | 130.184 | — | 48 | $80^{14}$ | EtOH | B4, 156 |
| C-Nitrosoaromatics | | | | | | |
| Nitrosobenzene | 107.11 | Rhomb., monoclin. | 67–68 | $57–59^{18}$ | Alc., ether | B5[2], 169 |
| 2-Nitrosotoluene | 121.14 | Prisms, needles | 72 | — | Alc., ether | B6[2], ether |
| 4-Nitrosophenol | 123.11 | Yellow needles | 126 d. | — | Alc., ether | B6[2], 205 |
| 4-Nitrosodiphenylamine | 198.23 | Yellow powder | 143 | — | Alc., benz. | B12[2], 122 |
| N-Nitrosoaromatics | | | | | | |
| N-Nitrosodiphenylamine | 198.23 | Yellow prisms | 66.5 | — | Benz. | B12[2], 310 |
| N-Nitrosophenylhydroxylamine | 138.13 | Needles | 59 | — | Alc., ether | B16[2], 344 |
| N-Nitroso-N-methylaniline | 136.15 | Yellow | 14.7 | $136^{13}$ | Alc., ether | B12[2], 309 |

[a] The number following the letter "B" indicates the volume of the main series of Beilstein, *Handbuch der organischen Chemie*; the superscript refers to the first, second, etc., supplement; and the following number is the page in the supplement.
[b] Abbreviations: d = decomposition; Alc. = alcohol; Alk. = alkalis; benz. = benzene; superscripts are pressures in mm of Hg.

compounds are sensitive to light, especially aromatic compounds which have —CH= in the position *ortho* to the nitro group; consideration should be given to the possibility that the composition of a sample may have been changed because of exposure to light (8,250), especially if oxygen is also present (498).

*N*-Nitrosoamines are readily decomposed by ultraviolet light; for example, simple dialkyl *N*-nitrosamines decompose in aqueous solutions of methanol or in pure methanol to release nitrite ion, but even the nitrite ion is destroyed by the uv light. If the photolysis takes place in alkaline solution, however, the released nitrite ion is not decomposed; indeed, the nitrite ion serves as a quantitative indicator of the amount of *N*-nitrosamine (133) (see below).

## B.  CHEMICAL PROPERTIES

### 1.  Reduction

From the viewpoint of an analytical chemist, the most important chemical reaction exhibited by nitro and nitroso groups is reduction to amines or hydroxylamines, that is,

$$R{-}NO_2 \xrightarrow{4H} RNHOH \xrightarrow{2H} RNH_2$$

The ease with which nitro and nitroso groups are reduced is determined in large measure by the nature of the atom or radical attached to the nitrogen. Most of the investigations concerning analytical reductimetric procedures for nitro and nitroso groups have involved aromatic $C{-}NO_2$ compounds; by comparison, only relatively few systematic studies have been reported on other nitro and nitroso group functions.

All nitro and nitroso groups, regardless of attachment, can be more or less quantitatively reduced by a combination of concentrated hydriodic acid (57%) and red phosphorus at 150–200°C in a sealed tube:

$$CH_3NO + 2\,HI \rightarrow CH_3NHOH + I_2$$

$$C_6H_5NO + 4\,HI \rightarrow C_6H_5NH_2 + H_2O + 2\,I_2$$

The combination is extraordinarily powerful, for it also can convert alcohols to hydrocarbons and aromatic hydrocarbons to hydroaromatic derivatives, and this must be taken into consideration, along with the possibility that nitrogen will be split off. Milder reductions take place in a refluxing solvent at atmospheric pressure (470,750). A sealed-tube method in which only hydriodic acid is used has been developed by Aldrovandi and De Lorenzi (4) for quantitative determination of nitro groups as well as a variety of other nitrogen functional groups [see Section VI.A.1.b]. The HI reduction procedure often yields unexpected products; for example, $p$-nitrosodimethylaniline is reduced and demethylated to $p$-aminoethylaniline (232). Reduction of aromatic nitroso compounds with hydriodic acid (149,150) may yield azoxy compounds by reaction of unreduced nitroso compound with the phenylhydroxylamine which is first produced:

$$C_6H_5NO + C_6H_5NHOH \rightarrow C_6H_5N{=}N(O)C_6H_5 + H_2O$$

Reduction of nitro and nitroso groups to an amine is seldom straightforward, for the route of the reduction is governed largely by the nature of the compound bearing the functional group. As a result, no single procedure can be recommended for the quantitative analysis of all $C$-nitro and $C$-nitroso compounds; the performance of any method must be examined methodically as applied to a given compound. However, hydriodic acid reduction in a sealed tube at an elevated temperature is about as general as can be expected, and thus may be used for reduction of occasional samples of relatively unknown structures. Since the procedure has not been studied sufficiently, little more can be said about its general effectiveness or limitations at this time.

Reduction of the $C$-nitro group usually proceeds through formation of a $C$-nitroso group as a first intermediate, which may have a transient existence or may linger long enough to participate in other reactions. In acidic media, reduction of the nitroso group occurs rapidly and forms a hydroxylamine derivative, and this, in turn, yields the expected amine. In alkaline media, an azoxy compound is often formed by condensation of the $C$-nitroso group compound and the corresponding hydroxylamine:

$$C_6H_5NO + C_6H_5NHOH \rightarrow C_6H_5NNOC_6H_5 + H_2O$$

Moreover, in alkaline solution, phenylhydroxylamine disproportionates into aniline and azoxybenzene:

$$3\, C_6H_5NHOH \rightarrow C_6H_5NH_2 + C_6H_5NNOC_6H_5 + 2\, H_2O$$

Reductions of $C$-nitro groups in neutral media favor formation of $C$-nitroso compounds and hydroxylamines, but yields of amines are generally quite poor. Primary and secondary aliphatic $C$-nitro compounds usually yield $C$-nitroso compounds when treated with an acidic reductant such as stannous chloride, presumably because of rapid isomerization to an oxime, but hydroxylamines can also be formed (263). Hydrogen sulfide in aqueous solutions forms sulfur with nitroso compounds.

Strong reducing agents such as titanous, chromous (682), and vanadous (30,216) ion can react with all forms of nitro and nitroso groups, and the anticipated amine is obtained quantitatively for a sur-

## TABLE IV
### Typical Reductions of Nitro and Nitroso Groups

| Class | Reductant and conditions | Products and remarks | Ref. |
|---|---|---|---|
| All | HI | $N_2 + I_2$ | 4 |
| | Zn + alcohol + $CaCl_2$ | Hydroxylamines | 507,716 |
| Aliphatic C—$NO_2$ | Zn + $NH_4Cl$; electrolysis | C—NHOH $\xrightarrow{[H]}$ C—$NH_2$ | 558 |
| | Nitronate + HI | Oxime | 362 |
| | Zn + $NH_4Cl$; $H_2$ + Raney catalyst; electrolysis | C—NO | 480 |
| | Fe + $H^+$; $H_2$ + catalyst; $LiAlH_4$; $TiCl_3$ | C—$NH_2$; no reduction by $NaBH_4$ | 263,480 |
| | Electrolysis | C—$NH_2$ | 480 |
| Aromatic C—$NO_2$ | Fe, 200°C | C—NO, low yield | 582 |
| | Zn + $NH_4Cl$ | C—NHOH | 338,471 |
| | Metals + $H^+$; $TiCl_3$; $CrCl_2$; $SnCl_2$ | C—$NH_2$ | 84,395,564,714 |
| | $Na_2S_2$; $NH_4HS$; $H_2S$; $(NH_4)_2S$ | C—$NH_2$ | 54,231,586 |
| | $NaHSO_3$; $Na_2S_2O_4$ | C—$NH_2$ plus sulfonated, sulfamated products | 191,570,631 |
| | Glucose | —C—N=N—C, good yield $\downarrow$ O | 214 |
| | Electrolysis | Chiefly C—$NH_2$ | 558 |
| | $N_2H_4$ + Raney Ni or Pd | C—$NH_2$ | 659 |
| | Zn + NaOH | C—$NH_2$ | 466 |
| | Catalyst + $H_2$ | C—NHOH (—control $H_2$ amount) | 64a |
| | $FeSO_4$ + $NH_4OH$; catalyst + $H_2$ | C—$NH_2$ | 1,310 |

| Substrate | Reagent | Product | Ref. |
|---|---|---|---|
| Aliphatic C—NO | Zn + HAc | Dimer → C—NH—NH—C $\xrightarrow{Zn+H^+}$ C—NH$_2$ | 616 |
| | Metals + H$^+$; H$_2$ + catalyst | C—NH$_2$ | 21 |
| | Electrolysis | C—NH$_2$ | 290 |
| Aromatic C—NO | H$_2$ + catalyst; metal + H$^+$; SnCl$_2$ | C—NH$_2$ | — |
| | NH$_4$SH | C—NH$_2$ | 388 |
| | NaHSO$_3$ | C—NH$_2$ and sulfanilate; —SO$_3$H substit. | 399 |
| | Na$_2$S$_2$O$_4$ | C—NH$_2$ | 122 |
| | Alkaline arsenite | —R—N=N=N—R and other products (→O) | 246 |
| | N$_2$H$_4$ + Raney Ni | C—N=N=C and —C—N=N—C (→O) | 297 |
| | Mg, MgI$_2$ | R—N—N—R (H H) | 23a |
| Aromatic or aliphatic R—N—NO (R) | Zn + HAc | —N—NH$_2$, good yield; —N—NH$_2$ | 708; 269,708 |
| | H$_2$ + catalyst; metal + H$^+$; TiCl$_3$ | —NH$_2$ | 708 |
| Aliphatic R—NH—NO | [H] | RNHNH$_2$ | |
| RNH—NO$_2$ | Zn + NH$_4$Cl | R—NHNO, e.g., nitroguanidine | 480 |
| | Zn + HAc | RNH—NH$_2$, very poor yields | |
| | Metals + H$^+$; TiCl$_3$ | RNH$_2$, poor yields | |
| | Electrolysis | RNH—NH$_2$, e.g., nitrourea | 304 |
| R$_2$N—NO$_2$ | TiCl$_3$; metals + H$^+$ | RNH$_2$ (infrequent) | |
| | Zn + HAc; electrolysis | R$_2$N—NH$_2$, poor yields | |

prising number of aromatic nitro and nitroso compounds. In altogether too many instances, however, conversion is incomplete when the nitro or nitroso group is attached to an aliphatic radical or to an atom other than carbon.

Trivalent molybdenum (215) may be used to reduce $C$-nitro and $C$-nitroso groups to amines; in general, a two- to threefold excess of reductant is added to the sample under an atmosphere of carbon dioxide, and then the excess is titrated with ferric alum, using methylene blue as indicator.

Table IV summarizes the effects of various reducing agents on nitro and nitroso groups.

### a. $C$-NITROPARAFFINS

The nitroparaffins are in general more resistant to reduction than the aromatic nitro compounds; unfortunately, the nitroparaffins seldom give rise to one product when treated with reducing agents. However, nitronic acids often can be readily reduced to oximes in excellent yields, and nitroolefins and bromonitroalcohols also may form oximes on reduction (145).

Dinitroparaffins such as 2,2-dinitropropane readily form nitrite on reduction in alkaline solution; in contrast, the mononitroparaffins seldom form nitrite when reduced in strongly alkaline solutions (634), but nitrite can be formed simply by the action of hot concentrated alkali on mononitroparaffins (56); see Section VI.A.3.b(1)(e).

### (1) Reduction by Metals

Aliphatic $C$-nitro compounds are best reduced to corresponding amines by iron and hydrochloric acid (263); hydrogenation with Raney nickel is also effective (520). To obtain good yields of amines, it is necessary to use strong reducing agents at low temperatures in order to minimize the formation of derivatives such as isoxazoles by interaction of the amines with unreduced primary and secondary nitroparaffins (430).

Aliphatic $C$-nitro compounds also form $N$-alkyl hydroxylamines, as well as amines, with mild reducing agents, and it is possible to control conditions so as to obtain high yields of the hydroxylamines, especially when metals or amalgams are used to effect reduction. Thus, although stannous chloride and even tin and hydrochloric acid (a favorite couple in laboratory reductions) can provide good yields of $N$-alkyl hydroxylamines, better conditions can be established with zinc dust and water, as well as with solutions of salts such as ammonium chloride. Catalytic reduction (palladized barium sulfate) in aqueous or alcoholic oxalic acid gives high yields of hydroxylamine derivative without amine formation (611).

Reduction of aliphatic C-nitro compounds in alkaline media may take place largely through attack on the nitronate ion with formation of an oximate ion:

$$RCH_2NO_2 \xrightleftharpoons{H^+} RCH=N \begin{array}{c} O^- \\ \diagup \\ \diagdown \\ O \end{array} \overset{[H]}{\rightleftharpoons} RCH=N \begin{array}{c} O^- \\ \diagup \\ \diagdown \\ OH \end{array} \xrightarrow{[H]} RCH=NO^-$$

Nitronic acids can be reduced by aluminum amalgam, sodium amalgam, or zinc and alkali to the corresponding oximes (367), but reduction to the amine seldom takes place (other than as an incomplete side reaction). Reduction of *gem*-dinitroparaffins usually yields oximes; vigorous reduction yields amines.

### (2) Reduction by Hydriodic Acid

Reduction of primary and secondary saturated nitroparaffins for analytical work can be effected by hydrogen iodide; for this purpose, a mixture of potassium nitronate and an iodide is acidified (362):

$$RCH=N \begin{array}{c} O^-K^+ \\ \diagup \\ \diagdown \\ O \end{array} \xrightarrow{H^+} RCH=NO_2H \xrightarrow{HI} RCH=NOH + I_2$$

The preparation of the nitronate and the subsequent acidulation must be performed with care and at low temperatures in order to minimize decomposition reactions (e.g., Nef reaction; see Section III.C.3.a). For analysis of nitro compounds, the liberated iodine is titrated by thiosulfate (see Section VI.A.1.b).

### (3) Electrolytic Reduction

Electrolytic reduction of aliphatic C—$NO_2$ groups generally occurs in two irreversible steps:

$$RNO_2 \xrightarrow{4e} RNHOH \xrightarrow{2e} RNH_2$$

Nitromethane, nitroethane, and 1-nitropropane are reduced at nickel cathodes to hydroxylamines at 15–20°C in a 10–15% alcoholic solution acidulated by sulfuric acid; it is customary to use platinum anodes in porous pots containing dilute sulfuric acid. Conversions up to 80% are observed, but at temperatures near 70°C, comparable yields of amines are obtained (263). Cathodes of lower overvoltage (copper, tin, nickel) usually permit partial reduction to hydroxylamines; cathodes of higher overvoltage, such as mercury and lead, favor complete reduction to the amine. The electrolytic reduction of poly-C-nitroalkanes has been studied (226).

The polarographic reduction of aliphatic C—NO$_2$ compounds in strongly alkaline solutions occurs in one step. In strongly acidic solutions, the second wave appears at about 1.54 V (vs. SCE), but it is obscured by a hydrogen wave; strongly buffered media must be used to maintain constant pH [e.g., sodium acetate or 0.05 $M$ H$_2$SO$_4$ (141)]. For samples of limited solubility in water, solvents such as a methanol-water mixture containing lithium chloride or a 0.1 $M$ solution of LiCl in dimethylformamide may be used (see Section VI.A.4).

Aliphatic poly-C-nitro compounds also are reducible polarographically; if a *gem*-dinitro group also has a hydrogen on the same carbon, one nitro group is reduced mainly through conversion into the *aci* form without C—N fission. If no proton is on the same carbon, C—N fission appears to occur first with formation of nitrite ion and an *aci*-nitro compound. At pH 6 and pH 2, nitroform yields dihydroxyguanidine (12-electron reduction), and at pH 12, hydroxyguanidine is formed (14 electrons). Similarly, at pH 6 and pH 2, hydroxylamidoxime, CH$_3$C=NOH(NHOH), is formed from 1,1,1-trinitroethane by a 6-electron reduction; at pH 12, there is formed an amidoxime, CH$_3$C=NOH(—NH$_2$), by an 8-electron reaction (520).

Coulometric reductions of aliphatic C-nitro compounds are conveniently performed and have high sensitivity; usually, four electrons are required for each nitro group (formation of hydroxylamine) at pH 2 (see also Section VI.A.5).

### b. C-NITROAROMATICS

Aromatic C-nitro compounds are readily and more or less quantitatively converted by strong reducing agents to the corresponding amines. Of course, the route of the reaction and the ease of reduction of the C-nitro group are determined in large measure by the nature of the substituents on the aryl radical, and when multiple nitro groups are involved, one group is usually more readily and more rapidly reduced than the others.

### (1) Reduction by Metals

Reduction with metal-acid couples takes place readily and, ordinarily, essentially quantitatively, but Koniecki and Linch (376) reported some difficulty in the zinc reduction of *o*-nitrobiphenyls, *o*-nitrotoluene, and nitronaphthalene at the low levels encountered in assays for these compounds in urine.

Reduction of an aromatic $C—NO_2$ group on the industrial scale is usually accomplished by iron filings and a small amount of hydrochloric acid; the chlorides of iron are largely hydrolyzed to hydroxides (or hydrated oxides), and the liberated acid attacks a fresh quantity of metal (451). This method of reduction is not very useful for quantitative work, but it may be used for qualitative procedures; for example, hydrated ferrous oxide in ammonium hydroxide (310) provides a method of detection for the nitro group (and other oxidants) by a change in color of the reaction mixture [see Section V.B.1.a(1)].

Zinc dust is used extensively in the laboratory for the reduction of the aromatic $C—NO_2$ group in acidic media, but a variety of reduction products can be obtained simply by varying the pH; for example, zinc dust in sodium hydroxide (466) and in calcium chloride (392) converts $C—NO_2$ to $C—NH_2$, whereas zinc dust in ammonium chloride reduces nitrobenzene to $\beta$-phenylhydroxylamine (5). The combination of zinc dust and sodium hydroxide in alcoholic media often is used to form azo or hydrazo compounds.

The use of zinc dust and ammonium chloride for quantitative estimation of aromatic nitro and nitroso compounds was described as early as 1898 by Green and Wahl (239). Weighed amounts of assayed zinc dust were used, and the reaction was completed at the boiling point; the zinc dust that remained was oxidized by ferric sulfate, and aliquots of the resulting solution were analyzed for ferrous ion by titration with permanganate. The amount of zinc dust consumed in the reduction of the nitro or nitroso compound was then computed. The method is not accurate.

A useful analytical method was developed by Vanderzee and Edgell (713); it is a modification of several older methods in which the loss of weight suffered by metallic tin in contact with hydrochloric acid and an aromatic $C—NO_2$ compound is used to determine the amount of nitro compound. Unlike the earlier methods, the procedure of Vanderzee and Edgell is stoichiometric, but it also suffers interferences from other reducible functional groups (see Section VI.A.2). The kinetics of the tin reduction has been studied in some detail (714). As can be anticipated, other metal-acid couples have been found suitable for the indirect gravimetric determination of aromatic nitro groups; for example, cadmium and copper (see below). Procedures using tin, cadmium, or copper metal-acid couples (335) seem to be applicable to mononitro- as well as polynitroaromatics, and they may also be applied to aliphatic and aromatic $C—NO$ groups; $N—NO$ and $N—NO_2$ groups also are reduced by these couples.

Amalgams have been found useful for reduction of aromatic $C$-nitro compounds. For example, electrolytic cadmium in a Jones reductor appears to be effective for nitrobenzene, *m*-nitrotoluene, *o*-nitrobenzoic acid, and *m*-nitrobenzenesulfonic acid (436). Similarly, zinc amalgam has been found as satisfactory (546) as zinc in acid (99) for most simple nitroaromatic compounds; it is nearly impossible to ascertain whether the lack of quantitative conversion as reported by various workers with metal-acid systems ($\pm 0.1\%$ of theor.) is due to inefficiencies in the reduction step or in the methods used to measure the amount of amine formed.

An amalgam of zinc at the 3% level (239) appears to be an effective reducing agent for most nitro compounds; however, compounds such as nitrobenzaldehyde give low yields of amines (condensation side reactions). Ammonium amalgam (prepared electrically) gives a variety of products with nitrobenzene, including azoxybenzene and phenylhydroxylamine as well as aniline; in sulfuric acid solution, *p*-aminophenol is the main product (676). Copper sponge and monosodium phosphite reduce alcoholic solutions of nitro compounds; hydrogen is evolved (452).

### (2) Reduction by Titanous Ion

The kinetics of reduction of nitrobenzene and its substitution products by titanous ion has been examined by Newton et al. (515). The reaction proceeds through the following steps:

$$RNO_2 \rightarrow RNO \rightarrow RNHOH \rightarrow RNH_2$$

The active species appear to be $[RNO_2H]^+$ and a hydrolyzed form of $Ti^{3+}$ that is present as a minor constituent. The rate of reduction is proportional to

$$\frac{[Ti^{3+}][RNO_2]}{[H^+]^2}$$

The condensation of nitrosobenzene and phenylhydroxylamine (both are formed during the reduction sequence) apparently is not rapid enough to form azoxybenzene in sufficient quantities to divert significant amounts of nitrobenzene through the following sequence of reduction steps (662):

$$R{-}N{=}N{-}R \rightarrow R{-}N{=}N{-}R \rightarrow R{-}N{-}N{-}R \rightarrow NH_2{-}R{-}R{-}NH_2 + RNH_2$$

Moreover, the rate of reduction of nitrosobenzene and phenylhydroxylamine is much greater than the rate of reduction of nitrobenzene; the interaction of these two compounds to give azoxybenzene is extremely fast. The rate of reduction of azoxybenzene is only about one seventh as fast as the rate for nitrobenzene at the same molar concentration (515).

Because acidity plays an important role in establishing the redox potentials of strong reducing agents such as the titanous ion, the experimental procedure employed determines in large measure the extent to which an aromatic C-nitro or C-nitroso group is attacked and the route of the reduction reaction. Thus, for example, in strong HCl media, Earley and Ma (151,447) found that azo and aromatic C—NO groups can be reduced completely, whereas aromatic C-nitro and hydrazino groups are not; it is important to recognize that aromatic C—NO groups and aromatic C—NO$_2$ groups can be determined when in admixture [see Section VII.F].

In spite of irregularities, titanous ion remains an attractive reagent for determination of nitro groups from the point of view of reaction rates, sensitivity to atmospheric oxidation, and selectivity, especially when titanous sulfate solutions are used as titrants in buffered media, for example, citrates (40). Titanous reductions can be performed success- fully on the submicroscale with samples of aromatic nitro compounds of the order of 50–75 $\mu$g, provided that an excess of reagent is added and then back-titrated. Aromatic C—NO groups sometimes can be titrated directly at 40–60°C, but a 50–100% excess of reductant and a reaction time of 3–5 min are required; 4 equiv. of titanium(III) are required for each C—NO and 6 equiv. for each C—NO$_2$ group.

### (3) Reduction by Stannite Ion

The more powerful reducing couple stannite-stannate has received little attention. The formation of alkali-metal stannite probably occurs because hydrated tin(II) oxide behaves as follows in the presence of hydroxide ions (398):

$$[Sn(OH)_2(H_2O)] \rightleftharpoons H[Sn(OH)_3]$$
$$H[Sn(OH)_3] + H_2O \rightleftharpoons [Sn(OH)_3]^- + H_3O^+$$
$$[Sn(OH)_3]^- \rightleftharpoons H_2O + HSnO_2^-$$

for which the dissociation constant is:

$$K = \frac{[Sn(OH)_3^-][H^+]}{[Sn(OH)_2]} = 4 \times 10^{-10}$$

The normal potential of the stannite-stannous couple indicates that it is a strong reducing agent (398):

$$HSnO_2^- + 3 OH^- + H_2O \rightleftharpoons Sn(OH)_6^- + 2 e \qquad E_0 = 0.93$$

Goldschmidt and Eckardt (230) made a detailed study of the kinetics of reduction of aromatic nitro compounds by sodium stannite; in the course of their work, they determined the nature of the products formed and gathered evidence that indicated $HSnO_2^-$ is the reducing species involved in the reaction. The effect that substituents on the aryl nucleus have on the rate of reaction is essentially the same as for reductions involving stannous salts in acidic media, probably because the chief species in both reduction processes are anions. The excess of stannite was back-titrated with a standard iodine solution; in some instances, iodine reacted with the products of reduction. Only four of the compounds studied required 6 equiv. for reduction (to amines): o- and p-nitrophenol and o- and p- nitroaniline. On the other hand, a large group of nitro compounds required 2 equiv. for reduction, and could be divided into two categories according to whether they formed hydroxylamine derivatives or azo and azoxy derivatives. By far the greater number of the compounds fell into the second category, and only o-nitrobenzoic acid, o-nitrobenzaldehyde, and o-nitrosobenzoic acid were found to form hydroxylamine derivatives.

Stannite solutions are not stable and are seldom employed as volumetric reducing agents, in part because of their unpredictable action on nitro compounds, but largely because reductions usually are not quantitative. There are no definitive studies on the use of stannite solutions in quantitative analysis, other than a comparatively short investigation by Sato (605) which indicated that under certain conditions alkaline stannite solutions can reduce nitro groups, nitrato groups, and hydroxylamine to ammonia.

### (4) Reduction by Hydrosulfite, Sulfides, and Thiosulfate

Sodium hydrosulfite, $Na_2S_2O_4$, in alkaline media is especially useful for reduction of C-nitro and C-nitroso groups to amine groups (122,574,639). On the other hand, reductions with this agent may also sulfonate aromatic structures; for example, 5-nitrouracil is transformed to 5-aminouracil-4-sulfonic acid, and 3-nitro-4-hydroxypyridine forms 3-amino-4-hydroxypyridine-2-sulfonic acid (191). This behavior is not much different from the action of bisulfites and sulfites; with these reagents, however, the products are often N-sulfonated amines, $RNHSO_3H$ (570,631).

Sodium thiosulfate in strongly alkaline solutions also reduces nitro compounds; yields are generally low, seldom over 75% (329), but it is of interest that nitro compounds can react with this important titrant $(Na_2S_2O_3)$.

Reductions of aromatic $C—NO_2$ groups can also be effected by hydrogen sulfide, alkali-metal sulfides (57,287,301), and ammonium sulfide (221,508). These reagents have the specific property of reducing one nitro group of a polynitro compound much more rapidly than the remaining groups (302), but the specificity does not appear to be sufficient for quantitative purposes. Hodgson (286) notes that these reducing agents preferentially attack beta nitro groups in naphthalenes, whereas stannous chloride attacks alpha groups.

### (5) Electrolytic Reduction

The polarographic and coulometric reduction of $C$-nitroaromatics is discussed in detail in Section VI.A.4 and VI.A.5.

Electrolytic reduction of the aromatic nitro group occurs readily at platinum cathodes and nearly any other metallic electrode (558); graphite electrodes may also be used. Although the products formed by electroreduction may be influenced by the nature of the cathode, the pH of the catholyte appears to be the most important factor which controls the route of reduction. Reduction to amines occurs in acidic solutions, but in very strongly acidic catholytes, benzidines may be formed by the acid-catalyzed benzidine rearrangement of intermediate hydrazo compounds. In weakly acidic or neutral media, aromatic hydroxylamines and nitroso compounds are formed in preponderance. However, the aromatic hydroxylamines are not readily reduced at a platinum cathode in dilute sulfuric acid; often they rearrange to $p$-aminophenols:

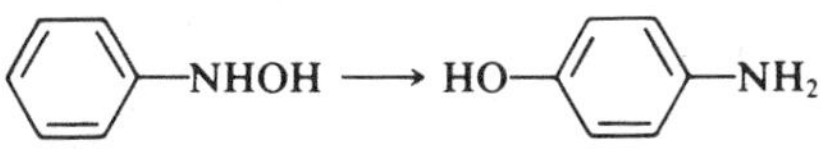

Other reactions can also take place; thus 3-nitrophthalic acid is reduced to the corresponding hydroxylamine, but as rapidly as the hydroxylamine is formed, it reacts with the carbonyl group to give benzisoxazolone-4-carboxylic acid (225).

Reduction of nitroaromatic compounds in basic media leads to the formation of bimolecular products in which the azo, azoxy, and hydrazo compounds predominate, because of the base-catalyzed condensation of intermediate products such as the nitroso- and hydroxylamine compounds. Cathodes of low hydrogen overvoltage (such as nickel) in alkaline solutions nearly always lead to the formation of azoxy compounds; the low hydrogen overvoltage prevents further reduction of the azoxy compound. On the other hand, if alcohols or aromatic sulfonic acids are included in the electrolyte and the reduction is made to take place near the boiling point and at high current densities, azo compounds are formed. Hydrazo compounds are formed at low current densities.

### (6) Reduction with Stannous Chloride

Stannous chloride has been found to be an effective agent for aromatic C-nitro compounds, and in acidic media has been used for quantitative determinations (418); see Section VII.A. The rates of reduction of o-nitrobenzenes by stannous chloride were determined by Manabe and Hiyama (458); reduction appears to occur by simultaneous reaction of $SnCl_2$ and $HSnCl_3$ with R—$NO_2$.

Reduction of complex aromatic C-nitro compounds often leads to ring closures, and, for example, stannous chloride reduction of 2-oxo-1-(2-nitrophenylacetyl)cyclohexane in ether solution gives a 90% yield of 2-(2-oxocyclohexyl)indole along with other ring structures (522):

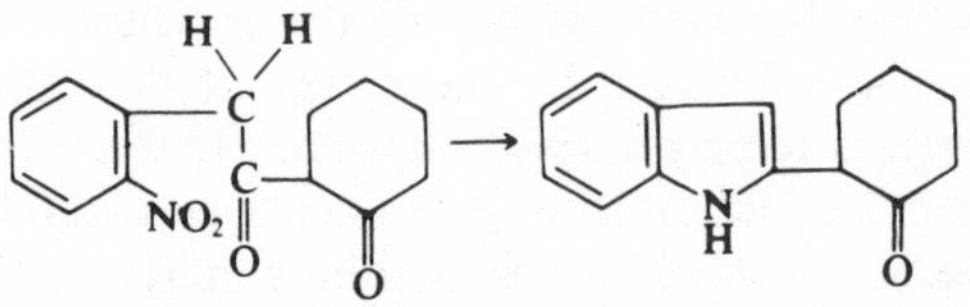

### (7) Miscellaneous Reducing Agents

Glucose in alkaline solution reduces many simple aromatic C-nitro compounds such as o-nitrobenzoic acid to the corresponding azoxy compounds in good yield (214); di-C-nitro compounds such as m-dinitrobenzene or 3,5-dinitrobenzoic acid do not form azoxy compounds, but the quinoid structures that are formed may be used to detect such compounds (57). It is of interest to note that a few C-nitroaromatic compounds, notably nitrobenzene, can be converted by alcoholic potassium hydroxide to azoxy compounds; this reaction is highly restricted in its application.

Phosphine gas in alkaline solutions reduces aromatic C—$NO_2$ compounds to azoxy compounds (86).

Reduction of aromatic C-nitro compounds can also be accomplished with lithium aluminum hydride, sodium hydride, and sodium borohydride [see Section V.B.1.a(3)(a)].

### c. N-NITRO COMPOUNDS

### (1) Reduction by Titanous Ion

The need for rapid analyses of explosive compositions containing the primary nitramine, nitroguanidine, and the cyclic secondary trinitramine, RDX (cyclonite), has resulted in a series of investigations on the behavior of these compounds with titanous chloride. It was found that nitroguanidine is readily reduced in 15 min when treated with an excess of an acidic solution of titanous chloride at the boiling point, but only from 3.8 to 4.2 equiv. are consumed per mole of the nitramino group. This erratic behavior [discussed more fully in Section VI.A.1.a(1)] was also observed with nitrourea, nitroaminoguanidines, and substituted nitroguanidines (747). In contrast, when similar conditions are employed with RDX (a cyclic secondary nitramine), results are low (60% of the expected 12 equiv.). Ferrous chloride itself is scarcely without action on RDX, but a combination of predetermined amounts of ferrous chloride and a large excess of titanous chloride (boiling for 30 min!) provided the desired 12-equiv. reduction to the extent of 98–99% (384). Subsequently, it was found that nitroguanidine can also be made to react with consumption of 6 equiv. of titanous chloride in the presence of fixed amounts of ferrous chloride; with large amounts of ferrous chloride, as much as 8 equiv. can be consumed. Obviously, aliphatic $N—NO_2$ groups such as occur in nitroguanidine require more rigorous reduction conditions; the method of Sternglanz et al. (664) accomplishes this smoothly. The procedure involves dissolution in dilute acetic acid, addition of large amounts of citrate and titanous solution (200% excess), and a reaction time of at least 10 min; results generally tend to be about 3% low (6 equiv.).

Nitrourea consumes only 2 equiv. of titanous chloride regardless of the presence of ferrous iron, and it thus appears that the unstable nitrosourea is formed (747):

$$NH_2CONHNO_2 \rightarrow NH_2CONHNO \xrightarrow{H^+} N_2 + CO_2 + NH_4^+$$

Studies of the reaction of titanous chloride with nitroguanidine suggest that nitrosoguanidine is an intermediate in the initial 4-equiv. reduction (65). Additional details are presented in Section VI.A.1.a(1)(c). Nitrosoguanidine is somewhat stable in dilute acidic solutions (703) but rapidly decomposes in 1:1 HCl (65); the redox system nitrosoguanidine $\rightleftharpoons$ nitroguanidine appears to be reversible (646).

### (2) Electrolytic Reduction

Electrolytic reduction of nitramines usually results in conversion to an amine; for example, a 75–80% yield of aminoguanidine is obtained from nitroguanidine with a tin cathode and stannous chloride in an electrolyte of 5% sulfuric acid; nitrosoguanidine was identified as an intermediate reduction product (657). The polarographic reduction of nitroguanidine in 1:1 HCl appears to be useful for quantitative determinations; the wave occurs at $E_{1/2} = -0.24$ V versus the mercury pool (65).

### (3) Miscellaneous Reducing Agents

Reduction of primary nitramines, $RNHNO_2$, by vigorous agents ruptures the N—N linkage to form amines and ammonia. Milder reducing agents give poor yields of a wide variety of products.

### d. C-NITROSO COMPOUNDS

In general, reduction of C-nitroso groups occurs readily, and the methods ordinarily used for reduction of C-nitro groups are frequently employed.

Polarographic reduction of aliphatic and aromatic C—NO groups occurs readily in methanol-water solutions in a fashion similar to that for C—NO$_2$ groups (290). The reactions often are irreversible; because reductions occur at positive potentials, chloride ion interferes and the usual calomel anode must be replaced with a mercuric sulfate-mercury anode (291). The reduction forms hydroxylamines (two-electron step) and is at maximum sensitivity below pH 4; reduction at higher pH values forms an amine by a four-electron step in buffered solutions (pH 4–5, acetate; pH 7, phosphate), but the potential is sufficiently negative to permit use of the normal mercury-chloride pool anode. As can be anticipated, the half-wave potential is a function of pH; however, the wave height is nearly independent of pH. Dimeric nitroso compounds require six electrons for formation of hydrazines (610):

$$\begin{array}{c} R \\ \diagdown \\ \diagup \\ O \end{array} N{=}N \begin{array}{c} \diagup O \\ \diagdown \\ R \end{array} \rightarrow \begin{array}{c} R \\ \diagdown \\ H \end{array} N{=}N \begin{array}{c} \diagup H \\ \diagdown \\ R \end{array}$$

The kinetics of the reduction of nitroso compounds has been studied (389).

In certain instances it is possible to reduce C-nitroso compounds preferentially in the presence of C-nitro compounds, owing to the great reactivity of the C-nitroso group; for example, in Section III.B.1.b(2) it was noted that the reduction potential of titanous ion can be controlled sufficiently to permit selective reduction of aromatic C-nitroso compounds in the presence of aromatic C-nitro compounds and aromatic hydrazines.

Aromatic and aliphatic C-nitroso compounds can be reduced by hydrogen iodide to hydroxylamine derivatives (149,150):

$$R{-}NO + 2\,HI \rightarrow RNHOH + I_2$$

The reaction is quite rapid with monomeric nitroso compounds; dimeric compounds will not react directly. However, the aromatic hydroxylamine derivatives react very quickly with the original nitroso compounds to form an azoxy compound, for example, with p-chloronitrosobenzene:

$$ClC_6H_4NO + 2\,HI \rightarrow ClC_6H_4NHOH + I_2$$

$$\xrightarrow[\text{ClC}_6\text{H}_4\text{NO}]{} ClC_6H_4N\!\!=\!\!\overset{\overset{\text{O}}{\uparrow}}{N}C_6H_4Cl$$

Additionally, the aromatic C-nitroso compounds can be reduced to amines by hydriodic acid:

$$C_6H_5NO + 4\,HI \rightarrow C_6H_5NH_2 + H_2O + 2\,I_2$$

Naturally, consideration must be given to the possibility that iodine will be taken up by other reactive functional groups (e.g., unsaturated bonds).

Aromatic C-nitroso compounds often are readily reduced to amines by zinc dust in acetic acid, but the reduction is seldom quantitative; it is frequently used for qualitative detection of aromatic C—NO groups by diazotizing the resulting amines and coupling to form colored products.

Nitrosobenzene is reduced initially to azobenzene and finally to hydrazobenzene by an excess of a mixture of magnesium and magnesium iodide, but the yields are too poor for use in quantitative procedures (23a).

### e. N-NITROSO COMPOUNDS

As a general rule, reduction of nitrosamines in acidic solution by chemical agents yields mostly the hydrazines. On the other hand, a compound like N-nitrosodiphenylamine forms diphenylamine as the main product when reduced by LiAlH$_4$ (609).

The results of titrations of the aromatic N-nitroso compounds cupferron, N-nitroso-N-phenylbenzylamine, and N-methyl-N-nitrosoaniline with titanous chloride in a medium buffered with sodium acetate have been reported by Ma and Earley (447). Cupferron could not be determined since reduction did not stop when the nitroso group had been reduced, but continued by splitting the N—N bond; the splitting could not be made quantitative. The other two compounds were treated with a 100% excess of titanous chloride; a 10-min reaction period was permitted before back titration. Results were of the same order of accuracy and precision as with C-nitroso compounds.

Reduction of N-nitroso compounds with alkaline sodium hydrosulfite may not always yield amines; for example, reduction of N-nitrosodibenzylamines at 60° in basic ethanolic solutions results in quantitative nitrogen evolution and formation of hydrocarbons instead of the anticipated amines, but $N$-nitrosobenzylphenylamine forms chiefly 1-benzyl-1-phenylhydrazine (536) and does not release nitrogen.

As a rule, reduction of aromatic nitrosamines, N—NO, with mild reducing agents such as zinc and acetic acid initially yields diaryl hydrazines; stronger reducing agents yield amines:

$$R_2NNO \rightarrow R_2NNH_2 \rightarrow R_2NH + NH_3$$

Hydriodic acid also reduces $N$-nitroso compounds; the reaction involves denitrosation in acidic media and subsequent reaction of nitrosonium ion with iodide ion:

$$R_2N—NO + H^+ \rightleftharpoons R_2NH + NO^+$$
$$NO^+ + I^- \rightarrow \tfrac{1}{2}I_2 + NO$$

The rate of acidic denitrosification controls the overall release of iodine, and by careful selection of conditions it is possible to obtain quantitative release of iodine within acceptable time intervals (213,661). Of course, the possibility of reaction of iodine with other functional groups must be taken into consideration.

## 2. Action of Bases

As can be anticipated, very strong solutions of alkali decompose most organic compounds; the action of hot 70% potassium hydroxide on aromatic $C$-nitro compounds has been investigated, and azobenzenes have been found to be among the products (438). Mono-$C$-nitroparaffins form nitrite when treated with hot alkalies, for example, boiling in 60% KOH (56) or autoclaved at 120°C in pH 9.5 buffer (476). Milder solutions of alkalies also produce results that are of interest to analytical chemists (414).

Table V summarizes the effects of bases on typical nitro and nitroso functional groups.

### a. $C$-NITROPARAFFINS

Primary and secondary $C$-nitroparaffins exhibit $aci$-nitro tautomerism because the shift of an atom of hydrogen from the $\alpha$-carbon to an oxygen of the nitro group makes possible the simultaneous existence of three forms:

$$R_2CHNO_2 \underset{}{\overset{slow}{\rightleftharpoons}} R_2C=N\!\!\begin{array}{c}OH\\O\end{array} \underset{H^+}{\overset{rapid}{\rightleftharpoons}} R_2C=N\!\!\begin{array}{c}O^-\\O\end{array} + H^+$$

Nitro form          Nitronic acid          Nitronate ion
                     $aci$ form

TABLE V

Typical Reactions of Nitro and Nitroso Groups with Bases

| Class | Base and conditions | Products and remarks | Ref. |
|---|---|---|---|
| Aliphatic C—NO$_2$ | Prolonged action by strong bases (conc.) | Extensive decomposition; $CH_3NO_2 \rightarrow NaOOCCH=NO_2Na$ ($NO_2^-$ formed) | 263 |
| Nitroalcohols | | Decomposition; HCHO + C—NO$_2$ | 56 |
| Nitroolefins | | Nitroalkanes, etc. | 263 |
| Aliphatic C—NO$_2$ 1° or 2° | Alkali-metal hydroxides | Salts of nitronic acids insoluble in NaHCO$_3$ (nitronic acids are) | 517 |
| Aromatic C—NO$_2$ | Moderately conc. | Mononitro = stable; polynitro—HNO$_2$; colors | 56 |
| | Hot, strong conc. | Extensive decomposition | 438 |
| Aliphatic C—NO | Dilute bases | Decomposition | Section III.B.2.e |
| Aromatic C—NO | Dilute caustic | Stable | 414 |
| Aliphatic N—NO | Dilute caustic | Decomposition | 394 |
| Aromatic Ar—N—NO R | Dilute caustic | Function of aryl substitution | 394 |
| Aliphatic N—NO$_2$ } Aromatic N—NO$_2$ } | 10–20% NaOH, 100°C | 1°—most are stable, soluble, form salts 2°—decompose → RNH$_2$ + HNO$_2$, or N$_2$ | 327,394 |
| Aliphatic C(NO$_2$) or Aromatic C(NO$_2$) | (Pseudonitroles) dilute alkali | Primary isomerize to nitrolic acids. Secondry usually insoluble | |
| Aliphatic C=NOH | (Nitrolic acids) dilute alkali | Red solutions | |

85

The nitronate ion in solution is readily evident because of its strong yellowish color [and by strong absorption in the ultraviolet (188,217)], developed when primary and secondary aliphatic nitro compounds are dissolved in solutions of strong bases. Moreover, the nitro forms give no color with ferric chloride, but the *aci* forms (the enolic isomers) produce a red color typical of enolic compounds (263); this reaction forms the basis of the Konovalov test [Section V.B.1.a(2)]. The nitro form is slowly soluble in strong base; the *aci* form is instantly soluble in sodium bicarbonate. The corresponding nitronic acids are formed and can be isolated from alkaline solutions of the nitroalkanes by careful neutralization with acids stronger than the nitronic acid (usually, $pK_a = 4$–6); most nitronic acids (and their salts or esters) decompose more or less slowly into the corresponding nitroalkanes. The rates of transformation of nitronates to the nitroalkane form can be followed polarographically (16,494); see Section VII.A.4.a.

In contrast to other nitroparaffins (which simply form nitronates), nitromethane is very sensitive to alkalies. Sodium or potassium hydroxide converts it to salts of methozonic acid (263), that is, nitroacteldehyde oxime, and eventually to salts of nitroacetic acid:

$$HON{=}CHCH{=}NO_2K \rightarrow [N{\equiv}CCH{=}NO_2K] \rightarrow KOOCCH{=}NO_2K \xrightarrow{[H^+]} HOOCCH_2NO_2$$

Actually, the sodium salt of nitroacetic acid can be obtained simply by dropping nitromethane into a 50% aqueous solution of sodium hydroxide held at 50°C and then boiling for 10 min; when cooled, the crystalline salt precipitates.

The heavy metal salts (e.g., silver, mercury, copper) of nitronic acids are insoluble, essentially covalent compounds that are best prepared by metathesis from the alkali-metal salts. However, it is important to recognize that a dangerous compound, mercuric fulminate, is formed when the sodium salt of *aci*-nitromethane is treated with mercuric chloride:

$$CH_2{=}NO_2Na \rightarrow [(CH_2{=}NO_2)_2Hg] \rightarrow Hg(ONC)_2$$

The stable salts formed by 2-nitro-1,3-indandione are useful for characterization of amines (711); of great interest is the fact that nitromethane at pH 8 nitrates tyrosine quantitatively (583).

The *aci* form of a primary or secondary *C*-nitro compound is readily hydrolyzed by 25% sulfuric acid into carbonyl compounds and nitrous oxide (the Nef reaction):

$$2\,RCH{=}NOOK + 2\,H_2SO_4 \rightarrow 2\,RCHO + N_2O + 2\,KHSO_4 + H_2O$$

$$2\,R_2C{=}NOOK + 2\,H_2SO_4 \rightarrow 2\,R_2CO + N_2O + 2\,KHSO_4 + H_2O$$

The Nef reaction takes place rather rapidly; as a result, it is nearly impossible to neutralize alkaline solutions of nitronates without loss of nitrogen as $N_2O$. A variety of other side reactions attend the acidification (380,521). See also Section III.C.3.a.

The ionization constants of many nitronic acids and nitroalkanes in water are listed, and the detailed chemistry of nitronic acids and their esters are presented by Nielsen (517). Some nitronates can be determined acidimetrically in aqueous solution (356), for example, nitrocyclohexane.

Nitroparaffins with at least one hydrogen atom on the same carbon atom as the nitro group can be directly titrated as acids with strong bases in nonaqueous media (207); for example, sodium methoxide can be used to titrate nitroparaffins in a basic solvent such as butylamine, dimethylformamide, or pyridine, with thymol blue as an indicator.

The *gem*-dinitroparaffins with a hydrogen on the carbon bearing the nitro groups are strong acids, for example, dinitromethane, $pK_a \approx 4$.

The *vic*-dinitroparaffins in which at least one of the carbon atoms bearing a nitro group also carries a hydrogen atom split off nitrous acid readily; in some instances, the formation of nitrous acid takes place spontaneously or on boiling in ethanol, but the action of dilute alkali is more positive.

As can be anticipated from consideration of the structure of a nitronic acid, the $C{=}N$ grouping undergoes many typical addition reactions; in fact, a nitro compound adds bromine slowly, but its nitronate adds bromine instantly:

$$R_2C{=}NO_2H + Br_2 \rightarrow R_2CBrNO_2 + HBr$$

The rapid reaction of nitronates with bromine was used by Meyer and Wertheimer (489) to determine the amount of the *aci*-nitro form in a mixture of the two tautomers. A small amount of ferric chloride is added to the sample, and a standard bromine solution is run in rapidly until the color of the ferric complex disappears. The amount of bromine consumed is equivalent to the nitronate content of the sample.

The nitronic acids are oxidants; iodine is liberated and an oxime is formed when a mixture of an alkali-metal nitronate and potassium iodide is acidified:

$$R_2C{=}NO_2H + 2HI \rightarrow R_2C{=}NOH + I_2 + H_2O$$

### b. *C*-NITROAROMATICS

Aromatic mono-*C*-nitro compounds do not have an $\alpha$-hydrogen; consequently, they do not exhibit *aci*-nitro tautomerism and are usually indifferent to cold aqueous alkali, but derivatives of *o*- and *p*-nitroaniline develop colors under certain conditions. However, when two or more nitro groups are attached to an aryl nucleus, reactivity to alkali in acetone is almost always observed [the Janovsky reaction (313)]. For example, the red color developed by 1,3,5-trinitrobenzene in alcohol containing a strong base is attributed to an additive compound believed to be of the following structure (481):

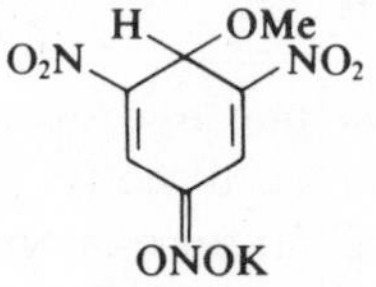

There is evidence, however, that the reaction of alkali with poly-*C*-nitroaromatics is extremely complex. The absorption spectrum of the 1,3,5-trinitrobenzene adduct resembles that of tetryl; the visible spectrum has been reported (198). Nitrobenzene and $\alpha$-nitronaphthalene do not form colored compounds; $\beta$-nitronaphthalene does form colored addition compounds. Most dinitro- and trinitroaromatic compounds form colored solutions with strong bases and alcohol; acetone (272), other ketones, and aldehydes may be used in place of alcohol. The color reaction serves for identification and estimation of polynitro compounds [see Sections V.B.1.a(3) and VI.A.3.b(2)].

Dilute alkali at about 100°C removes a nitro group which is *ortho* or *para* to another nitro group on an aryl nucleus; nitrite is formed, and a hydroxyl group replaces the dislodged nitro group. The displacement of nitro groups in *m*-dinitro compounds is only partial (56).

Aromatic *C*-nitro compounds often may be titrated as acids in nonaqueous media (207); for example, trinitrophenol can be titrated potentiometrically as a triprotic acid in ethylenediamine with sodium colaminate (salt of ethanolamine) as titrant. After 2 mol of the base is added, a yellow precipitate appears; however, a marked potential change does not appear until 3 mol has been added (73). Mixtures of nitro compounds often can be differentially resolved by nonaqueous titrations.

c.  *N*-Nitro Compounds

*N*-Nitro compounds are not basic; the primary compounds are markedly acidic, and the secondary are very weakly acidic. For example, R—NHNO$_2$ is about 5% as strong as the analogous compound R—COOH, and *N*-nitrourethane is stronger than formic acid (394).

Nearly all primary nitramines are quite stable in alkali, being converted into their salts; the —NHNO$_2$ group withstands attack by 10–20% aqueous solutions of sodium hydroxide at 100°C. Formation of the salts (isonitramines) of the primary nitramines betrays the tautomerism:

$$RNHNO_2 \rightleftharpoons RN=NOOH$$

The secondary aliphatic nitramines are decomposed with varying degrees of ease by hot aqueous sodium hydroxide. In general, the decomposition with base first forms nitrous acid:

$$R—N(NO_2)CH_2R' \rightarrow R—N=CHR' + HNO_2$$

and subsequently the amine and an aldehyde:

$$R—N=CHR' \rightarrow RNH_2 + R'CHO$$

Detection of nitrous acid provides a sensitive colorimetric test for secondary aliphatic nitramines.

The secondary nitramides, —CON(R)NO$_2$, resist cold sodium bicarbonate solutions, but they are readily decomposed by dilute aqueous ammonia or sodium hydroxide; hydrolysis often occurs on boiling in water.

Many nitrimines are soluble in sodium bicarbonate or alkalies, but as a rule their instability in aqueous media leads to hydrolytic reactions which yield nitrous oxide.

### d. N-NITROSO COMPOUNDS

Aqueous sodium hydroxide and other strong bases convert N-nitrosoamides (such as N-nitrosoacetanilide) to syn-diazotates at room temperature:

$$C_6H_5N(NO)COCH_3 + 2 NaOH \rightarrow C_6H_5N=NONa + CH_3COONa + H_2O$$

If the reaction is performed in the presence of a coupling agent such as β-naphthol, azo dyes will be formed.

A similar reaction occurs with N-nitrosomethylurethane, CH$_3$N(NO)COOC$_2$H$_5$; the diazotate CH$_3$N=NOK is first formed, but since it is unstable (especially in the presence of moisture), diazomethane is formed. Diazoethane is formed from nitrosoethylurethane; other diazoparaffins are formed from analaqous urethanes.

### e. C-NITROSO COMPOUNDS

Aromatic C-nitroso compounds do not release nitrite when hydrolyzed with alkali. Nitrosophenol forms benzoquinone oxime (285);

similarly; *p*-nitrosoaniline and *N*-methyl-*p*-nitrosoaniline are hydrolyzed to quinone monoxime:

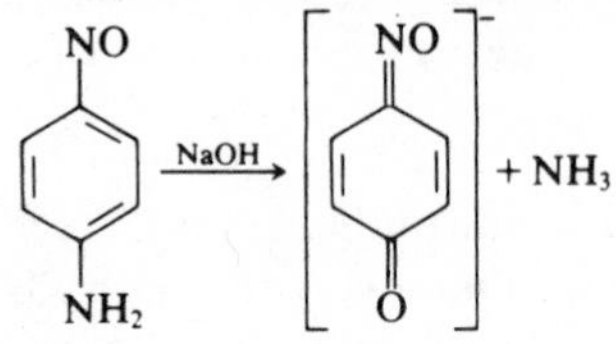

In very dilute alkali, where hydrolysis does not take place, *p*-nitrosoaniline dissolves to form a cation:

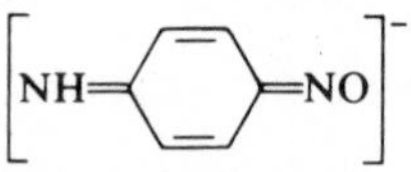

The *C*-nitrosoparaffins readily rearrange to oximes; for example, aqueous or alcoholic solutions of monomeric nitrosomethane rapidly isomerize to form formaldoxime:

$$CH_3NO \rightarrow CH_2NOH$$

The isomerization of many *C*-nitrosoparaffins is catalyzed by alkali as well as acid.

### 3. Action of Acids

#### a. *C*-NITROPARAFFINS

Aliphatic primary *C*-nitro groups resist attack by strong acids of moderate concentrations. In concentrated solutions (about 50% $H_2SO_4$), the primary *C*-nitroparaffins undergo rearrangement and form chiefly hydroxamic acids; as anticipated, the yield of hydroxamic acid is at maximum in the absence of water:

$$RCH_2NO_2 \rightleftharpoons RCH=NO(OH) \xrightarrow{H^+} RC(=NOH)OH$$

When water is present, hydrolysis of the hydroxamic acid takes place and hydroxylammonium salts are formed:

$$RC(=NOH)OH + H_2O + 2\,HCl \rightarrow RCOOH + (NH_3OH)^+Cl^-$$

Hydroxylamine is produced commercially by this reaction (431). The analysis of primary *C*-nitroparaffins can be accomplished by determination of the hydroxylamine formed by acid hydrolysis; for example, hydroxylamine can be oxidized to nitrite by iodine and then determined in the usual manner by diazotization-coupling reactions [see Section VI.A.3.b(1)(e)].

Concentrated sulfuric acid splits off nitrous acid from most secondary C-nitroparaffins; some tertiary C-nitroparaffins also are cleaved to nitrous acid (323).

The salts of primary and secondary aliphatic C-nitro compounds, the nitronates, can be reconverted to their parent compounds by careful neutralization, but they are readily converted to aldehydes and ketones by rapid acidification with dilute sulfuric or hydrochloric acid at an elevated temperature (Nef reaction):

$$2\,RCH{=}NO_2Na + 2\,H^+ \rightarrow 2\,RCHO + 2\,Na^+ + N_2O + H_2O$$

$$2\,R_2C{=}NO_2Na + 2\,H^+ \rightarrow 2\,R_2CO + 2\,Na^+ + N_2O + H_2O$$

A transient blue or green coloration is often observed, and the products formed include nitrous oxide; unfortunately, however, the reaction is not quantitative because oximes, hydroxamic acids, and nitrosated derivatives (pseudonitroles and nitrolic acids) are also formed by side reactions. See also Section V.B.1.a(2).

The action of acids on aliphatic polynitro compounds is determined largely by the number and position of the nitro groups on the hydrocarbon skeleton. As a general rule, the polynitroalkanes are stable to hydrolysis by water or mineral acids, but there are many exceptions. The compounds that are hydrolyzable are attacked by hot, constant-boiling hydrochloric acid (a condition which might readily be found in titanous chloride reductions according to Knecht and Hibbert; see below). The following listing summarizes the usual actions of hot mineral acid on polynitro compounds (336a,520):

$$R_2C(NO_2)_2 \rightarrow \text{no reaction}$$
$$RCH(NO_2)_2 \rightarrow RCOOH$$
$$R_2CHC(NO_2)_3 \rightarrow \text{no reaction (CH not activated)}$$
$$R_2CHC(NO_2)_3 \rightarrow R_2C{=}O \text{ (CH activated)}$$
$$RC(NO_2)_3 \rightarrow RCOOH$$

It is of interest to note that the nitronate salts of compounds containing electron-withdrawing groups (e.g., p-nitrophenylnitromethane) and those with the gem-dinitromethyl group withstand the action of acid, that is, the familiar Nef reaction (conversion of a nitronate salt of mononitroparaffin to a carbonyl compound on acidification) does not take place (520,521).

The action of nitrous acid on nitroparaffins is not of the same nature as the reactions noted above, but it serves the purposes of analytical chemistry very well, for it permits distinguishing between primary, secondary, and tertiary derivatives of the nitroparaffins. The reactions were first noted by Meyer and Locher (491); the following listing summarizes the actions of nitrous acid:

Primary:   $R-CH_2NO_2 \rightarrow RC \begin{smallmatrix} \nearrow NOH \\ \searrow NO_2 \end{smallmatrix}$   Nitrolic acids; solution of salt is red

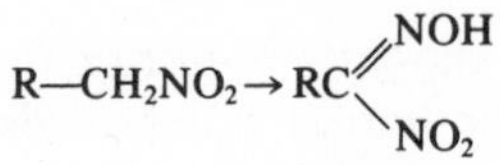

Secondary:  $R_2CHNO_2 \rightarrow R_2C \begin{smallmatrix} \nearrow NO \\ \searrow NO_2 \end{smallmatrix}$   Pseudonitroles; solution is blue

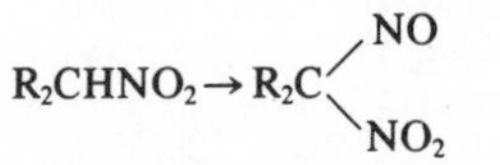

Tertiary:   $R_3CNO_2 \rightarrow$ no reaction   (Colorless solution)

Although the nitrolic acids are usually weak acids and colorless, their salts are bright red in solution; they are colored when solid and are violently explosive. When neutralized by acids, nitrolates form polymeric substances believed to be *n*-mers of nitrile oxides ($R-C \equiv N \rightarrow O$). Solutions of the blood-red alkali salts of nitrolic acids slowly become colorless when kept warm or in sunlight.

The pseudonitroles are usually pungent-smelling solids that are relatively insoluble in dilute aqueous solutions of acids or alkalies and are colorless in the solid state (dimeric); like true nitroso compounds, however, they become colored when melted or put into solution (monomer):

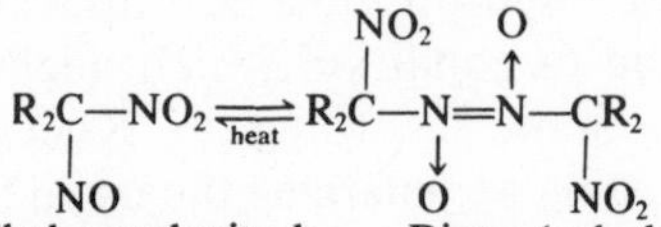

$$R_2C-NO_2 \underset{heat}{\rightleftarrows} R_2C-N=N-CR_2$$

Dialkyl pseudonitrole   Dimer (colorless)
(blue)

### b.  C-NITROSOPARAFFINS

The aliphatic C-nitroso compounds readily rearrange to oximes:

$$R_2CHNO \rightarrow R_2C=NOH$$

The oxime group is called the isonitroso group in older literature because of its relationship to the C-nitroso group. The reaction leading to the formation of the oxime group is seldom reversible, and the isomerization is catalyzed by strong acids as well as bases; polar solvents also may induce isomerization. As can be anticipated, the molecular weight of the alkyl radical influences the conversion to an oxime, and the heavier radicals increase the resistance of the group to isomerization; when monomeric C-nitroso compounds are present in small concentrations (i.e., the dimer is in preponderance), oxime formation is favored.

### c. C-NITROSOAROMATICS

The aromatic C-nitroso compounds often undergo condensation reactions in concentrated acids. For example, nitrosobenzene forms nitrosodiphenylhydroxylamine, $NOC_6H_4N(C_6H_5)OH$, in concentrated sulfuric acid; other strong acids also catalyze the condensation (62). The *ortho*- and *meta*-substituted nitrosobenzenes form analogous products, but a variety of products are obtained by the more complicated reactions that occur when substituents are *para* to the nitroso group (28).

Both aliphatic and aromatic compounds can form adducts with acids; for example, salts are formed with both hydrogen bromide and hydrogen chloride (587).

### d. N-NITRO COMPOUNDS

The aliphatic primary nitramines, $—NHNO_2$, are of themselves acidic substances, but they can be converted into alcohols, with quantitative release of $N_2O$, by action of dilute (about 2%) aqueous sulfuric acid. The secondary nitramines, $=N—NO_2$, are also decomposed very rapidly by treatment with concentrated acids, but they are more stable than the primary compounds. For example, a dialkyl nitramine generally resists attack by 40% sulfuric acid (cf. primary nitramines, above). In contrast, a secondary isonitramine, e.g., butyl butylisonitramine, $C_4H_9—N=NOOC_4H_9$, yields $N_2O$ when treated with 40% sulfuric acid. The aliphatic nitramines such as ethylnitramine, nitrourea, and nitroguanidine also release nitrous oxide when treated with 40% sulfuric acid (or stronger), but as a rule at least part of the nitro group is released as nitrous or nitric acid.

The action of concentrated sulfuric acid on many N-nitro compounds such as nitroguanidine and cyclotrimethylenetrinitramine (RDX) is of special interest in analytical chemistry: nitric acid is formed. Thus these compounds can be determined in a nitrometer (see Section VI.A.7.a), or the nitric acid can be determined by the Bowman-Scott method (see Section VI.A.1.g). Stalcup and Williams (660) have found that the nitric acid released in concentrated sulfuric acid can nitrate salicylic acid (197,503), and the nitrate ester can then be determined readily by titration with titanous solution, whereas the original $N—NO_2$ compounds give grossly inaccurate results when used directly in titanous reductions.

Most aryl nitramines isomerize in the presence of acids; for example,

$N$-nitroaniline (phenylnitramine) forms a mixture of $o$- and $p$-nitranilines. The isomerization occurs readily by action of concentrated sulfuric acid in glacial acetic acid or simply with concentrated sulfuric acid, but it also takes place on prolonged boiling of an aryl nitramine in dilute aqueous mineral acid or in acidic reagents. Unfortunately, the isomerization is attended by a series of side reactions; for example, primary aryl nitramines form diazonium compounds (by action of released nitrous acid), the amounts depending on the other substituents that are on the aryl nucleus. Nitrosamines may also be formed. Tetryl (trinitrophenylmethylnitramine) does not appear to isomerize; consequently, nitric acid is liberated quantitatively (see Section VI.A.1.g).

e. $N$-NITROSO COMPOUNDS

The nitrosamines,

$$\begin{matrix} R \\ \phantom{a} \\ R \end{matrix} \!\!\! \diagdown \!\!\! N\!\!-\!\!NO,$$

are decomposed to nitrous acid and the parent secondary amine by action of strong acids such as $1:1$ HCl or concentrated sulfuric acid; since nitrous acid is readily detected by Liebermann's reaction [see Section V.B.1.a(4)], this sensitive reaction is used to identify nitrosamines and to distinguish them from the great majority of $C$-nitroso compounds, which do not release nitrous acid when treated with acids. Aromatic nitrosamines that have the *para* position open often form $p$-nitrosoarylamines with acidic reagents, especially when treated with dry hydrogen chloride in alcohol (the Fischer–Hepp rearrangement). The fundamental reaction of acids with $N$-nitroso compounds is denitrosation, leading to the reactive nitrosonium ion:

$$R_2N\!\!-\!\!NO + H^+ \rightleftharpoons R_2NH + NO^+$$

The $N$-nitrosoamide group is readily hydrolyzed by acids to the nitrosamine group and thereby is decomposed as indicated above, especially since the primary nitrosamine group is itself quite unstable:

$$R\!\!-\!\!N(NO)COR \rightarrow R\!\!-\!\!NH(NO)$$

Acid-catalyzed reactions in water or alcohols destroy some nitrosamides (213):

$$R'CON(R)NO \xrightarrow{\text{R''OH}} R'COOR'' + R\!\!-\!\!N\!\!=\!\!N\!\!-\!\!OH$$

Table VI summarizes the effects of acids on typical nitro and nitroso groups.

TABLE VI

Typical Reactions of Nitro and Nitroso Groups with Acids

| Class | Acid and conditions | Products and remarks | Ref. |
|---|---|---|---|
| Aliphatic C—NO$_2$ | R$_1$R$_2$C=NOONa + H$^+$— | R$_1$R$_2$C=O + N$_2$O (Nef reaction); no reaction with gem-dinitro | 316,520 |
| (see Section III.B.3.b) | Secondary + conc. H$_2$SO$_4$ | Form HNO$_2$ | 323 |
| | RCH$_2$NO$_2$ + conc. H$_2$SO$_4$ | Hydroxamate → RCOOH + NH$_2$OH | 263 |
| | Polynitro + mineral acid | Generally stable | 520 |
| | Nitrous acid | Red, white, and blue reaction | Section V.B.1.a.(2) |
| Aromatic C—NO$_2$ | Mineral acids | Highly resistant | |
| Aliphatic C—NO | Mineral acids | Variety of products, including oximes | 20 |
| | | Adducts or salts | 587 |
| Aromatic C—NO | Conc. mineral acids | Nitrosodiarylhydroxylamine and other products | 28 |
| | | Adducts or salts | 587 |
| Aliphatic N—NO | Conc. mineral acids | Regenerate amines; denitrosation (HNO$_2$) | 213 |
| | Dilute mineral acids | May form salts; denitrosation (HNO$_2$) | 213,600 |
| Aromatic N—NO | Dilute mineral acids | Fischer–Hepp rearrang. → NO—R—NH$_2$; | 213,480 |
| H | Dilute mineral acids | R—NH$_2$ + HONO | |
| Aromatic N—NO | Conc. mineral acids | Form HNO$_3$; NO—Ar—NH$_2$ (Fischer–Hepp rearrang.) | 480 |
| Aliphatic RNHNO$_2$ | Strong mineral acids | ROH + N$_2$O; HNO$_3$, etc.; stable in dilute acids | 327,394 |
| R$_2$NNO$_2$ | Mineral acids | Relatively stable; conc. acids → HNO$_3$ | 172,327 |

## 4. Action of Oxidants

Table VII summarizes the reactions of typical nitro and nitroso functional groups with oxidants.

### a. *C*-NITRO GROUP

The *C*-nitro group resists oxidative attack except under conditions in which the C—N bond is attacked; no further oxidation of the nitrogen atom can take place. The oxidative cleavage of the C—NO$_2$ group, however, is of interest to analytical chemists, for it permits quantitative determination of C—NO$_2$ and C—NO groups by the amount of nitric acid formed. One of the first uses of chromic acid for determination of nitro groups was described by Friedemann (204). The excess (about 120%) of chromic acid was determined iodometrically. More recently, Jurecek et al. (334,527) have shown that chromic acid in dilute sulfuric acid oxidizes carbon and hydrogen to carbon dioxide and water, respectively, and nitro and nitroso groups (attached to carbon, nitrogen, or oxygen) to nitric acid. If two nitrogen atoms are attached to each other as in azo compounds, hydrazines, pyrazoles, and so on, elementary nitrogen is split off; aromatic amino groups are converted to ammonia. Chromic acid oxidizes iodine to iodic acid, but chlorine and bromine are liberated as the elements; all groups containing sulfur are converted to sulfuric acid. The nitric acid formed can be determined as ammonia after reduction with Devarda's alloy.

Hydrogen peroxide in alkali under specified conditions of concentration and temperature quantitatively converts primary and secondary *C*-nitroparaffins to nitrite ion; the reaction appears to be a combination

TABLE VII
Typical Reactions of Nitro and Nitroso Groups with Oxidants

| Class | Oxidant and conditions | Products and remarks | Ref. |
|---|---|---|---|
| All | CrO$_3$ | → HNO$_3$ | 334 |
| C—NO$_2$ | Resists KMnO$_4$ | | |
| C—NO$_2$ | H$_2$O$_2$ | NO$_2^-$ | |
| C—NO | HNO$_3$; H$_2$O$_2$: KMnO$_4$ | → C—NO$_2$ | 29 |
| C—NO | Auto-oxidation | → CNO$_2$ | 411 |
| N—NO | Mild oxidants | → N—NO$_2$ | |
| N—NO$_2$ | HOCl | Symmetrical —N=N— | 748 |
| $\diagup\!\!\!^{NO}_C\diagdown_{NO_2}$ | Mild oxidant | $\diagup\!\!^C\diagdown$C(NO$_2$)$_2$ | 748 |

of oxidation and reduction before or after hydrolysis. Mononitroaromatics, amides, aromatic amines, most aliphatic amines, aliphatic nitriles, derivatives of hydrazine, and nitrogen-containing heterocycles such as pyridine do not ordinarily yield nitrite ion. Polynitroaromatics, nitrite esters, and some derivatives of hydroxylamine may release nitrite ion under certain experimental conditions (674). Trinitromethyl compounds except esters of trinitroethanol are quantitatively converted to the corresponding 1,1-dinitro anion (227).

Primary and secondary hydroxamic acids are rapidly cleaved to a carboxylic acid and an oxidized derivative of hydroxylamine (162).

### b. C-NITROSO GROUP

The C-nitroso group can be oxidized by almost any strong oxidant; typically, hydrogen peroxide, permanganate, persulfate, and chromic acid are employed. The oxidation is limited, however, to tertiary aliphatic and aromatic nitroso derivatives; often a variety of products are formed.

### c. N-NITRO GROUP

The primary nitramines (those which can be prepared) are stable to nonacidic oxidizing agents, but they are completely hydrolyzed by acids into an alcohol and $N_2O$. On the other hand, since the secondary nitramines are more stable to acids, attack at the N—N bond can take place; however, cleavage requires oxidants of such a nature that the C—N bond is often preferentially ruptured.

### d. N-NITROSO GROUP

Nitrosamines that are not readily hydrolyzable may be oxidized to nitramines by the usual oxidizing agents; however, when hydrogen peroxide is used, nitrosamines may revert to amines (72). Trifluoroacetyl peroxide is a superior oxidizer (163).

### 5. Miscellaneous Reactions

### a. C-NITROPARAFFINS

Primary and secondary C-nitroparaffins condense in basic media with formaldehyde (and other alkyl aldehydes and ketones) to form nitrohydroxy compounds; in turn, these compounds can be reduced to aminoalcohols (commercial preparation). For example, nitroethane reacts with acetaldehyde to form 3-nitro-2-butanol:

$$CH_3CH_2NO_2 + CH_3CHO \rightleftharpoons CH_3CH(NO_2)CHOHCH_3$$

Similarly, 2-nitro-2-methyl-1-propanol is formed from formaldehyde and 2-nitropropane:

$$CH_3C(NO_2)CH_3 + HCHO \rightleftharpoons CH_3\!-\!\underset{\underset{CHOH}{|}}{\overset{\overset{CH_3}{|}}{C}}\!-\!NO_2$$

The position of the equilibrium depends on the nature of the aldehyde and the nitroparaffin; with lower members of both homologous series, reaction is rapid and there is a very great tendency to continue condensation to polyhydroxynitro compounds. For example, the 3-nitro-2-butanol formed as indicated above can condense with another molecule of aldehyde:

$$CH_3CH(NO_2)CHOHCH_3 + CH_3CHO \rightleftharpoons CH_3C(NO_2)(CHOHCH_3)_2$$

The nitroalcohols are decomposed by alkali; thus 2-nitro-2-methyl-1-propanol prepared as indicated above will regenerate formaldehyde and the parent nitro compound. On the other hand, 2-nitro-2-methyl-1,3-propanediol, 2-nitro-2-ethyl-1,3-propanediol, and tris(hydroxymethyl)nitromethane yield two molecules of formaldehyde (324).

Condensation of nitroparaffins with aromatic aldehydes forms conjugated nitroalkenes; for example,

$$C_6H_5NO + CH_3CH_2NO_2 \rightarrow C_6H_5CH\!=\!C(NO_2)CH_3 + H_2O$$

The condensation reactions of C-nitroparaffins with aldehydes and ketones are of little value for analytical determinations; however, these reactions often interfere in the determination of small amounts of nitroparaffins. For example, alcoholic solutions often contain enough aldehydes to react with traces of nitroparaffins, especially under alkaline conditions.

The condensation of nitromethane (601) with 1,2-naphthoquinone-4-sulfonic acid in alkaline media yields a water-soluble p-quinoidal compound that is specific for nitromethane (185):

Compounds containing mobile $NH_2$ and $CH_2$ groups also react with 1,2-naphthoquinone-4-sulfonic acid.

Primary $C$-nitroparaffins as their nitronate salts or as nitronic acids react with diazonium salts to form $\alpha$-nitroaldehyde hydrazones in high yields (319,490), and quantitative colorimetric methods for the determination of nitroparaffins by this reaction have been reported (114,693):

$$C_6H_5CH\!=\!NO_2^- + C_6H_5N\!\equiv\!N^+ \rightarrow C_6H_5\!-\!\underset{\underset{NO_2}{|}}{C}\!=\!N\!-\!NHC_6H_5$$

The nitroaldehyde hydrazones are soluble in alkalies and are usually of a red to red-orange color.

### b. $C$-NITROSOAROMATICS

The condensation of aromatic $C$-nitroso compounds with primary aromatic amines to form azo compounds is an important reaction from the viewpoint of analytical chemistry:

$$RNO + RNH_2 \rightarrow R\!-\!N\!=\!N\!-\!R + H_2O$$

The formation of azo compounds is seldom uniformly realizable; thus, although $m$- and $p$-nitroaniline form the expected azo compounds, $o$-nitroaniline forms $o$-nitro-$p'$-nitrosodiphenylamine (7). Sometimes, unexpected products are formed (e.g., $\alpha$-naphthylamine with nitrosobenzene). The reactions of alkyl amines with nitrosoaromatics are also irregular; for example, dilute aqueous ethylamine yields 1-nitroso-2-$N$-ethylaminonaphthalene when treated with 1-nitroso-2-naphthol. Secondary amines in general are not reactive with nitrosoaromatic compounds, but when reaction takes place the products are complex.

Similarly, condensation of a $C$-nitrosobenzene with hydroxylamine in the presence of sodium carbonate yields a $syn$-diazotate, provided that the nitroso derivative does not first isomerize into an oxime:

$$RNO + H_2NOH \rightarrow R-N=N-OH$$

Nitrosopyrroles and nitrosophenols do not react this way with hydroxylamine; however, the primary and secondary amino derivatives of aromatic nitroso compounds do not readily isomerize, and so they are diazotized by hydroxylamine.

Condensation of an aromatic hydroxylamine and an aromatic nitroso compound in acidic, neutral, or basic solution leads to formation of azoxy compounds. Generally, mixed azoxy compounds are formed owing to an equilibrium between the nitroso and hydroxylamino compounds. For example (9,529):

$$p\text{-}XC_6H_4NO + C_6H_5NHOH \rightleftharpoons p\text{-}XC_6H_4NHOH + C_6H_5NO$$

and on condensation the following products are formed:

$$
\overset{O}{\underset{\uparrow}{C_6H_5N}}=NC_6H_5 \quad
\overset{O}{\underset{\uparrow}{XC_6H_4N}}=NC_6H_5 \quad
\overset{O}{\underset{\uparrow}{C_6H_5N}}=NC_6H_4X \quad
\overset{O}{\underset{\uparrow}{XC_6H_4N}}=NC_6H_4X
$$

Diazonium salts can also be formed directly by action of nitric acid on a solution of a $C$-nitrosobenzene in an inert solvent such as chloroform or by action of nitrous acid on an acetic acid solution of a $C$-nitrosoaromatic.

Nitrosoanilides react directly with aniline to form diazoamino compounds and with phenols to yield azo compounds.

### 3. GENERAL REFERENCES

The following references provide detailed information on the synthesis and reactions of nitro and nitroso compounds.

*Sidgwick's The Organic Chemistry of Nitrogen*, 3d ed., revised and rewritten by I. T. Millar and H. D. Springall, Clarendon Press, Oxford, 1966. A comprehensive, systematic survey of all nitrogen compounds, including nitro and nitroso compounds.

Rodd, E. H. (Ed.), *Chemistry of Carbon Compounds*, Vol. III, Part A: "Aromatic Compounds," Elsevier, Amsterdam, 1954. A review of the properties of nitro and nitroso derivatives of benzene is given in Chapter III; derivatives of amines, sulfonic acids, anilines, and phenols are discussed in Chapters IV, V, VI, and VIII.

Coffey, S., *Rodd's Chemistry of Carbon Compounds*, 2nd ed., Vol. I, Part B: "Aliphatic Compounds," Elsevier, Amsterdam, 1965. A review of the properties of nitro and nitroso compounds is given in Chapter 6; references.

Kirk, R. E., and D. F. Othmer, *Encyclopedia of Chemical Technology*, Vol. 9, The Interscience Encyclopedia, Inc., New York, 1950, pp. 375–403, 428–461. A cursory summary; good for orientative purposes.

Urbanski, T., *Chemistry and Technology of Explosives*, Vol. I, Macmillan, New York, 1964. An excellent summary of synthetic methods, properties and reactions of nitro compounds, and status of the technology of explosives.

Urbanski, T. (Ed.), *Nitro Compounds*, Macmillan, New York, 1964.

Feuer, H. (Ed.), *The Chemistry of the Nitro and Nitroso Groups*, Part I, Wiley-Interscience, New York, 1969. A comprehensive compilation; a detailed survey of the literature, written largely for organic chemists.

Noble, P., Jr., F. G. Borgardt, and W. L. Reed, "Chemistry of the Aliphatic Polynitro Compounds and their Derivatives," *Chem. Rev.*, **64**(1), 19 (1964). A systematic, detailed review of synthesis, reactions, and physical properties.

Millar, I. T., and H. D. Springall, *A Shorter Sidgwick's Organic Chemistry of Nitrogen*, Clarendon Press, Oxford, 1969. An updated, abridged version of Sidgwick's original work.

Davis, T. L., *Chemistry of Powder and Explosives*, John Wiley, New York, 1943.

## IV. TOXICITY AND INDUSTRIAL HYGIENE

All nitro and nitroso compounds should be handled as hazardous materials in well-ventilated areas, and contact with the person should be avoided.

Nitroparaffins are moderately toxic, of the order of petroleum naphtha (263). However, their halogenated derivatives must be handled with caution. For example, chloropicrin, $NO_2$—$CCl_3$, induces edema and is lethal in high concentrations; on the other hand, bromopicrin and dichlorodinitromethane have one-tenth the activity of chloropicrin (477). Salts of alkyl nitro compounds (nitronates) are explosive when dry.

Aromatic nitro compounds are to be treated as insidiously toxic materials; the degree of toxicity is considered to be at a maximum with nitrobenzene derivatives, decreasing through nitrotoluenes to the nearly nontoxic nitronaphthalenes (366). Dinitrobenzene is especially toxic; like many solid or liquid aromatic nitro compounds, it is absorbed through the skin.

Vapors of nitro compounds are readily absorbed by the lungs. Contact with aromatic nitro compounds, especially nitrobenzene, invariably leads to the production of methemoglobin in the blood; the central nervous system is affected, producing headaches, dizziness, visual disturbances, abdominal cramps, and convulsions (650).

Typically, the symptoms of cyanosis include a characteristic bluish tinge of the lips, mucous membranes, ear lobes, fingernail beds, and tongue. The cyanosis is the result of the conversion of hemoglobin in red blood cells to methemoglobin, which is incapable of transferring oxygen from the lungs to body tissues; this causes varying degrees of tissue anoxia and asphyxia. With prolonged exposure to nitro compounds, headache and drowsiness set in; occasionally, severe spasms of nausea and vomiting attend a general feeling of fatigue. Should the methemoglobin content of the blood continue to rise, unconsciousness results, and at this stage the illness has progressed to the point where death may occur. Complete rest and relaxation are mandatory; administration of oxygen is advisable; no alcoholic stimultants should be used. Orange juice, lemon juice, or sugar water may be given. Proper medical care must be sought.

Damage to the liver is also frequent. Studies have shown that living cells reduce dinitrobenzenes to extremely toxic nitrophenylhydroxylamines (281,432).

Halogenated nitroaromatic compounds are highly irritating to the skin; after an initial contact, subsequent exposures often lead to severe dermatitis (pruritis and vesication) and an eventual hypersensitivity to traces of the compounds. Hexyl (hexanitrodiphenylamine) and trinitrotoluene (TNT) are typical examples of toxic industrial materials; even waste water must be treated for removal of such nitro compounds, for as little as 0.1 ppm of hexyl is lethal to fish. Treatment of waste effluents with chlorine offers only a partial solution because most nitro compounds are resistant to oxidation, and the nitro group is not destroyed. Certain biological methods employing strains of *Actinomycetes* (502), *Pseudomonas* (640), or *Corynebacterium* (314) may offer practical solutions for the detoxification of nitro group compounds. Absorption of nitro compounds on charcoal seems to be an economical method for treating effluents, and collection on lime offers an acceptable alternative, in terms of cost (387), in some instances.

Many nitro compounds are potential explosives, and samples submitted for analysis should not be subjected to shock, attrition, or pressure. Since the nitro group is an oxidant, all nitro compounds should be kept apart from combustible materials. It is reported that the explosive decomposition of mixtures of sulfuric acid and dinitrotoluene or dinitrobenzene is catalyzed by lubricating oil (80). On the other hand, most compounds are desensitized when grossly diluted; for example, tetranitromethane is highly explosive in the presence of many organic materials but is relatively safe in dilute solution (e.g., chloroform).

Aliphatic polynitro compounds are hazardous materials, particularly when present as dry nitronate salts; distillation of any polynitro compound is especially dangerous and must be performed on a small scale at reduced pressure and behind barricades (520). The lower aliphatic nitroalcohols may detonate when distilled; the hazard is not removed by distillation at about 1 mm of pressure.

The N-nitro and N-nitroso compounds are also hazardous. For example, N-nitroso-N-methylurethane has been shown to be carcinogenic (612a); there is a possibility that activity is due to release of diazomethane, which has also been found to be carcinogenic. Other

$N$-nitroso compounds too are carcinogenic (724); they are reported to produce persistent inflammation of the eye (391) and may disrupt the olfactory sense for long periods of time (299). $N$-Nitrosodimethylamine has been shown to be carcinogenic; on the other hand, diphenyl-nitrosamine and $N$-nitroso-$N$-methyl-$p$-toluenesulfonamide are essentially noncarcinogenic (724).

A sensitive gas chromatographic procedure for the determination of nitro compounds in air has been described by Camera and Provisani (101).

## V. QUALITATIVE DETERMINATIONS

### A. PREPARATION OF SAMPLES

Nitro and nitroso compounds often may be used directly in qualitative determinations; occasionally, they are first dissolved in ethanol, glacial acetic acid, or other solvents, but this must be performed with caution since the compounds may be strong oxidants. Moreover, since nitro compounds may be explosive, they must be handled with caution; only small amounts should be placed in a sample bottle for use in analyses. Solutions and unused samples should be discarded by transferral to suitable containers and treated as explosive and inflammable wastes (destroyed in open fires).

Many nitro and nitroso compounds can be detonated by heat or friction; samples should not be teased by spatulas or subjected to pressure. Since detonations of polynitro compounds may be violent, unknown samples should not be tested for combustibility or thermal stability without benefit of sturdy safety shields and remote-handling equipment. Determination of the nitrogen contents of unknown samples of nitro and nitroso compounds by the Dumas or other combustion procedures should be avoided or performed cautiously behind heavy shields; the possibility of the complete destruction of expensive glassware must always be taken into consideration. Mixing the sample with clean, ignited sand usually modifies the rate of burning so that combustion analyses can be performed without mishap, but detonable samples may still present problems (5). Shea and Watts recommend calcium carbonate in place of sand (633); Heron (275) urges better control of the flow of carbon dioxide and gives modifications which are alleged to make a C + H train operate better with nitro compounds.

Inhalation of vapors and contact of samples with the skin must be avoided. Heating or boiling solutions of unknown nitro and nitroso compounds in organic solvents entails risk; even known compounds should be regarded with suspicion. Should heating be necessary, apparatus should be arranged behind safety shields and steam baths used whenever possible.

It is good practice to wear a face shield when handling samples of unknown nitro and nitroso compounds (or those known to be hazardous) and to keep the hands well away from bottles, test tubes, and the like by using long tongs and holders. One should avoid grasping the sample bottle containing an unknown in one hand while digging in with a spatula to remove a portion; deflagrations or detonations are often accidentally triggered by friction, although damage to the person will usually be at a minimum as long as miniscule samples are handled, and the hands and face, protected by gloves and shields, are kept as far away as possible. Distance is the key to minimization of personal hazard.

## B. DETECTION

### 1. Chemical Methods

#### a. COLOR REACTIONS

#### (1) General Tests

The positive detection of nitro and nitroso group functions by chemical methods is still an unsolved problem. Unfortunately, the remarks made by Mulliken and Barker (507) in 1899 are still true: "There are few radicals of equal importance whose detection in organic compounds by direct qualitative tests is so inadequately provided for as is that of the nitro group." The only significant chemical property common to all nitro or nitroso group compounds is oxidizing power. As a result, the important analytical reactions of the nitro and nitroso group compounds exploit oxidizing powers directly, or indirectly in reactions with reducing agents, but oxidation cannot be used to detect the nitro and nitroso groups with certainty. The problem is further complicated because the reactivity of the nitro and nitroso groups is governed to a large extent by the atomic aggregates attached to them; thus, in broad outlook, the reactivities of the alkyl nitro and nitroso compounds differ markedly from those of compounds in the aromatic series, although in each class there is a gradation of properties, a circumstance that leads to considerable overlap between the properties of the alkyl and aromatic compounds.

The following general test for nitro and nitroso groups is predicated on their oxidizing properties; light green ferrous hydroxide is oxidized to brown ferric hydroxide (106,635).

A 5% solution of ferrous ammonium sulfate is prepared in air-free distilled water containing 0.4 ml of concentrated sulfuric acid/100 ml. About 10 mg of the sample is put into a tube; 7 ml of ferrous ammonium sulfate solution and then 0.5 ml of an alcoholic solution of potassium hydroxide are added (30 grams KOH dissolved in 30 ml $H_2O$ and mixed with 200 ml 95% ethanol). The air in the tube is displaced with an inert gas, and then the tube is stoppered and shaken vigorously. The reaction occurs within a few seconds; after 5 min, the test is considered terminated. A positive test is indicated by the formation of a red-brown to brown precipitate of ferric hydroxide.

A light green precipitate is a negative test, but in some instances the precipitate may appear to be a darker green because of slight oxidation. Hydroxylamine, quinones, and —O—NO and —O—$NO_2$ compounds, as well as other oxidants, also give a positive test. A list of the colors formed in this test by some nitro and nonnitro compounds is given by Hearon and Gustavson (271). Not all aliphatic C-nitro compounds show a positive reaction; for example, nitroethane and 1-nitropropane do not oxidize ferrous hydroxide in the above test (106).

Compounds containing nitrogen attached to oxygen (nitro, nitroso, azoxy, hydroxylamines, etc.) can by pyrolyzed to release nitrous acid (and perhaps $N_2O_3$ or $N_2O_4$). According to a procedure given by Feigl and Amaral (181), the sample is taken just to dryness in a micro test tube; then the tube is covered with a disk of filter paper on which is absorbed a drop of a solution prepared by mixing equal volumes of a 1% solution of sulfanilic acid in 30% acetic acid and a 0.1% solution of $\alpha$-naphthylamine in 30% acetic acid. The residue in the tube is then pyrolyzed by heating over a microburner for 1–2 min; a red or pink coloration on the paper indicates that the unknown contains nitrogen attached to oxygen.

Nitrous acid is also released when certain C-nitro compounds are treated with hot, concentrated alkali solution; Bose (56) has suggested the following procedure:

Heat 10–50 mg of the sample for not more than 2 min with 1 ml of a solution made by dissolving 10 grams of KOH in 6 ml of water. Cool; add 1 ml of water, and pour a few drops of the alkaline solution into a test tube. Acidify with 1 ml of 50% acetic acid solution; the resulting mixture should be straw-yellow. If necessary, add more water to form a light-colored solution. Then add about 0.5 ml of Griess–Ilosvay reagent [see Section V.B.1.a(3)(b)]; a rose-red color indicates a positive test.

Nearly all C-nitro compounds except the aromatic mononitro compounds decompose under the experimental conditions just described to yield nitrous acid. The aliphatic nitro compounds (and esters of nitrous acid) give a positive reaction even when the alkali is allowed to react in the cold; it is thus possible to distinguish nitroparaffins (or nitrites) from aromatic polynitro compounds (require heat for decomposition) and aromatic mononitro compounds (no reaction). Of great interest is the fact that p-nitrosophenol and p-nitrosodiphenylamine do not form nitrite with hot alkali (56).

Another general test for the nitro and nitroso groups involves reduction to a substituted hydroxylamine by boiling in 80% ethanolic solution with zinc dust (716) and then testing the resulting solution for its reducing power by a silver mirror test (507), Tollen's reagent, or Fehling's solution. It is claimed that calcium chloride materially assists the reduction process (740). The reduction is performed as follows:

Dissolve a few milligrams of the sample in 3 ml of hot 80% ethanol, and add 6–7 drops of 10% calcium chloride solution and about 50 mg of zinc dust. Heat the test tube in a steam bath; after 20–30 min, cool the solution and filter. Check the reducing power of the resulting solution with Tollen's reagent or by the silver mirror test.

Tollen's reagent is prepared as follows. Add drops of 1:10 ammonium hydroxide to 3 ml of 5% silver nitrate until the precipitate that first is formed just dissolves; add 1 drop of 10% sodium hydroxide solution and, if a precipitate of silver oxide forms, cautiously add drops of dilute ammonium hydroxide until the precipitate dissolves. Add a few drops of the sample solution, and warm slowly in a hot-water bath; separation of silver occurs if the zinc dust reduction produced a hydroxylamine or other reducing compound.

The silver mirror test is performed on an ammoniacal solution of silver nitrate prepared as indicated above, without addition of sodium hydroxide. The test tube in which the test is to be performed should be cleaned by boiling in it a strong solution of sodium hydroxide and then rinsing thoroughly. As before, a few drops of the solution of reduced sample should be introduced and the tube warmed gently. Deposition of a silver mirror on the walls of the tube constitutes a positive test. The test should also be applied to the unreduced material; nitro or nitroso groups gain the power to reduce ammoniacal silver nitrate after treatment with zinc dust. Azo and azoxy groups respond similarly.

The filtrate from the zinc reduction in ethanol may also be tested as follows. Add one drop of benzoyl chloride and warm the mixture; then add a drop of concentrated hydrochloric acid and a drop of 10% ferric chloride solution. The formation of a wine-colored ferric hydroxamate signifies that the sample contained a nitro or nitroso compound.

Fehling's solution may also be used for testing reducing power, but the silver reagents are more sensitive.

Veibel (716) recommends oxidizing the hydroxylamine with aqueous ferric chloride to a nitroso compound; an ether extract of the nitroso compound usually exhibits the green or blue color characteristic of a monomeric C—NO compound.

### (2) C-Nitroparaffins

Primary and secondary nitroparaffins may be distinguished from tertiary nitroparaffins and aromatic C-nitro compounds by their solubility in alkalies, but they are usually detected by the action of nitrous acid in alkaline solutions: primary nitroparaffins form the reddish amber color characteristic of a nitrolic acid; secondary nitroparaffins yield the sky-blue color of a pseudonitrole; tertiary nitroparaffins do not react (106,263,635). This test is sometimes known as the red, white, and blue reaction, first described by Meyer and Locher (491).

Dissolve about 0.1 gram of the nitroparaffin in 2 ml of a 10% solution of sodium hydroxide in water (or a 40% solution in methanol), and add 1 ml of 10% sodium nitrite solution; slowly add drops of 10% sulfuric acid, making certain each drop is mixed well. A red or orange-red color appears just before complete neutralization if a primary alkane is present; the color disappears on further addition of acid but can be regenerated when the solution is again made alkaline. If secondary nitroparaffins are present, the solution will be blue or blue-green. If the acidified solution is shaken with chloroform or toluene, the solvent layer extracts the pseudonitrole and becomes blue or blue-green.

The secondary nitroparaffins are not as sensitive as the primary ones, and the test fails completely with high molecular nitroparaffins [7 (263) to 16 (283) carbon atoms]. Moreover, nitrous acid gives colored reaction products with many other types of structures (e.g., phenols and aromatic tertiary amines).

The ferric chloride test used for detection of enolic compounds may also be applied to nitronic acids; the characteristic red color is formed with aqueous or alcoholic solutions of nitronic acids.

Because the reaction of ferric chloride occurs just with nitronates, it is obvious that only primary and secondary nitroparaffins will respond; this reaction was first studied by Konovalov (377) and will give positive results with primary and secondary nitroparaffins of much higher molecular weight than respond to the red, white, and blue reaction described above.

Treat a small amount of the nitro compound with a few drops of concentrated potassium hydroxide solution, and let stand for about 15 min. Dilute with a few milliliters of water, and then remove some of the clear supernatant solution and place it in a separate tube. Add a few milliliters of ether; add a strong solution of ferric chloride drop by drop, shaking the mixture well after each addition. At first, there will be formed a precipitate of ferric hydroxide, but eventually the ether will assume a red or reddish brown color.

The following procedure for developing ferric chloride colorations with primary mono-$C$-nitroparaffins has been developed by Scott and Treon (619) into a sensitive quantitative method:

Prepare a solution of from 1 to 20 mg of the nitro compound in about 10 ml of water; add 1.5 ml of 20% sodium hydroxide, and let the mixture stand for 15 min (occasional agitation). Acidify the solution with 6 ml of 1:7 hydrochloric acid, and immediately add 0.5 ml of 10% ferric chloride. The characteristic brown-red color of ferric chloride will be converted into a deep red ($\lambda_{max} \approx 500$ nm).

The final pH of the solution should be about 1.25, and maximum sensitivity for the test will be obtained if the volumes of sodium hydroxide solution and 1:7 hydrochloric acid are controlled so as to obtain the requisite pH. Nitromethane does not react; 1-nitropropane, 1-nitrobutane, and nitroethane form stable color complexes, but the colors formed with 2-nitropropane and 2-nitrobutane fade rapidly. Green (254) or brown (738) colors sometimes appear, possibly because of formation of ferric nitronate salts similar in nature to those formed between ferric ions and phenols.

The Liebermann test [see Section V.B.1.a(4)], which depends on the liberation of nitrous acid by secondary and some tertiary nitro compounds when heated in sulfuric acid, may be used as a means of detection of $C$-nitroparaffins (323); the nitrous acid is conveniently detected by resorcinol, but it has been found best to add the resorcinol after treatment of the sample with sulfuric acid (323):

Place about 1 mg of the sample in a test tube, and add 5–10 ml of concentrated sulfuric acid. Heat in a boiling-water bath for about 5 min; then cool to room temperature. Stratify 2–3 ml of a 1% aqueous solution of resorcinol over the sulfuric acid, and mix carefully. Return the tube to the boiling-water bath for at least 15 min; cool the contents to room temperature. A red-blue or purple color constitutes a positive test.

The procedure can detect a few micrograms of secondary or tertiary aliphatic compounds (323); reagent blanks should be run. Halogenated nitroalkanes with the nitro group and halogen on the same carbon respond to the test, but nitroalcohols do not. Oxidizing and reducing agents must be absent. It is believed that nitrosylsulfuric acid is first formed by decomposition of the nitroparaffin to nitrous acid; subsequently, nitrosylsulfuric acid combines with resorcinol to form a $p$-nitrosophenolic compound. Hydroxybenzene reagents other than resorcinol may be used; phenol, orcinol, pyrogallol, and pyrocatechol also give a red-blue color. If the hydroxybenzene compound is put into the sulfuric acid at the same time as the nitro compounds, a transitory coloration is obtained.

The coupling of primary $C$-nitroparaffins with diazonium salts may be used as a means of detection. The reaction has been used by Turba et al. (693) with diazotized sulfanilic acid for determination of nitroparaffins. Cohen and Altshuller (114) have also made a study of the procedure [see Section VI.A.3.b(1)(d)].

Specific spot tests for nitromethane and nitroethane were developed by Feigl and Goldstein (184); the tests are based on the Nef degradation of nitromethane to formaldehyde and of nitroethane to acetaldehyde.

A drop of an ethanolic solution containing nitromethane is shaken with a drop of a 2% sodium hydroxide solution; after several minutes, 3 or 4 drops of a suspension of chromotropic acid in concentrated sulfuric acid are added, and the mixture is warmed in a water bath. A violet color indicates nitromethane. The limit of identification is 2.5 $\mu$g, and is positive in the presence of nitroethane.

The test for nitroethane is performed as follows. The sample is dissolved in methanol, and a few drops of the solution are shaken with a drop of 2% sodium hydroxide solution. Three drops of 1:1 sulfuric acid are combined with a drop of the acidulated sample solution in a test tube. The mouth of the test tube is covered with a filter paper moistened with a reagent consisting of equal volumes of 20% aqueous piperidine or morpholine and 5% aqueous sodium nitroprusside. The tube is warmed in a water bath; a blue stain appears on the filter paper if nitroethane is present, Since ethanol may give a faint blue color in this test, methanol must be used; the limit of detection is 8 $\mu$g.

$\alpha$-Nitroalcohols are decomposed by strong alkali into the starting materials, formaldehyde and a nitroparaffin; hence detection of formaldehyde can be used to indicate the presence of an $\alpha$-nitroalcohol (see Section III.B.5.a). Jones and Riddick (324) have developed a procedure based on the detection of formaldehyde by the violet-colored complex formed with chromotropic acid (1,8-dihydroxynaphthalene-3,6-disulfonic acid) in sulfuric acid; a few micrograms of nitroalcohol can be detected.

Mix the sample with about 2 ml of 0.50 N sodium hydroxide, and adjust the volume to about 4 ml. About 5 min later, add 2 ml of a 2% aqueous solution of chromotropic acid, and then bring to a volume of 25 ml with concentrated sulfuric acid; mix well and place in a hot-water or steam bath for 10 min. A violet color is a positive test.

A sensitive and specific method for the detection of nitromethane in the presence of nitroethane was developed by Feigl and Goldstein (185); it is based on the condensation of nitromethane with 1,2-naphthoquinone-4-sulfonic acid in alkaline solution to give a blue-violet, water-soluble $p$-quinoidal compound (601,693); see Section III.B.5.a. One drop of a 5% solution of the quinone is mixed with 1 drop of the sample dissolved in alcohol, several milligrams of calcium oxide are added, and the mixture is shaken; the limit of detection is 0.6 $\mu$g. The same reaction was applied for quantitative determination of nitromethane by Jones and Riddick (326). Aminoalcohols inhibit the color-forming reaction; however, if the sample is refluxed with sodium bisulfite and alkali for 30 min, and the test is performed on an aliquot of the resulting solution, interference is avoided (324).

A blue color is obtained by reaction of secondary and tertiary aliphatic nitro compounds, nitrates, nitrites, and many nitrosamines with diphenylamine in concentrated sulfuric acid (in the absence of other oxidizing agents, which also give blue colors).

The procedure is as follows (238). Dissolve 10 $\mu$l of the liquid sample or 10 mg of the solid in 10 ml of sulfuric acid; add 10 $\mu$l of this solution to 1 ml of freshly prepared diphenylamine reagent (20 mg pure diphenylamine in 100 ml warm 29 $N$ sulfuric acid solution). Warm the mixture in a boiling water bath (at least 98°C); a blue color will develop within 1 hr if the test is positive. The lower limit of sensitivity seems to be about 1 $\mu$g. In some instances, the test is best performed by heating the sample in sulfuric acid while passing through it a gentle stream of air, which then is passed through warm diphenylamine reagent.

The performance of this test for a series of pure aliphatic $C$-nitro compounds was studied by Grebber and Karabinos (238); the following table summarizes their findings:

Diphenylamine Test for Aliphatic Nitro Compounds

| Positive | Negative |
|---|---|
| NO$_2$<br>\|<br>R—CH—R' | R—CH$_2$NO$_2$ |
| R''<br>\|<br>R—C—NO$_2$<br>\|<br>R'' | NO$_2$<br>\|<br>R—C—R<br>\|<br>NO$_2$ |
| | R'''—C(NO$_2$)$_3$ |

R = alkyl; R' = alkyl, NO$_2$, X; R'' = alkyl, X; R''' = alkyl, NO$_2$.

It was shown that the nitrous acid or the oxide of nitrogen released by these compounds is the oxidant that produces the blue color; the colored material is believed to be a mixture of phenazine and benzidine derivatives. The test can also be performed with diphenylbenzidine as reagent (13). Isonitroso compounds (the oximes) do not react.

A method for differentiating between $C$-nitroparaffins and mono-$C$-nitroaromatics is given in the following section.

Manzoff (459) has developed a colorimetric test for nitromethane which involves reaction with vanillin in the presence of ammonia; $p$-hydroxybenzaldehyde reacts similarly. A mixture of nitromethane, ammonia, and vanillin remains colorless at room temperature; when warmed, it assumes a red color that becomes deeper at higher temperatures, but gradually disappears when cold. As expected, the reaction is not suitable for quantitative work (450).

### (3) C-Nitroaromatics

### (a) Mononitro Compounds

The most commonly used general test for the C-nitro group on an aryl nucleus involves reduction to an amine, diazotization, and coupling to form an azo dye. The reduction is best performed by metal and acid, but it may also be accomplished by catalytic hydrogenation. The following is a suitable procedure for reduction of nitro groups to amino groups (also with other reducible functional groups):

About 0.2 gram of the compound and 1 gram of zinc dust are mixed in a 125-ml Erlenmeyer flask, and then 5 ml of dioxane and 1 ml of glacial acetic acid are added. The mixture is refluxed for 30–60 min; then it is cooled and filtered (365). The acidic filtrate contains the amine; any test for amines can be used.

Catalytic reduction with hydrogen can be accomplished in a variety of ways; however, for general laboratory work, the hydrogenation can be performed readily at atmospheric pressure. For example, hydrogen generated by acid and metal and dried by sulfuric acid can be passed directly into an agitated, warm solution of the unknown in which is suspended a catalyst (106). For additional details on hydrogenation, see Section V.C.1.c of "Azo Group," Vol. 15; it should be recognized that, since the reduction is exothermic, addition of hydrogen must be made cautiously.

A typical test for aromatic nitro compounds which involves formation of an azo dye is given by Wolthuis et al. (741):

A drop or two of liquid (or a comparable amount of solid) is dissolved in 3 ml or more of glacial acetic acid, with the aid of heat if necessary; 1 ml of concentrated HCl and 3 ml of water are added. About 1 gram of zinc dust is added carefully and the mixture is boiled for about 5 min (until nearly colorless). The solution is filtered and cooled (10–20°C); then 3–5 drops of 0.1 N sodium nitrite are added, and the fluids are mixed well. The resulting solution is spotted on a piece of filter paper and allowed to diffuse into a spot of 2% R-salt (sodium 2-naphthol-3,6-disulfonate) in 5% sodium carbonate solution.

The formation of a color, usually reddish, shows coupling and indicates that the original substance was probably a nitro compound. Mononitro compounds give a positive test, but polynitro compounds often fail to show the typical coupling because of side reactions during diazotization. Of course, amines will interfere, as will compounds containing nitroso, azo, azoxy, hydrazo, and other groups that can be reduced or cleaved to amines. This test obviously cannot be used to show the presence of the nitro group in a primary nitroamine since the parent compound as well as its reduction product respond to the diazo test.

Nystrom and Brown (528) have shown that reduction of nitrobenzene, p-nitrobromobenzene, and nitromesitylene by lithium aluminum hydride (LAH) yields the corresponding symmetrical azo compounds, and suggest that this general reaction constitutes a simple and positive test for the nitro group. Gilman and Goreau (224) checked the following procedure with over 65 aromatic compounds which contained at least one nitro or nitroso group:

A saturated solution of LAH in dry ether or ethylene glycol diethyl ether is used as a reagent, and is added drop by drop to 1 ml of a saturated solution of the compound in ethylene glycol diethyl ether (or dry ether). The appearance of any definite color or color change in the solution (or in the precipitate, if one is initially formed) constitutes a positive test for aromatic nitro and nitroso groups.

The colors, various shades of brown, yellow, orange, or red, are the result of the formation of azo compounds. The test is confirmed by pouring the reaction mixture into 3 ml of water, and adding aqueous 50% sulfuric acid until all salts and hydroxides are dissolved; then 2 ml of the acid is added in excess, and the resulting solution is compared with a blank prepared by acidulating a saturated solution of the sample in ethylene glycol diethyl ether (the preferred solvent). A colored solution differing from the color of the blank is confirmation of a positive test. Compounds of limited solubility may give a negative test; for example, sodium 2,4-dinitrobenzenesulfonate gives no indication of the presence of the nitro group, but a positive result is obtained with the free acid or its lithium salt (224). Although aliphatic nitro and nitroso compounds may be reduced by LAH, they do not form the characteristic colorations of the aromatic azo compounds. Nitrates will not interfere.

Nelson and Laskowski (514) recommend the use of solid lithium aluminum hydride for reduction of nitro groups to azo compounds:

About 100 mg of the unknown dissolved in 5 ml of anhydrous ethyl ether is treated with about 10 mg of LAH. A positive test is indicated if within 5 min there is a change (but not disappearance), in the color of the solution, the formation of a colored precipitate, or both.

Twenty-six aromatic nitro compounds representing a wide variety of functional groups were tested; only 2-chloro-5-nitrobenzenesulfonic acid failed to give a positive test. It is worthy of note that hydroxylamine, amyl nitrate, amyl nitrite, benzoquinone, hydroquinone, nitroethane, and nitrocyclohexane gave negative tests, for many of these compounds do not respond to other tests for aromatic nitro compounds. Phenyl-acetonitrile gives a positive test, presumably because of a reaction involving its $\alpha$-hydrogen, since octanenitrile and benzonitrile give negative results. When performing qualitative tests with LAH to detect aromatic $C$-nitro compounds, it should be recognized that aromatic nitroso, aromatic hydrazo, and azoxy compounds, and probably any aryl compound that has functional groups in which an N—O group is attached directly to the aryl nucleus can be converted to azo compounds. The test may also be performed by refluxing the sample in ethanol containing LAH.

Aromatic mononitro compounds, as well as some polynitro compounds, can be reduced by sodium hydride (539). About 10 mg of the sample dissolved in 0.1–0.2 ml of acetone is treated with 0.05–0.10 gram of sodium hydride. Evolution of hydrogen and formation of a color within 3 min constitute a positive test; a blank should be run concurrently.

Reduction of $C$-nitroaromatics can be made to take place at ambient pressure and room temperature by a carbon-supported platinum catalyst prepared with the aid of sodium borohydride (77,376). Koniecki and Linch (77) have found thiourea dioxide (formamidine sulfinic acid) a particularly suitable reducing agent for aromatic $C$-nitro compounds at levels of 5–50 $\mu$g/ml. The chloro derivatives of nitrobenzenes and nitro-toluenes, nitronaphthalene, nitrobiphenyls, alkoxynitrobenzenes, and nitroacetanilides are reduced by thiourea dioxide in 10–15 min at 50°C in an aqueous alkaline buffer (10% sodium carbonate); the excess thiourea dioxide can be destroyed by heat (50°C) in acidic media.

A general test predicated on the oxidizing power of the nitro group in C—$NO_2$ aromatic compounds has been described by Mulliken and Barker (507), but it is not used very much at this time; in the presence of iron and acid, "aniline-red oil" (equal parts of aniline and $o$- and $p$-toluidine) is changed by nitro compounds and other oxidants to the highly colored triphenylmethane dye, magenta red ($p$-rosaniline; Basic Red 9; C.I. 42510). The procedure must be performed carefully.

Three or four drops of aromatic nitro compound (or an equal weight of solid) are boiled for 2 or 3 min in a test tube with a mixture of 2 ml of nearly colorless aniline-red oil, 2 ml of water, 2 ml of hydrochloric acid (sp. gr., 1.20), and about 1 gram of iron filings. A large excess of acid prevents the appearance of the color, while too little diminishes its intensity and purity. The red color of the rosaniline is best observed by pouring a few drops of the reaction mixture into 10–20 ml of dilute acetic acid; a blank test should be performed.

The test is difficult to perform on volatile compounds, which escape from the hot reaction mixture before they can demonstrate their oxidizing qualities. Dinitro compounds produce a redder dye of less strength than mononitro compounds; trinitro compounds give dirty reddish or greenish browns. Many compounds that have groups where oxygen is joined to nitrogen will form the red color, for example, azoxybenzene, nitrosobenzene, and ethyl nitrate and nitrite. Iodoso- and iodobenzene give purplish solutions, but oxidants such as chloranil, quinone, and inorganic nitrates and nitrites do not form a red color.

A great number of aromatic C—NO$_2$ compounds give an intense orange-red color in benzene solution when treated with an excess of anhydrous aluminum bromide; the color is destroyed by water (535). More recently, the color reaction between anhydrous aluminum chloride and aromatic compounds in chloroform has been reexamined; functional groups can be classified by the colors that are formed (678).

Yellow or red solutions are formed by most aromatic nitro compounds when dissolved in concentrated sulfuric acid; the intensity of the coloration increases almost threefold as the acid concentration is increased from 97 to 100% (249).

A procedure for characterization of nitrobenzene derivatives by the colors formed when the compounds are dissolved in ammonium hydroxide—alcohol and sodium hydroxide—acetone was reported by Rudolph (595). Similarly, a transient pink color is sometimes obtained when nitrobenzene is dissolved in ethanol and then made alkaline (553); the color is formed by nitrothiophene, since pure nitrobenzene does not give a color (573). A quantitative procedure for mono- and di-nitrothiophene based on this color reaction is given in Section VI.A.3.b(2)(b). Ohkuma (532) indicates that m-dinitrophenyl compounds yield a violet, blue, or red coloration when spots of their ethanolic solutions are treated with 2 N sodium hydroxide and acetone; about 1 μg can be detected, but p-nitrobenzaldehyde semicarbazone, and p-nitrophenylhydrazine also give similar colors.

The colors formed by a series of more than 65 aromatic C-nitro compounds when each is treated with tetraethylammonium hydroxide in dimethylformamide solution have been reported by Porter (559); the wavelengths of maximum absorption may serve as a method of identification [see Section VI.A.3.b(2)(b)].

Aromatic nitro compounds can be reduced electrolytically to nitroso compounds in neutral or alkaline solution with nickel and lead electrodes or by zinc dust and calcium chloride; if the reduction takes place in the presence of sodium pentacyanoamminoferroate, sometimes green, and sometimes violet, colors are formed (180,531,533). A list of nitro compounds, colors resulting from this test, and limits of detection are given by Feigl (180). Aromatic thioaldehydes, pyridine, and some thioketones react to give blue colors; interference by aromatic hydrazines which give red or violet colors may be prevented by addition of formaldehyde. Similar colorations are produced by sodium nitroprusside (531). In these reactions, the reagent species is $Fe(CN)_5NH_3^{-3}$, and the nitroso group replaces the ammonia molecule in the complex; as a result, the original yellow color of the reagent is replaced by a color that depends on the compound subjected to the test. Sunlight often is helpful in accelerating the reaction.

The reagent is prepared as follows. A mixture of sodium nitroprusside and about three times its volume of concentrated aqueous ammonia is allowed to stand at 0°C for 24 hr. The mixture is diluted with ethanol and the solid is filtered off by suction and washed with alcohol. An aqueous solution of about 1% strength is used for tests. In general, a few drops of the sample are mixed with a few drops of the reagent. Colorations develop after a short time.

Nagasawa (511) reported that, when aromatic nitro compounds are reduced with zinc in warm neutral or acidic solution and acidified with 10% sulfuric acid after cooling, there results a color which is stable for 24 hrs; mono- and dinitro compounds give a positive test, but trinitro compounds show no response.

Aromatic nitro compounds give a green color with "bindone" (anhydrobisindandione) after reduction to a primary amine (zinc, acetic acid); amines from aliphatic nitro compounds produce a violet color, and the reaction is more sensitive than with aromatics. However, the test is not specific because any easily reducible nitrogen group (i.e., nitroso, azoxy, hydrazino, or hydrazo) will also give the same reaction (710). A few polynitro compounds, 3-nitroalizarin, 1,8-dinitronaphthalene, and a few ortho-substituted nitro compounds (which form rings on reduction) fail to show positive reactions.

A specific test for m-diamines has been applied by Bunton et al. (88) to the detection of m-C-nitrophenols and m-C-nitrophenolic ethers; the nitro group is reduced to an amino group and then reacted with glycerol and oxalic acid in the presence of zinc or calcium chloride to give a fluorescent 2,8-diaminoacridine.

### (b) Polynitro Compounds

Polynitroaromatic compounds form colored derivatives when dissolved in alkali alcoholates; 2,4-dinitrotoluene, 4,6-dinitro-m-xylene, and 2,4,6-trinitro-m-xylene give colors with EtOK, but the polynitro derivatives of mesitylene do not (700).

Polynitroaromatic compounds, as well as aliphatic $C$-nitro compounds, may be detected in the presence of aromatic mono-$C$-nitro compounds by the following test, which depends on formation of nitrite ion:

Heat 10–50 mg of the substance with 1 ml of KOH solution (10 grams in 6 ml of water) for not more than 2 mins. The color of the solution darkens during boiling; usually a yellow to dark brown color is formed. Cool under water; then add 1 ml of water and pour a few drops of the alkaline solution into a test tube. Acidify the solution with about 1 ml of 50% acetic acid; a straw-yellow solution should result, but it may be necessary to add more water to obtain a pale color. Then add about 0.5 ml of Griess–Ilosvay reagent;* a rose-red color indicates a positive reaction (56).

The test does not distinguish aliphatic $C$-nitro, from aliphatic $O$-nitro compounds because both are decomposed by cold dilute alkali; thus it is possible to distinguish these types of compounds from the aromatic mono-$C$-nitro compounds because the latter resist attack by strong alkali. Ordinarily, the mono-$C$-nitroaromatics can be distinguished from the aromatic poly-$C$-nitro compounds, for a nitro group in the *ortho* or *para* position to another nitro group is readily replaced with —OH by dilute alkali at 100°C; on the other hand, the $m$-dinitro groups are very resistant to attack. In the experimental procedure described above, however, enough of the $m$-dinitro groups are dislodged to provide a positive reaction. The aromatic nitroso compounds do not release nitrous acid on treatment with alkali; $p$-nitrosophenol and $p$-nitrosodiphenylamine give negative results.

The number of nitro groups in benzene derivatives can usually be estimated by the color developed when the compounds are dissolved in acetone or alcohol solutions of alkali [the Janovsky reaction (313); see Section III.B.2.b]; as reported by Bost and Nicholson (58), for example, mono-$C$-nitro compounds develop no color, dinitro compounds generally form a purplish blue color, and trinitro compounds form a deep red color when about 0.1 gram of the sample is dissolved in acetone and mixed with 3 ml of 5% sodium hydroxide solution. On the other hand, trinitroresorcinol gives no color with acetone and alkali (53); in general, trinitro compounds with an amino, substituted amino, or hydroxyl group form no color; and even richly substituted polynitrobenzenes yield no color (e.g., 2,4-dinitromesitylene and 2,4,6-trinitromesitylene). It is of interest to note that 2-nitro- but not 3-nitro-9-

*Dissolve 0.5 gram of sulfanilic acid in 150 ml of $2\,N$ acetic acid; prepare a colorless solution of 0.1 gram of $\alpha$-naphthylamine in 20 ml of boiling water; mix both solutions, and add 150 ml of $2\,N$ acetic acid.

aminoacridine gives a purple color in alcoholic alkali; rearrangement of bonds is possible with the 2-isomer (2):

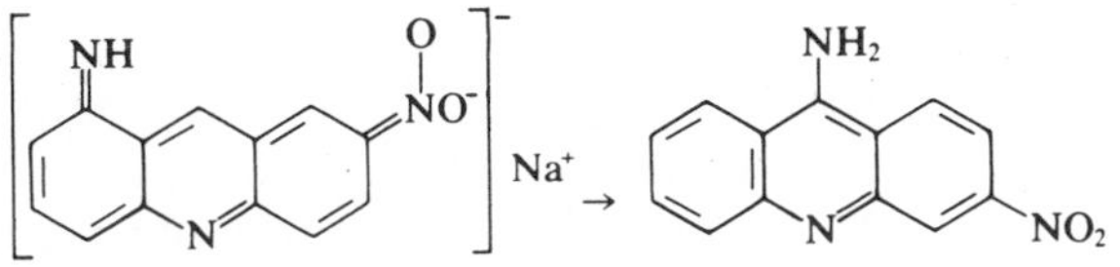

Amino or hydroxy groups on the benzene nucleus interfere; often the Janovsky reaction colors fade rapidly (106,519,578,635). The work of Bost and Nicholson should be consulted for the colors formed with a wide variety of compounds and especially for corroborative colorations obtained when the test solution is diluted, made alkaline, and so on. Porter (559) observed that dimethylformamide solutions of certain mononitro compounds, principally the derivatives of $o$- and $p$-nitroaniline and of $p$-nitrotoluene, develop orange, red, or purple colors when tetraethylammonium hydroxide is added. A detailed comparison was made of the colors formed by the Janovsky reaction (alkali in acetone or alcohol) and by tetramethylammonium hydroxide in dimethylformamide; the wavelength of maximum absorption, the absorbance of the colored solution containing 10 $\gamma$ of the compound/ml, and the appearance of the color to the eye have been recorded for 67 compounds. Similarly, a spectrophotometric study of the reactions of $p$-nitrotoluene in tert-butanol with potassium tert-butoxide (red colors) has been reported by Miller and Pobiner (495). Cavett and Heatis found that greater color stability can be obtained by use of diethylamine and dimethyl sulfoxide (104); the diethylamine had to be refluxed for 48–72 hrs with 40 grams/liter of sodium or potassium fluosilicate in order to stabilize the color. Most colorations are purple ($\lambda_{max}$, 540–575 nm), but there are exceptions. For example, 2,4-substituted compounds show $\lambda_{max}$, 640–695 nm; thus 2,4-dinitrophenyl isothiocyanate gives a yellow-green color ($\lambda_{max}$, 430 nm), and 2,2'-4,4'-tetranitrobiphenyl a pink color ($\lambda_{max}$, 640 nm). Compounds such as 3,5-dinitrobenzoic acid and 2,6-dinitro-4-chloraniline do not form colors.

A spot test for the detection of polynitrobenzene derivatives described by Sawicki and Stanley (606) consists of adding a drop of the compound dissolved in dimethylformamide to a mixture of a drop of 25% aqueous tetraethylammonium hydroxide and a drop of 5% fluorene in dimethylformamide. Brilliant blue, green-violet, and sometimes red colors are obtained for a number of polynitro compounds, but some mononitro compounds and polynitrobenzene homologs fail to respond; compounds containing strong electronegative groups other than nitro groups can give the color reaction (e.g., 1,3-bisphenylazobenzene). The polynitroaryl compounds react in alkaline solution with compounds containing the cyclopentadiene $CH_2$ group to give blue to green dyes. Dialkyl and aralkyl ketones react with polynitroaryl compounds to give violet and blue dyes. Thus, in the above procedure, a 50% solution of butanone can often be substituted for the fluorene solution. The colors formed by these reactions are somewhat unstable and cannot be used for colorimetry unless photometric measurements are made very soon after the mixing of reagents.

The formation of a red-brown to violet precipitate in an alkaline cyanide solution is specific for *m*-dinitro compounds; mononitrobenzene and *o*-dinitrobenzene derivatives do not form this coloration. Colorations and limits of detection are given by Feigl (180) and by Anger (12) for a series of *m*-dinitro compounds. The color is permanent in dilute acids; this distinguishes *m*-dinitro compounds from the nitrophenols. Guilbault and Kramer (244) found that *p*-nitrobenzaldehyde, *o*-dinitrobenzene, and cyanide ion react to form an intense color; the reaction is recommended as a sensitive, specific test for cyanide. The dinitrobenzene is reduced to *o*-nitrophenylhydroxylamine.

The purple color formed by derivatives of 3,5-dinitrobenzamides with dimethylformamide and methylamine has been used by Smith (647) to determine 3,5-dinitro-*o*-toluamide (used to control coccidiosis in chickens). A study of the color-forming reaction indicated that only the following structures form shades of purple with dimethylformamide and methylamine; other dinitro compounds produced colors varying from red through blue in the visible spectrum:

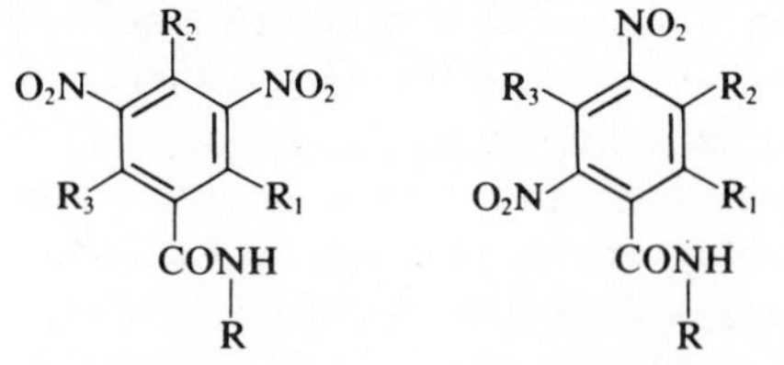

R = H or alkyl
$R_1$ = H or alkyl
$R_2$ = $R_3$ = H

R = H or alkyl
$R_1$ = H *and* $R_2$ = alkyl
*or* $R_1$ = alkyl *and* $R_2$ = H
$R_3$ = H

The carboxamide group must be present; 3,5-dinitrobenzoic acid does not form a purple complex. The original reference includes spectral characteristics of colors obtained from more than 50 dinitro compounds.

Prepare an acetone solution of sample so that the final concentration is 500 $\gamma$/ml. Prepare a solution of 3,5-dinitro-o-toluamide of similar concentration (for a standard). Pipet 1 ml of each solution into separate 50-ml volumetric flasks, and add 24 ml of reagent-grade dimethylformamide which has been cooled to 20°C. Add 25 ml of 40% aqueous solution of methylamine (cooled to 3–5°C). Mix thoroughly; transfer each solution to a cuvette, and obtain absorption within 3 min of mixing (use triplicates) at 390 and 550 nm. The color gradually fades.

Feigl and Gentil (183) have developed a spot test for acidic poly-C-nitroaromatic compounds such as dipicrylamine, picric acid, picrolonic acid, 5-nitrosalicylic acid, Martius Yellow (alkali salt of 2,4-dinitrophenol), Naphthol Yellow (alkali salt of 2,4-dinitro-1-naphtholsulfonic acid), 2,4-dinitrophenol, and p- or m-nitrophenol.

### (4) C-Nitroso Compounds

The Liebermann test (416,417) is often employed for the detection of nitroso compounds even though it is far from being specific [see Section V.B.1.a(2)]. In essence, the sample is thermally decomposed in a mixture of phenol and concentrated sulfuric acid; nitrous acid formed by decomposition of the C—NO group reacts initially with phenol to produce a red color, and when the reaction products are made alkaline, an intense blue color develops. Secondary and tertiary nitroparaffins, nitrosohydroxylamine, aromatic C—NO compounds, aliphatic O—NO compounds, and N-nitroso compounds give a positive test. The oxime group (=NOH) does not give a positive test. There are many variations of the test (712); however, the following procedure is generally applicable:

Place about 1 mg of the sample in the bottom of a small tube, together with one or two small crystals of phenol (10–20 mg); heat carefully over a small Bunsen flame until the phenol melts. Then cautiously add a few drops of concentrated sulfuric acid, and continue heating very gently to promote reaction. If the sample contains the nitroso group, the melt is red. When cool, the melt is treated cautiously with enough water to dissolve it; the solution is then made alkaline with sodium hydroxide. A blue color confirms the presence of the nitroso group.

The nitrous acid released by decomposition of the nitroso group nitrosates phenol, which reacts with more phenol to form a red indophenol; when an excess of sodium hydroxide is added, the deep blue salt of indophenol is formed:

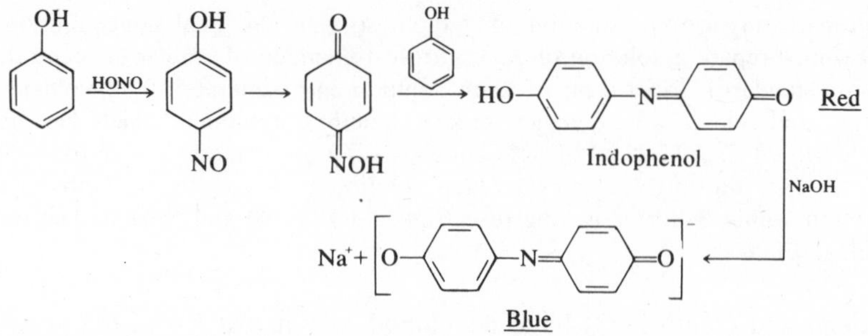

If the volume of the phenol-sulfuric acid mixture used for the decomposition of the nitroso compound is considerable, the indophenol dissolves and the resulting solution, containing indophenolate ion in the reaction mixture, is colored green to blue, as occurs in the following test procedure:

Place about 1 mg of the sample and a few crystals of phenol in the bottom of a small tube; add 1 ml of concentrated sulfuric acid, and warm gently. A green or blue color develops. Pour the acid mixture into 5 ml of water; the color will change through violet to reddish violet and then to red. On being made alkaline, the mixture becomes green or blue.

Compounds in which the NO group is strongly bound do not give a positive Liebermann test, for example, 3-nitrosopyridine and the nitrochloride of tetramethylethylene. Conversely, certain nitro compounds can release nitrous acid under the conditions of the Liebermann test and give a positive result, for example, 2-nitro-1-phenylethylene, phenylnitromethane,    2-nitro-2-phenyl-1,3-indanedione,    and    1-nitromethyl-1-methoxyphthalide (712). It is also necessary to note that, when the Liebermann test procedure is applied to certain C-nitro compounds, spurious colors are formed; in these instances, it may be difficult to detect positive color formation. For example, when heated with phenol and sulfuric acid and then diluted with water, phenylnitromethane forms a dirty red color with a green irridescence, o-tolylnitromethane produces an intense violet color, m-tolynitromethane yields a brilliant blue, and β-naphthylnitromethane colorations are red, but they go over to dark blue on addition of sodium hydroxide (212,556).

The Liebermann test is sufficiently sensitive to be able to detect small amounts of nitroso compounds which are formed as intermediates when nitro compounds are reduced electrolytically (182). The following typical procedure requires only a small amount of the test solution:

Place a drop of an alcoholic or aqueous solution of the sample in a microcrucible, together with a drop of alkali or sodium sulfate solution (to impart conductivity). Electrolyze for at least 10 min but not longer than 0.5 hr with a 4-V source, using a nickel wire cathode and a lead wire anode. Use one drop of the solution for the Liebermann test; alternatively, use the ammonia pentacyanoferrate test [Section V.B.1.a(3)(a)].

Highly colored monomeric nitroso compounds can be decolorized by dropwise addition of a solution of hydrazine or hydrazine hydrate. Moreover, monomeric C-nitroso compounds (as well as N-nitroso compounds) react with acidulated potassium iodide (HI) to liberate iodine. Often, aromatic C-nitroso compounds will produce typical colorations of azoxy compounds owing to the reaction of the initial C-nitrosoaromatic with the phenylhydroxylamine formed by reduction.

The formation of a red nitrosopalladium precipitate in alcoholic solution is a selective test for p-nitrosoaromatic amines (180); unfortunately, many amines and amine-phenols react with palladous chloride to give a yellow color which can obscure the red color formed by small amounts of the nitroso-substituted aromatic amine. Interfering nitrosonaphthols may be removed by treatment with aqueous sodium hydroxide and extraction of the nitrosoaromatic amine with ether.

The reduction of p-nitrosoaromatic amines such as p-nitrosodimethylaniline, p-nitrosodiethylaniline, and p-nitrosodiphenylamine with ascorbic acid in the presence of p-dimethylaminobenzaldehyde yields a red Schiff base of an intensity proportional to the amount of the nitrosoaromatic amine (186); other C-nitroso compounds do not give a positive test.

One drop of an alcoholic solution of the sample in a tube is treated with a 1% solution of p-dimethylaminobenzaldehyde in concentrated acetic acid; a few milligrams of ascorbic acid is added, and the mixture is heated in a steam or water bath. A violet or pink color will form immediately or within 1–2 min if a p-nitrosoaromatic amine is present.

Primary aromatic amines and compounds with active $CH_2$ groups condense with p-nitrosodimethylaniline or p-dimethylaminobenzaldehyde in alcoholic or acetic acid solutions to yield colored azo products; but if no color forms in parallel tests with an alcoholic solution of the sample and these compounds, the test can be considered valid. Sulfanilic acid may be used in aqueous solutions. Hydrazine salts interfere because the base forms a red acid-resistant aldazine with p-dimethylaminobenzaldehyde.

Aromatic C-nitroso compounds form benzenediazohydroxides on reaction with hydroxylamine, but since diazohydroxides are unstable, they are converted to azo dyes by coupling reactions. Thus an alcoholic solution of the nitroso compounds is treated with α- or β-naphthol and then with an aqueous solution of hydroxylamine hydrochloride; when a dilute solution of sodium carbonate is added dropwise, a color change takes place (usually from green to brown or red). Finally, the mixture is diluted with water, whereupon the azo dye separates as a voluminous precipitate. The dye may be separated and crystallized from benzene for further characterization.

Aromatic C-nitroso compounds are converted to intensely yellow nitrosodiaryl hydroxylamines by concentrated sulfuric acid; these compounds usually form red solutions in alkalies. However, *para*-substituted nitrosobenzenes are not condensed by sulfuric acid.

A great number of C-nitroso and nearly all N-nitroso compounds give the familiar "brown ring" test for nitrates (concentrated sulfuric acid and ferrous sulfate).

The reaction of nitroso compounds with sodium pentacyanoamineferroate produces a green color, but a violet color is seldom observed (180). The test is described in Section V.B.1.a(3)(a).

C-Nitroso compounds can be detected by the green, blue-green, or violet color formed with 1 drop each of sample solution, 1% potassium ferricyanide, molar mercuric chloride, and a molar acetate buffer of pH 4 (386); a muddy violet color is formed by p-nitrosophenol, but this can be converted to blue or a blue-green by addition of 1 drop of dilute ammonium hydroxide. Mercuric ions catalyze the reaction of C-nitroso compounds with pentacyanoamineferroate ions.

Nitroso compounds form a blue color when dissolved in a solution of 10 mg of diphenylbenzidine in 10 ml of 85% sulfuric acid (13). The detection limit for a number of nitroso compounds ranges from 0.2 to 30 $\mu$g; isonitroso compounds (oximes) do not react, but oxidants and compounds that release nitrous acid or oxides of nitrogen will also form the blue color (238). Diphenylamine in concentrated sulfuric acid also forms a blue color when heated with nitroso compounds; other oxidants respond similarly [Section V.B.1.a.2].

Sunlight acting on a solution containing an aromatic nitroso compound and potassium ferrocyanide first gives a red color, followed by formation of complex compounds of the type $K_3[Fe(CN)_5 \cdot RNO]$ (602).

Crystalline complexes, possibly useful for microscopic identification, are formed in alcoholic solutions by nitrosobenzene and cadmium iodide (552) and by bismuth trichloride and p-nitroso-N,N-dimethylaniline (715). Stannic chloride and titanic chloride form yellow amorphous complexes with nitrosobenzene (576).

A procedure for detecting nitrosated aromatic compounds in smokeless powders by the intense violet color formed in a 1% acetic acid solution of $\alpha$-naphthylamine is applicable also to nitroso derivatives and nitrosamines (445).

### (5) N-Nitroso Compounds

Nitrosoamines may be detected by the Liebermann test described in the preceding section. The test generally is positive with N-nitroso compounds, but not with all C-nitroso compounds (106).

The detection of N-nitroso compounds by release of nitrous acid or nitrogen oxides was studied by Feigl and Neto (187); Griess reagent or sulfamic acid was used to detect nitrite. Detection is made more certain when decomposition of the N-nitroso group is forced by heating with manganese or zinc sulfate; three tests are given below:

1. One drop of Griess reagent and 1 drop of hydrochloric acid (1:1) are added to 1 drop of the test solution, which is then warmed in a water bath. An intense red-violet color appears at once or within several minutes. The Griess reagent is prepared by mixing equal volumes of a 1% solution of sulfanilic acid in 30% acetic acid with a 0.1% solution of $\alpha$-naphthylamine in 30% acetic acid.

2. Sulfamic reagent is prepared by dissolving 5 grams of barium chloride dihydrate and 5 grams of sulfamic acid in 100 ml of 1:1 dioxane-water; 1 drop of this reagent is mixed with 1 drop of aqueous or alcoholic test solution, and the solution gently warmed. A precipitation or turbidity of $BaSO_4$ results if nitrosamines are present, because nitrous acid released from the nitrosamines reacts with sulfamic acid to form sulfuric acid.

3. A tiny bit of the solid compound is placed with several centigrams of hydrated manganese (or zinc) sulfate in a small test tube. The tube is covered with filter paper moistened with Griess reagent; upon heating to about 200°C, a red-violet stain appears on the paper if N-nitroso compounds are present. The dehydration of the sulfates produces a superheated steam; the nitrosamine is hydrolyzed and caused to form nitrous vapors.

The release of iodine from acidulated solutions of potassium iodide may also be used to detect N-nitroso (as well as C-nitroso) compounds; of course, other oxidizing substances interfere, as do substances or groups that take up iodine readily.

### (6) N-Nitro Compounds

The primary and secondary aliphatic nitramines and nitramides respond to Franchimont's test (199); in acetic acid solution, a green color is developed with dimethylaniline after addition of zinc dust. With $\alpha$-naphthylamine, a pink color is formed; other amines can also be used.

Dissolve 0.5–1 mg of the sample in several drops of glacial acetic acid; add several drops of a 1% solution of the amine in glacial acetic acid. Then add a tiny amount of zinc dust; too much is deleterious. Generally, a green color (with dimethylaniline) is formed with nitramines.

The performance of this test with a variety of classes of nitroguanidine compounds is indicated in the following table (480):

| | Color with | |
| Class | Diethylaniline | Dimethylaniline |
| --- | --- | --- |
| N-Alkyl-N'-nitroguanidines* | Light green | None |
| N-Aralkyl-N'-nitroguanidines | Light green | None |
| N-Aryl-N'-nitroguanidines | None or light green | Pink |
| 1-Substituted 2-nitroamino-<br>2-imidazolines* | Pink | Pink |
| 1,3-Dinitro-1,3-diaza-<br>2-cycloalkanones | Deep green | Deep green |
| 1-Nitro-2-amino-2-imidazolines | Deep green | Deep green |
| Nitrimines | — | Fugitive green |

Aromatic primary nitramines react with nitrous acid to form diazonium compounds (27):

$$ArNHNO_2 + HNO_2 \rightarrow [ArN{\equiv}N]^+NO_3^- + H_2O$$

Consequently, after treatment with nitrous acid, azo colors can be developed by coupling with compounds such as the naphthols. The diazotization is performed by cautious addition of a solution of sodium nitrite to a solution of the sample in 50% acetic acid. This test obviously cannot detect the aromatic primary nitramine group in the presence of amino groups or other groups that can be diazotized.

### b. Oxidation

The effect of nitrobenzene on the rate of reaction of metals with weak acid has been used to indicate its presence in benzene and the artificial musks used for perfumes (560). The sample is heated in 80% acetic acid solution in the presence of a piece of tin foil; with more than 10% of nitrobenzene present, the foil disappears almost immediately, with 1% it disappears in 10 min, with 0.1% it turns black after 10 min, and with 0.01% it turns gray-black overnight. Other nitro compounds behave similarly; for example, a sulfuric acid solution containing the sodium salt of a sulfonated aromatic nitro compound is used to strip nickel plated on steel, and certain aromatic nitro compounds may be employed in cyanide solutions to promote the solubility of noble metals (707).

Oxidation of N-substituted hydroxamic acids and N-alkyl hydroxy-

---

*Nitrate esters, for example, 1-(β-nitroxyethyl)-2-nitramino-2-imidazoline and N-(β-nitroxyethyl)-N'-nitroguanidine, generally give deep green colors with both dimethyl- and diethylaniline.

lamines by periodic acid yields a nitroso dimer which can be detected by the appearance of an intense absorption band in the ultraviolet at 270–290 nm [see Section V.B.2.b(4)].

### c. Gas Evolution

The evolution of a gas may often be used to detect $N$-nitro and $N$-nitroso groups or to confirm the results of other tests.

The nitrosamines, in contrast to $C$-nitroso compounds, are decomposed by hydrazoic acid (or a mixture of sodium azide and a strong acid):

$$=N-NO + HN_3 \rightarrow =NH + N_2O + N_2$$

Apparently, a nitrosamine establishes a rapid hydrolytic equilibrium in aqueous solution:

$$=N-NO + H_2O \rightleftharpoons =N-H + HNO_2$$

The nitrous acid reacts with hydrazoic acid, thus leading to complete decomposition of the nitrosamine (655):

$$HNO_2 + HN_3 \rightarrow H_2O + N_2O + N_2$$

The test is very simply performed by placing a nitrosamine sample in a dilute solution of hydrochloric acid and adding sodium azide; the evolution of gas is readily detected.

The secondary amine from which a nitrosamine has been formed can be regenerated by action of hot, concentrated hydrochloric acid in the presence of ferrous chloride or cuprous chloride. Nitric oxide is formed, and its evolution may be taken to indicate the presence of a nitrosamine. However, nitramines also liberate nitric oxide.

### 2. Physical Methods

Infrared spectroscopy is a more useful tool for detecting the presence of nitro and nitroso groups in compounds than are the chemical methods discussed above, but in many instances it will be necessary to use ultraviolet spectroscopy as well as chemical tests for confirmation. Brief summaries of ir absorption bands useful for detecting the presence of nitro and nitroso compounds are given in Tables VIII and IX.

### a. Infrared Spectrophotometry

#### (1) C—N Absorptions in C-Nitro Compounds

The C—N stretching vibration frequency in aromatic C—NO$_2$ and C—NO compounds is centered at about 850 cm$^{-1}$ (11.75 $\mu$). The aliphatic

TABLE VIII

Characteristic Stretching Frequencies of the $NO_2$ Group in the Infrared Spectra of Typical C—$NO_2$ Compounds

| Compound | Solvent or phase | Asymmetrical | | Symmetrical | | Ref. |
|---|---|---|---|---|---|---|
| | | $(cm^{-1})$ | $(\mu)$ | $(cm^{-1})$ | $(\mu)$ | |
| 2-Nitropropane | Liq. | 1553 | 6.42 | 1361 | 7.35 | 266 |
| Nitropropane | Liq. | 1553 | 6.42 | 1385 | 7.22 | 266 |
| Nitromethane | Liq. | 1567 | 6.38 | 1379 | 7.25 | 266 |
| Nitroethane | Liq. | 1550 | 6.45 | 1368 | 7.31 | 266 |
| Nitrobutane | Liq. | 1550 | 6.45 | 1381 | 7.24 | 79 |
| Nitropentane | Liq. | 1550 | 6.45 | 1379 | 7.25 | 266 |
| Nitroethylene | Liq. | 1527 | 6.55 | 1353 | 7.39 | 266 |
| 1,2-Dichloro-1-nitrobutane | CHCl₃ | 1580 | 6.33 | 1351 | 7.40 | 266 |
| Chloropicrin | — | 1610 | 6.21 | 1307 | 7.65 | 266 |
| 2,2-Dinitropropanol | CHCl₃ | 1577 | 6.34 | 1330 | 7.52 | 699 |
| 2,2,2-Trinitroethanol | KBr | 1603 | 6.24 | 1309 | 7.64 | 699 |
| Nitrobenzene | CHCl₃ | 1534 | 6.52 | 1348 | 7.42 | 124 |
| 2,4-Dinitrotoluene | CHCl₃ | 1540 | 6.49 | 1348 | 7.42 | 124 |
| 2,4,6-Trinitrophenol | CHCl₃ | 1554 | 6.44 | 1347 | 7.42 | 124 |
| p-Nitrophenol | CHCl₃ | 1524 | 6.56 | 1346 | 7.43 | 79 |
| p-Nitroanisole | CHCl₃ | 1515 | 6.60 | 1346 | 7.43 | 79 |
| P-Chloronitrobenzene | CHCl₃ | 1527 | 6.55 | 1350 | 7.41 | 79 |
| 1-Nitronaphthalene | KBr | 1524 | 6.56 | 1343 | 7.44 | 194 |
| 2-Nitroanthraquinone | KBr | 1527 | 6.55 | 1331 | 7.52 | 194 |
| p-Nitroacetanilide | — | 1506 | 6.64 | 1348 | 7.42 | 566 |
| p-Nitroaniline | KBr | 1505 | 6.64 | 1335 | 7.49 | — |
| N-(1-Nitro-2-methyl-2-propyl) hydroxylamine | Liq. | 1556 | 6.43 | 1373 | 7.28 | 79 |
| Anilinium-5-nitroaminotetrazole | Oil mull | 1543 | 6.48 | 1330 | 7.52 | 415 |
| Di(diethylammonium)-5-nitro-aminotetrazole | Oil mull | 1562 | 6.40 | 1300 | 7.70 | 415 |
| Tetranitromethane | Liq. | 1618 | 6.18 | 1266 | 7.89 | 420 |
| 2,2-Dinitropropane | — | 1575 | 6.35 | 1330 | 7.52 | 699 |

compounds have a wider variation, but the region 830–920 cm⁻¹ (10.9–12.05 $\mu$) encompasses very nearly all compounds studied. Hence absorption at about 850 cm⁻¹ (11.75 $\mu$) can be assumed to be indicative of the C-N bond; with characteristic absorptions in other regions (see below), identification of C—$NO_2$ and C—NO can be assured.

It is to be noted, however, that the intensity of absorption at about 850 cm⁻¹ (11.75 $\mu$) is markedly decreased when the 2-, 4-, and 6-positions of phenyl C-nitroso compounds are occupied; the band is also weak in 2-nitrodiphenylamine. Weak absorption (probably due to phenyl sub-

TABLE IX

Characteristic Frequencies of the NO Group in the Infrared Spectra of Typical Compounds

| Compound | Solvent or phase | Monomer (cm⁻¹) | Monomer (μ) | Dimer (cm⁻¹) | Dimer (μ) | Ref. |
|---|---|---|---|---|---|---|
| C-Nitroso | | | | | | |
| Trifluoronitrosomethane | Gas | 1595 | 6.27 | | | 472 |
| 2-Chloro-2-nitrosopropane | — | 1585 | 6.31 | | | 446 |
| Bis(1-nitro-2-nitrosobutane) | Solid | | | 1393 | 7.18 | 79 |
| Bis(1-nitro-2-nitrosocyclohexane)(cis) | Solid | | | 1399 | 7.15 | 79 |
| Bis(1-nitro-2-nitrosocyclohexane)(trans) | Solid | | | 1397 | 7.16 | 79 |
| 5-Nitroso-4,6-diamino-2-methyl-pyrimidine | Nujol | 1650 | 6.06 | | | 81 |
| Nitrosobenzene | Solid | | | 1487 | 6.73 | 229 |
| | CCl₄ | 1506 | 6.64 | | | 229 |
| | CCl₄ | 1513 | 6.61 | | | 268 |
| Nitrosomesitylene | — | 1490 | 6.71 | 1391 | 7.19 | 229 |
| | — | 1497 | 6.68 | 1475 | 6.78 | 512 |
| p-Methylnitrosobenzene | Solid | | | 1480 | 6.76 | 512 |
| | Solution | 1521 | 6.58 | | | 512 |
| p-Dimethylaminonitrosobenzene | Solid | 1538 | 6.50 | 1506 | 6.64 | 512 |
| 1-Nitroso-2-naphthol | — | 1630–1490 | 6.1–6.7 | 1530–1390 | 6.5–7.2 | 52 |
| | Solution | 1630–1490 | 6.1–6.7 | 1530–1390 | 6.5–7.2 | 52 |
| N-Nitroso | | | | | | |
| Dimethylnitrosoamine | CHCl₃ | 1399 | 7.15 | | | 267 |
| | — | 1453 | 6.88 | | | 268 |
| Dipropylnitrosoamine | — | 1350 | 7.40 | | | 267 |
| | — | 1460 | 6.85 | | | 268 |
| Nitrosopiperidine | Liq. film | 1429 | 7.00 | | | 267 |
| Nitrosodiphenylamine | CHCl₃ | 1480 | 6.76 | | | 735 |
| Nitrosoethylaniline | CHCl₃ | 1478 | 6.77 | | | 735 |

stitution) is often exhibited by aromatic $N$-nitroso compounds in this region.

### (2) $NO_2$ Absorptions in $C—NO_2$ Compounds

The infrared and Raman spectra of nitro compounds have been studied extensively within the last 5 years, and it is now possible to assign characteristic frequencies in vibrational spectra to asymmetric and symmetric stretching modes of the $NO_2$ group.

Absorption bands for $C$-nitro compounds occur in the infrared at about $1530 \, cm^{-1}$ ($6.54 \, \mu$) for the $—NO_2$ asymmetrical stretching and at about $1350 \, cm^{-1}$ ($7.41 \, \mu$) for the $—NO_2$ symmetrical stretching mode. Displacement of these generalized bands is a function of the type of substituent; groups such as $—CN$, $—COOEt$, and $—SO_3H$ tend to shift the bands to higher frequencies. In aromatic compounds, the presence of electron-donating groups such as amino groups *ortho* or *para* to the $—NO_2$ group distorts the spectrum; the $1530$-$cm^{-1}$ band and the $1350$-$cm^{-1}$ band are shifted to lower frequencies and increased in intensity. The absorption for $p$-nitroaniline is at $1475 \, cm^{-1}$ ($6.78 \, \mu$) and at $1310 \, cm^{-1}$ ($7.64 \, \mu$); see Fig. 1. However, the band centered near $1350 \, cm^{-1}$ is not excessively affected by the nature of the compound, and thus it is a more reliable indicator of the presence of nitro groups. Studies of a large number of compounds permit the following generalizations regarding the location of the characteristic infrared absorption bands for the $NO_2$ group in aliphatic and aromatic compounds:

Aromatic:    $1530 \pm 50 \, cm^{-1}$ and $1350 \pm 30 \, cm^{-1}$    (42,79,121,124,566)

Aliphatic    $1570 \pm 40 \, cm^{-1}$ and $1350 \pm 40 \, cm^{-1}$    (42,79,121,266,382)

Kornblum et al. (382) claim that primary and secondary $C$-nitroparaffins absorb at $6.45 \pm 0.01 \, \mu$ ($1550 \pm 5 \, cm^{-1}$) and tertiary nitroparaffins at $6.51 \pm 0.01 \, \mu$ ($1536 \pm 4 \, cm^{-1}$); primary $C$-nitro compounds also absorb strongly at $7.25 \pm 0.02 \, \mu$ ($1379 \pm 8 \, cm^{-1}$), secondary at $7.25$–$7.30 \, \mu$ ($1379$–$1370 \, cm^{-1}$), and tertiary at $7.40$–$7.45 \, \mu$ ($1351$–$1342 \, cm^{-1}$). These ranges of frequencies apply to compounds in the solid state as well as in solution in a variety of solvents. Generalities also seem to indicate that the asymmetric stretching mode usually occurs at $1500$–$1545 \, cm^{-1}$ ($6.47$–$6.66 \, \mu$) for mononitro compounds, $1535$–$1555 \, cm^{-1}$ ($6.43$–$6.52 \, \mu$) for dinitro compounds (203), and $1555$–$1580 \, cm^{-1}$ ($6.33$–$6.43 \, \mu$) for trinitro compounds.

The sodium salts of some nitroalkanes have been studied (189), and the frequencies of the carbonitronate ion are assigned to the regions $1200$–$1320 \, cm^{-1}$ ($8.33$–$7.58 \, \mu$) and $1040$–$1175 \, cm^{-1}$ ($9.61$–$8.51 \, \mu$). The

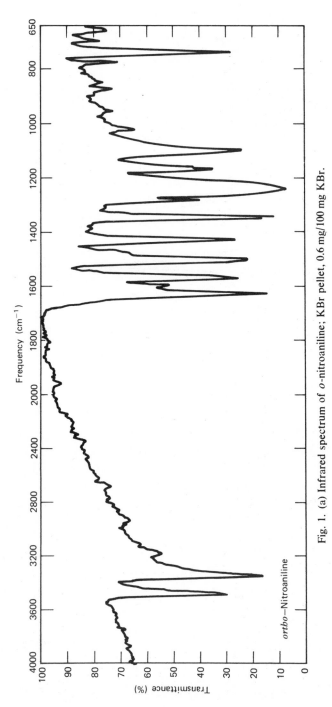

Fig. 1. (a) Infrared spectrum of *o*-nitroaniline; KBr pellet, 0.6 mg/100 mg KBr.

*ortho*−Nitroaniline

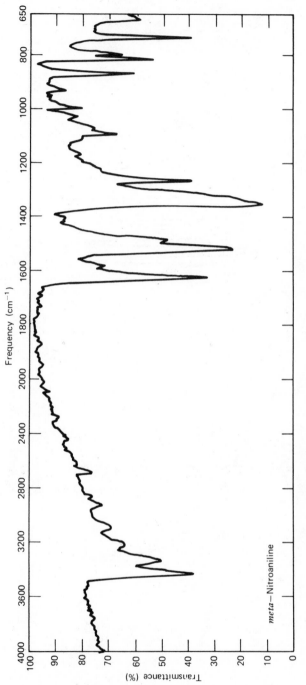

Fig. 1. (Continued) (*b*) Infrared spectrum of *m*-nitroaniline; KBr pellet, 0.6 mg/100 mg KBr.

*meta*−Nitroaniline

130

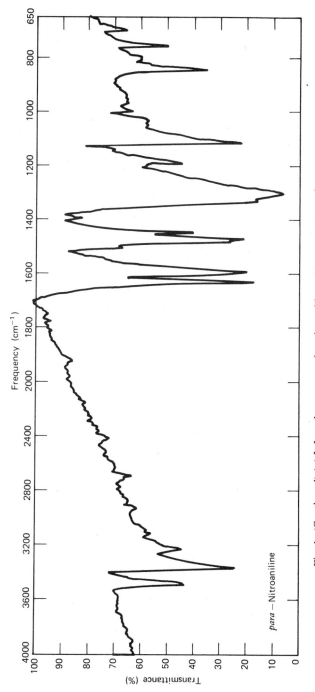

Fig. 1. (Continued) (c) Infrared spectrum of *p*-nitroaniline; KBr pellet, 0.6 mg/100 mg KBr.

*para* – Nitroaniline

131

—CN frequency occurs in the region 1585–1605 cm$^{-1}$ (6.31–6.23 $\mu$). The nitronic acids betray structural similarity to oximes:

$$R—CH{=}N\overset{OH}{\underset{O}{\diagdown}}$$

by showing the infrared C=N absorption characteristic of this grouping in oximes. Thus absorption occurs at 1620–1680 cm$^{-1}$ (6.17–5.95 $\mu$); conjugation shifts absorption to slightly lower frequencies. The nitronic esters absorb intensely at 1610–1660 cm$^{-1}$ (6.21–6.02 $\mu$).

### (3) C—NO Absorptions

Because of the ease with which C-nitroso compounds dimerize (tertiary) or pass into oximes (primary and secondary), assignment of characteristic infrared bands for N=O stretching is difficult; however, there is general agreement that the N=O stretching frequency for C-nitroso monomers lies in the region 1625–1475 cm$^{-1}$ (6.15–6.78 $\mu$); the aliphatic nitroso monomers absorb at 1625–1535 cm$^{-1}$, and the aromatic at 520–1475 cm$^{-1}$ (42,121,234,551). The C—N stretching vibrations usually occur in two bands, one near 1100 and the other near 800 cm$^{-1}$ (9.1 and 12.5 $\mu$) in aromatic C-nitroso compounds.

The following tabulation summarizes the main infrared absorption bands of the dimeric C-nitroso compounds:

|  | Cis | Trans |
|---|---|---|
| Aliphatic: | 1420–1330 cm$^{-1}$ (7.04–7.52 $\mu$) 1350–1320 cm$^{-1}$ (7.41–7.58 $\mu$) | 1295–1175 cm$^{-1}$ (7.72–8.51 $\mu$) |
| Aromatic: | 1420–1400 cm$^{-1}$ (7.04–7.14 $\mu$) 1400–1385 cm$^{-1}$ (7.14–7.22 $\mu$) | 1300–1250 cm$^{-1}$ (7.69–8.00 $\mu$) |

### (4) N—NO$_2$ Absorptions

The N—N frequencies are not readily identifiable in the spectra of N—NO and N—NO$_2$ compounds; however, certain N-nitro compounds show a band of medium intensity at about 2800 cm$^{-1}$ (3.57 $\mu$), and the asymmetric NO$_2$ frequency occurs at essentially the same place as with C—NO$_2$ compounds. The symmetric absorption is displaced toward lower frequencies. Thus nitramines and other N-nitro compounds such as nitroamides and N-nitrocarbamates absorb, in general, in the regions 1630–1530 cm$^{-1}$ (6.14–6.54 $\mu$) and 1315–1260 cm$^{-1}$ (7.60–7.94 $\mu$) (121). The symmetric band centered near 1290 cm$^{-1}$ is reasonably constant; however, absorption at higher frequencies may be observed with salt forms in which the structure $N{=}\overset{+}{N}{=}O_2^-$ influences the spectrum.

The asymmetric band centered at $1600 \text{ cm}^{-1}$ is influenced by the nature of the substituents on the N—$NO_2$ group. Thus nitroguanidine and related compounds with alkyl substituents absorb between 1620 and $1660 \text{ cm}^{-1}$; aryl guanidines and nitrourea, between 1585 and $1570 \text{ cm}^{-1}$; and polynitramines and salts of 5-nitroaminotetrazole, between 1560 and $1545 \text{ cm}^{-1}$. The majority of polynitramines, however, show absorption at $1585–1525 \text{ cm}^{-1}$ and $1290–1260 \text{ cm}^{-1}$, as is quite characteristic of nitramines.

### (5) N—NO Absorptions

The infrared spectra of the nitrosamines have not been examined in sufficient detail to permit definite assignment of absorption bands. One of the difficulties in such studies is that the nitrosamines dimerize readily. Moreover, the strong bands appearing in the spectra of neat liquid nitrosamines generally decrease in intensity and shift to lower frequencies when diluted with solvents such as carbon tetrachloride. Earl et al (148) suggest that nitroso group absorption occurs in the general region of $1400 \text{ cm}^{-1}$ ($7.14 \mu$); however, from a study of 20 nitrosamines, Williams et al. (735) conclude that aliphatic nitrosamines have a band at $1460–1425 \text{ cm}^{-1}$ ($6.85–7.02 \mu$) associated with N=O stretching vibration and a strong band at $1150–1030 \text{ cm}^{-1}$ ($8.70–9.71 \mu$). Aromatic nitrosamines have N=O stretching bands at $1500–1450 \text{ cm}^{-1}$ ($6.67–6.90 \mu$) and bands at $1025–925 \text{ cm}^{-1}$ ($9.76–10.81 \mu$) and $1200–1160 \text{ cm}^{-1}$ ($8.33–8.62 \mu$), all showing solvent sensitivity. There is coupling between the N=O stretching vibration and C—H bending modes, as well as ring C—C vibrations.

### b. ULTRAVIOLET SPECTROPHOTOMETRY

Absorption maxima and extinction coefficients are given for typical nitro compounds in Table X, and for typical nitroso compounds in Table XI.

### (1) C-Nitroaromatics

C-Nitroaromatics absorb over a wide range of the ultraviolet spectrum, as suggested by Fig. 2; however, the intense absorption band at about 210 nm that is attributed to the nitro group and the weak band from its $n \rightarrow \pi^*$ transition at about 270 nm (569) are usually unrecognizably intermingled with the intense bands of the $\pi \rightarrow \pi^*$ transitions of the aromatic nucleus. Moreover, since the nitro group is strongly electron-withdrawing, the aromatic bands undergo marked bathochromic shifts; thus the exact locations of absorption maxima are dependent on the orientation of nitro groups and the presence of other substituents as well

TABLE X

Ultraviolet Absorbance Maxima for Typical Nitro Compounds

| Compound | Solvent | $\lambda_{max}$ (nm) | $\epsilon_{max}$ | Ref. |
|---|---|---|---|---|
| C-Nitroaromatic | | | | |
| Nitrobenzene | EtOH-H$_2$O | 269 | 7,770 | 124 |
| o-Dinitrobenzene | EtOH-H$_2$O | <210 | — | 124 |
| 1,3,5-Trinitrobenzene | EtOH-H$_2$O | 235 | 12,500 | 124 |
| o-Nitrotoluene | EtOH-H$_2$O | 265 | 5,500 | 124 |
| 2,4-Dinitrotoluene | EtOH-H$_2$O | <210 | 14,100 | 124 |
| 2,4,6-Trinitrotoluene | EtOH-H$_2$O | 232 | 18,260 | 124 |
| 2,3-Dinitro-1,4-dichlorobenzene | EtOH | 220 | 18,500 | 251 |
| 2-Nitro-1,4-dibromobenzene | EtOH | 223 | 17,400 | 251 |
| 2,4-Dinitroresorcinol | EtOH | 400 | 20,180 | 153 |
| 1-Nitro-2-naphthol | EtOH | 330 | 3,130 | 153 |
| 2-Nitrophenol | EtOH | 272 | 6,000 | 614 |
| 2,4,6-Trinitrophenol | Basic (NaOH) | 358 | 14,100 | 614 |
| 2,4-Dinitrophenylhydrazone of acetaldehyde | EtOH | 356 | 21,000 | 584 |
| 2,4-Dinitrophenylhydrazone of p-hydroxybenzaldehyde | EtOH | 395 | 28,700 | 584 |
| C-Nitroaliphatic | | | | |
| Nitromethane | EtOH | 278 | 20 | 266 |
| Chloropicrin | EtOH | 277 | 62 | 266 |
| 2,2-Dinitropropanediol | CH$_2$Cl$_2$ | 277 | 5.75 | 699 |
| Ethanenitrolic acid, CH$_3$C(NO$_2$)NOH | 0.1 N NaOH | 323 | 8,000 | 474 |
| 2-Hydroxymethyl-2-nitro-1,3-propanediol | MeOH | 275 | 5.76 | 699 |
| 1,1-Dinitromethane, potassium salt | Dilute KOH | 363 | 20,000 | 337 |
| 1,1-Dinitroethane | H$_2$O, pH 12 | 352 | — | 475 |
| 1,1-Dinitropropane, potassium salt | Dilute KOH | 382 | 17,000 | 337 |
| Trinitromethane | H$_2$O, pH 12 | 352 | — | 475 |
| Nitroamines | | | | |
| Nitroguanidine | EtOH | 265 | 15,300 | 614 |
| Dimethylnitramine | Dioxane | 240 | 6,300 | 327 |
| Dinitroethyleneurea | EtOH | 246 | 10,500 | 327 |
| Methylnitramine | 10$^{-3}$ HCl | 233 | 7,200 | 327 |
| Methylenedinitramine | EtOH | 226 | 12,530 | 327 |
| 1,3,5-Trinitro-1,3,5-triazacyclo-hexane (RDX) | EtOH | 233 | 14,900 | 327 |

as their orientation. For example, nitro-substituted benzenes and toluenes have intense single maxima over the range 210–280 nm, halogen-substituted benzenes absorb at 220–230 and 235–245 nm, mono- and dinitrophenols have two bands in general, at 230–310 and 270–350 nm, and dinitrophenylhydrazone derivatives show a band at 200–370 nm. Bathochromic shifts are large when the substituents are electron donors; thus, whereas nitrobenzene shows a band at about 257 nm, p-nitroaniline exhibits its principal band at about 370 nm (in alcohol). The bathochromic shift is also evident in the 2,4-dinitrophenylhy-

TABLE XI
Ultraviolet Absorbance Maxima for Typical Nitroso Compounds

| Compound | Solvent | $\lambda_{max}$ (nm) | $\epsilon_{max}$ | Ref. |
|---|---|---|---|---|
| *C*-Nitroso | | | | |
| 2-Nitroso-2-nitropropane | EtOH | 230 | 12,180 | 153 |
| Nitrosobenzene | EtOH | 280 | 9,000 | 89 |
| Methylnitrosobenzene | EtOH | 315 | 11,375 | 512 |
| *p*-Chloronitrosobenzene | EtOH | 313 | 11,880 | 512 |
| Nitrosodimethylaniline | EtOH | 421 | 23,800 | 89 |
| 4-Nitroso-*N*,*N*-diethylaniline | EtOH | 425 | 32,400 | 614 |
| *p*-Nitrosophenol | EtOH | 310 | 6,340 | 153 |
| 3-Chloro-4-nitrosophenol | Neutral | 299 | 6,785 | 285 |
| 2,4-Dinitrosoresorcinol | EtOH | 280 | 12,710 | 153 |
| 1-Nitroso-2-naphthol | EtOH | 260 | 13,200 | 153 |
| 2-Nitroso-1-naphthol | EtOH | 260 | 22,100 | 153 |
| Ethylazaurolic acid, $CH_3C\begin{smallmatrix}\diagup NH\\\diagdown N=O\end{smallmatrix}$ | 0.1 N–NaOH | 348 | — | 474 |
| *N*-Nitroso | | | | |
| Dimethylnitrosoamine | EtOH | 232 | 5,900 | 267 |
| Diethylnitrosamine | EtOH | 233 | 7,400 | 267 |
| *N*-Nitroso-4-nitro-*N*-ethylaniline | EtOH | 312 | 15,000 | 614 |
| *N*-Nitrosocarbanilide | EtOH | 231 | 18,300 | 614 |
| *N*-Nitroso-4-nitrodiphenylamine | EtOH | 318 | 13,700 | 614 |

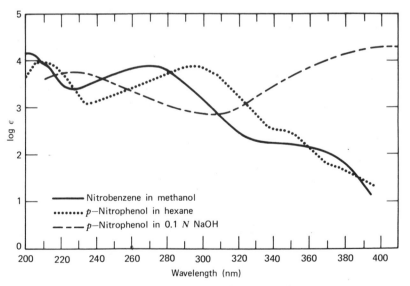

Fig. 2. Absorption spectra of typical *C*-nitroaromatics. [Data adapted from Schroeder et al. (614).]

drazones (often used as derivatives for identification of carbonyl compounds); the 2,4-dinitrophenylhydrazones of saturated aldehydes and ketones absorb strongly at about 360 nm, while those of the $\alpha,\beta$-unsaturated derivatives absorb at about 380 nm (550). Alkyl nitrobenzoates are often used to identify or determine alcohols (617); as expected in this instance, the maximum absorption of these compounds occurs at about 253 nm.

### (2) C-Nitroparaffins

The $C$-nitroparaffins show characteristic $n \rightarrow \pi^*$ low-intensity maxima in the region 265–285 nm (exhibiting hypsochromic shifts in polar solvents); in unsaturated nitro compounds, the $n \rightarrow \pi^*$ bands are generally masked by intense $\pi \rightarrow \pi^*$ absorption bands derived from conjugation of the double bond and the nitro group, and thus nitroolefins show unassigned bands at 220–250 nm (68,501).

A study of nitroparaffin absorption in the ultraviolet has established concentration ranges over which Beer's law holds (196); for example, Beer's law is obeyed by solutions of nitromethane, nitroethane, 1-nitropropane, and 2-nitropropane in water and in carbon tetrachloride over the ranges 0.010–0.050 $M$ and 0.12–0.30 $M$. It is of interest to note that the results of a detailed study of the ultraviolet spectra of nitromethane in a variety of solvents suggest that the compound may form a dimer (452a). Figure 3 includes a typical uv spectrum for a $C$-nitroparaffin.

The uv spectra of nitronic acids and esters closely resemble the spectra of the alkali-metal salts of nitroparaffin compounds [the nitronates which absorb in the region 220–250 nm (270); see Fig. 3]. A strong $\pi \rightarrow \pi^*$ band is usually found in aqueous or ethanolic solutions near 220–230 nm ($\epsilon \approx 10,000$) for simple aliphatic nitronic acids and at about 10 nm higher and with equal extinction coefficients for the corresponding nitronate anions (depending on structure) (517). Spectral data for nitronate anions have been published by Armand (16), Hawthorne (270), and Williams et al. (734). Conjugation of the nitro group with olefinic structures or aromatic rings produces the expected bathochromic shift; thus the nitronic acid derived from phenylnitromethane absorbs very strongly at 284 nm ($\epsilon = 20,000$), and the corresponding nitronate ion absorbs at 294 nm ($\epsilon = 25,000$) (201).

Kamlet and Glover (337) found the uv spectra of the nitronate salts of a series of compounds suitable for identification of the parent dinitro compounds. The 3-substituted 1,1-dinitro-2-propane salts absorb sharply at 313–326 nm and have a diffuse absorption band at 395–410 nm; saturated 1,1-dinitroalkane salts show maximal absorption at 379 ± 5 nm,

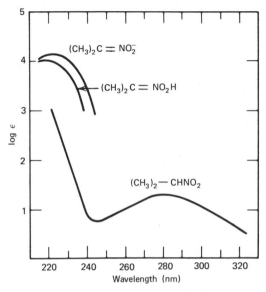

Fig. 3. Absorption spectra of typical *C*-nitroparaffins: nitroisopropane and the nitronic acid and nitronate ion derived from it. [Data adapted from Haszeldine (266).]

and salts with electron-withdrawing substituents in the 2-position have maximal absorption at $366 \pm 3.5$ nm.

### (3) *N*-Nitro Groups

Primary nitramines absorb in the region 225–235 nm with $\epsilon_{max}$ about 7000, and the secondary nitramines absorb slightly less strongly ($\epsilon_{max}$ about 5500) in the region 230–250 nm.

Secondary aliphatic nitramines containing more than one nitramine group per molecule usually show molecular extinction coefficients in ethanol or dioxane that are close multiples of 5500, and primary aliphatic nitramines show coefficients that are multiples of 7000 (there are many exceptions). If a primary and a secondary nitramine group are present on the molecule, the extinction coefficients appear to be additive and the value will be essentially 12,500 (327).

Figure 4 shows the spectra of typical primary and secondary nitramines; it is evident that these nitramines are rather difficult to distinguish. However, since the hydrogen atom attached to the nitrogen in the primary nitramine group is acidic, the spectrum of a nitramine compound is somewhat different in alkaline solution than it is in neutral solvent. Accordingly, as is shown in Fig. 4, in alkaline solution there is a small increase in intensity of absorption and the position of the maxi-

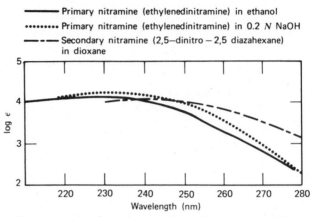

Fig. 4. Absorption spectra of typical primary and secondary nitramines. [Data adapted from Jones and Thorn (327).]

mum is shifted to longer wavelengths. Secondary nitramines are usually unstable in alkali, but for compounds that are sufficiently stable to enable the recording of spectra, the intensity of absorption is essentially the same as in neutral solution. In acidic media, the spectra of primary and secondary nitramines are the same as they appear in neutral media.

### (4)  C-Nitroso Groups

#### (a)  Nitrosoparaffins

The monomeric C-nitrosoparaffins (usually blue in color) show a low-intensity absorption maximum at 630–790 nm ($\epsilon$ from 1 to 60), a maximum in the region 270–300 nm ($\epsilon$ about 80), and an intense band about 220 nm ($\epsilon$ about 5000); both short-wavelength bands appear to originate from $n \rightarrow \pi^*$ transitions, but the one at 220 nm arises from $\pi \rightarrow \pi^*$ transitions (234). The long-wavelength band is common to all C-nitroso compounds (monomers), and since dimerization involves band formation by lone-pair electrons, the extinction for this band varies with compound structure, temperature, and concentration, as well as any other condition affecting dimer–monomer relationships. The band at 270–300 nm is also used in studies of dimeric nitroso compounds (336). Absorption in the 630–800 nm region will disappear if the nitroso group is oxidized to the nitro group. Also, it is important to recognize that the bands below 300 nm may be submerged in strong aromatic absorptions if the C-nitrosoparaffin structure also is associated with an aromatic group. Additionally, when a dimeric C-nitrosoparaffin is formed, a new $\pi \rightarrow \pi^*$

transition appears in the region about 270 nm ($\epsilon$ about 10,000) (235) and the color of the compound changes from blue to yellow.

Nitrosoacetylenes are blue-green in the monomeric state and show absorption maxima in the region 640–654 nm (588). Jander and Haszeldine (311) have recorded data on a number of halogenated $C$-nitrosoparaffins.

The $\pi \to \pi^*$ transitions of the *cis* dimers of $C$-nitrosoparaffins are always of lower wavelengths than the *trans* dimers, and the *trans* forms tend to show larger solvent hypsochromic shifts than do the *cis*, as is indicated by the following tabulation of $\lambda_{max}$ values for $C$-nitroso-*i*-butane (235):

|        | $H_2O$ | EtOH | $Et_2O$ | $CCl_4$ |
|--------|--------|------|---------|---------|
| *Cis*: | 270    | 276  | 294     | 298     |
| *Trans*: | 285  | 291  | 294     | 298     |

### (b) Nitrosoaromatics

Monomeric $C$-nitrosoaromatics (usually green in color) show the weakly absorbing band characteristic of all $C$-nitroso monomers in the visible region (630–800 nm) which is affected by heating or dilution; the position of the band is influenced by substituents and exhibits hypsochromic shifts in polar solvents (see Fig. 5). Only the long-wavelength $n \to \pi^*$ transition is seen distinctly in aromatic $C$-nitroso compounds; the lower wavelength $n \to \pi^*$ transitions are usually submerged in the much stronger $\pi \to \pi^*$ transitions of the aromatic nucleus.

The dimers of aromatic $C$-nitroso compounds show absorption bands at 260–315 nm, except for compounds containing substituents such as amino groups which absorb at 310–425 nm.

The *cis* form of aromatic $C$-nitroso compounds is more stable, but in *ortho*-substituted dimers the *trans* form is more stable because the aryl groups are twisted out of plane. As was noted for $C$-nitrosoparaffins, the wavelength of maximum absorption is lower for *cis* than for *trans* isomers.

The ultraviolet spectra of a series of substituted *p*-nitrosophenols and a comparison with the spectra of the corresponding tautomers (*p*-benzoquinoneoximes) has been recorded by Hodgson (285).

### (5) N-Nitroso Groups

The spectra of nitrosamines resemble those of nitrites and exhibit an absorption maximum around 360 nm which is of low intensity ($\epsilon \approx 100$) and has a fine structure in nonpolar ($n \to \pi^*$ transition); there is also a

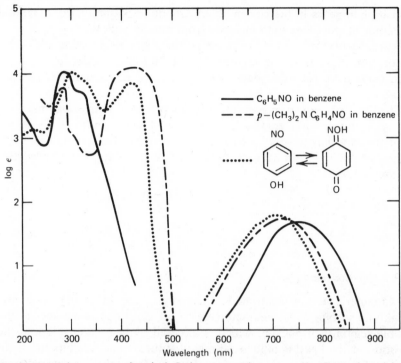

Fig. 5. Absorption spectra of typical *C*-nitrosoaromatic compounds. [Data adopted from Holleck and Schindler (292), Burawoy (89), and Kessler and Luttke (355).]

relatively intense absorption band ($\epsilon \approx 6500$) around 235 nm from the $\pi \to \pi^*$ transition. The band at 360 nm shows hypsochromic shifts in polar solvents with loss of fine structure and appears to be characteristic of the monomeric nitrosamines (see Fig. 6); consequently, the position and intensity of the band are influenced by temperature, concentration, and other factors which affect dimerization (268). Incidentally, the marked hypsochromic shift and loss of fine structure when the solvent is changed from light petroleum to ethanol readily distinguish a nitrosamine from a nitrite (267).

## C. IDENTIFICATION

### 1. Chemical Methods

#### a. REDUCTION TO AMINES

#### (1) *C*-Nitroaromatic Compounds

The principal chemical method used for identification of a *C*-nitro compound is reduction to an amine which can be readily identified when

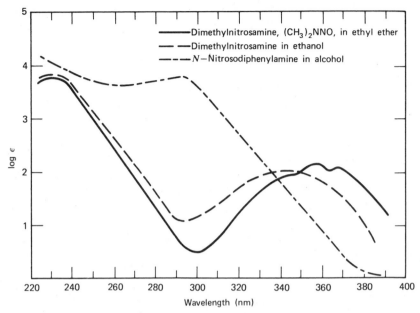

Fig. 6. Absorption spectra of typical $N$-nitroso compounds. [Data adapted from Haszeldine and Jander (267), Haszeldine and Mattinson (268), and Schroeder et al. (614).]

converted to a substituted thiourea, an acetamide, an aryl sulfonamide, an $N$-substituted benzamide, or other tractable derivative; amines may also be identified as hydrochlorides by $X$-ray diffraction methods (71).

The reagents commonly used for reduction of $C$-nitro compounds are tin, zinc, or zinc amalgam in acidic media, but catalytic hydrogenation is preferable for small samples; on the other hand, stannous, titanous, or hydrosulfite reductions are in general use. Details for reduction procedures are given in Sections V.C.1.b and V.C.1.c of "Azo Group," Vol. 15. Hydrogenation of aromatic nitro compounds is often possible at ambient conditions (77). After reduction, the solution is rendered alkaline and the amine is extracted with a solvent such as ether. In any event, it is important to bear in mind that other functional groups in the compound may be altered or removed during reduction of the nitro group. An elemental analysis or even a molecular weight determination of the original nitro compound should be available for comparison with the identified reduction product.

In many instances, samples contain an unknown nitro compound that is quite pure. However, the analytical chemists's lot is seldom simple, and it is best to assume that after reduction two or more amines will be at hand. Thus, for aromatic nitro compounds, it is appropriate to use the general procedures (or suitable modifications) given in Sections V.C.1

and V.C.2 of "Azo Group," Vol. 15, for identification of reduction products via benzamides (if amines), and via S-benzylthiouronium derivatives (if sulfonic acid groups are on the aniline nucleus). Infrared analyses and gas chromatographic separations may help to establish the identities of amines.

### (2) C-Nitroparaffin Compounds

Aliphatic primary mono-C-nitro compounds are readily reduced to primary amines by zinc and hydrochloric acid; other methods are given in Section V.C.1 of "Azo Group," Vol. 15. Mild reduction such as is indicated in Section V.C.1.b(5) of "Azo Group" yields aldoximes with the primary nitroalkanes, and ketoximes with the secondary. N-Alkyl hydroxylamines are formed from nitroalkanes by reduction with stannous chloride (as in Section V.C.1.b(1) of "Azo Group"). However, all these compounds can be reduced to amines. Catalytic hydrogenation is the best procedure for reduction of an unknown aliphatic C-nitro compound; Raney nickel at room temperature and elevated pressure is suitable (316). Methanolic solutions of nitroalcohols are readily reduced to aminoalcohols by hydrogen at 3000 psi (30°C) in the presence of Raney nickel. Lithium aluminum hydride can be used to reduce aliphatic primary mono-C-nitro compounds.

Reduction products of aliphatic C-nitro compounds (the corresponding alkyl amines) are best characterized as phenylthioureas (74) and p-toluenesulfonamides as well as benzamides. After reduction, the solution containing reduction products is rendered alkaline and distilled; the distillate is collected in dilute acid. Special precautions must be taken because alkyl amines of low molecular weight are volatile liquids or gases.

The polynitroparaffins do not behave systematically on reduction (176,520,608); hence reduction products are of only indirect assistance in the identification of these compounds.

### (3) C-Nitroso Compounds

Aromatic C-nitroso compounds can be reduced easily to amines with zinc, tin, or stannous chloride, just as are the C-nitroaromatics, and the resulting aromatic amines can be identified in the usual fashion. Catalytic hydrogenation is an especially attractive procedure for reduction of C-nitrosoaromatics because the high concentrations of hydrochloric acid sometimes used in acid-plus-metal reductions may lead to the production of significant amounts of chloro-substituted aryl amines in addition to the expected product. Nevertheless, the tin-hydrochloric acid reducing system is commonly used for converting C-nitrosoaromatics to their corresponding amines.

### (4) N-Nitroso Compounds

Aromatic N-nitroso compounds [the secondary nitrosamines, ArN(NO)R] are best reduced to secondary amines by tin and hydrochloric acid; the amount of hydrochloric acid must be kept at a minimum to limit formation of chloro-substituted aryl amines. Titanous sulfate [see Section V.C.1.b(3) of "Azo Group," Vol. 15] is an especially elegant reagent for reduction; the amine is extracted by ether from an alkaline solution.

The reduction routes of N-nitrosoalkanes have not been examined sufficiently so that positive statements can be made about the products. In the few cases examined, reduction of nitrosoalkane dimers has given rise to a wide variety of products in poor yield (394).

### (5) N-Nitro Compounds

The nitroamines (and nitrosoamines) may be reduced to the corresponding hydrazines, which in turn are readily converted to hydrazones by reaction with carbonyl compounds and identified, but not many hydrazones of the substituted hydrazines are described in the literature. Moreover, yields of hydrazines from these compounds are usually poor; therefore, even if hydrazines might be obtained on first trial, it is best to attempt energetic reduction to the corresponding secondary amine by zinc and hydrochloric acid and derivatization of the amine.

### b. NITRATION

Aromatic mono-C-nitro compounds (nitro group on nucleus) often can be nitrated further to dinitro compounds which of themselves can serve as derivatives for identification and occasionally can be confirmed by conversion to naphthalene addition compounds (see next section). However, the nitration procedure often alters other functional groups (e.g., the aryl C-nitroso group); thus this method is used only for corroboration of results obtained from procedures involving reduction of an amine.

The introduction of another nitro group is accomplished with 99–100% nitric acid (white fuming nitric acid, WFNA (106):

About 200 mg of the compound to be nitrated is put in an 8-in. test tube, and 2 ml of concentrated sulfuric acid is added. The tube is immersed in cold water; 2 ml of fuming nitric acid is added a drop at a time, shaking and rotating the tube constantly to mix the contents well and to cool them. After all the nitric acid has been added, the tube is placed in a water bath at 45–55°C and allowed to remain for 15–20 min (occasional agitation). Then the tube is plunged into water; when the contents are cool, they are poured into a 25-ml beaker containing 8–10 ml of cold water. The diluted mixture is returned to the test tube and cooled by immersion in water. The nitro compound that separates is filtered and washed with three or four 1-ml portions of cold water and then returned to the reaction test tube, where it is dissolved in a small amount of boiling methanol or ethanol. The hot solution is filtered, if necessary, and then drops of water are added until a permanent cloudiness appears. The solution is warmed until it is clear and then cooled to allow crystals to form; the purification is repeated until two successive crystallizations show a variation of no more than 0.5–1° in the melting point.

Very low yields of product are obtained when aryl $C$-nitroso compounds are nitrated directly. Ingold's method, utilizing phorphorus pentoxide and dinitrogen pentoxide, increases yields but still leaves much to be desired (306).

### c. OXIDATION

$C$-nitroaromatic compounds are generally stable enough to resist oxidation. Thus, by subjecting such compounds to the usual oxidants, side chains and sensitive functional groups can be altered. A simple example is the oxidation of 2,4,6-trinitrotoluene to 2,4,6-trinitrobenzoic acid by chromic acid. The oxidation product can be used to characterize an unknown.

Still another example is afforded by the $C$-nitrosoaromatic compounds in which the functional group can be readily oxidized to yield the corresponding nitro compound; hydrogen peroxide, nitric acid, or permanganate (29) can be used to accomplish the transformation. Although the resulting nitro compound itself can serve to identify the progenitor, corroboration is readily possible, for after reduction the amine can be converted to identifiable derivatives.

### d. CONDENSATION

The 3-arylamino-2-phenylindolenines (anils) and substituted bromoazobenzenes were found suitable by Levy and Campbell (412) for identification of $C$-nitroso compounds. The anils are prepared by dissolving 0.5 gram of the nitroso compound and 0.5 gram of 2-phenylindole in alcohol, cooling the solution, and adding a few drops of alcoholic potassium hydroxide; a violent reaction occurs. The derivative is crystallized from alcohol, and the melting point is determined. Bromoazobenzenes are obtained from $p$-bromoaniline according to the procedure of Ingold (306), involving the condensation of a $C$-nitrosoaryl with an aminoaryl (see Section III.B.5.b). Nitrophenols can be identified as aryl oxyacetic acids (243).

## 2. Physical Methods

### a. INFRARED SPECTROPHOTOMETRY

Identification of nitro compounds requires exact location of the characteristic —NO$_2$ frequencies and other corroborating spectral features; a library of reference spectra is necessary. An example of a method of approach to the determination of unknown structures can be seen in the work of Lothrop et al. (442); comparison of the infrared spectra of polynitro compounds with the spectra of other nitro compounds containing substituent groups such as —NH, —NH$_2$, and —CH$_3$ supported evidence that a so-called pentanitrophenylmethylnitramine was in fact a tetranitrophenylmethylnitramine. This work also provides reference spectra for pentanitroaniline and picramide, as well as the following nitramines: tetranitrophenylmethylnitramine, 2,3,4,6-tetranitrophenylmethylnitramine, and tetryl (2,3,6-trinitrophenylnitramine). The ir spectra of a group of nitramides, nitramidines, and other nitrogen compounds have been reported by Lieber et al. (415); the spectra of some secondary aromatic nitramines are given by Lothrop et al (442). The identification of HMX and RDX in mixtures has been reported by Bedard et al. (37).

Brown (79) has correlated symmetrical and unsymmetrical stretching of the —NO$_2$ group with electron-releasing substituent groups attached to the group, negative groups on the carbon holding the nitro group, and the conjugation of the nitro group; a tabulation of the main bands in the ir spectra of more than 100 aromatic and aliphatic compounds provides a valuable source for analytical laboratories.

An estimation of the number of nitro groups in aliphatic polynitro compounds as made possible by measurement of the absorption in the 6,3-$\mu$ region was demonstrated by Grabiel et al. (236). A comprehensive compilation of the ir spectral characteristics (with references through 1963) of aliphatic polynitro compounds has been published by Noble et al. (520).

Additional information on the ir spectra of nitro and nitroso compounds is given in Section V.B.2.a.

### b. ULTRAVIOLET SPECTROPHOTOMETRY

Although ultraviolet spectrophotometry is generally considered incapable of establishing the identity of nitro or nitroso compounds, it can provide corroborative evidence. Thus Kamlet and Glover (337) have shown that uv absorption permits identification of salts of nitroparaffins with greater certainty than is possible with melting or decomposition points [see Section V.B.2.b(2)]. Haszeldine (266) has also demonstrated that nitroparaffins are more readily distinguished from the isomeric nitrites by their uv spectra than by their ir spectra. The nitrites absorb at much longer wavelengths: 275–285 nm for nitroalkanes and 310–385 nm for nitrites.

Ultraviolet absorption has been used by Hodgson (285) to demonstrate the tautomerism of the benzoquinoneoxime–$p$-nitrosophenol system. The difference of absorption in acidic and alkaline solutions may serve as a means for identification of nitrosophenols; for example, $p$-nitroso-phenol exhibits maxima at 302 nm in acidic solution and 397 nm in alkaline solution.

The following references contain compilations of uv maxima and extinction coefficients or spectra which may be of assistance to the analyst in the identification of nitro and nitroso compounds: dinitro-phenylhydrazone derivatives (67,584); an extensive series consisting of all classes of nitro compounds and derivatives in ethanol solution (614); nitrosobenzenes (512); examples of the spectra of $C$-nitro and $C$-nitroso counterparts (93); aliphatic nitrosamines (153,267); nitrobenzenes and toluenes (124); picrylamines (641); halogen-substituted nitrobenzenes (251); nitramines and nitrosamines (327,614); and polynitroalkanes (520).

### c. MASS SPECTROSCOPY

Mass spectroscopy provides one of the most certain and rapid means of identification of nitro compounds, particularly in mixtures. Since a large number of nitro compounds have a vapor pressure of about 10 $\mu$ at room temperature, most of these compounds can be readily introduced into a mass spectrometer inlet system. Although the molecular-ion peaks of nitroparaffins of molecular weights higher than nitromethane are weak or not evident, characteristic peaks at $m/e$ 30 ($NO^+$) and $m/e$ 46 ($NO_2^+$) are present; reference spectra for nitroparaffins through $C_3$ are given by Collin (117) and may also be found in *Index of Mass Spectral Data* (303). The spectrum of an aliphatic nitro compound exhibits peaks attributable to hydrocarbon fragments up to the parent ions minus the —$NO_2$ group.

Aromatic nitro compounds characteristically give a strong molecular-ion peak ($M$), and the major peaks are formed by elimination of the —$NO_2$ radical (e.g., molecular ion minus 46) and of a neutral NO molecule with a rearrangement to form the phenoxy cation (molecular ion minus 30). For many compounds, loss of $CH{=}CH$ from the molecu-lar-ion-minus-46 fragment leads to a strong peak at molecular ion minus 72; the molecular-ion-minus-30 fragment also readily loses CO to form a peak at molecular ion minus 58. Of course, the characteristic $NO^+$ peak at $m/e$ 30 betrays the presence of an oxygenated nitrogen compound, but more often the fact that the molecular-ion peak has an odd $m/e$ value can be taken as indication of the nitro group (along with $NO^+$).

Reference spectra for $o$-, $m$-, and $p$-chloronitrobenzenes are given by Momigny (500); the molecular-ion peaks at $m/e$ 157 with the isotopic peaks at $m/e$ 159 clearly indicate the presence of chlorine; major fragmentation peaks occur at $M - 46$ ($NO_2$) and $M - (46 + 35)$ ($NO_2 + Cl$). Significant peaks at $M - 30$ arise from the splitting out of NO, a rearrangement process that is characteristic of aromatic nitro compounds and is summarized by Beynon (47).

Reference spectra for various nitro derivatives of benzene, toluene, aniline, phenol, xylene, toluidine, chlorobenzene, and aminochlorobenzene are given by Beynon et al. (48). All of these compounds have molecular-ion peaks which permit immediate molecular weight assignments. Distinguishing features of these spectra are given in Table XII; isomeric species are easily identified by accurate measurements of peak ratios when the sample consists essentially of one compound.

TABLE XII
Characteristics of Mass Spectra of Aromatic Nitro Compounds

| Compound | Molecular weight | Relative intensity | | | | |
| --- | --- | --- | --- | --- | --- | --- |
| | | Molecular ion | $M - 17$ | $M - 30$ | $M - 46$ | 100% peak |
| Nitrobenzene | 123 | 48.6 | — | 10.4 | 100.0 | $m/e$ 77 |
| $o$-Nitrotoluene | 137 | 25.1 | 100.0 | 1.4 | 82.4 | $m/e$ 120 |
| $m$-Nitrotoluene | 137 | 71.2 | 0.5 | 10.0 | 100.0 | $m/e$ 91 |
| $p$-Nitrotoluene | 137 | 72.5 | 0.2 | 10.3 | 100.0 | $m/e$ 91 |
| $o$-Nitroaniline | 138 | 100.0 | 4.2 | 4.6 | 16.7 | $m/e$ 138 |
| $m$-Nitroaniline | 138 | 100.0 | 0.2 | 7.7 | 23.3 | $m/e$ 138 |
| $p$-Nitroaniline | 138 | 47.0 | — | 51.8 | 39.1 | $m/e$ 65 |
| $o$-Nitrophenol | 139 | 100.0 | 7.6 | 8.1 | 12.7 | $m/e$ 139 |
| $m$-Nitrophenol | 139 | 90.0 | 0.1 | 4.0 | 54.0 | $m/e$ 65 |
| $p$-Nitrophenol | 139 | 90.9 | 0.1 | 70.3 | 30.9 | $m/e$ 65 |
| Nitro-$p$-xylene | 151 | 43.3 | 100.0 | 2.0 | — | $m/e$ 134 |
| 3-Nitro-$p$-toluidine | 152 | 100.0 | 2.8 | 2.2 | 60.7 | $m/e$ 152 |
| $p$-Dinitrobenzene | 168 | 69.8 | — | 2.0 | 28.4 | $m/e$ 30 |
| 2,3-Dinitrotoluene | 182 | 18.2 | 100.0 | 1.1 | 6.1 | $m/e$ 165 |
| 2,4-Dinitrotoluene | 182 | 9.0 | 90.3 | 0.6 | 1.7 | $m/e$ 89 |
| 2,5-Dinitrotoluene | 182 | 10.1 | 100.0 | 2.0 | 1.9 | $m/e$ 165 |
| 2,6-Dinitrotoluene | 182 | 1.6 | 100.0 | 0.5 | 1.6 | $m/e$ 165 |
| 3,4-Dinitrotoluene | 182 | 56.6 | — | 5.2 | 3.8 | $m/e$ 30 |
| 3,5-Dinitrotoluene | 182 | 86.8 | 0.3 | 1.7 | 36.2 | $m/e$ 89 |
| $p$-Nitrochlorobenzene | 157 | 100.0 | — | 40.5 | 92.8 | $m/e$ 157 |
| 1-Amino-2-nitrochlorobenzene | 172 | 100.0 | 3.6 | 9.2 | — | $m/e$ 172 |

Prominent mass spectral features are given for a number of dinitro-phenylhydrazone derivatives by Kleipool and Heins (360). Molecular-ion peaks are easily detectable, and characteristic peaks appear at $m/e$ 224 and 206 for aldehydes, $m/e$ 228 and 178 for methyl ketones, and $m/e$ 252 and 178 for ethyl ketones.

Reference spectra for dialkyl nitrosamines, nitrosomethane, and nitrosobenzene are given by Collin (118). These materials show molecu-lar-ion peaks and are characterized by peaks at $m/e$ 42 ($CH_2NCH_2^+$), $m/e$ 44 ($N—N\!=\!O^+$), and the anticipated $M - 30$ intense peak cor-responding to loss of $m/e$ 30 ($NO^+$); the peak at $m/e$ 30 ($NO^+$) also indicates splitting off of NO.

### d. ADSORPTION CHROMATOGRAPHY

Chromatographic separations of nitro compounds appear to be best accomplished on columns packed with silicic acid adsorbents or with 2 : 1 mixtures of silicic acid and Celite (689); alumina and kaolin ad-sorbents are also effective. As might be anticipated, each packing exhibits wide batch-to-batch variations in adsorption qualities (457); uniform adsorptive properties and minimum spectrophotometric back-grounds can be obtained by prewashing with acetone–ether, ether, and ligroin (bp, 60–70°).

Streak reagents for chromatographic identification of nitro compounds on columns have been developed by LeRosen et al (408); the reagents were selected for use on silicic acid adsorbent with benzene develop-ment, but they may have wider applicability. Stokes's reagent is recommended for detection of $C$-nitro compounds; a red color is for-med. The reagent is prepared as follows:

Dissolve 1.5 grams of ferrous sulfate and 1.0 gram of tartaric acid in 50 ml of water. Just before use, take a known volume of the ferrous-tartaric solution and add drops of a molar solution of ammonium hydroxide until the precipitate that first forms is just dissolved. Then add approximately the same volume of 50% potassium hydroxide as the ferrous-tartaric solution taken initially; shake well.

When adsorbed on silica, nitro compounds readily give a red zone on a greenish background with the reagent, but the whole turns red on standing. Nitrobenzene, nitropropane, and $m$-dinitrobenzene were the only ones tested.

The chromatographic separation of analogous $C$-nitro and $C$-nitroso compounds of phenol, resorcinol, naphthol, and propanol has been reported by Edwards and Tate (153). The nitro and nitroso analogs are dissolved in benzene or petroleum ether at dilutions of less than 0.01 $M$; the adsorbent is ether silicic acid or silicic acid–Celite mixture packed in columns, and the developer is benzene or a mixture of benzene with acetone or petroleum ether. The recovery of solute from the mixtures ranges from 85 to 97%, with the recovery of nitro compounds about 2–10% higher than that of nitroso compounds. Preliminary identification of the compounds was made by measurement of the zone position or the use of streak reagents such as 6 N sodium hydroxide solution, 0.005 M potassium permanganate in 0.125 M sodium hydroxide, and diphenylamine in sulfuric acid. Further identification and quantitative estimations of the compounds were made by determination of the ultraviolet absorptivity of an ethanolic eluate of isolated column zones. In a subsequent paper, Edwards et al. (152) include the chromatographic separation of other nitroaromatics, nitramines, and azo and azoxy compounds. Some three-component mixtures were separated satisfactorily; recovery of single solutes ranged from 96 to 97%.

Teague et al. have found columns packed with silicic acid or kaolin to be effective adsorbents for di-, tri-, and tetranitrostilbenes (680).

The chromatographic resolution of isomeric quinone oximes formed by nitrosation of phenol has been reported by Gullstrom et al. (245); samples are dissolved in acetone and adsorbed on a column of activated alumina. Elution with 1% acetic acid in acetone followed by 5% methanol in acetone provides distinct bands which can be separated for quantitative assay.

Angell (11) separated small amounts of m-dinitrobenzene (1%) in nitrobenzene by adsorption from a light petroleum solution on an alumina column and elution with ethanol. The eluate is treated with acetone and some sodium hydroxide solution; a pink color indicates a small concentration of the dinitro compound, and the actual amount can be determined spectrophotometrically. A red color indicates a greater concentration of solute; this amount is best determined polarographically by the step height of the wave at −0.45 V (vs. SCE).

Mixtures of o-, m-, and p-nitroanilines and many of their substituted derivatives can be separated more or less completely on activated alumina columns. In a typical procedure (397), a solution of about 1 gram of the nitranilines in 75 ml of benzene is applied to a 2.5 × 30 cm column of alumina; the chromatogram is developed with a 1:4 mixture of ethyl acetate and benzene. The *para* compounds are adsorbed strongly and remain near the top of the column; the *ortho* compounds move rapidly down the column and are readily eluted; the *meta* compounds, however, are intermediately situated as a yellow band.

Mixtures of aliphatic 2,4-dinitrophenylhydrazones have been separated on bentonite from ether or hexane (727) and on alumina from benzene (83).

Chromatographic separation of the isomers of nitrophenols, nitroanilines, chloronitrobenzenes, and nitrotoluenes by means of columns packed with clathrates was investigated by Kemula and Sybilska (354); aqueous solutions of $\gamma$-picoline with potassium thiocyanate in different proportions were used as solvents and as eluents. Nitrobutanes and nitropropanes were also separated.

The chromatographic properties of hexogen (hexahydro-1,3,5-trinitro-$s$-triazine), a high explosive, and of a series of related polynitramines have been reported by Malmberg et al. (457). The adsorbent used was Celite–silicic acid (1:2), and a modification of the Franchmont test and the Schryver test for the formaldehyde obtained by reaction with concentrated sulfuric acid served as streak reagents.

Thin-layer chromatographic (TLC) procedures developed for nitramines by Harthon (256) were extended to compounds such as hexahydro-1,3,5-trinitro-$s$-tetrazine (RDX) and octahydro-1,3,5,7-tetranitro-$s$-tetrazine (HMX) by Fauth and Roecker (177); the general techniques employed were described by Russell (596). Parallel TLC strips of alumina spotted with Griess's reagent are used to locate the positions of the RDX and HMX. Scrapings of the sample strips were eluted with acetonitrile for spectrometry at 228 nm. A more extensive series of aliphatic cyclic and linear nitramines was examined by Bell and Dunstan (41) on silica gel TLC plates. The dried plates were sprayed with 1% diphenylamine in ethanol and irradiated with uv light (115).

The derivatives of diphenylamine formed in the aging of double-base powders (munitions) have been separated, identified, and determined quantitatively by Schroeder et al. (615); a number of $N$-nitroso compounds and nitrodiphenylamines were isolated on silicic acid–Celite adsorbents and, after elution, were determined spectrophotometrically at appropriate wavelengths between 280 and 425 nm. Procedures for separation of the diphenylamine derivatives from other components in the double-base powders also have been developed.

Colman (119) has described a paper chromatographic procedure by which any of eighteen 2-, 4-, 6-substituted 1,3,5-trinitrobenzenes can be identified rapidly.

### e. Gas Chromatography

The lower boiling nitroparaffins (to $C_4$) can be separated from each other and identified or determined quantitatively in $4.6 \times 1300$ mm glass columns packed with firebrick and substrates such as Apiezon T, Lubriseal, or transformer oil (substrate/brick weight ratio = 0.35); helium at a rate of 30 ml/min is used with column temperatures of 65° or 98°, and 20-$\mu$l samples are introduced through an injection port maintained at about 210° (46).

Increased resolution in the separation of $C_1$–$C_4$ nitroparaffins can be obtained in 6-ft lengths of $\frac{1}{4}$-in. (o.d.) copper tubing packed with a mixture consisting of 10 grams of Armeen SD (745), 10 grams of Apiezon N, and 100 grams of Johns-Manville Type C-22 firebrick; before use, the packing is maintained for 24 hr at 150°C in a stream of air. The chromatographic separation can be effected isothermally at 50°C with a hydrogen flow rate of 90 ml/min (28°C) or with a linear temperature programming rate of 2.9°C/min (start 40°C) at a hydrogen flow rate of 60 ml/min; in either instance, a thermal conductivity detector is used, and 4-$\mu$l samples are injected into a port section maintained at about 200°C (45).

Fear and Burnet (179) applied gas chromatography to the analysis of vapor-phase butane nitration products; separations of $C_1$–$C_4$ mononitroparaffins, alcohols, aldehydes, and paraffins were satisfactory on 20-ft columns (0.125-in. o.d.) containing 3 grams of bis(2-ethylhexyl)adipate on 100 grams of Chromosorb W operated first at 25° and 60°C and then at a linear rate of temperature increase (1.3°/min) until all components were eluted (helium flow at 6 ml/min; flame ionization detector). Water in the sample was removed by calcium carbide before chromatography. Hydrogen, oxygen, nitrogen, nitric oxide, methane, and carbon monoxide were analyzed in a separate column.

Essigman and Issenberg (173) have described the results of a government-sponsored study on the determination and identification of volatile nitrosamines (nanograms per gram) in foods contaminated with nitrates and nitrites. Distillation—extraction procedures are used to concentrate $N$-nitrosamines before chromatography; the procedures leave much to be desired, especially since the simultaneous presence of aldehydes and nitrites in foods may induce formation of nitrosamines in neutral solutions during prolonged distillation. It is contended that vanishingly small amounts of $N$-nitrosamines ingested by human beings over many years may be a source of carcinoma.

Evered and Pollard (175) have found 30–60 mesh firebrick with 15–20% squalane or dinonylphthalate at 80–111°C suitable for separation of nitroparaffins; the carrier gas is a 3:1 mixture of hydrogen and nitrogen. A hydrogen flame detector was used with a stream of oxygen to facilitate the burning of the carrier gas (see also Section V.C.2.e).

The gas chromatographic separation of aromatic nitro compounds at 200°C on firebrick (C-22, of 30–60 mesh) impregnated with DC-710 silicone fluid has been found adequate for quantitative separations (321). Samples up to 25 $\mu$l can be flash-volatilized at about 260°C and led into

an 8-ft column (0.25-in. o.d.) packed with 20 grams of firebrick impregnated with 8 grams of the silicone; helium flow rate is maintained at
20 ml/min. The silicone fluid "bleeds" from the column; as a result,
separation of isomers is no longer possible after columns have been
operated for about 56 hr at 200°C. Silicones on firebrick have also been
found to be quite useful for separations of nitroxylenes (143) and
isomers of trinitrotoluene (218,593). It is of interest to note that RDX (a
nitramine) can be separated from HMX by a silicone-packed column
(594). Nitrophenols are separated by columns packed with Chromosorb
W impregnated with 5% diethylene glycol adipate and 0.4% phosphoric
acid (112). Chromosorb W impregnated with Apiezon L has been found
best for the separation of mono- and dinitrotoluene isomers (538).

The determination of nitrobenzenes in cigarette smoke concentrates
has been described by Hoffman and Rathkamp (288); separations appear
to be best at 200°C by 5% OV-225 on Gas Chrom P. Camera and
Pravisani (101) have shown that electron-capture detectors are suitable
for detection of C-nitro compounds in air.

Nitroso compounds can be separated on a column of sodium chloride
impregnated with polyoxyethylene glycol (651). Aliphatic N-nitrosamines which have been extracted with methylene chloride can be
separated by use of a 6-ft × 0.125-in. column packed with 15% diethylene
glycol stearate on Chromosorb W operated at 120–150°C and a helium
flow of about 35 ml/min.

### f.  X-Ray Diffraction

X-ray diffraction techniques are perhaps among the most useful
methods for identifying crystalline nitro and nitroso compounds. Unfortunately, the number of X-ray diffraction patterns for nitro and
nitroso compounds that have been published is presently so small that
use of X-ray diffraction methods for identification is sharply limited.

Soldate and Noyes (652) were among the first to recognize that X-ray
diffraction methods are of considerable aid for the rapid identification of
the many crystalline compounds that are used in explosive compositions; diffraction pattern data have been recorded for compounds
such as 2,4-dinitrotoluene, 2,4,6-trinitrotuluene (TNT), picric acid,
guanidine picrate, ethylene dinitramine, nitroguanidine, RDX (hexogen),
HMX, N-nitrosodiphenylamine, 2-nitrodinitrodiphenylamine, 2,4′-dinitrodiphenylamine, 4,4′-dinitrodiphenylamine, dihydroxyethylnitramine
dinitrate, and octahydro-1-acetyl-3,5,7-trinitro-s-tetrazine.

Burkardt (90) has reported data on the nitroguanylhydrazone of
acetaldehyde, cinnamaldehyde, and acetophenone, as well as on a series
of nitroguanidine derivatives. The crystal forms of 2,4,6-trinitrotoluene

as obtained from X-ray studies have been discussed by Burkardt and Bryden (92). Furthermore, Burkardt (91) has shown that molecular complexes (adducts; see below) of 2,4,6-trinitrotoluene with compounds such as naphthalene, anthracene, 2,4-dinitroanisole, 2,4-dinitromesitylene, phenanthrene, and 2-iodo-3-nitrotoluene have powder diffraction patterns that are useful for characterizations.

## g. ADDUCTS

The adducts of aromatic polynitro compounds with naphthalene were studied by Dermer and Smith (138); the following general procedure was used:

As nearly equimolecular amounts as possible of naphthalene and the unknown are fused together and then cooled. The melt is recrystallized from alcohol, and the melting point of the derivative is determined.

Obviously, the molecular weight of the polynitro compound must be known or closely estimated; hence adducts are useful only for confirmation of results obtained with other methods. Of 17 polynitro compounds which are commercially available in large quantities or are in themselves analytical reagents, all but 4—2,4-dinitrophenetole, ethyl 3,5-dinitrosalicylate, 3,5-dinitroanisole, and 2,4,2',4'-tetranitrobiphenyl—could be recrystallized without decomposition. Adducts are formed with other hydrocarbons such as anthracene, aniline, and substituted anilines, but at least two nitro groups must be on an aromatic ring if addition compounds are to be formed. There are, however, many peculiarities; for example, polynitrophenols and polynitrophenolic ethers usually give stable adducts, but polynitroamines rarely form compounds with naphthalene. Similarly, 2,4-dinitromesitylene and 1-*tert*-butyl-3,5-dimethyl-2,4,6-trinitrobenzene appear to be inactive because of the hindering effects of methyl groups situated between nitro groups. Nitro groups *ortho* to each other impede in the formation of adducts; thus *o*-dinitrobenzene, 3,4-dinitrotoluene, 2,3,4,6-tetranitrophenol, 2,3,4-tetranitroaniline, and polynitroveratroles fail to yield complexes. The generalization is not rigorous, however, because 2,3,4-trinitrotoluene, 4,5,6-trinitro-*o*-cresol, and several polynitrochlorobenzenes readily form adducts.

Mercury nitroform, $Hg[C(NO_2)_3]_2$, forms adducts with nitrobenzene and substituted nitroaromatics that have an open position *meta* to a nitro group (520). Tetranitromethane forms complexes with many aromatic hydrocarbons.

A list of characteristic adducts of polynitro compounds is given in Table XIII.

TABLE XIII
Characteristic Adducts of Polynitro Compounds (138,523)

| Polynitro compound | Second component | Adduct | |
| --- | --- | --- | --- |
| | | Melting  point (°C) | Color |
| 1,3-Dinitrobenzene | Naphthalene | 53 | Colorless |
| 1,3-Dinitrobenzene | Aniline | 42 | Red |
| 2,4-Dinitrotoluene | Naphthalene | 60 | — |
| 2,6-Dinitrophenol | Naphthalene | 58 | — |
| 2,4-Dinitrophenetole | Naphthalene | 41 | — |
| 2,4-Dinitroanisole | Naphthalene | 50 | — |
| 3,5-Dinitroanisole | Naphthalene | 69 | — |
| 2,4-Dinitroresorcinol | Naphthalene | 165 | — |
| 3,5-Dinitroguaiacol | Naphthalene | 94 | — |
| Ethyl 3,5-dinitrobenzoate | Naphthalene | 75 | — |
| 1,3,5-Trinitrobenzene | Naphthalene | 152 (156) | Colorless |
| 1,3,5-Trinitrobenzene | Phenanthrene | 158 | Orange |
| 1,3,5-Trinitrobenzene | Anthracene | 174 | Scarlet |
| 1,3,5-Trinitrobenzene | Dimethylaniline | 107 | Dark violet |
| 2,4,6-Trinitrotoluene | Naphthalene | 97–98 | Colorless |
| 2,4,6-Trinitrotoluene | Aniline | 84 | Red |
| Hexanitroethane | Naphthalene | d. | Red |
| 2,4,6-Trinitrobenzaldehyde | Naphthalene | 136–137 | — |
| 2,4,6-Trinitrophenetole | Naphthalene | 39 | — |
| 2,4,6-Trinitroanisole | Naphthalene | 69–70 | — |

## VI.  QUANTITATIVE DETERMINATIONS

The determination of nitrogen in any nitro and nitroso compound can be accomplished by the Dumas method (first introduced in 1826); however, since the results represent total nitrogen, whenever the nitrogen content in the form of the nitro or nitroso group is to be determined, resort must be made to procedures that depend upon a specific reactivity of these groups. Characteristically, the procedures in prevalent use are predicated on reduction to convert nitro or nitroso groups to an amine. The most widely used reducing agents for titrimetry are stannous chloride and titanous chloride; reducing agents such as chromous salts and metals are not as popular in spite of their superiority.

## A. NITRO GROUP

### 1. Titrimetric Procedures

#### a. TITRATION WITH REDUCING AGENTS

In principle, since the nitro group is an oxidant, it should be possible to titrate it directly with strong reducing agents such as titanous salts, stannous salts, chromous or vanadous salts, and sodium hydrosulfite. However, reduction of most nitro compounds takes place too slowly for direct titration; reactions often need to be driven to completion by a large excess of reductant under special conditions that must be established for each combination of reducing agent and nitro compound. Although general procedures are cited in subsequent sections, a careful study must be made of each method before it can be applied with confidence to a given compound. In practice, it has been found that only powerful reducing agents will suffice; unfortunately, such agents must be protected from oxygen at all times, but special equipment such as that developed by Stone provides a convenient method for storage and titration (668).

#### (1) Reduction with Titanous Salts

Six equivalents of titanous chloride solutions are required for reduction of each nitro group to an amino group; the course of reduction proceeds through at least three steps:

$$-NO_2 \rightarrow -NO \rightarrow -NHOH \rightarrow NH_2$$

Each step involves an exchange of two electrons, for an overall consumption of 6 equiv. of reducing agent. When an amine is the product,

$$6\,Ti^{3+} + R-NO_2 + 6\,H^+ \rightarrow RNH_2 + 6\,Ti^{4+} + 2\,H_2O$$

or 4 equiv. when substituted hydroxylamines are formed:

$$4\,Ti^{3+} + R-NO_2 + 4\,H^+ \rightarrow RNHOH + 4\,Ti^{4+} + 2\,H_2O$$

Hinshelwood et al. (515,662) found that the reduction of nitrobenzene by titanous chloride in aqueous alcohol is a first-order reaction with respect to each reactant and, over a certain range, is approximately inversely proportional to the square of the hydrogen ion concentration. As already indicated in Section VI.B.1 of "Diazonium Group," Vol. 15, the active species are suggested to be $C_6H_5NO_2H^+$ and a hydrolyzed form of titanous ion. In the sequence of reactions indicated above, reduction of the nitrosoaryl compound and the hydroxylamino compound proceeds many times faster than the reduction of nitroaryl to nitrosoaryl compound; however, condensation forms an azoxy compound, and this diversion of a considerable part of the main stream of reaction results in an overall decrease in reaction rate. It is suggested that the rate-determining step is the transfer of electrons from a titanous complex to the nitro compound, inasmuch as electron-attracting substituents on the aryl ring increase, and electron-repelling substituents decrease, the rate of reduction.

It is necessary at all times to take into consideration the interaction of other functional groups with titanous salts during reduction of nitro groups. Moreover, the route of reaction of titanous salts with C-mononitro compounds is different from that with N-nitro compounds, and the nature of substituents, as well as aromaticity or resonance, often plays an important role in determining the products of reduction, the extent of reduction, and the rate of reduction [see Section III.B.1.b(2)].

### (a) C-Nitroaromatics

The use of standard titanous chloride solutions for quantitative reduction of the C-nitro group was first described by Knecht and Hibbert (363); a general procedure involved a 5-min boiling of the nitro compound with an excess of titanous chloride solution in the presence of a large amount of hydrochloric acid, and then a back titration of excess titanous ion in the cold with ferric alum solution, using ammonium thiocyanate as indicator. In the original procedures, C-nitro compounds that are insoluble in water or acids (e.g., nitrobenzene) are first sulfonated with fuming sulfuric acid, and compounds that are neither soluble in water nor easy to sulfonate (e.g., trinitrotoluene) are dissolved in alcohol. In the light of recent work, however, solubilization by sulfonation is not recommended, especially since pure samples may be degraded by the drastic conditions used in typical sulfonation procedures and, more often, sulfonation procedures are not quantitative. It is of interest to note that in 1919 van Duin (146) had observed that titanous solutions are markedly decomposed on heating (147); however, even to this day, little attention is given to this source of error. Results within 0.4% accuracy were obtained by Knecht and Hibbert for o-nitroaniline, nitrobenzene, picric acid, m-dinitrobenzene, dinitrotoluene, and trinitrotoluene, but poor results were obtained with nitroanisole and α-nitronaphthalene; on the other hand, β-nitronaphthalene behaved normally.

Various modifications of the Knecht–Hibbert method for reduction by titanous chloride were used by English (171) in a critical study. The strong acidity required for the titanous reduction was obtained by use of sulfuric acid instead of hydrochloric acid because the former contained fewer reducible impurities; boiling time was reduced to 5 (from 10) min, and a constant volume was maintained during boiling. The indicator was added toward the end of the back titration rather than at the beginning. Results from 99.8 to 100.1% of theoretical were obtained with *m*-nitroaniline, *o*-nitrophenol, 2-nitro-4-methylaniline, 3-nitro-4-methylaniline, *m*-dinitrobenzene, 2,4-dinitrotoluene, picric acid, trinitrotoluene, and picramic acid. Low results were obtained with mononitro compounds such as nitrobenzene, nitrotoluene, nitroxylene, nitrochlorobenzenes, and the nitronaphthalenes; high results, with 2,4-dinitrochlorobenzene. In view of the low results, English concluded that aromatic mononitro compounds were resistant to reduction by titanous salts; it was noted that the presence of positive or negative substituents (with the exception of chloride) facilitated reduction, and that the orientation of substituent groups with respect to the nitro group had no appreciable effect.

Careful work by Callan and Henderson (98) revealed that the conclusion drawn by English about the resistance of mononitro compounds to reduction by titanous chloride was in error; low values were shown to be caused by steam distillation of the samples during the prolonged boiling needed to accomplish complete reduction. Furthermore, it was found that titanous chloride in strong hydrochloric acid solutions appears to induce chlorination of aryl nitro compounds, especially in the presence of alcohol; thus it was suggested that in some instances low results are obtained because the hydrogen atom displaced by the entering halogen atom acts as a reducing agent. Accordingly, the use of titanous sulfate solution rather than titanous chloride was proposed by Callan et al. (99) as a means of preventing chlorination of the nitro compound undergoing reduction, particularly in alcoholic solutions. With titanous sulfate in a sulfuric acid reduction medium and a reflux condenser to prevent loss of sample by steam distillation, excellent results were obtained with easily halogenated compounds such as *α*-nitronaphthalene and *p*-nitroanisole; these compounds could not be determined satisfactorily by the method of Knecht and Hibbert. Callan and Henderson (98) presented tables of results obtained with mono-*C*-nitro hydrocarbons, *C*-nitromonochloro compounds, *C*-nitrophenols, and so on under various conditions in sulfuric acid media and also cited comparative results for reductions in hydrochloric acid media. Twelve results obtained with five pure mono-*C*-nitro hydrocarbons were within 99.8–110.0% of theoretical. More recently, Musha (509,510) has obtained good results with titanous sulfate solutions that were kept reduced by zinc amalgam. In spite of the advantages realizable with titanous sulfate solutions, titanous chloride solutions continue to be used widely; perhaps the ready availability of concentrated titanous chloride solutions is a detriment to the use of titanous sulfate. The simplicity of preparation of titanous sulfate solutions from titanium hydride, however, should be an inducement for discarding titanous chloride as a titrant (718).

The reduction potential of the $Ti^{3+} - Ti^{4+}$ couple increases as the acid content decreases; consequently, titanous solutions are more powerful reducing agents in basic than in acidic media. Moreover, it appears that reaction kinetics may be more favorable in alkaline media because Kolthoff and Robinson (375) were able to reduce several C-nitro compounds at room temperature with only a slight excess of 0.05 N titanous chloride in a buffered alkaline medium (sodium bicarbonate and sodium citrate); the slight excess of titanous ion was determined potentiometrically with ferric alum. Results within ±0.3% absolute were obtained for m- and p- nitroaniline, picric acid, p-nitrophenol, α-nitrophenol, α-nitronaphthalene, and nitrobenzene, and the potentiometric results were found to be within 0.3% of the values obtained when the back titration was performed in the usual way with thiocyanate as indicator. Dachselt (131) showed that a direct potentiometric titration of o-nitroaniline with titanous chloride solution in an inert atmosphere is possible when the determination is performed at an elevated temperature (50–80°C) in an alcoholic solution containing tartaric acid, but since most nitro compounds are difficultly reducible and the reaction must be driven by an excess of titanous reagent, Dachselt's procedure cannot be applied to the great majority of C-nitro compounds; application to C-nitroso compounds has not been reported.

The citrates, tartrates, and acetates of alkali metals are effective buffers for titanous reductions and prevent precipitation of basic salts; since complexes are formed with titanium ions, the reduction potential of the titanous–titanic couple is increased slightly over the amount ascribable to pH effects, and this also facilitates reduction of nitro compounds. In strongly alkaline solutions, titanous ion can release hydrogen by reaction with water (368a); buffered solutions (generally below pH 10) minimize the probability of decomposition of titanous titrants. Several workers state a preference for acetate (94,591) over citrate or tartrate, but others (663) select the salt that is most soluble in the reaction medium; no definitive studies have been made to permit identification of the best salt, although it is to be expected that salts which complex titanium in the alkaline region will be the best. On the other hand, Earley and Ma (151) developed a micromethod in which an acetate buffer not only maintains an optimum pH for titration of nitro compounds but also appears to limit or prevent nuclear chlorination. (It is not indicated whether chlorination is prevented by acetate or whether chlorination simply does not occur at high pH.) Additionally, the authors report that *C*-nitro group compounds will not react with titanous chloride in concentrated hydrochloric acid whereas the majority of azo and *C*-nitroso compounds are reduced; moreover, mono-*C*-nitro compounds are seldom reduced unless the reactivity of the nitro group is increased by substitution. The procedure for titration of azo compounds in the presence of *C*-nitro compounds and *C*-nitroaryl hydrazines follows (151):

Weigh 3–8 mg of sample, and transfer to a reaction flask containing 5–10 ml of water or 95% ethanol; connect the flask to a microburet containing titanous chloride. Agitate the solution or stir it mechanically until the sample dissolves; then introduce 4 ml of 12 *M* hydrochloric acid. After purging the vessel and contents with nitrogen for 5 min, add about 6 ml (100% excess) of titanous chloride; the reduction is complete after about 3 min. Back-titrate with a solution of ferric alum and 2 ml of 2.5 *M* ammonium thiocyanate to a pink end point that lasts for at least 1 min. Perform blank determinations in the same way.

In a sodium acetate buffer (371), nuclearly attached nitro groups can be titrated on nitroaryl hydrazines together with the hydrazine group.

The usual factors in titanous reductions which are varied in order to induce complete reduction are the time the titanous salt is allowed to remain in contact with the nitro compound, the temperature, the quantity of titanous titrant used in excess, and the acidity of the reaction medium. As expected, the more refractory nitro compounds are subjected to extremes, and a large number of papers published on the use of titanous reduction methods are mere records of studies of the behavior of one compound or a class of compounds under such extremes (10). For example, the classic Knecht–Hibbert procedure involving a 5-min boiling period does not reduce a number of aliphatic *N*-nitro compounds such as nitroguanidines to amines; however, Kouba et al. (384) were able to obtain satisfactory results (even though only 4 equiv. were consumed by nitroguanidine) when the hydrochloric acid concentration originally specified by Knecht and Hibbert was brought up to 1:1 in the reaction medium, and the time of boiling was prolonged to 15 min. Similarly, Zimmerman and Lieber (747) were able to obtain a consumption of 6 equiv. of reductant by inclusion of empirically determined amounts of ferrous iron along with an excess of titanous ion, but the procedure is not recommended for accurate assays of nitroguanidine. The work of Earley and Ma (151) is another example of forced reduction (or inhibition) by use of extreme conditions; also in this category is the work of Sternglanz et al. wherein a six-electron reduction of nitroguanidine is achieved by use of a 200% excess of titanous chloride in a medium buffered at about pH 4 by potassium citrate (591,664).

After initial work by Kolthoff and Robinson (375), there was an increased effort to use buffered media with titanous solutions in the analysis of difficultly reducible nitro compounds. By employing buffered media and a large excess of titanous titrant, compounds that gave unsatisfactory results with the original Knecht–Hibbert procedure were successfully analyzed by titanous titrations; for example: nitroguanidines (593,663), cyclotrimethylenetrinitramine (RDX) (384), and dinitrotoluene (94). Roth and Wegman (591) suggest that the order of addition of samples and reagents may be critical (e.g., titanous chloride and buffer must be in the titration flask before a nitroguanidine sample is added); unfortunately, the effect of order of reagent addition in titanous reductions has not been studied in sufficient detail to permit the drawing of conclusions.

A micromethod and apparatus for the determination of aromatic $C$-nitro groups by titanous chloride reduction were described by Maruyama (468). In accordance with Kolthoff and Robinson (375), the compound is dissolved in glacial acetic acid and mixed with sodium citrate solution; a small amount of sodium bicarbonate is added to help displace dissolved oxygen. Under an atmosphere of carbon dioxide, the solution is treated with an excess of titanous chloride solution, heated to 50–60°C, and titrated with ferric alum.

A similar method was described by Tiwari and Sharma (686); titanous sulfate reductions are performed in a concentrated citrate buffer, and it is reported that aromatic nitro compounds can be reduced within 3 min at room temperature. A 10–25 mg sample is dissolved in 3–5 ml of glacial acetic acid; then 10 ml of potassium citrate solution (w/v) is added (under nitrogen atmosphere), followed by 20 ml of 0.05 N titanous sulfate solution. The mixture is shaken for 1 min, warmed on a hot plate for 1 min, and then cooled rapidly. Sulfuric acid is added (3 ml of 50% v/v), and the excess titanous sulfate is titrated with 0.05 N ferric sulfate, adding potassium thiocyanate indicator near the end point.

The micromethod described below for the determination of nitro compounds according to Ma and Earley (447) prescribes only sodium acetate as buffer [not with acetic acid as specified by Maruyama (468)]; the pH of the reaction medium is higher, and therefore it is not necessary to apply heat to bring the reduction to completion.

A 3–8 mg sample of nitro compound is dissolved in alcohol and treated with sodium acetate solution; after a 5-min flushing of the apparatus with nitrogen, the solution is titrated with titanous chloride solution to a deep violet color. Then concentrated hydrochloric acid is added, and the solution is titrated with ferric alum, using ammonium thiocyanate as an indicator; a blank is run concurrently (447).

All aromatic nitro compounds tested by Ma and Earley were readily reduced (without nuclear chlorination) and determined with a relative error of about ±0.03%. Nitroethane, 2-nitro-2-methyl-1-propanol, and nitrostyrene were not quantitatively reduced.

Each nitro group of poly-C-nitroaromatic compounds apparently is reduced smoothly with consumption of 6 equiv. of titanous ion; thus Fauth and Roecker (176) reported quantitative reduction of all nitro groups in 2,4,6-trinitromesitylene by a procedure very similar to that of Knecht and Hibbert; Butts et al. (94) claimed rapid and complete reduction of dinitrotoluene by citrate-buffered titanous chloride, and Ma and Earley (447) have shown similar results for a variety of dini-troaromatic compounds in an acetate-buffered medium. C-nitro and di-(C-nitro)phenylhydrazines can be reduced with titanous chloride in sodium acetate by the procedure developed by Earley and Ma (151) (see above); both the nitro and the hydrazine group of aryl hydrazines are reduced under the experimental conditions.

Macbeth and Price (449) studied the action of titanous chloride on p-nitro- and 2,4-dinitrophenylhydrazones in titrations performed ac-cording to the classical method of Knecht and Hibbert. Because of the strongly acidic media, the phenylhydrazones are first hydrolyzed to the corresponding phenylhydrazines and carbonyl compounds; then the phenylhydrazines are attacked by titanous ion. Cleavage of p-nitro-phenylhydrazine to ammonia takes place, with consumption of 8 equiv. of titanous ion; on the other hand, 2,4-dinitrophenylhydrazine is not cleaved, and the formation of diaminophenylhydrazine requires 12 equiv.

### (b) C-Nitroparaffins

The reduction of C-nitroparaffins by titanous salts does not appear to take place with the same facility as that of C-nitroaromatics. Thus, with the sodium acetate method of Ma and Earley (447), nitroethane, 2-nitro-2-methyl-1-propanol, and nitrostyrene are not reduced quantitatively (75% max.); in contrast, the C-nitroaromatics are reduced cleanly (see the preceding section). When an excess of titanous chloride is used with glacial acetic, C-nitroparaffins seldom are reduced quantitatively, and in some instances the reduction scarcely takes place (176). Obviously, reduction of C-nitroparaffins by titanous salts can be applied as an analytical procedure only after an extensive series of experiments have shown that quantitative results can be obtained.

On reduction, C-polynitroparaffins readily lose all but one nitro group attached to a carbon; this clearly indicates that the reduction of the

*C*-monoaminomononitro group is difficult, but it is emphasized that this type of nitro group may be reduced with yields less than theoretical. Thus *gem*-dinitroparaffins consume the anticipated 6 equiv. of titanous ion/mol in a Knecht–Hibbert reduction, but the number of equivalents consumed per mole tends to be higher than 6 in an acetate-buffered medium, indicating that these conditions induce partial attack on the remaining nitro group (176). The end products of the reaction of titanous ion with aliphatic *gem*-dinitro compounds have not been isolated. As expected, aliphatic *C*-trinitro compounds like 2,2,2-trinitro-ethanol lose two nitro groups in a Knecht–Hibbert reduction. The actions of titanous chloride and other reducing agents on some substituted nitroparaffins have been summarized by Henderson and Macbeth (273).

### (c) N-Nitro Compounds

Reduction of the *N*-nitro group with titanous salts seldom forms a single species, for several steps are involved and the stability of some of the intermediate products is such that further reduction may not take place or other reactions take place competitively:

$$RN-NO_2 \rightarrow RN-NO \rightarrow RNNHOH \rightarrow RNH + NH_2OH \rightarrow RNH + NH_3$$

For example, the reduction of nitrourea proceeds with the consumption of essentially 2 equiv. of titanous ion; Zimmerman and Lieber (747) interpret this behavior as the result of formation of extremely unstable nitrosourea:

$$NH_2\overset{O}{\underset{}{C}}-NHNO_2 \overset{2H^+}{\longrightarrow} NH_2\overset{O}{\underset{}{C}}-NHNO \rightarrow N_2 + CO_2 + NH_3$$

Nitroguanidine and other nitroammonocarbonic acids also do not give reproducible values in reduction with titanous salts [see also Section III.B.1.c(1)]. Thus Kouba et al. (384) found that about 4 equiv. of titanous chloride are consumed per mole of nitroguanidine when refluxed for 15 min in a hydrochloric acid medium. Zimmerman and Lieber (747) suggest that in this instance the nitrosoguanidine that is first formed is sufficiently stable to be further reduced to a "hydroxylamine," but the "hydroxylamine" is unstable and decomposes:

$$NH_2\overset{NH}{\underset{}{C}}-NHNH(OH) \rightarrow N_2 + H_2O + CO_2 + NH_3$$

However, the degradation is not clean, and other decomposition products are also produced. Nevertheless, 4 equiv. are involved in the reaction.

As has been noted in prior discussions, the presence of impurities such as iron and zinc in titanous solutions may direct the path of the reduction; addition of large amounts of substances such as ferrous salts may provide conditions in which unstable products may linger long enough to be attacked by titanous reducing species. For example, Zimmerman and Lieber (747) showed that nitroguanidine, when reduced in the presence of 0.70 equiv. of ferrous iron/mol of sample, will consume very closely the theoretical value of 6 equiv. for each nitro group. However, since values as high as 7.6 equiv. per nitro group can be obtained in the presence of larger amounts of iron, the iron concentration must be rigidly controlled in order to obtain reproducible and accurate results as well as a desired equivalency; the procedure is not recommended for general use, although it is suitable for routine analyses. The results of later studies suggest that the theoretical value for reduction of nitroguanidine in the presence of iron may be 8 equiv. of titanous chloride/mol of compound.

In an effort to explain peculiarities in the reduction of nitroguanidine, an aliphatic nitroamine, Brandt et al. (65) begin by suggesting [like Zimmerman and Lieber (747)] that the reduction involves formation of a hypothetical, unstable hydroxylaminoguanidine for which only 4 equiv. of titanous ion are required:

$$\underset{\underset{\displaystyle NH_2-C-NHNO_2}{\parallel}}{NH} \xrightarrow{4\,equiv.} \underset{\underset{\displaystyle NH_2-C-NHNHOH}{\parallel}}{NH}$$

It is assumed that in the usual titanous reduction procedures the hydroxylamino compound rapidly decomposes without further attack by titanous ion; on the other hand, it is suggested that large amounts of ferrous iron stabilize the hydroxylamino compound to the point where it is made available for reductive cleavage and subsequent reduction of hydroxylamine by 4 additional equiv. of titanous ion:

$$\underset{\underset{\displaystyle NH_2-C-NHNHOH}{\parallel}}{NH} \xrightarrow{2\,equiv.} \underset{\underset{\displaystyle NH_2-C-NH_2}{\parallel}}{NH} + NH_2OH \xrightarrow{2\,equiv.} \underset{\underset{\displaystyle NH_2C-NH_2}{\parallel}}{NH} + NH_3$$

The behavior of RDX (cyclotrimethylenetrinitramine, or hexahydro-1,3,5-trinitro-$s$-triazine) on reduction by titanous chloride in the presence of ferrous chloride appears to follow a similar reaction mechanism, for only 12 equiv. of titanous ion are consumed per mole of RDX (384). Thus, since RDX is an $N$-nitro compound, it may undergo reduction after a fashion similar to that postulated by Brandt et al. (see above) or by Zimmerman and Lieber (747), that is, the formation of a hydroxylamine-type product requiring only 4 equiv. of reductant for conversion of each N—NO$_2$ group.

At this point, it is of interest to note that the role of metal ions in the erratic behavior of nitrate esters (specifically, nitroglycerine) on reduction with titanous ions was studied by Fainer (175a); no significant effect was observed.

### (2) Reduction with Stannous Chloride

A method for quantitative reduction of the nitro group by stannous chloride was first described by Limpricht in 1878 (418,656,744); however, the method is treacherously simple, for although it merely requires heating a nitro compound with an excess of stannous chloride and back titration with iodine, it is plagued by inaccuracies arising from air oxidation, nuclear chlorination, loss of sample by volatilization, and nonstoichiometric reduction [see Section III.B.1.a.(6)]. The reduction in acidic solutions is as follows:

$$RNO_2 + 3\,SnCl_2 + 6\,HCl \rightarrow RNH_2 + 3\,SnCl_4 + 2\,H_2O$$

The following procedure is representative of the earlier methods:

A solution of stannous chloride is prepared by dissolving 150 grams of tin in concentrated hydrochloric acid; the clear supernatant is diluted with 50 ml of the concentrated acid and then made up to 1 liter with water. About 0.2 gram of the sample is placed in a glass-stoppered flask together with an excess of the stannous chloride reagent (total of about 10 ml). The reaction mixture is warmed for about 30 min, cooled, and diluted to exactly 100 ml; a 10-ml aliquot is used for determining the excess of stannous chloride. The aliquot is diluted with water, and enough basic solution is added to dissolve the precipitate which first forms; 0.1 N iodine solution is used to back-titrate with starch as indicator. The basic solution consists of 90 grams of sodium carbonate and 120 grams of sodium potassium tartrate dissolved in sufficient water to make 1 liter of total volume.

In the light of present-day knowledge, Limpricht's method appears primitive, and, as expected, it was found to yield erratic and low results. A number of workers within a span of 40 years suggested modifications to eliminate apparent sources of error, but the elimination of one error usually introduced others. For example, low results were first attributed to the fact that the reduction is not quantitative; as a result, modified procedures inflicted more vigorous conditions for longer times to force completion, but altogether too many workers in the early years simply refused to recognize that prolonged treatment emphasized oxidation of the reagent by atmospheric oxygen and that loss of sample occurs during heating (because many nitro group compounds are volatile or are readily carried over with water vapor). Moreover, although it was common knowledge when Limpricht introduced the method that nuclear chlorination often occurs during stannous reductions in strong hydrochloric acid solutions (especially if alcohol is present), attempts to eliminate chloride ion were not made before 1920 (99); the first practical method developed expressly to minimize substitution by halogen was reported in 1939 by Hinkel et al. (284).

Florentin and Vandenberghe (195) made a systematic study (1920) of the general Limpricht procedure and noted that the end point is difficult to establish because of interaction of iodine with alkali and concomitant oxidation of amines by hypoiodite. To avoid this difficulty, the authors suggested oxidation of excess stannous ion by ferric chloride and determination of the ferrous ion formed by direct titration with permanganate in a solution of the neutrality established by solid calcium carbonate and diammonium citrate. These authors also found, however, that the excess of stannous chloride can be titrated directly with standard iodine under cover of an inert atmosphere.

As has already been indicated, the large amount of hydrochloric acid present in stannous chloride reductions often leads to nuclear chlorination, and the results will be low because only 4 equiv. of hydrogen are consumed instead of the 6 expected for formation of an amine:

$$R—NO_2 \xrightarrow{4H} R—NHOH \xrightarrow{2H} RNH_2$$
$$\Big|\xrightarrow{HCl} RNHCl \rightarrow RClNH_2$$

Replacement of a substantial fraction of the hydrochloric acid used in the reduction by sulfuric acid effectively eliminates chlorination, but the rate of reduction is lowered (284).

The procedure given in Section VII is essentially that of Hinkel et al. (284). Within the limitations of the method, a large variety of compounds are reduced quantitatively ($\pm 2$–3 parts per thousand). Samples that are quite volatile may not be trapped by the reflux condenser and will be carried away by the stream of carbon dioxide used to exclude air. In these instances, it is appropriate to use the sealed tube method originally employed by Spindler (656) and also by Florentin and Vandenberghe (195). Essentially, the sample is placed in a small tube or ampul and weighed; the sample container is transferred to a larger tube to which is also transferred the usual aliquot of stannous chloride solution; the open end of the larger tube is fused shut after displacing air with carbon dioxide. The sealed tube is heated in a water bath for several hours with occasional agitation of the contents. When cold, the tube is opened under cover of an inert atmosphere (such as carbon dioxide) and the contents are quantitatively transferred and diluted before titration with standard iodine solution. When reduction products are highly colored and interfere with detection of the end point, potentiometric methods may provide a solution. In many instances, however, titration with permanganate is a feasible alternative.

### (3) Reduction with Chromous Salts

Reduction of nitro compounds by a chromous solution followed by titration of the excess chromous ion with ferric solution was reported by Someya (654); Terent'ev and Goryacheva (682) titrated m- and p-nitroaniline directly with chromous chloride solutions, using Neutral Red (C.I. 50040) as indicator. However, these procedures were not considered very satisfactory; moreover, very frequent standardization of the chromous solution was required.

Using to advantage the procedure of Lingane and Pecsok (426) for preparing and maintaining a standard chromous chloride solution, Bottei and Furman (59) were able to perform satisfactory (±5 parts per thousand) determinations on several aromatic nitro compounds after carefully establishing proper conditions of acidity, buffering, temperature, and time of reaction for each compound. In brief, the sample is dissolved in water or glacial acetic acid, concentrated hydrochloric acid is added, and then the solution is diluted with water. Carbon dioxide is initially passed through, and then over, the solution. An excess of chromous chloride (200–400%) is added, and after a reaction period of 1 or 2 min the reaction mixture is back-titrated with standardized ferric alum solution; the end point is determined potentiometrically (SCE; Pt).

The direct potentiometric titration of aromatic C-nitro compounds with chromous sulfate was studied by Jucker (330). It was demonstrated that isomers such as o- and p-nitrophenol could be determined in admixture since, in many instances, two "breaks" in the curve of emf versus volume of titrant were exhibited. The results in nearly every case were from 95 to 99+% of theoretical. On the other hand, whereas polynitro compounds such as picric acid and trinitrotoluene show only one end-point break, at pH = 7, m-dinitrophenol shows two distinct breaks in pyridine sulfate solution. Other typical titration media include NaAc—HAc buffers and a mixture of dimethylformamide, sulfuric acid, and water.

Tandon (679) titrated satisfactorily (95% average) several aromatic C-nitro compounds directly with chromous sulfate solution under a carbon dioxide atmosphere, using Neutral Red (C.I. 50040), Phenosafranine (C.I. 50200), or p-ethoxychrysoidine as indicator. Alternatively, the same compounds were analyzed by reduction with an excess of chromous sulfate solution and back titration with ferric alum solution, using ammonium thiocyanate as indicator.

The possibility of nuclear chlorination when chromous chloride is used in strongly acidic (HCl) solutions must be taken into consideration. The use of chromous sulfate in sulfuric acid and a reaction medium maintained acidic by sulfuric acid is recommended particularly when alcohol is used as a solvent, inasmuch as Callan et al. (99) and Hinkel et al. (284) have shown that chlorination is especially facilitated by ethanol.

### (4) Reduction with Other Metal Salts

Vanadous sulfate was used by Gapchenko and Sheintis (216) to reduce solutions of milligram quantities of aromatic C-nitro compounds in acetone. A 200–300% excess of vanadous reagent was allowed to react with the sample for 5 min at room temperature; ferric alum was used to back-titrate the excess reducing agent in the presence of Safranine-O (C.I. 50240) as indicator. Banerjee (30) used vanadous sulfate to reduce nitro compounds at the boiling point; ferric alum with thiocyanate as indicator was employed for back titration. Difficultly soluble compounds were either sulfonated or dissolved in ethanol, and it is claimed that results compare favorably with those obtained with titanous sulfate reduction. Witry–Schwachtgen (739) described the preparation of vanadous sulfate by zinc amalgam; a number of aromatic nitro compounds were determined by reduction with vanadous sulfate and back titration with sodium dichromate, using Phenosafranine (C.I. 50200) as indicator.

Experimental results for the reduction of picric acid and cupferron by trivalent molybdenum are cited by Gapchenko (215); the excess molybdenum was titrated with ferric alum, using methylene blue indicator.

### b. REDUCTION BY HYDRIODIC ACID

The combination of concentrated hydriodic acid (57%) and red phosphorus is a powerful reducer when employed in a sealed tube at 150–200°C, and it has been frequently used to convert alcohols into hydrocarbons (see Section III.B.1); however, phosphorus is not needed for reduction of nitro, nitroso, and diazo compounds, as demonstrated by Aldrovandi and De Lorenzi (4).

For the determination, 1 ml of 57% hydriodic acid is added to 20–30 mg of nitro compound in a small glass tube, which then is sealed by fusion. The tube is placed in an oil bath at a temperature of 100–300°C for a suitable period of time; when cool, the tube is opened and the contents are diluted with water and titrated with 0.1 N sodium thiosulfate solution, using starch indicator. Blank determinations are performed concurrently. Naturally, groups that consume iodine will interfere.

The oxidizing property of nitronic acids has been used by Klimova and Zabrodina (362) to liberate iodine from hydriodic acid in a micromethod for primary and secondary nitroparaffins. The nitronate is first formed by dissolving 3–5 mg of the sample in 3 ml of a 2% alcoholic potassium hydroxide solution and allowing the solution to stand for 10 min in a flask equipped with a ground-glass stopper. Then 3 ml of 50% potassium iodide solution and 5 ml of 15% hydrochloric acid solution are added to the sample solution, and the stoppered flask is kept in the dark for 10 min. Finally, the sample solution is diluted with 50 ml of water and titrated with 0.01 $N$ sodium thiosulfate, using starch indicator; results ranging from 99.9 to 98.8% of theoretical are reported for nitroethane, 1-, and 2-nitropropane, and nitrocyclohexane. The *aci* form of the alkyl nitro compounds reacts with hydrogen iodide to form an oxime and iodine (see Section III.B.2.a):

$$R_1R_2-C{=}NOOH + 2 H \rightarrow R_1R_2C{=}NOH + I_2 + H_2O$$

## c. REDUCTION BY METALS AND AMALGAMS

Reduction of nitro compounds with metallic zinc and subsequent determination of the amine was suggested by Callan et al. (99) as an alternative to the titanous reduction method of Knecht and Hibbert, which fails when nuclear chlorination takes place. It is of interest to note that a few years later the authors found that titanous sulfate reductions were satisfactory and thus there was no need for zinc reduction.

The sample (0.05 mol) is dissolved in excess hydrochloric acid solution, zinc dust is added in considerable excess, and the reaction is allowed to proceed, with occasional heating, for 1 hr; the reaction mixture is then cooled and filtered. The filtrate is diluted, ice is added, and the amine is diazotized by direct titration with 0.5 $N$ sodium nitrite solution, using starch-iodide paper to detect the end point.

Since diazotization titrations are often employed in the control of dyestuff manufacture, this procedure for determination of nitro groups is of practical importance. Moreover, it is immaterial whether nuclear halogen is removed by the reduction step or whether chlorinated aromatic amines are formed during reduction, for such amines also react with nitrous acid to form diazo compounds; additionally, since exceptionally vigorous reducing conditions can be brought to bear on the nitro compound being analyzed, quantitative conversion to an amine is possible more frequently than in reductimetric titrations (e.g., titanous, stannous, or chromous) where stoichiometry of reduction must be maintained. Conversion to an amine and titrimetric determination via diazotization is recommended as a check on other methods for determining the nitro group. Wolthuis et al. (741) suggest that the titration need not be performed at a low temperature; data are offered to show that 15–20°C is sufficient. A procedure for determining nitro groups by reduction—diazotization is given in Section VII.

The titration of amines with sodium nitrite is not as widely applicable as is the reduction of aromatic $C$-nitro compounds by metals and acids. For example, nitroanilines, chloranilines, and any negatively substituted aryl primary amine very readily form insoluble diazoamino compounds during diazotization, and it is necessary to increase the mineral acid concentration and to titrate with nitrite as rapidly as possible at a lower temperature than usual. Iodoaryl compounds (e.g., $o$- and $p$-iodonitrobenzene) are sometimes reduced to iodide, and, in the subsequent nitrite titration, iodine is liberated and immediately reacts with the starch of the indicator paper. Dinitro compounds in which both nitro groups are on the same ring form phenylenediamines which react incompletely with nitrite or are subject to a variety of side reactions. The $o$-diamines react incompletely to form triazoles and other products: $m$-diamines form dyes of the general structure of the Bismarck Browns, and these highly colored products often obscure the starch-iodine end point. The direct titration of $p$-diamines with nitrite is seldom quantitative unless a large excess is added (741); in this instance, the amine can be determined indirectly by using the diazotized amine as a titrant in a coupling with $\beta$-naphthol.

Per'e and Lobunets (544) determined $p$-nitrophenol by reduction with zinc amalgam; the amine was titrated at ice temperature with sodium nitrite solution, and potassium bromide was added before diazotization as an accelerator (697). Results within ±0.6% absolute were obtained with $p$-nitrophenol, and 0.1–0.2% with nitrobenzene (545). Similarly, nitronaphthalene (546) and $o$-, $m$-, and $p$-nitrotoluene were determined with an accuracy of 0.2–0.3% (543). Lobunets determined $m$-nitrobenzenesulfonic acid with an accuracy of 0.3% absolute, but results were only within 3% for nitronaphthalenesulfonic acid (433). The isomers of nitrobenzoic acid, and $p$- and $m$-nitrocinnamic acid were determinable with an accuracy of 0.3–0.5%; $o$-nitrocinnamic acid could not be determined because of the abnormal course of the reduction process (434), and only $m$-nitrobenzaldehyde responds quantitatively (435).

Ruzhentseva and Goryacheva (599) claimed excellent results for the determination of many aromatic nitro derivatives by reduction with a skeletal nickel catalyst of the Raney type, prepared by treatment of

Al–40% Ni alloy with sodium hydroxide (600); the catalyst is stored under water or alcohol.

About 3 grams of moist nickel catalyst paste and 25 ml of 0.5 N alcoholic potassium hydroxides are added to 0.5–1 mmol of nitro compound. The mixture is boiled for 1–2 hr and filtered into 10 ml of concentrated hydrochloric acid. The catalyst is washed with 50 ml of ethanol, and the combined filtrates are evaporated. The residue is then treated with 10 ml of concentrated hydrochloric acid and 3 grams of potassium bromide, diluted to 100 ml, and titrated at ice temperature with 0.1 N sodium nitrite solution, using starch-iodide paper to detect the end point.

Reduction with the Al–40% Ni alloy should be applicable to aliphatic C-nitro compounds, provided that the resulting amines are determined by appropriate methods.

Zinc dust was used to reduce p-nitrophenol after its separation from parathion (an agricultural insecticide); the resulting amine was determined with sodium nitrite (534).

Another variation of the method for determining nitro compounds by titrating the amine formed on reduction has been introduced by Lobunets (436); nitro compounds are reduced with electrolytic cadmium in a Jones reductor and the resulting amines determined by the Day and Taggart (136) adaptation of Koppeschaar's (378) original bromination method. Briefly, the amine is treated in an iodine flask with excess bromide–bromate solution, acidified with hydrochloric acid, and allowed to stand in the dark for a period of time determined by the reactivity of the amine. The resulting mixture is cooled, potassium iodide is added, and the free iodine is titrated with thiosulfate, using starch as indicator (136). Although the procedure is by no means general, results are said to be satisfactory with the amines obtained from m-nitrotoluene, o-nitrobenzoic acid, m-nitrobenzenesulfonic acid, and nitrobenzene (713).

Ordinarily, the action of bromine is one of substitution, but many aryl C-nitroamino compounds have activating substituents, and the nitro group may be displaced, or bromination occurs incompletely at certain sites. In fact, the nitro group in nitroanilines may be partly dislodged; the displacement of the nitro group may be attributed to free bromine or to hypobromous acid, but certain compounds may react with both species:

$$ArNO_2 + Br_2 + H_2O \rightarrow ArBr + HBr + HNO_3$$
$$Br_2 + H_2O \rightarrow HOBr + HBr$$
$$ArNO_2 + HOBr \rightarrow ArBr + HNO_3$$

Johnson et al. (317,318) have modified the Koppeschaar method for use with mixtures of picric acid with nitrophenols and nitroanilines, where the displacement of a nitro group from picric acid would introduce inaccuracies.

Cadmium metal is used as a reductant in a procedure for aromatic $C$-nitro compounds as described by Budesinsky (87); the resulting cadmium is titrated with EDTA:

Five disks of metallic cadmium (8-mm diameter), which have been activated by agitation in 0.25 $N$ HCl for 30 min under an atmosphere of flowing carbon dioxide, are added to about 0.8 mg-equiv. of the sample dissolved in methanol. Five milliliters of 0.7 $N$ hydrochloric acid solution is added to the sample solution, and the mixture is agitated for 1 hr. Then 5 ml of buffer solution ($NH_4Cl$—$NH_4OH$, pH 10) and Eriochrome Black T is added to the mixture, and the solution is titrated with 0.05 $M$ EDTA; a blank is run concurrently. When determining picric acid and $p$-nitrophenol, 0.5 gram of sodium sulfite must be added before the solution is made alkaline with the buffer. Aliphatic nitroso and azo compounds cannot be determined.

### d. HALOGENATION OF NITROPARAFFINS

Primary or secondary nitroparaffins can be chlorinated by sodium hypochlorite, and the excess of reagent determined iodimetrically. The procedure involves replacement of hydrogen atoms alpha to the nitro group by chlorine (274) or, more rapidly, by bromine (332). In the method as proposed by Jones and Riddick (325), the chlorination reagent for primary nitroparaffins is a 0.25 $N$ hypochlorite solution made from sodium hydroxide and a commercial laundry bleach; a stronger solution is used for secondary paraffins (2.0 $N$). Excess hypochlorite is added to the nitroparaffin sample, and the mixture allowed to stand at room temperature for 10 min. Then the solution is cooled in an ice bath and acidified with acetic acid; potassium iodide is added, and after 10 min the liberated iodine is titrated with 0.1 $N$ sodium thiosulfate solution, using starch as an indicator. A standard solution of hypobromite may also be used. Of course, all compounds that can be halogenated in alkaline solution or that in any way consume halogen will interfere. The method is accurate to $\pm 0.1\%$; a suitable procedure is described in Section VII.

Compounds with terminal dinitromethyl groups are readily brominated in basic solution to form the corresponding bromodinitro derivative. Klager (359) found that the bromodinitro derivatives are readily purified and that their halogen contents can be easily determined inasmuch as methanolic solutions of the derivatives liberate iodine quantitatively when treated with potassium iodide:

$$
\begin{array}{c}
NO_2 \\
| \\
R{-}C{-}Br + 2\,KI \rightarrow
\end{array}
\begin{array}{c}
NO_2K \\
\| \\
R{-}C \\
\diagdown \\
NO_2
\end{array}
+ KBr + I_2
$$

The reaction of dibromodinitromethane with potassium iodide yields an alkaline solution which must be neutralized before titration with sodium thiosulfate solution:

$$C(NO_2)_2Br_2 + 4\,KI + H_2O \rightarrow NO_2CH\!\!=\!\!NO_2K + 2\,KBr + KOH + 2\,I_2$$

### e. Acidimetric Procedures

The nitro group of primary and secondary C—NO$_2$ aliphatics can enolize to yield the *aci* form, but the resulting acidic functions are rarely strong enough to be titrated in aqueous media. On the other hand, the nitro compounds themselves can be titrated in nonaqueous media, and their salts can be titrated as bases. Thus a nonaqueous titration procedure with a pH meter (465) as indicator has been used by Feuer and Vincent (190) for the determination of a series of salts of nitro compounds. The salt (e.g., lithium-1-propane-nitronate), is dissolved in glacial acetic acid and titrated with 0.1 $M$ perchloric acid solution.

An interesting approach to the acidimetric determination of primary nitroparaffins has been described by Reynolds and Underwood (579). The primary nitroparaffin is converted to its *aci* form with base and then treated with nitrous acid to form the corresponding nitrolic acid. The resulting nitrolic acid is titrated photometrically (beginning at pH 5) to its basic form by addition of standard sodium hydroxide; changes in absorbance at 420 m$\mu$ are recorded until a plateau is reached.

The unsubstituted aromatic nitro compounds are not acidic; but when several nitro groups or other negative substituents are *meta* to each other, the acidity function is enhanced and direct titration in nonaqueous media is possible. For example, *m*-dinitrobenzene can be titrated in two steps, and trinitro-*m*-cresol can be titrated first as monobasic and then as a tribasic compound. Since the salts of polynitro compounds are markedly colored, it is necessary to determine the end point potentiometrically (73).

Fritz and Lisicki (207) first studied the titration of several nitro compounds as acids in nonaqueous solvents, using both visual and potentiometric end points. The titration of *m*-nitrobenzoic acid and 3,5-dinitrobenzoyl chloride in benzene–methanol solution was performed by the procedures ordinarily used for determination of a carboxylic acid and an acid chloride, that is, with thymol blue as an indicator and sodium methylate as the titrant. The results for nitromethane with butylamine as solvent (which may have caused partial decomposition) were somewhat erratic and several percent low with the visual end point; a potentiometric titration of nitromethane in butylamine (pH meter and calomel–antimony electrodes) indicated that the thymol blue indicator changes slightly before the end point. Other basic solvents such as dimethylformamide, pyridine, and ethylenediamine can be used in place of butylamine; titrants such as tetraalkylammonium hydroxides are also satisfactory. The first color change of thymol blue (red to yellow) can serve to determine stronger acids such as the carboxylic acids, and the change from yellow to blue is used to determine the nitro compounds. Very weak acids such as phenols, polynitroamines, and nitriles interfere with the determination of nitro compounds, as will be indicated below.

The method of Fritz and Lisicki was used successfully by Kaye (345) for the determination of RDX in mixtures containing motor oils or waxes; dimethylformamide served as solvent, sodium methoxide in benzene–methanol as titrant, and azo violet [4-($p$-nitrophenylazo) resorcinol] as indicator.

Because an electron-withdrawing group such as —$NO_2$ increases the acidity of phenols, and the position of the —$NO_2$ group further influences the acidity, that is, $p$-$NO_2 > o$-$NO_2$, Fritz and Yamamura (209) found that several nitrophenols could be titrated as weak acids in acetone, using tetraalkylammonium hydroxide in benzene–methanol as titrant, and that a differentiating titration of mixtures of such weak acids was possible if the titrations were followed potentiometrically. Results of better than 98.5% were obtained for $o$-, $m$-, and $p$-nitrophenol and 2,4-dinitrophenol; also, these nitrophenols could be titrated in the presence of many other phenols.

Potentiometric titrations of nitroaromatic amines as weak acids in pyridine, using triethyl-$n$-butylammonium hydroxide as titrant, were carried out by Fritz et al. (208); although the results were satisfactory, (100 + %), they were not so good as those obtained with the nitrophenols, and the probability that one hydrogen of the amino group may be acidic was noted.

Although 2,4-dinitrophenylhydrazine can be titrated as a base in glacial acetic acid, Sensabaugh et al. (629) also titrated it as a weak acid with tetrabutylammonium hydroxide. Subsequently, it was found that the 2,4-dinitrophenylhydrazones of a number of aldehydes and ketones could be quantitatively titrated with an 0.01 $N$ solution of tetrabutylammonium hydroxide (in contrast to the 0.1 $N$ solution employed for phenols and amines); the titration provides a rapid method for identification of carbonyl derivatives. A table of neutralization equivalents for the 2,4-dinitrophenylhydrazone derivatives of a number of aldehydes and ketones is given by these authors (629). An extension of this technique is the application of a nonaqueous titration procedure to the analysis of nitro derivatives used for the characterization of organic compounds. Smith and Haglund (645) titrated the dinitrophenyl derivatives of esters, ethers, amines, and amides; acetone was found to have more general utility as the solvent than pyridine in potentiometric titrations with tetra-$n$-butylammonium hydroxide.

Nonaqueous titrations of explosives and certain of their ingredients are of sufficient accuracy to be recommended as elegant replacements for cumbersome nitrometer procedures and titanous chloride titrations. Methods have been developed for ammonium and sodium nitrate (differentially titrated as bases in acetic acid–chloroform) (82); ammonium nitrate (206); trinitrotoluene (97); hexogen (345); and mononitrotoluene, dinitrotoluene, and pentaerythritol tetranitrate (604). Nonaqueous determinations with apparatus for automated differential titrations have been employed by Sarson (604) for analysis of finished explosives and the ingredients for their manufacture. Trinitrotoluene, dinitrotoluene, and RDX (hexahydro-1,3,5-trinitro-$s$-triazine, or hexogen) respond as acids in methyl isobutyl ketone. Nitrotoluene can be titrated as an acid in dimethylformamide. If titrations are performed in methyl isobutyl ketone or dimethylformamide, trinitrotoluene, dinitrotoluene, and nitrotoluene can be resolved. It is of interest to recognize that extraction of explosive mixtures with methyl isobutyl ketone separates nitro compounds from inorganic nitrates (which can also be titrated in nonaqueous media).

### f. AQUAMETRIC METHODS

Aromatic C—$NO_2$ groups can be determined by procedures which are based on the measurement of water produced by reaction of the oxygen in the nitro group. For example, water is formed when a nitro compound is condensed with an aromatic amine or when it is reduced under appropriate anhydrous conditions; Kissin and co-workers have used aquametry for this purpose (597,598), but the method has not been applied to a sufficient number of compounds to establish its general utility.

### g. FERROUS SULFATE TITRATION

Ferrous sulfate of itself is not able to reduce nitro groups; however, aliphatic and aromatic N—$NO_2$ groups are decomposed by rather strong mineral acids into nitric acid (see Section III.B.3.d), which then reacts with ferrous sulfate. Mitra and Srinivasan (499) were able to assay tetryl (trinitrophenylmethylnitramine) alone or in admixture with trinitrotoluene by decomposing it in cold, concentrated sulfuric acid and subsequently determining nitric acid by the ferrous sulfate titration devised 30 years earlier by Bowman and Scott (61). The reaction of nitric acid with ferrous sulfate takes place in sulfuric acid media where the water content does not exceed 25% and the temperature is below 60°C:

$$HNO_3 + 2\,FeSO_4 + 2\,H_2SO_4 = HNOSO_4 + Fe_2(SO_4)_3 + 2\,H_2O$$

Nitrite may interfere, mainly because of the formation of nitrate in the strongly acidic medium:

$$3\,HNO_2 = HNO_3 + 2\,NO + H_2O$$

Johnson (315) has modified the Bowman–Scott method for determining nitrates by maintaining very high concentrations of sulfuric acid during titration and employing a special apparatus to minimize air oxidation. Because C—NO$_2$ groups do not form nitric acid upon treatment with sulfuric acid, the ferrous sulfate method appears to be useful for determination of N—NO$_2$ compounds in admixture with aromatic C-nitro compounds. Cottrell et al. (128) found that the visual end point in the determination of nitroguanidine by the sulfuric acid–ferrous sulfate method was difficult to detect; however, they were able to obtain accuracies of the order of $\pm0.2\%$ absolute with a potentiometric end point (tungsten–platinum). It is to be noted that nitrate esters, nitrite esters, and N-nitroso- as well as N-nitroamines react with ferrous sulfate in concentrated sulfuric acid.

## 2. Gravimetric Procedures

An indirect gravimetric procedure for the determination of an aromatic nitro group was first proposed by Vanderzee and Edgell (713,714). The procedure involves reduction of the nitro group by a weighed amount of tin in aqueous hydrochloric acid solution and subsequent weighing of undissolved tin. The determination is conducted in an inert atmosphere (carbon dioxide), and a sufficiently large sample is employed so as to consume 0.6–0.18 gram of tin; a minimum amount of methanol is used as solvent, and the molarity of the hydrochloric acid is carefully adjusted so as to minimize side reactions involving atmospheric oxidation of tin and evolution of hydrogen. The authors claim an accuracy of $\pm0.5\%$ and note that p- and m-iodonitrobenzene and 1,3,5-trinitrobenzene cannot be determined by this method.

The method of Vanderzee and Edgell, with minor modifications, is essentially as follows:

Clean reagent-grade, 30-mesh tin by shaking it for 10 min in a solution consisting of 50 ml of methanol, 50 ml of 0.75 N hydrochloric acid, and about

1 gram of nitrobenzene (or a few milliliters of nitric acid). Decant most of the solution, transfer the tin to a coarse-fritted glass filter, and wash with methanol; dry the metal in an oven at 110°C for a few hours.

Transfer about 10 grams of clean tin to a medium-porosity sintered-glass crucible. Heat the crucible and its contents at 75°C for several hours, allow to cool in a desiccator, and then weigh.

Prepare a reaction apparatus consisting of a 250-ml single-neck flask (standard taper) with a side gas-inlet tube that conducts carbon dioxide well into the liquid, and a reflux condenser that can be connected to the flask. Alternatively, a simple flask can be used and carbon dioxide can be conducted through a long glass tube which passes through the central tube of the condenser and projects nearly to the bottom of the flask.

Place in the flask (without loss) nearly all the weighed tin, and transfer an accurately weighed sample of the nitro compound (0.2–0.4 gram) with the aid of 5–25 ml of reagent-grade methanol; the quantity of solvent should be just enough to dissolve the nitro compound. Connect the condenser; pour 35 ml of 0.75 $N$ hydrochloric acid through the condenser, and follow with enough water to make a total volume of 75 ml in the reaction flask. Boil the contents of the flask for about 1 hr; cool and then transfer the remaining tin to the original weighed fritted-glass crucible by means of water. Wash the tin residue well with water and then with methanol. Dry the crucible and its contents at 75°C; reweigh, and determine the loss of weight. Each gram of dissolved tin corresponds to 0.03934 gram of nitro nitrogen (3 Sn = $NO_2$).

In view of the attractive features of a quantitative procedure that does not require preparation and careful storage of standard solutions of reductants, and the ideal conditions supplied by the tin method, Juvet et al. (335) devised an improved gravimetric method based on the use of copper; the method is not as critical of acid concentration and solvent volume as the tin method, largely because of the position of copper in the electromotive series. The procedure uses ethanol or ether as the solvent, sulfuric acid, and pure copper; the reaction

$$3\,Cu + ArNO_2 + 7\,H^+ \rightarrow 3\,Cu^{2+} + ArNH_3^+ + 2\,H_2O$$

gives results with a precision and accuracy of $\pm 0.5\%$. Aldehydes, ketones, carboxylic acids, alcohols, phenols, amines, and nitriles do not interfere in this determination; however, inorganic oxidants, nitroso compounds, phenylhydrazines, and complexing ions such as the halides will interfere. A laboratory procedure is described in Section VII.

### 3. Spectrometric Methods

Spectrometric methods are widely used for quantitative determinations of $C$-nitro and $N$-nitro compounds. In particular, the infrared methods are pressed into service whenever it is possible to obtain standards of reasonably purity and all other methods of determining the nitro compounds are inapplicable; mixtures of the standards are prepared and their absorption spectra examined for bands that are diagnostic of the desired component (518), but more often it is necessary to plot absorption at a given wave number as a function of the desired component. As noted above, the ir methods are used because other methods are incapable of determining the desired component in the simultaneous presence of analogous nitro compounds. Thus in the next section are cited examples that demonstrate the application of ir methods for analysis of mixtures of closely related components; it is of interest to note that the bands assigned to C—NO$_2$ absorption are not used [see Section V.B.2.a(2)].

The ultraviolet-visible spectrophotometric methods are especially useful for determination of small amounts of nitro compounds, but they are more often used as quality control methods for determining the minor components in a product.

## a. INFRARED PROCEDURES

An infrared quantitative procedure has been developed by Pristera (561) for the determination of up to 10% of $\beta$- and $\gamma$-trinitrotoluene and 2,4-dinitrotoluene in $\alpha$-trinitrotoluene; values obtained for these impurities were of the order of 0.1–0.2% in error, and for the $\alpha$-TNT about 1%. Absorbance measurements of a 25% solution of the sample in dry benzene are made with an 0.1-mm cell at 9.23 $\mu$ (1083 cm$^{-1}$) for $\alpha$-TNT, 9.36 $\mu$ (1068 cm$^{-1}$) for 2,4-DNT, 11.57 $\mu$ (864.3 cm$^{-1}$) for $\gamma$-TNT, and 12.44 $\mu$ (803.9 cm$^{-1}$) for $\beta$-TNT. The value of each component is determined by the method of successive approximations, calculating in the sequence $\alpha$-, $\gamma$-, and $\beta$-TNT and 2,4-DNT. Three approximations are usually sufficient. ($\alpha$-TNT is the 2,4,6-isomer, $\beta$-TNT is the 2,3,4-isomer, and $\gamma$-TNT is 2,4,5-trinitrotoluene.)

In a later publication, Pristera and Halik (562) describe a similar procedure for the analysis of mixtures of $o$-, $m$-, and $p$-mononitrotoluene (NT) and 2,4-dinitrotoluene (DNT); the absorbance of a 20% solution of the sample in dry cyclohexane is measured in an 0.1-mm cell at 8.70 $\mu$ (1149 cm$^{-1}$) for $o$-NT, 12.49 $\mu$ (800.6 cm$^{-1}$) for $m$-NT, 8.49 $\mu$ (1178 cm$^{-1}$) for $p$-NT, and 9.39 $\mu$ (1065 cm$^{-1}$) for 2,4-DNT.

Schwartz and Mark (616a) have developed an infrared spectrophotometric procedure for analysis of up to 8% of hexahydro-1,3,5-trinitro-$s$-triazine (RDX) in admixture with octahydro-1,3,5,7-tetranitro-$s$-tetrazine (HMX). The RDX content of mixtures is extracted by dimethyl sulfoxide, which is saturated with HMX; measurements of the extract are made in the region 787–803 cm$^{-1}$ (12.71–12.45 $\mu$), and it is necessary to use a spectrophotometer that provides ordinate expansion. Grindlay (242a) used HMX-saturated 1,2-dichloroethane to extract RDX and found it possible to perform the analysis with a simple ir spectrophotometer; absorption at 1590 cm$^{-1}$ (6.29 $\mu$) (using HMX-saturated solvent as a reference) provides a measure of the RDX content of a sample (up to 25%). To increase accuracy, it is also suggested that cyclohexanone be added as internal standard and that ratios of the absorption of RDX at 1590 cm$^{-1}$ (6.29 $\mu$) to the carbonyl absorption maximum in the vicinity of 1715 cm$^{-1}$ (5.83 $\mu$) be used to establish calibration plots.

### b. ULTRAVIOLET-VISIBLE SPECTROPHOTOMETRY

#### (1) Nitroparaffins

A short time after the development of processes for the vapor-phase nitration of paraffins (701,702), it became of interest to develop analytical procedures for quantitative determinations of the primary, secondary, and tertiary nitroparaffins which are simultaneously produced. The determination of one nitroparaffin in the presence of another is difficult because of closely related properties; moreover, the color reactions of the primary nitroparaffins and halogenated nitroparaffins of low molecular weight which were known before about 1920 were not specific or sufficiently sensitive for the determination of secondary compounds such as 2-nitropropane and 2-nitrobutane or of tertiary compounds such as 2-nitro-2-methylpropane in the range of concentrations useful for toxicity studies or for the detection of traces in residues after fumigation or fruit sprayings. Most of the colorimetric procedures for determination of aliphatic nitro compounds that were disclosed before about 1952 are listed by Jones and Riddick (322–324,326).

The ultraviolet absorption spectra of a considerable number of polynitroalkanes have been summarized by Noble et al. (520). Characteristically, they show a high-intensity band near 210 nm and a band of lower intensity near 280 nm. In a number of instances, the maximum at 280 nm appears as a shoulder on the higher intensity band. Dinitroparaffins have molar absorptivities that are about twice those of the corresponding mononitro compound. The nitro groups appear to absorb independently; thus there is hardly any distinction between geminal and vicinal positions. In fact, the value of molar absorptivity in hexane at 280 nm may be used to estimate the number of nitro groups present; mononitroalkanes, 25; dinitro-, 60; trinitro-, 100; tetranitro-, 140 (520,642).

### (a) Colors Formed by Reaction with Nitrous Acid

As was noted in Section III.B.3.a, Meyer and Locher (491) made use of the color formed with nitrous acid to distinguish qualitatively between the three types of aliphatic nitro compounds: primary compounds form nitrolic acids, the salts of which are red; secondary compounds form pseudonitroles; and tertiary compounds do not react and therefore form no color. Turba et al. (693) developed the pseudonitrole reaction into a quantitative method, but it is not sensitive enough to detect small amounts of secondary nitroparaffins in the presence of other types.

A more sensitive procedure for primary and secondary nitroparaffins and nitroalcohols has been developed by Altshuller and Cohen (6): aqueous solutions of a sample are treated with potassium nitrite and sulfuric acid and extracted with diethyl ether; the ether extract is treated with sodium hydroxide solution, and the absorbance of the aqueous layer containing alkylnitrolate (red-orange to red-brown) is read at 330 nm. The color system is found to provide sensitivities at least an order of magnitude greater than does the method of Scott and Treon (see below).

### (b) Color Formation with Ferric Chloride

One of the first colorimetric methods for the determination of nitro-ethane, 1-nitropropane, and 1-nitrobutane in air and biological samples was developed by Scott and Treon (619). It depends on the fact that a nitro group attached to a primary or secondary alkane carbon exists in equilibrium with its tautomeric *aci* form and the equilibrium is shifted quantitatively to the right by alkali:

$$R_2CHNO_2 \rightleftharpoons R_2C{=}N\underset{\searrow O}{\overset{\nearrow OH}{}}$$

The *aci*-nitroalkane forms a colored complex with ferric chloride, as is characteristic of enolic compounds [see Section V.B.1.a(2)]. From 0.5 to 15 mg/ml can be determined by the procedure; the sample is dissolved in sodium hydroxide solution, acidified with dilute hydrochloric acid solution, and treated with 10% ferric chloride solution. The pink color which develops is measured at 500 nm; nitroethane can be determined in air within 1%, and in biological tissue within 5%. An extension of this procedure was reported by Jones and Riddick (322); the color is developed in organic solvents instead of water in order to permit detection of compounds that are not soluble in water.

### (c) The Liebermann Reaction

Although the Liebermann reaction is used to detect aliphatic nitro compounds [see Section V.B.1.a(4)], it has not been employed very much in quantitative determinations because it lacks specificity. However, Jones and Riddick (323) developed a sensitive method which is superior to the Liebermann test; notable improvement was realized by substituting resorcinol for phenol, because it was found that most secondary (and some tertiary) $C$-nitroparaffins react with resorcinol in strong sulfuric acid to give a red-blue color suitable for photometry. Nitroparaffins decompose to form nitrous acid; nitrosylsulfuric acid then forms and immediately reacts with resorcinol to yield a colored $p$-nitrosophenolic compound. The absorbance of the colored solution, stable for 24 hr, is measured at 560 nm; the limit of detection is 1 $\mu$g with an accuracy of $\pm 5\%$. Primary $C$-nitroparaffins form fatty acids and salts of hydroxylamine when warmed with sulfuric acid; thus they do not interfere with the determination of other nitroparaffins. The method has been adapted to the determination of halogenated nitroparaffin fumigant residues on grains and to the determination of 2-nitropropane in air.

### (d) Coupling Reactions

Primary nitroparaffins, in the absence of 2-nitro-2-alkyl-1,3-alkenediols, have been determined in the range of 1–50 $\mu$g/ml by Cohen and Altshuller (114); secondary nitroparaffins and 2-nitro-2-alkyl-1-alkanols do not interfere. In accordance with a procedure described by Turba et al. (693), nitroparaffins are coupled with $p$-diazobenzenesulfonic acid in a buffered phosphate solution of pH 4.3; the absorbance of nitromethane is determined at 440 nm, and that of higher nitroparaffins at 395 nm.

### (e) Formation of Nitrous Acid

Conversion of $C$-nitroparaffins to nitrous acid or nitrite ion and subsequent determination of nitrite by colorimetric methods often is a useful quantitative technique. For example, alkaline solutions of hydrogen peroxide convert primary and secondary $C$-nitroparaffins quantitatively to nitrite ion, which can be readily detected by the Griess–Ilosvay reagent (see Section III.B.4.a); (674); however, conditions must be closely controlled. The reagent system was employed by Glover (227) for the assay of trinitromethyl compounds and esters of trinitroethanol and 2,2-dinitroalcohols; reaction with hydrogen peroxide and sodium hydroxide forms 1,1-dinitro anions. Trinitromethyl compounds are reduced quantitatively; esters are saponified and demethy

lolated quantitatively with sample solutions of about 0.02 $M$, and recovery is from 90 to 100%. Solutions of trinitromethyl compounds are
added to hydrogen peroxide–water–methanol solution and then to a
0.1 $N$ sodium hydroxide solution; solutions of trinitromethyl esters in
methanol are added to a solution of hydrogen peroxide and sodium
hydroxide. Absorbance is read over the range 320–400 nm, and quantities are estimated by reference to calibration curves.

On reduction, $C$-dinitroalkyl compounds readily lose one nitro group
as nitrite; the convenience of reduction by zinc amalgam was invoked by
Shinozuka and Stock (634) for the determination of 2,2-dinitropropane
alone or in the presence of the corresponding mononitro compound.
Sample solutions $10^{-4}$ to $10^{-3}$ $M$ in 0.1 $M$ sodium hydroxide were passed
through a Jones reductor containing zinc amalgam, but the reduction
could also be effected by shaking a buffered (0.1 $M$ NaOH) solution of
the sample with a liquid zinc amalgam. Cadmium and lead amalgams
were slower in action (i.e., more than 2 min required), and bismuth
amalgam was ineffective. Nitrites can be removed from acidified solutions as methyl nitrite (192,698). Griess–Ilosvay reagent was added to the
reduced sample solution and absorbance measured at 520 nm. In alkaline
solution, 2-nitropropane is converted to the nonreducible *aci* form and
so will not interfere with the determination of *gem*-dinitroalkanes; on
the other hand, since mononitro compounds may also form nitrite in
dilute alkali, it is expected that these compounds will interfere (56), but
when reduced in strong alkali, 1-nitroalkanes yield no nitrite. Primary
$C$-nitroparaffins couple with Griess–Ilosvay reagent under weakly acidic
conditions, but the colored product scarcely absorbs at wavelengths
greater than about 500 nm.

Gwatkin (247) has made use of the rather specific reaction of primary
and secondary aliphatic nitro compounds to form hydroxylamine on
hydrolysis with sulfuric acid as the basis for a colorimetric procedure.
The hydroxylamine is oxidized by iodine to nitrous acid (55) and then
determined via diazotization and coupling.

A dilute solution of the sample containing from 10 to 70 $\mu$g of nitro
nitrogen/ml is used; 1 ml of the sample is pipetted into a tube with a mark at the
10-ml level. Then 1 ml of sulfanilic acid solution (0.5 gram in 35% acetic acid)
and 1 ml of 6 $N$ sulfuric acid are added, and the tube is warmed at 100°C (water
bath) for 15 min. The contents of the tube are cooled, and 1.8 ml of acetate
buffer is added (35 grams of sodium acetate in 600 ml of water); after thorough
mixing, 0.4 ml of iodine solution is added (1.3 grams iodine in 100 ml glacial
acetic acid). The contents of the tube are mixed well by shaking and allowed to
stand for 4 min; 1 ml of 2% sodium arsenite solution is then added to remove
excess iodine. Finally, 1 ml of naphthylamine solution (0.2 gram $\alpha$-naphthyl-

amine hydrochloride in 100 ml 35% acetic acid) is added, and the volume is brought to 100 ml with water. Twenty minutes is allowed for formation of the magenta dye resulting from the coupling of diazotized sulfanilic acid with $\alpha$-naphthylamine; the intensity of the color is measured in a photometer at 550 nm.

The following primary aliphatic nitro compounds were tested at a concentration of 5 $\mu$g or less: nitroethane, 1-nitropropane, 4-chloro-$\beta$,2-dinitrostyrene, and ethyl $\beta$-carbethoxy-$\delta$-nitrovalerate. The secondary compounds found to respond to the method were diethyl $\gamma$-nitropimelate, 1-phenyl-2-nitropropane, and ethyl $\beta$-hydroxy-$\gamma$-nitrocaproate. Aromatic compounds such as chloramphenicol, nitrobenzaldehyde, p-nitrophenylacetic acid, p-nitrophenylserinol, and dinitrophenol do not react. Similarly, tertiary aliphatic nitro compounds do not react, tested were 2-nitro-2-methyl-1-propanol and tris($\beta$-carboxyethyl)nitromethane.

### (f) Nitroalcohols

None of the methods discussed thus far can be satisfactorily applied to nitroalcohols; however, aliphatic nitroalcohols in the range of 1–100 $\mu$g can be determined with 2% accuracy by a procedure developed by Jones and Riddick (324) in which a nitroalcohol (but not an aminoalcohol) is decomposed to yield formaldehyde by refluxing with alkali. Loss of formaldehyde is prevented by sodium bisulfite. Subsequently, a colored complex with formaldehyde is formed by addition of chromotropic acid and sulfuric acid (154); absorbance is read at 580 nm.

### (g) Determination of Nitromethane

A large number of methods have been developed for the determination of nitromethane, but most of them are not specific. However, a specific determination for nitromethane in the presence of other nitroparaffins was developed by Jones and Riddick (326) and found to provide an accuracy of ±2% and a precision of ±1%. The procedure depends on a reaction of nitromethane disclosed by Turba et al. (693). Nitromethane (1–25% of mixture) is dissolved in a buffered solution (pH 9.5) and condensed with a freshly prepared solution of sodium 1,2-naphthoquinone-4-sulfonate to form a violet complex that is stable for at least 1 hr in amyl alcohol; then isoamyl alcohol (saturated with pH 9.5 buffer) is added, and the phases are separated by centrifugation. A portion of the isoamyl alcohol layer is transferred to a Corex cell, and absorbance of the complex is read at 585 nm. Because large amounts of other primary nitroparaffins inhibit formation of the colored complex, nitromethane is separated from higher nitroparaffins by an azeotropic distillation with methanol, inasmuch as the higher nitroalkanes do not form azeotropes with methanol. The azeotropic separation of nitromethane finds use in other analytical procedures.

### (h) Determination of 3-Nitropropanoic Acid in Natural Products

Matsumoto et al. (476) have developed a spectrophotometric procedure for the determination of 3-nitropropanoic acid, a naturally occurring nitroparaffin. The aforementioned methods for determining aliphatic nitro compounds were found unsuitable for measuring the amount of 3-nitropropanoic acid in plant extracts, and Bose's method (56) (release of nitrite in hot alkali) as applied by Cooke (125) yielded only a fraction of the nitrite expected; moreover, plant pigments give rise to interfering colorations. Creeping indigo leaf (*Indigofera endecaphylla* Jacq.) is digested with 0.1 N HCl; the digest is cooled, adjusted to pH 6, and brought to a known volume. Solids are allowed to settle (or centrifuged); an aliquot of the supernatant is mixed with formaldehyde and 0.05 M borax–NaOH buffer (pH = 9.5) and then is autoclaved at 120°C for at least 45 min to convert the nitro group to nitrite. After dilution and cooling to ice temperature, Griess–Ilosvay reagent and dilute hydrochloric acid are added, followed by 2 M sodium acetate to establish a pH of 2–2.5. The solution is made to volume with 95% ethanol, and the absorbance is measured at 525 nm against a reagent blank. The addition of formaldehyde in this procedure enhances the displacement of the nitro group and prevents the reaction of interfering amino acids (found in natural products) with nitrite ions when the solution is acidified to promote reaction with Griess–Ilosvay reagent. The recovery of from 0.08 to 0.4 mg of 3-nitropropanoic acid added to alfalfa meal was found to be from 94 to 97%.

### (i) Determination of N-Nitro Compounds

Methods for the determination of aliphatic and aromatic N-nitro compounds by colorimetric procedures are essentially methods for the determination of the nitric acid which can be formed by hydrolysis in acidic media or the products formed in alkaline media. Thus Laccetti et al. (393a) adapted a modification of the colorimetric ferrous sulfate method described in Section VI.C.2.d of "Nitrate and Nitrite Esters Groups" principally for determination of nitrite esters, although it was found that the procedure is equally applicable to most N-nitro (nitramine) groups; however, RDX (hexahydro-1,3,5-trinitro-s-triazine) and HMX (octahydro-1,3,5,7-tetranitro-s-tetrazine) did respond quantitatively. In subsequent studies, Semel et al. (627) found that the red color formed by the reaction of nitric oxide generated from the nitric acid split off by the hydrolysis of RDX and HMX in concentrated sulfuric acid–ferrous sulfate solutions obeyed Beer's law, but the color formed by RDX was less intense than would be expected for qualitative release of the —NO$_2$ functional fragment and the color formed by HMX was still less intense. It was suggested that formaldehyde formed in the hydrolysis of RDX and HMX could be an inhibitor for the colorimetric system:

$$(—CH_2N—NO_2)_n + 2n\ H_2O \rightarrow n\ NH_3 + n\ CH_2O + n\ HNO_3$$

For RDX, $n = 3$; for HMX, $n = 4$.

### (2) Aromatic Nitro Compounds

The most common colorimetric method for the determination of C-nitroaromatics involves reduction to an amine, diazotization, and coupling to form a highly colored azo dye; unfortunately, the colored azo dyes formed by these methods seldom can provide differentiation between similar C-nitroaromatics.

Data suitable for quantitative analysis by ultraviolet and visible absorption in ethyl alcohol have been recorded by Schroeder et al. (614) for a variety of nitramines, nitrosamines, and nitro compounds. Data for the nitramines include the usual high explosives; other data include nitrotoluenes, nitrophenol, nitro- and N-nitrosoanilines (also tetryl), nitro derivatives of urea such as nitrocentralites (centralites are 1,3-dialkyl-1,3-diphenylureas), nitrodiphenylamines, and nitrocarbazoles.

The solvent must be specified for definitive results by spectrophotometric methods; this observation is especially significant when substances are eluted from columns and it is necessary to change the eluent to a solvent for which spectral data are on hand, because simple evaporation of one solvent and dissolution of the residue in another is seldom possible owing to the volatility of nitro and nitroso group compounds. It is better to allow the solvent in a given chromatographic zone to evaporate and then to elute the zone with the required solvent. For substances that are volatile or decompose on prolonged contact with the adsorbent, about 100 mg of a high-boiling liquid compound such as diethylene glycol (diluted with solvent) is added to the wet adsorbent before evaporation; when the volatile solvent has evaporated, the sensitive compound will remain dissolved in the high-boiling solvent, correspondingly reduced in vapor pressure. Of course, the high-boiling liquid must not interfere with the spectrophotometric determination of the desired substance (613).

Most troublesome are mixtures of analogous nitro and nitroso compounds such as may be formed by oxidation of nitrogenous substances, especially since their properties are quite similar and ordinary separations are, in consequence, ineffective. A procedure for the spectrophotometric, quantitative estimation of nitro compounds subsequent to column chromatographic separation from analogous nitroso compounds was developed by Edwards and Tate (153) and is described more fully in Section VII.H.

Thin-layer chromatography was used by Kovac and Sohler (385) to isolate $O,O$-dimethyl-$O$-(3-methyl-4-nitrophenyl) thiophosphate residues in fruits and vegetables. The thiophosphate compound can be extracted with petroleum ether and then treated with dyes and acetonitrile, separated from the dyes by TLC on $Al_2O_3$, and oxidatively split by hydrogen peroxide and base to a 3-methyl-4-nitrophenolate; absorbance is then measured at 400 nm.

### (a) Reduction—Diazotization—Coupling

A large number of methods are available for determining aromatic $C$-nitro compounds colorimetrically by reduction—diazotization—coupling sequences. The following procedure, designed to determine nitrobenzene in biological materials, is novel in that it uses formamidine sulfinic acid as reducing agent. The procedure was developed by Koniecki and Linch (376); the nitro compound at a concentration of 5–50 ppm in 5-ml samples (such as urine) is reduced to a primary amine in alkaline solution ($Na_2CO_3$) with formamidinesulfinic acid (thiourea dioxide), diazotized with sodium nitrite reagent, and coupled with Chicago acid (1-amino-8-naphthol-3,4-disulfonic acid). The color that develops is measured at the wavelength appropriate for the particular nitro derivative, generally in the region 520–560 nm. Recovery of known quantities of nitro derivatives added to fresh urine ranged from 50 to 90%; since recovery of 95–100% of corresponding amines added under the same conditions was possible, it was felt that the diazotization and coupling procedures were correct, but *in vivo* hydroxylation promoted by enzymes may be the cause of lower results. Since glassware cleaned in the usual manner leaves much to be desired, nitrite-absorbing contaminants and residual azo dyes should be removed by soaking in 5% trisodium phosphate for several hours and washed away by distilled water. Corex cuvettes should be soaked in 37% nitric acid.

In many instances, the separate steps involved in reduction—diazotization—coupling procedures can be eliminated by use of reducing agents, such as sodium borohydride and lithium aluminum hydride, which convert 2 mol of the $C$-nitroaromatic directly to an azoic coloring matter [see Section V.B.1.a(3)(a)]. Thus a procedure for determining 2,4-dinitrophenylamino groups (DNP) in DNP-acids described by

Ramachandran (565) is based on the intense red color developed when sodium borohydride is added to aqueous solutions of DNP-acids in the presence of sodium bicarbonate. Absorbance is measured at 420 nm; the detectable concentration is 0.01–0.06 $\mu$ mol/ml with 2% precision.

### (b) Colorations in Alkaline Media

The colorations formed by poly-$C$-nitroaromatics in alkaline solutions (i.e., the Janovsky reaction) are used extensively in quantitative procedures (see also Section III.B.2.b).

The fugitive colors formed by reaction of di-($C$-nitro) aromatics in a specially prepared sodium hydroxide solution were utilized by English (169) for the determination of these compounds when present in amounts ranging from 0.05 to 5% in their mononitro counterparts. The caustic reagent was prepared from sodium monoxide instead of sodium hydroxide to ensure generation of a color sufficiently stable for colorimetry; erratic and inaccurate results were obtained when the reagent was prepared from reagent-shelf sodium hydroxide. The method is of limited use, however, because the amounts of sample, reagent, conditions, and so on must be determined empirically.

The colorimetric determination of nitrobenzene, azobenzene, aniline, and benzidine in air contaminated by a process for production of benzidine has been described by Bykhovskaya (95). The air is passed through nitrating acid to form nitro compounds. $p$-Nitroazobenzene formed from azobenzene is reacted with alkaline glucose to give a blue coloration; the dinitrobenzene obtained from nitrobenzene forms a violet color with acetone and sodium hydroxide; benzidine trapped by passing air through water produces a blue color with ferric chloride and sodium hydroxide. Aniline forms a violet color with sodium hypochlorite and phenol.

A study of glycolaldehyde in enzymatic oxidation prompted development of a procedure for glyoxal-bis(2,4-dinitrophenylhydrazone) by Banks et al. (31). The nitro derivative in freshly prepared alkaline acetone absorbs at 600 nm and is distinctly different from the phenylhydrazones of other carbonyl compounds, which absorb at about 440 nm. Concentrations of 5–45 $\mu$g are detectable with a standard error of 0.002.

Liebmann and Woods (407) have shown that dinitrothiophene can be determined colorimetrically in amounts of 5–500 ppm in nitrobenzene by mixing a sample with 0.1 $N$ sodium hydroxide in ethanol and measuring the absorbance at 540 nm. It is possible to determine mononitrothiophene after its nitration with fuming nitric acid to dinitrothiophene; dinitrobenzene does not interfere.

Although strong bases such as potassium hydroxide and sodium hydroxide are usually used in water, alcohol, or other solvent systems, strong organic bases may sometimes provide advantages; thus colorations formed with ethylenediamine before and after addition of ethanol have been used to determine small amounts of polynitroaromatics in urine and serum (168). Smith (647) and Smith and Swank (648) used ethylenediamine as well as other diamines for the analysis of 3,5-dinitro-o-toluamide and related dinitrobenzamides in dimethylformamide. The nitroaromatic complexes of ethylenediamine appear to be quite stable (607). Glover and Kayser (228) have developed quantitative spectrophotometric procedures for compounds such as 1,3,5-trinitrobenzene, 2,4,6-trinitrotoluene, and 1,4,5,8-tetranitronaphthalene; red products are formed with ethylenediamine in dimethyl sulfoxide. In general, the spectra of most of the compounds are too similar to serve for identification; however, chromatographic separations may be used effectively. The procedure is useful for the determination of TNT in mixtures of RDX. Cruse and Mittag (129) made chromatographic separations on magnesium oxide and then determined 1,3,5-trinitrobenzene by the red complex which it forms with sodium methylate in an alcohol-benzene medium, and m-dinitrobenzene by its absorption in the ultraviolet.

As was noted in Section III.B.2.b, certain monoC-nitroaromatic compounds form colorations in alkaline solutions. Porter (559) studied the formation of colors when mono-C-nitroaromatic compounds in dimethylformamide solution are treated with tetraethylammonium hydroxide; the study was designed to establish whether these colorations can be used for quantitative determinations of various compounds. The wavelength of maximum absorption, color, and absorbance at maximum absorbance are recorded for over 65 compounds; similar tabulations are given for colors formed in acetone and sodium hydroxide (Janovsky).

### (c) Adducts

The general absorptivity exhibited by trinitrophenol (TNP), trinitrobenzene (TNB), or trinitrofluorene (TNF) in molecular adducts with naphthalene and its derivatives is such that useful absorbance measurements of their solutions in glacial acetic acid ($10^{-3}$ to $10^{-4}$ M) can be made over the range 360–400 nm (for TNF) and 330–370 nm (for TNB and TNP); Klemm et al. (361) have shown that determination of the trinitroaromatics can be achieved with an error of about 1.5%.

### (d) Miscellaneous Color Reactions

The indophenol reaction has been used for p-nitrophenol in blood and urine by Lawford and Harvey (403). The p-nitrophenol is first extracted with a solvent mixture of hydrochloric acid, amyl alcohol, ethyl ether, and petroleum ether. The extract is made ammoniacal, reduced with zinc dust, and centrifuged to remove solids. Then aqueous o-cresol reagent is added, and absorbance is measured at 620 nm. Average recovery of p-nitrophenol from urine was 92% and from blood 98%; detection levels are of the order of 1–15 $\mu$g.

The method of Auerbach (23) for tetrachloronitrobenzene, an agricultural dust, is based on the formation of an intensely colored nitroquinoid ion on reaction of the compound with acetone in a freshly prepared tetraethylammonium hydroxide solution in acetone; the purplish red color that develops is measured at about 548 nm.

## 4. Polarographic Methods

The polarographic reduction of nitro compounds is often used as a quantitative method for the determination of specific compounds at low levels of concentration. Unfortunately, many other functional groups are also reducible at the dropping mercury electrode, and so it is necessary to isolate the nitro compound before analysis or to make a systematic investigation in order to ensure that no other reducible or interfering substances are present, should it be desired to introduce the sample directly into the suspension medium.

### a. POLAROGRAPHIC BEHAVIOR OF C-MONONITROPARAFFINS

A short time after the first production of nitroparaffins on an industrial scale in this country, DeVries and Ivett (141) studied the polarographic reduction of a series of commercially available mononitroparaffins ranging from nitromethane to the nitrobutanes. A single quantitative step (four electrons) was observed in the polarogram of each compound in a 0.05 $M$ sulfuric acid supporting medium; diffusion currents were found proportional to concentration in the range of 0.0005–0.167 M, but sulfuric acid and sodium sulfate were the only supporting media investigated systematically. It has now been established that quantitative determination of nitroparaffins is best performed in 0.05 $M$ sulfuric acid; however, as can be seen from the half-wave potentials for the series of aliphatic nitro compounds recorded in Table XIV, it is not possible to distinguish the various members because the compounds are reduced at essentially the same potential.

A few years later, the polarographic behavior of some nitroparaffins as a function of pH in buffered media (Britton–Robinson) was studied by Petru (548). In an acidic medium, only one wave was found in the vicinity of $-0.7$ V versus SCE, much as DeVries and Ivett had observed; starting at pH 4.5, a second wave at about $-1.5$ V becomes visible (obscured by the hydrogen wave in strongly acidic media), but it disappears in strongly alkaline media (pH $\geqslant 10$). The variation of the half-wave potential of a given nitroparaffin changes only slightly with pH, and the diffusion currents are readily reproducible in acidic buffers with a step height which decreases as the molecular weight is increased; however, at pH $\geqslant 7$, the diffusion current of a freshly prepared solution decreases on standing to an equilibrium value owing to a slow transformation of the nitro form to the unreducible *aci* form. In nearly every instance, nitroparaffins show no polarographic activity in 0.1 $N$ sodium hydroxide solutions (141) that have stood for some time.

The reduction of mononitroparaffins is irreversible and involves four electrons in the first step to produce a hydroxylamine derivative; the second step involves only two electrons and results in formation of the amino group (or an alkyl aminium ion in acidic supporting solutions):

$$R-NO_2 \xrightarrow{4e} R-NHOH \xrightarrow{2e} R-NH_2$$

Studies of the polarographic reduction of $N$-methylhydroxylamine and hydroxylamine verify the overall mechanism of the behavior of mononitroparaffins; a second reduction wave at about $-1.5$ V is also observed with the hydroxylamines. According to Petru (548), however, the four-electron irreversible reduction step involves formation of a free radical which immediately couples with a hydrogen ion to form $RNH^+$, the cation of the alkyl hydroxylamine; the second step, exactly one-half the height of the first, is a two-electron reduction of the alkyl hydroxylamine cation to the corresponding alkyl ammonium ion.

As already indicated, at pH values greater than about 7 the diffusion current of freshly prepared solutions of nitroparaffins decreases to an equilibrium value because of transformation of the nitro form to the corresponding *aci* form, which is not reducible in alkaline solution. Thus, by introducing an alkyl nitro compound into a buffered solution of a given pH and establishing the decrease of diffusion current with time, the rate of enolization can be determined (16,494).

The polarographic half-wave potential of a nitroparaffin in aqueous solution is influenced by the nature of the suspending medium. For example, Stewart and Bonner (665) found that the diffusion current for each of a series of nitro compounds remains constant regardless of the buffer employed, but that the nitroparaffins are more readily reduced in citric acid–disodium phosphate and acetate buffers than in benzoate or phthalate buffers. Although the concentration of buffer need not be exactly specified for quantitative work, because the effects of minor

## TABLE XIV
### Polarographic Data for Some Aliphatic Nitro and Nitroso Compounds

| Compound | Supporting medium | $E_{1/2}$ vs SCE | Maximum suppressor | Ref. |
|---|---|---|---|---|
| Nitromethane | 0.05 $M$ $H_2SO_4$ | −0.681 | Gelatin | 141 |
| Nitroethane | 0.05 $M$ $H_2SO_4$ | −0.659 | Gelatin | 141 |
| 1-Nitropropane | 0.05 $M$ $H_2SO_4$ | −0.613 | Gelatin | 141 |
| 1-Nitrobutane | 0.05 $M$ $H_2SO_4$ | −0.561 | Gelatin | 141 |
| 2-Nitropropane | 0.05 $M$ $H_2SO_4$ | −0.673 | Gelatin | 141 |
| 2-Nitrobutane | 0.05 $M$ $H_2SO_4$ | −0.641 | Gelatin | 141 |
| Nitroethane | MeOH–benzene (1:1), 0.3 $M$ LiCl | −1.16 | None | 563 |
| 1-Nitropropane | MeOH–benzene (1:1), 0.3 $M$ LiCl | −1.16 | None | 563 |
| 1-Nitrobutane | MeOH–benzene (1:1), 0.3 $M$ LiCl | −1.22 | None | 563 |
| 2-Nitropropane | MeOH benzene (1:1), 0.3 $M$ LiCl | −1.31 | None | 563 |
| 2-Nitrobutane | MeOH–benzene (1:1), 0.3 $M$ LiCl | −1.31 | None | 563 |
| 1,3-Dinitropropane | MeOH–benzene (1:1), 0.3 $M$ LiCl | −1.16 | None | 563 |
| 2,2-Dinitropropane | MeOH–benzene (1:1), 0.3 $M$ LiCl | −0.86 | None | 563 |
| 1,5-Dinitropentane | MeOH–benzene (1:1), 0.3 $M$ LiCl | −1.17 | None | 563 |
| 2,2-Dimethyl-1,3-dinitropropane | MeOH–benzene (1:1), 0.3 $M$ LiCl | −1.22 | None | 563 |

| Compound | Conditions | Value | Indicator/Additive | Ref. |
|---|---|---|---|---|
| Nitromethane | HAc (gl), 1 $M$ NH$_4$Ac | −1.125 | Gelatin | 44 |
| Nitroethane | HAc (gl), 1 $M$ NH$_4$Ac | −1.100 | Gelatin | 44 |
| 1-Nitropropane | HAc (gl), 1 $M$ NH$_4$Ac | −1.100 | Gelatin | 44 |
| 2-Nitropropane | HAc (gl), 1 $M$ NH$_4$Ac | −1.140 | Gelatin | 44 |
| Ethane nitrolic acid, CH$_3$C(NO$_2$)NOH | 0.1 $N$ NaOH | −0.816 | — | 474 |
|  |  | −1.355 | — | 15 |
| Acethydroxamic acid oxime, CH$_3$C(NHOH)NOH | 0.1 $N$ NaOH | −0.468 | — | 474 |
|  |  | −1.268 | — | 15 |
| Acetamidoxime, CH$_3$C(NH$_2$)NOH |  | N.R. | — | 474,15 |
| Dihydroxyguanidine | pH 6 | −0.382 |  | 475 |
| Tetranitromethane | 10% EtOH, pH 12 (first wave) | −0.155 |  | 475 |
| Cyclotrimethyltrinitramine (RDX) | pH ≈ 8, borate–methanol | −0.811 | Acid fuchsine | 328 |
| Nitroguanidine | pH 2.4, Britton and Robinson | −0.73 | Methyl red | 731 |
|  | pH 2.4, Britton and Robinson | −1.13 | Methyl red | 731 |
|  | pH 5.5, Britton and Robinson | −1.26 | Methyl red | 731 |
| Nitrosoguanidine | pH 5.7, Britton and Robinson | −0.90 | Methyl red | 731 |
|  | pH 5.7, Britton and Robinson | −1.20 | Methyl red | 731 |
| N-Nitrosodimethylamine | 0.3 $M$ HCl, pH 0.8 | −0.9 | Gelatin | 170 |
| N-Nitrosodimethylamine | pH 2.4, McIlvaine | −0.953 | Gelatin, Triton X-100 | 732 |
| N-Nitrosodiethylamine | pH 2.4, McIlvaine | −0.918 | Gelatin, Triton X-100 | 732 |
| N-Nitrosodi-n-propyl amine | pH 2.4, McIlvaine | −0.871 | Gelatin, Triton X-100 | 732 |
| N-Nitrosodiisopropyl-amine | pH 2.4, McIlvaine | −0.939 | Gelatin, Triton X-100 | 732 |

variations in suspension medium concentrations resulting from dilution factors will not be readily evident, the nature of the anion used in the supporting electrolyte may be of importance, as indicated above and also by the results of studies by Breiter et al. (69), which suggest that iodide ion influences the kinetics of reduction of nitromethane. Additionally, the cation may have some effect, for it has been reported that aliphatic nitro compounds give more stable waves in the presence of lanthanum salts (493). Some insight as to the effect of lithium ions on polarographic systems may be gained by study of the results of an investigation of the reduction of pyridinium systems as reported by Tsuji and Elving (692).

The effects of pH on the reduction of vic-hydroxynitrobutanes and some of their ethers and esters were studied by Seagers and Elving (620,621); reduction steps in acidic solutions form N-hydroxylalkyl hydroxylamines or N-alkyl hydroxylamines and were found to occur over a wide range of pH values [McIlvaine's buffers were used in studies involving pH 2–8 and boric acid–sodium hydroxide buffers (Sørensen) for values greater than 8]. The vicinal hydroxyl group apparently does not affect the nitro group because these compounds are reduced at half-wave potentials seldom differing by more than 0.1 V at any pH value.

Polarography of nitroparaffins in nonaqueous solvents has also proved adequate for quantitative analysis. Radin and DeVries (563) have found that a 1:1 methanol–benzene solvent which is 0.3 $M$ in lithium chloride is especially useful because small amounts of water (less than 1%) have little effect on half-wave potentials and diffusion currents ($\pm 3\%$); Cellosolve or ethylcellulose may be used as a maximum suppressor. The polarograms do not show waves as well defined and as regular as those usually obtained in aqueous solutions; on the other hand, Bergman and James (44) recommend a molar solution of ammonium acetate in glacial acetic acid because well-defined polarograms are obtained when this suspension medium is used with 0.01% gelatin as a maximum suppressor.

The di-($C$-nitro)paraffins in which the two $NO_2$ groups are not on the same carbon [vic-di-($C$-nitro)paraffins] reduce at about the same potentials as the mononitroparaffins in nonaqueous solvents; in other words, it appears that nitro groups distributed across an alkane chain have no influence on each other (563).

### b. gem-Di-($C$-nitro)paraffins

The polarographic reduction of gem-di-($C$-nitro)paraffins has been investigated in aqueous media by Stock (667) and by Masui and Sayo

(473,474). The reduction processes for 2,2-dinitropropane, $H_3C$—$C(NO_2)_2CH_3$, have been fairly well established by isolation of the products; the compound cannot assume an *aci*-nitro form because it does not have a hydrogen on the carbon carrying the nitro groups. In acidic suspension media, the polarographic reduction appears as two waves, each involving two electrons, and representing a sequence of an electrochemical reduction followed first by a reaction of the products with hydrogen ion, and then by another electrochemical reduction (ECE):

$$R_2C(NO_2)_2 \xrightarrow{2e} (R_2C{=}NO_2)^- + NO_2^- \xrightarrow{2H^+} R_2C(NO_2){\cdot}NO \xrightarrow[2H^+]{2e} R_2C{=}NOH + HNO_2$$

The products are acetoxime and nitrous acid (473); other *gem*-di-(*C*-dinitro)paraffins should yield the corresponding oximes and nitrous acid.

In alkaline suspension medium, a single two-electron wave is observed; the half-wave potential is almost independent of pH. The reduction products are the *aci* forms of 2-nitro-2-nitrosopropane and nitrite ion:

$$(CH_3)_2C(NO_2)_2 + 2e \rightarrow [(CH_3)_2C{=}NO{\cdot}O]^- + NO_2^-$$

The polarographic reductions of 1,1-dinitroethane and dinitromethane also were investigated by Masui and Sayo (474). These compounds have a hydrogen on the carbon bearing the two nitro groups; hence they can assume an *aci*-nitro form. Attempts to elucidate the reduction mechanism included isolation of the products formed by controlled-potential electrolysis of the *gem*-dinitro compounds. The polarograms are best explained by assuming that the following equilibrium forms are rapidly interchangeable:

I                    II                    III

Acidic ⟵——————⟶ Basic

Ionic form III, the *aci*-form, exists in very alkaline solution and exhibits a strong absorption band at 382 m$\mu$ (R=CH$_3$), and form I exists in very acidic media; as is expected, in acidic solutions the band at 382 m$\mu$ is markedly diminished in intensity and the weak absorption of the nitro form appears at 273 m$\mu$. The negative charge on the nitro group in form III causes the reduction potential to be more cathodic; in contrast, the inductive effect of one nitro group on the other in form I does not make the reduction potential as positive as the resonance effect in form II, and thus form II, the H-aci form, is the more easily reducible.

Dinitromethane and 1,1-dinitroethane exhibit three polarographic waves in a very strong acid, two in acidic media, one in neutral, and two in strongly alkaline solutions. Studies of the products of controlled-potential reductions with a mercury pool cathode suggested that in acidic solutions six electrons are taken up by dinitromethane to yield formhydroxamic acid oxime; 1,1-dinitroethane forms acethydroxamic acid oxime; in alkaline solutions, eight electrons are involved in the formation of formamidoxime and acetamidoxime:

Acidic solutions

$$
\left.\begin{array}{c} H_2C(NO_2)_2 \\ \\ \\ CH_3CH(NO_2)_2 \end{array}\right\} \xrightarrow{6e} \left\{\begin{array}{c} HC\diagup^{NHOH}_{\diagdown NOH} \\ \\ CH_3-C\diagup^{NHOH}_{\diagdown NOH} \end{array}\right.
$$

Alkaline solutions

$$
\left.\begin{array}{c} H_2C(NO_2)_2 \\ \\ \\ CH_3CH(NO_2)_2 \end{array}\right\} \xrightarrow{8e} \left\{\begin{array}{c} HC\diagup^{NH_2}_{\diagdown NOH} \\ \\ CH_3-C\diagup^{NH_2}_{\diagdown NOH} \end{array}\right.
$$

Because a number of intermediate products with a variety of half-lives were shown to be involved in the reduction at constant potential, it is obvious that the waves observed in polarograms are not necessarily representative of the actual reduction stages; it is more probable that the waves are the cumulative result of the presence of the intermediates.

The polarographic reduction of *gem*-dinitroparaffins has been investigated in nonaqueous supporting media by Radin and DeVries (563); the solvent systems employed were methanol, methanol–benzene (1:1), methanol-1,4-dioxane (1:1), methanol–glycerol, isobutanol, ethylene glycol, and glycerol. Other nonaqueous supporting media useful for polarographic analysis of *gem*-dinitro compounds are acetonitrile (289), dimethylformamide (289), and *N*-methylacetamide (625).

### c. *gem*-TRI-(*C*-NITRO)PARAFFINS

Masui and Sayo (475) also studied polarographic and controlled-potential reductions of *gem*-tri-(C-nitro)paraffins such as trinitromethane, 1,1,1-trinitroethane, and tetranitromethane. As implied above in the discussion of *gem*-di-(*C*-nitro)paraffins, compounds that can assume an *aci*-nitro form will undergo reduction by a different route from those that do not have a hydrogen atom on the carbon attached to the nitro groups and, therefore, cannot give rise to an *aci*-nitro form. Characteristically, the reduction of compounds with no hydrogen atom on the carbon atom attached to the nitro groups is initiated by fission of the C-N bond to produce nitrite ion and the *aci*-nitro form of a compound with one less nitro group than the starting compound; the nitrite may then enter into the reduction process, especially in acidic media. It appears that fission of the C-N bond takes place more readily (more positive potential) as the number of nitro groups on the carbon is increased, but no simple relationship has been found [see also Elving and Markowitz (161)].

The polarographic reduction of trinitromethane forms dihydroxyguanidine at all pH values, and 12 electrons are required. On the other hand, 1,1,1-trinitroethane undergoes C-N bond fission (first wave, 2 electrons) to form 1,1-dinitroethane and nitrite ion; the second and third waves of 1,1,1-trinitroethane thus correspond to the first and second waves of 1,1-dinitroethane, and the product of reduction in solutions greater than pH 6 accordingly is acetamidoxime, while in acidic solutions it is acethydroxamic acid oxime. The polarographic routes of reduction of tetranitromethane are not sharply defined, owing to reaction with mercury:

$$C(NO_2)_4 + 2 Hg \rightarrow C(NO_2)_2 : NO_2^- + NO_2^- + Hg_2^{2+}$$

As a result, polarograms in neutral and alkaline media are essentially similar to those obtainable from a mixture of trinitromethane and nitrite; in acidic media, nitrite reduction also takes place.

### d. MIXED AROMATIC–ALIPHATIC *C*-NITRO COMPOUNDS

Compounds that are quasi-aromatic in nature usually are reduced at potentials that are more negative than those for the corresponding substituted benzenes. For example, Hartley and Visco (261) found that nitroferrocene is a typical quasi-aromatic nitro compound; it absorbs in the infrared at $1507 \, cm^{-1}$ $(6.64 \, \mu)$, as is typical for aromatic nitro compounds. In 25% aqueous ethanol solutions that are $0.08 \, M$ in total buffer and $0.20 \, M$ in potassium chloride, it shows only one wave (believed to be reduction to aminoferrocene) over the pH ranges where aromatic compounds usually exhibit a change from two waves in acidic media. In freshly prepared solutions it is found that the diffusion current of the initial wave decreases rapidly, while a second, more cathodic wave appears and grows with time. However, solutions prepared and examined in diminished light show no change of diffusion current;

additionally, no maxima are observed in nitroferrocene reductions, as is typical for aromatic compounds. The formation of a free radical (electron spin resonance) has been observed; hence the first reduction step may involve one electron, as is characteristic for reduction of a nitro group.

As can be anticipated, a compound such as 5-aryl-4-nitro-1-cyclohexane behaves like a simple aliphatic nitro compound (233).

### e. N-NITROPARAFFINS

Relatively few aliphatic N-nitro compounds have been examined polarographically; one of the first to be analyzed with the polarograph is cyclotrimethyltrinitramine (RDX). Jones (328) found that RDX in buffered methanolic solution gives at least three polarographic waves between 0 and $-2$ V versus SCE in a supporting medium consisting of a borate buffer of pH 8.0 (Clark and Lubs), made with potassium rather than sodium hydroxide and methanol in place of water; 0.5 $M$ LiCl was the supporting electrolyte, and a methanolic solution of acid fuchsine (C.I. 42685) was used as a maximum suppressor. Although RDX undergoes slow solvolysis in methanol to produce nitrite, and alkaline treatment of RDX causes degradation (also to nitrite) such that all three polarographic waves decrease in height and the first can decline completely to zero, it is of extreme interest (for analytical purposes) that the buffered supporting electrolyte did not lead to perceptible destruction of RDX over a period of several weeks at room temperature. The first wave is the RDX wave, and its height is found to be linear with concentration; the half-wave potential is $-0.811 \pm 0.001$ V versus SCE (328). Quite independently, Lewis (413) found that methanolic solutions buffered with borax and deoxygenated with sulfite are excellent suspension media for the determination of RDX.

Nitrourea is also among the first aliphatic N-nitro compounds to be examined polarographically, but only recently has a satisfactory reduction route for the parent compound and its derivatives been suggested. According to Laviron et al. (402), the products isolated from reductions of nitrourea at controlled potentials (identified by infrared absorption) suggest that, in strongly acidic media (pH $< 2$), the single six-electron step represents formation of semicarbazide:

$$\mathrm{RNHCONHNO_2} \xrightarrow[6\,\mathrm{H^+}]{6\,e} \mathrm{RHNCONHNH_2}$$

In alkaline media, the reaction mechanism leads to ill-defined polarographic waves; controlled-potential reductions with a quiescent mercury pool have not provided results which permit identification of products, but results of studies with other compounds suggest that the N—NO$_2$ bond is cleaved.

In strongly acidic media (HCl), $N$-nitro-$N$-methylurea is reduced after the fashion of its $C$-nitro analog, methylnitrourea; the product is methyl-2-semicarbazide:

$$\text{NH}_2\text{CON(R)NO}_2 \xrightarrow[6\,\text{H}^+]{6\,e} \text{NH}_2\text{CON(R)NH}_2$$

However, above pH 2, $N$-nitro-$N$-methylurea rapidly decomposes into $N$-nitromethylamine; consequently, polarograms at pH values greater than 2 are essentially the same as those of the $N$-nitroalkyl amine.

Nitroguanidine in strongly acidic media (about 2 $M$ HCl) exhibits a well-defined wave with a small second wave (731); with a decrease in acid concentration (to about pH 1), the first half-wave potential becomes more negative and the step height of the second wave increases. On the other hand, in a solution 6 $M$ in HCl, only the first wave is observed. It appears that the total reduction involves six electrons, that nitroso-guanidine is an intermediate, and that aminoguanidine is the final product (402); however, the reaction mechanism is quite complex, as will become evident in the following discussion (see also Section VI.A.5). Two waves are found for nitroguanidine in the acidity range below about pH 8 (731); only one wave is observed in solutions more alkaline than pH 9 and more acid than about 2 $M$. This suggests that the predominant species in solution shifts as a function of pH (731); for example, the amphoteric properties of nitroguanidine can be denoted simply by

Strongly acidic ⟵——————————⟶ Strongly basic

Thus the final polarogram reflects the presence of two or more forms (the above are only some of the resonant and equilibrium forms of nitroguanidine), each with a different half-wave potential and with a fast or slow rate of equilibria. Since the polarograms in strongly acidic or strongly basic media exhibit essentially only one wave, the existence of a preponderance of one of the forms is implied; accordingly, the appearance of another wave at intermediate values of pH suggests (among other possibilities) that at least two forms of the compound are now involved.

Reduction of nitroguanidine in basic media (above pH 8) shows one well-defined wave and a somewhat poorly structured second wave at a more negative potential. According to Laviron et al. (402), the first wave corresponds to the formation of nitrosoguanidine, and the second wave to the reduction of nitrosoguanidine to guanidine itself. The products of controlled-potential reductions of nitroguanidine and nitrosoguanidine are, respectively, nitrosoguanidine and guanidine; hence the two waves in basic media correspond to the following reactions:

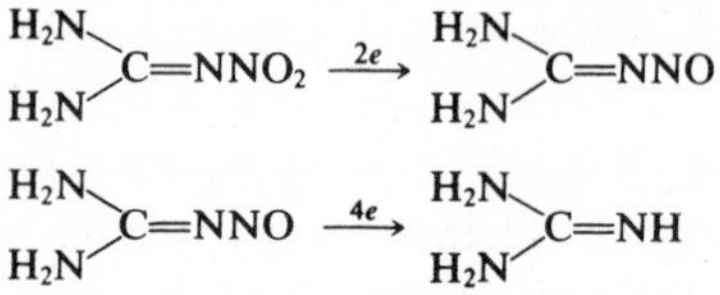

$$\begin{array}{c} H_2N \\ \diagdown \\ \diagup \\ H_2N \end{array} C{=}NNO_2 \xrightarrow{\ 2e\ } \begin{array}{c} H_2N \\ \diagdown \\ \diagup \\ H_2N \end{array} C{=}NNO$$

$$\begin{array}{c} H_2N \\ \diagdown \\ \diagup \\ H_2N \end{array} C{=}NNO \xrightarrow{\ 4e\ } \begin{array}{c} H_2N \\ \diagdown \\ \diagup \\ H_2N \end{array} C{=}NH$$

The results of a systematic study of the $N$-nitro derivatives of a variety of secondary aromatic and aliphatic amines were reported by Laviron and Fournari (400); the following summarizes the general behavior of $N$-nitro compounds:

*Acidic media*

$$RR'N{-}NO_2 \xrightarrow{\ 2e\ } RR'N{-}N{=}O \quad \left.\begin{array}{c} \\ \\ \end{array}\right\} \text{ one wave}$$

$$RR'N{-}N{=}O \xrightarrow{\ 4e\ } [RR'N{-}NH_2]H^+$$

*Alkaline and neutral media*

$$RR'{-}N{-}NO_2 \xrightarrow{\ 2e\ } RR'N{-}NO \quad \text{(first wave)}$$

$$RR'{-}N{-}NO \xrightarrow{\ 4e\ } RR'NNH_2 \quad \text{(second wave)}$$

Although $N$-nitrodiethylamine, $N$-nitropiperidine, and $N$-methyl-phenylnitramine respond as indicated above, $N$-nitropyrazole in either acidic or basic media shows a two-electron wave derived from rupture of the N-N bond and the formation of pyrazole and nitrite ion.

The polarographic reduction of the $N$-nitro compounds derived from the primary aromatic and aliphatic amines $n$-butylamine and aniline was studied by Laviron et al. (401). The salient feature in the polarograms of the $-NH-NO_2$ compounds is that the step height is relatively constant in acidic media and begins to diminish rapidly to zero after a pH of about 5; the diminution is attributed to formation of an ion which is not reducible at the dropping mercury electrode:

$$RNHNO_2 \rightleftharpoons RNNO_2H \underset{k_2}{\overset{k_1}{\rightleftharpoons}} [RNNO_2]^- + H^+$$

The tabulation summarizes data pertinent to the above equilibrium:

| Compound | $pK_a$ | $pK'$ | $k_2$ | $k_1$ |
|----------|--------|-------|-------|-------|
| $n$-Butylnitramine | 6.4 | 8.25 | $4.3 \times 10^9$ | $1.7 \times 10^3$ |
| Phenylnitramine | 4.8 | 7.25 | $2.1 \times 10^9$ | $3.4 \times 10^3$ |

In acidic or basic media, a six-electron reduction forms hydrazine derivatives:

$$R\text{—}NHNO_2 \xrightarrow{6e} RNHNH_2$$

### f. EXAMPLES OF DETERMINATIONS OF $C$-NITROPARAFFINS

The discussion in earlier paragraphs has already indicated that the proximity of the half-wave potentials observed in the polarographic reduction of a homologous series of aliphatic nitro compounds precludes the possibility of determining any one aliphatic mononitro compound in the presence of a similar compound, though exceptions certainly may be found and there are instances where it is possible to resolve mixtures of aliphatic and aromatic nitro compounds in appropriate suspension media or by ac polarographic methods. Thus the polarographic method of analysis is generally used to determine a nitro compound when it occurs alone or in admixture with other polarographically inert types of compounds. A typical polarographic procedure is given in Section VII.F.

One of the first examples of quantitative polarographic methods as applied to the direct determination of aliphatic nitro compounds was reported by Wilson and Hutchinson (736). The method was developed specifically for determination of concentrations of the order of 1–2 mg of nitromethane/liter of air. In brief, a sample of air is shaken with 50 ml of $0.2 N$ sulfuric acid; drawing air through the acidic solution was not found satisfactory. After about 30 min, the current–voltage curve of the sulfuric acid solution is obtained over the range $-0.3$ to $-1.3$ V versus SCE; about 95% of the nitromethane content of air is absorbed by the acidic solution. Nitromethane gives a well-defined quantitative step at a half-wave potential between $-0.7$ and $-0.8$ V versus SCE in $0.2 N$ sulfuric acid. A polarographic procedure for determination of nitroethane in blood and in air has been developed by Scott (618); the supporting electrolyte is a 1% solution of hydrochloric acid.

A rapid polarographic method for the determination of the naturally occurring toxic ingredient 3-nitropropanoic acid in the range of $3.5 \times 10^{-3}$ to $3.5 \times 10^{-5} M$ was developed by Frodyma et al. (210) to supplant the more tedious colorimetric procedure noted in Section VI.A.2.b(1). Plant material containing the nitro compound is digested in $0.1 M$ hydrochloric acid; after centrifugation and filtration, 20 ml of the extract is mixed with 25 ml of an acetate buffer and diluted to 50 ml. The buffer is 0.2 $M$ in acetic acid and 0.2 $M$ in sodium acetate (pH = p$K_a$) and provides a pH of 4.7 in the supporting medium. The polarographic reduction is irreversible, and the observed half-wave potential is somewhat dependent on the concentration of buffer; it occurs at values of $-0.76$ to $-0.80$ V versus SCE.

The determination of nitrofuranes in poultry feeds by dc polarography has been reported by Moore and Guertal (504); the ac polarographic determination of 3-(5-nitrofurfurylideneamino)-2-oxazolidone and 5-nitro-2-furaldehyde semicarbazone in poultry feed has been found to be quite specific in the presence of other nitroaromatic coccidiostats (132).

Substances that are not polarographically active are often rendered active by nitration. Thus aliphatic hydrocarbons can be nitrated and then used in polarographic procedures. However, by-products of the nitration and nitrogen dioxide may interfere in the polarographic determination of the nitroalkanes. An interesting method of preventing such interferences was employed by Ballod et al. (26); it was found that the reaction between nitroparaffins and the aldehydes in alkane nitration products could be arrested by making the solution alkaline and keeping it at 0°C. Additionally, the reaction of nitrous acid (from $NO_2$) with nitroparaffins to form nitrolic acids was avoided by dissolving the alkane nitration products in a neutral phosphate buffer.

### g. SEPARATION OF $C$-NITROPARAFFINS FOR POLAROGRAPHIC ANALYSIS

Chromatographic separations are widely used as a means of removing nitro compounds from materials that interfere with their determination by polarographic methods; however, chromatography also makes possible polarographic analysis of nitro compounds which have nearly the same half-wave potentials. Nearly any form of chromatographic separation can be applied, but the classical column form with alumina or silica as the adsorbent is preferred quite frequently because generous samples can be used and the column sections can be segregated and their extracts analyzed directly by polarography; alternatively, the column can be eluted continuously to provide incremental fractions for analysis. Excellent examples of column adsorption methods for separation of nitramine impurities in hexahydro-1,3,5-trinitro-$s$-triazine (RDX, trimethylenetrinitramine) are provided by Malmberg et al. (457); paper chromatography also may be used for separations, for example, the procedure described by Yasuda and Rogers (742a).

Kemula and Sybilska (354) found that mixtures of isomeric nitropropanes and nitrobutanes can be separated in columns packed with clathrate compounds (nickel thiocyanato–nitrogen base complex, where the base is a mixture of $\beta$- and $\gamma$-picoline and 2,6-lutidine) and eluted with an aqueous solution of potassium thiocyanate.

### h. POLAROGRAPHIC BEHAVIOR OR $C$-MONONITROAROMATIC COMPOUNDS

Polarographic reduction of aromatic $C$-nitro compounds is a nonreversible process, much as it is for the nitroparaffins; there is formed, via a four-electron reaction (one step), an intermediate aryl hydroxylamine which, in acidic media, can be reduced (additional step of two electrons) at more negative potentials to an amine:

$$RNO_2 \xrightarrow{4e} RNHOH \xrightarrow{2e} RNH_2$$

In basic media, the single-step four-electron reduction process involves two distinct one-electron steps in the initial stages, but the reduction does not proceed beyond the aryl hydroxylamine; the overall rate of reduction usually is determined by the second electron-transfer step:

$$R\!-\!NO_2 \underset{}{\rightleftharpoons} [R\!-\!\dot{N}O_2]^- \underset{slow}{\xrightarrow{e}} [R\!-\!NO_2]^{2-} \underset{fast}{\longrightarrow} [\text{reduction products}]$$

In fact, the results of oscillopolarographic studies of nitrobenzene in the presence of camphor suggest that the limiting current is decreased when there is retardation of the penetration reaction (497); this is strengthened by the finding that, in alkaline solutions, inhibitors such as camphor (343) and methylcellulose (342) can stop the reduction almost completely at the radical anion stage, that is, $[R\!-\!\dot{N}O_2]$ (340,341). Additionally, the entire reduction process is quite complicated because the results of ac polarographic studies of nitrobenzene reduction confirm the presence of intermediate processes which cannot be discerned from observations of dc polarography (70); for example, studies of the kinetics of reductions at a hanging drop electrode suggest that $p$-nitroaniline forms $p$-nitrosoaniline and the corresponding imine (347).

Radical anions have also been studied spectrophotometrically (349,351–353); the absorption spectra of the anion radical produced by the electrolytic reduction of nitrobenzene shows five peaks in the region between 390 and 450 nm (340).

Polarographic reductions of $C$-nitroaromatics are best performed in alcohol–water solutions buffered at low pH values or in aqueous solutions buffered at pH 7–10; the usual aqueous buffer solutions are among the most widely used, but often from 10 to 30% of methanol is added to facilitate the dissolution of nitro compounds. Nonaqueous supporting media have been found useful for determining the mechanism of reduction of aromatic compounds, but for quantitative determinations they seldom offer advantages over aqueous media (see Table XV). Although simple aqueous buffer solutions are most frequently used, a wide variety of supporting media have been investigated for analysis of aromatic nitro compounds; for example, an aqueous solution of sodium xylene sulfonate has been used for polarography of o- and p-nitrotoluene, chloronitrobenzenes, and p-dimethylaminobenzaldehyde (108). It is emphasized that the supporting medium often determines the kind of polarographic results that are obtained; thus alkaline solutions with dodecyltrimethylammonium chloride (555) show reduction waves of aromatic compounds that are shifted in reference to those observed in alkaline media containing alkali-metal hydroxides. High-purity quaternary ammonium salts must be used in polarographic procedures; of the many methods available to purify such compounds, recrystallizations from appropriate solvents seem to be the simplest (638). The nature of the supporting electrolyte also has been found to affect the polarographic behavior of *para*- and *meta*-substituted nitrobenzene compounds in nonaqueous solvents such as acetonitrile and dimethylformamide (289).

Supporting media consisting of anhydrous solutions of salts in liquid ammonia have been used for the polarography of nitro and nitroso aromatics. A number of anhydrous salts, such as $LiNO_3$, $NH_4SCN$, $NaSCN$, $NH_4I$, and $LiClO_4$, are liquefied by anhydrous ammonia owing to the formation of ammoniates of the general form $AB(NH_3)_x$; for example, at an ammonia pressure of 1 atm, the ammoniate of ammonium nitrate [Diver's fluid (144)] is a liquid at 25°C, while the ammoniate of ammonium thiocyanate is fluid up to 78°C (at higher temperatures, solid ammoniates deposit). Hubicki and Dabkowska (298) studied the ammoniate of lithium perchlorate; polarographic decomposition of the supporting medium begins at $-1.7$ V versus SCE. Sellers and Leonard (626) reported reductions of nitro- and nitrosoaromatics in the ammoniate of sodium iodide; a silver-silver iodide reference electrode is used, and gelatin was found to be satisfactory as a maximum suppressor. Typical half-wave potentials are as follows: m-nitroaniline, $-0.102$ and $-0.404$; p-nitroaniline, $-0.25$ and $-0.50$; o-nitroaniline, $-0.21$ and $-0.45$; p-nitroacetanilide, $-0.082$ and $-0.448$; p-nitrosoacetanilide, $-0.37$.

The polarographic reduction of C-nitroaromatics in dimethylformamide and acetonitrile shows a one-electron wave corresponding to formation of the anion radical and a second, three-electron wave. For

## TABLE XV
### Polarographic Data for Some Aromatic Nitro and Nitroso Compounds

| Compound | Supporting medium | $E_{1/2}$ vs. SCE | Maximum suppressor | Ref. |
|---|---|---|---|---|
| Nitrobenzene | 63:27:10 Ethanol-pyridine-$H_2O$ | −0.937 | 0.01% Polyvinyl pyrrolidone | 624 |
| m-Nitrotoluene | 63:27:10 Ethanol-pyridine-$H_2O$ | −0.960 | 0.01% Polyvinyl pyrrolidone | 624 |
| p-Nitrotoluene | 63:27:10 Ethanol-pyridine-$H_2O$ | −1.048 | 0.01% Polyvinyl pyrrolidone | 624 |
| o-Nitroaniline | 63:27:10 Ethanol-pyridine-$H_2O$ | −1.078 | 0.01% Polyvinyl pyrrolidone | 624 |
| m-Nitroaniline | 63:27:10 Ethanol-pyridine-$H_2O$ | −0.942 | 0.01% Polyvinyl pyrrolidone | 624 |
| p-Nitroaniline | 63:27:10 Ethanol-pyridine-$H_2O$ | −1.090 | 0.01% Polyvinyl pyrrolidone | 624 |
| m-Nitrobenzaldehyde | 63:27:10 Ethanol-pyridine-$H_2O$ | −0.830 −1.590 | 0.01% Polyvinyl pyrrolidone | 624 |
| p-Nitrobenzaldehyde | 63:27:10 Ethanol-pyridine-$H_2O$ | −0.652 −1.64 | 0.01% Polyvinyl pyrrolidone | 624 |
| p-Dinitrobenzene | 63:27:10 Ethanol-pyridine-$H_2O$ | −0.480 −1.58 | 0.01% Polyvinyl pyrrolidone | 624 |
| m-Nitroanisole | 63:27:10 Ethanol-pyridine-$H_2O$ | −0.899 | 0.01% Polyvinyl pyrrolidone | 624 |
| p-Nitroanisole | 63:27:10 Ethanol-pyridine-$H_2O$ | −1.035 | 0.01% Polyvinyl pyrrolidone | 624 |
| p-Nitrophenylhydrazine | 63:27:10 Ethanol-pyridine-$H_2O$ | −1.094 | 0.01% Polyvinyl pyrrolidone | 624 |
| m-Nitrodimethylaniline | 63:27:10 Ethanol-pyridine-$H_2O$ | −0.937 | 0.01% Polyvinyl pyrrolidone | 624 |
| p-Nitrodimethylaniline | 63:27:10 Ethanol-pyridine-$H_2O$ | −1.085 | 0.01% Polyvinyl pyrrolidone | 624 |

## TABLE XV (Contd.)
### Polarographic Data for Some Aromatic Nitro and Nitroso Compounds

| Compound | Supporting medium | $E_{1/2}$ vs. SCE | Maximum suppressor | Ref. |
|---|---|---|---|---|
| p-Nitrobiphenyl | 63:27:10 Ethanol-pyridine-H₂O | −0.891 | 0.01% Polyvinyl pyrrolidone | 624 |
| o-Nitrophenol | pH 3, McIlvaine, 10% EtOH | −0.305 −0.770 | Gelatin | 537 |
| m-Nitrophenol | pH 3, McIlvaine, 10% EtOH | −0.310 −0.885 | Gelatin | 537 |
| p-Nitrophenol | pH 3, McIlvaine, 10% EtOH | −0.405 −0.885 | Gelatin | 537 |
| o-Nitroanisole | pH 12, NaOH + Na₃PO₄ | −0.795 | Gelatin | 537 |
| m-Nitroanisole | pH 12, NaOH + Na₃PO₄ | −0.755 | Gelatin | 537 |
| p-Nitroanisole | pH 12, NaOH + Na₃PO₄ | −0.855 | Gelatin | 537 |
| o-Nitrotoluene | pH 5, KCl-HCl, 8% EtOH | −0.05 −0.20 | | 540 |
| m-Nitrotoluene | pH 5, KCl-HCl, 8% EtOH | −0.07 −0.16 | | 540 |
| p-Nitrotoluene | pH 5, KCl-HCl, 8% EtOH | −0.05 −0.22 | | 540 |
| 2,4-Dinitrotoluene | pH 5, KCl-HCl, 8% EtOH | −0.09 −0.18 | | 540 |
| 2,6-Dinitrotoluene | pH 5, KCl-HCl, 8% EtOH | −0.14 −0.23 | | 540 |

| Compound | Conditions | Value | |
|---|---|---|---|
| 2,3-Dinitrotoluene | pH 5, KCl-HCl, 8% EtOH | −0.07<br>−0.26 | 540 |
| 3,4-Dinitrotoluene | pH 5, KCl-HCl, 8% EtOH | −0.07<br>−0.25 | 540 |
| 2,4,6-Trinitrotoluene | pH 5, KCl-HCl, 8% EtOH | −0.05<br>−0.12<br>−0.18 | 540 |
| o-Nitrodiphenyl | pH 3.16, 0.2 $M$ NaAc-HCl, 60% EtOH | −0.450 | 219 |
| m-Nitrodiphenyl | pH 3.16, 0.2 $M$ NaAc-HCl, 60% EtOH | −0.386 | 219 |
| p-Nitrodiphenyl | pH 3.16, 0.2 $M$ NaAc-HCl, 60% EtOH | −0.380 | 219 |
| Nitrosobenzene | pH 4.0, McIlvaine | +0.04 | 649 |
| Nitrosobenzene | 1:1 Acetone, pH 7, $PO_4$ buffer | −0.20 | 295 |
| 4-Chloronitrosobenzene | 1:1 Acetone, pH 7, $PO_4$ buffer | −0.19 | 295 |
| 4-Methylnitrosobenzene | 1:1 Acetone, pH 7, $PO_4$ buffer | −0.24 | 295 |
| 2,6-Dichloronitroso-benzene | 1:1 Acetone, pH 7, $PO_4$ buffer | −0.08 | 295 |
| 2,4,6-Trichloronitroso-benzene | 1:1 Acetone, pH 7, $PO_4$ buffer | −0.07 | 295 |
| 2,4,6-Trimethylnitroso-benzene | 1:1 Acetone, pH 7, $PO_4$ buffer | −0.21 | 295 |

205

the reduction of nitrobenzoic acid and its ester in dimethylformamide or acetonitrile, the three waves that are observed are suggested to be the result of reductions of protonic hydrogen, formation of carboxylatonitronate dianions, and reduction to hydroxylaminobenzoate (14). Pyridine and diethylamine catalyze proton transition between acids and nitro compounds (453).

A four-electron step in the initial reduction of aromatic $C$-nitro compounds is often observed in essentially nonaqueous supporting electrolyte systems such as dimethylformamide containing sodium nitrate or tetraalkylammonium halide (294). On the other hand, in solutions where the dimethylformamide/water ratios are high, the reduction wave is often split; the first wave represents formation of radical ions, and the wave at more negative potentials represents reduction to a hydroxylamine (348). For example, $o$- and $m$-trifluoromethylnitrobenzene in dimethylformamide were shown to be reduced in two steps; the first step indicates a one-electron diffusion-controlled process which produces a stable anion radical. The second step involves a sequence of an electron-transfer reaction followed by a homogeneous chemical reaction that in turn is followed by another electron-transfer reaction (ECE); the combination of reactions produces a trifluoromethylnitrosobenzene anion radical (242,454). The formation of an anion radical by reduction of nitrobenzene in acetonitrile has been verified by electron spin resonance (222).

Extensive studies have indicated that substituents on a benzene ring can make the reduction of an aromatic nitro group more difficult; for example, substituents shift half-wave potentials to more negative values in the following order: $NO_2 < COR < COOH < OH < Cl < R$ (622). Electron-withdrawing groups, especially when *meta* to the nitro group, make the compound more readily reducible; electron-donating groups, especially when *ortho* or *para*, make reduction more difficult. *Ortho* groups which interfere with the normal resonance of the nitro group (usually by steric effects) make the compound reduce at half-wave potentials comparable to those for the aliphatic compounds. As can be anticipated from an equation expressing the reduction of the nitro group, the half-wave potential shifts with changes in pH; in most instances, the diffusion current is also affected by pH. Also, there appears to be a correlation of the observed half-wave potentials for reduction of aromatic nitro compounds in acetonitrile with molecular orbital energies (253,581). A plot of the longest wavelength bands in the absorption spectra of substituted nitrobenzenes versus the half-wave potentials is linear; the structural implications of this finding are discussed by Holleck and Schindler (291). Correlations of polarographic half-wave potentials with nuclear magnetic resonance chemical shifts have been reported by Bennett and Elving (43).

Page et al. (537) studied the polarographic reduction of a variety of substituted nitrobenzenes over a wide range of pH values; it was found that $m$- and $p$-nitrophenols produce two steps at low pH values. The total height of the two steps below pH 4 corresponds to a six-electron reduction, much as is the case for nitrobenzene. In alkaline solution, the reduction of $o$- and $p$-nitrophenols involves *six* electrons, but only the *para* compound exhibits a double step, which increases in height progressively as pH increases. In contrast, the reduction of $m$-nitrophenol in alkaline solution occurs in a single four-electron step (much as for nitrobenzene), but above pH 12 the single step is split into two. In the neutral region, $o$-nitrophenol differs from $p$-nitrophenol in that a four-electron reduction is obtained at a pH of 6, but in more acidic or more alkaline media the reduction becomes a six-electron step. Apparently the formation of a quinoneimine is catalyzed by acids or bases (666). The half-wave potentials for the $o$- and $m$-nitrophenols are very close (except at high pH values) and are less negative than those for the *para* isomer. The reduction of $o$- and $p$-nitrophenols has been studied by a multidrop mercury cathode (717).

Polarographic reductions of nitro and nitroso compounds often are sequential reactions consisting of an electrochemical reduction followed by a chemical reaction followed by an electrochemical reduction (ECE). Many polarographic ECE systems can be represented by the reaction sequence

$$A \underset{}{\overset{n_1 e}{\rightleftharpoons}} B \underset{k_2}{\overset{k_1}{\rightleftharpoons}} C \overset{n_2 e}{\rightleftharpoons} D$$

but since chemical equilibrium is usually fast, the sequence of reactions can be detected only in instances where chemical equilibrium is slow, as for $o$-nitrophenol and $p$-nitrosophenol (3). Interpretation of the reaction sequence and evaluation of the rate constants for the chemical reaction equilibrium occurring between the two electroreduction steps can often be made from chronopotentiometric potential–time curves (174,577,683,684). However, when the rate of a homogeneous chemical reaction is slow, the polarographic limiting current sometimes can be used to evaluate the reaction rate constant. Thus Nicholson et al. (516) developed mathematical relationships correlating the polarographic limiting current and the reaction rate constant; the results of a numerical solution of the integral equation applied to the polarographic reduction of $p$-nitrosophenol showed clearly that the intermediate chemical reaction involves acid–base catalyzed dehydration of $p$-hydroxyl-aminophenol:

$$\text{(p-nitrophenol)} \xrightarrow[\text{H}^+]{2e} \text{(p-hydroxylaminophenol)} \xrightarrow[-\text{H}_2\text{O}]{k} \text{(quinone imine)} \xrightarrow[2\text{H}^+]{2e} \text{(p-aminophenol)}$$

The rate constant for the dehydration is a function of pH; as is characteristic of an acid–base catalyzed reaction, it is large at both extremes of pH and only slightly influenced by pH at some central value such as pH 4 (for $p$-hydroxylaminophenol). The presence of gelatin (and other substances) has an effect on the rate constant. Cyclic chronopotentiometry is especially useful for measuring rate constants of chemical reactions at electrodes (32,276–279).

In summary, the first step of a polarographic reduction in protic and aprotic solvents is the reversible transfer of an electron to the aromatic $C$-nitro compound to form a free radical ion. In a protic solvent, the ion is neutralized with subsequent disproportionation to yield the hydrated nitroso compound $RN(OH)_2$. Very rapidly thereafter, there is a dehydration and the aryl hydroxylamine is formed by reduction. The dehydration of $RN(OH)_2$ to RNO may be quite slow. In aprotic solvents, the polarographic wave corresponding to the first reduction step often is distinctly separated from succeeding waves; in protic solvents, separation is seldom evident.

Proton donors in suspension media cause reduction of $o$-chloronitrobenzene and many other nitro compounds to shift to more positive values (312); an explanation of the effect of proton donors such as sulfonic acid and carboxylic acids on the reduction of nitro compounds in nonaqueous solvents has been offered (96).

The possibility of reduction of other functional groups must always be given consideration in procedures designed to determine the nitro group content. A typical example is the complex behavior of the nitrobenzaldehydes. For example, $p$-nitrobenzaldehyde in acidic solution shows three steps in polarographic reduction, the first two corresponding to reduction of the nitro group, and the third to reduction of the carbonyl group. At high pH values, a single step is observed for complete reduction of the carbonyl group; the step has a height that is about one half of its usual value in 0.01 $M$ sodium hydroxide and is connected with the growth of another step at a more negative potential, which can be seen best in a calcium hydroxide–calcium chloride buffer solution (18,137). The half-wave potentials for the three steps are more positive in this buffer solution than in the usual buffer solutions of the same pH value. The effect of calcium ions (554) on the polarographic behavior of nitrobenzaldehyde is not peculiar to this compound; as a rule, calcium ions shift the reduction waves of many aromatic compounds and appear to have, in general, a marked effect on other reduction processes [see Section V.B.1.a(1)] (740). The effects of salts on the half-wave potentials of nitro compounds in acetonitrile and dimethylformamide have been studied in detail (289).

Other unexpected reactions may take place; for example, Kitagawa et al. (358) have shown (hanging drop electrode—epr) that, in the polarographic reduction of iodonitrobenzene and bromonitrobenzene, the first polarographic wave corresponds to the removal of iodine to form nitrobenzene, the second to the formation of the nitrobenzene radical anion, and the third to further reduction of the nitrobenzene radical. Other studies of iodoaromatics and related compounds were made by Gergely and Iredale (219,220).

Detailed discussions and reviews of procedures for reduction of a number of aromatic nitro compounds are given by Kolthoff and Lingane (374) and by Milner (496); a complete bibliography of polarographic literature from 1922 to 1967 is published by Sargent-Welch Scientific Company (*Bibliography of Polarographic Literature*, Cat. No. S-29368). A listing of more recent studies of nitro compounds appears in *Analytical Reviews*, published every two years by the ACS periodical *Analytical Chemistry*. Additional details can be found in tables compiled by Semerano and co-workers of the Padua Center of Polarography (628).

### i. POLY-*C*-NITROAROMATICS

The polarographic method is also applicable to the determination of polynitroaromatic compounds such as picric acid; in alkaline solution (pH 11.7) three distinct steps are exhibited, but at pH values below 4 the three steps coalesce into one (513). Lingane (425) found that a $0.4 \times 10^{-3} M$ solution in $0.1 M$ hydrochloric acid gave a wave for which $n = 17.1$, suggesting that bis(3,5-diamino-4-hydroxylphenyl)hydrazine is the product:

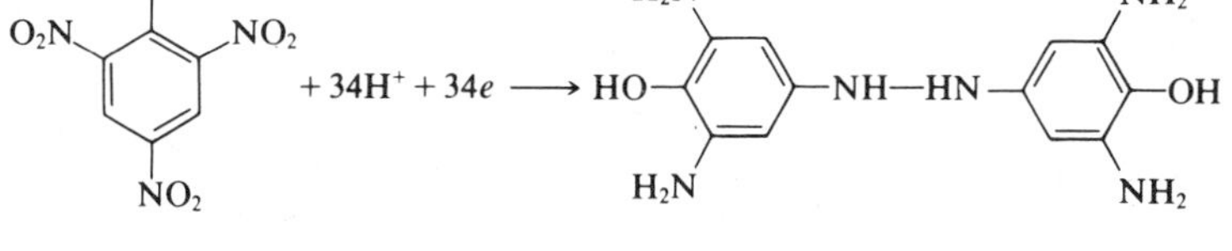

A study of the reduction at controlled potential ($-0.40$ V vs. SCE) in hydrochloric acid indicated that at concentrations below about $0.2 \times$

$10^{-3} M$ the product is indeed the expected 2,4,6-triaminophenol ($n = 18$), but at higher concentrations (e.g., $0.4 \times 10^{-3} M$) the value of $n$ is about 17; at still higher concentrations, $n$ is less than 17 (486). Clearly, the products of partial reduction are involved in side reactions with each other or with picric acid itself.

The effect of substituents on a variety of polynitroaromatic compounds has also been studied by Perret and Holleck (547), and a detailed investigation of the polarographic behavior of simple mono-, di-, and trinitro compounds (benzenes, toluenes, phenols, and resorcinols) has been made by Pearson (540–542) and by Page et al. (537).

### j.  *N*-Nitroaromatics

Studies of the polarographic reduction of aromatic nitramines are not numerous; recently, Laviron and Fournari (400) made a systematic study of some nitramines derived from secondary amines. Four compounds were studied, among which was *N*-methylphenylnitramine; the polarographic reaction in acidic medium proceeds by a two-electron reduction involving formation of the *N*-nitroso compound:

$$C_6H_5(CH_3)N\text{—}NO_2 \xrightarrow{2e} C_6H_5(CH_3)N\text{—}NO$$

which is followed immediately by a four-electron reduction to an asymmetric hydrazine:

$$C_6H_5(CH_3)N\text{—}NO \xrightarrow{4e} C_6H_5(CH_3)N\text{—}NH_3^+$$

The two steps are indistinguishable; consequently, only one wave is observable. Two waves are observed in neutral or alkaline media. The first wave involves two electrons and represents reduction to the nitroso derivative; the second wave, involving four electrons, is indicative of the reduction of the nitroso compound to the amine:

$$C_6H_5(CH_3)NNO_2 \xrightarrow[2\,H^+]{2e} C_6H_5(CH_3)NNO,$$

$$C_6H_5(CH_3)NNO \xrightarrow[3\,H_2O]{4e} C_6H_5(CH_3)NH + N_2O + 4\,OH^-$$

No protons are involved in the second reaction; thus the half-wave potential should be nearly independent of pH, and this is indeed observed.

The polarographic reduction of nitramines derived from primary aromatic amines follows the same routes as that for the aliphatic nitramines (see Section VI.A.4.e).

### k. Examples of Determinations of Nitroaromatics

It is best to isolate nitro compounds before attempting their determination by polarography, in order to eliminate interfering electroreducible substances or nonreducible materials which may interfere by increasing or depressing diffusion currents (126). However, polarographic procedures for quantitative determination of specific nitroaromatic compounds can often be applied directly to samples; for example, the direct quantitative determination of parathion (O,O-diethyl-O-p-nitrophenyl thiophosphate) has been reported by Bowen and Edwards (60), as well as others. A half-wave potential of $-0.39$ V versus SCE is obtained in the suspension medium, resulting from application of the following procedure:

About 1 gram of a dust formulation containing parathion is suspended in sufficient acetone to yield a solution that contains about 1 mg of parathion/ml; after 1 hr, during which the solution is occasionally shaken, the mixture is centrifuged. An aliquot of the centrifugate containing about 10 mg of parathion is removed and mixed with enough acetone to bring the volume to 50 ml; then there are added a solution consisting of 0.6 gram of acetic acid in 25 ml of water, and enough solution of gelatin to make its final concentration 0.01%. Finally, the mixture is brought to a total of 100 ml, and a portion is placed in an H-type cell, deaerated, and electrolyzed. A saturated calomel electrode is used as reference. The most probable contaminant of parathion, p-nitrophenol, does not interfere since it is reduced at a more negative potential ($-0.68$ V). An accuracy of $\pm 1\%$ is claimed, and the detection limit appears to be about 20 $\mu$g/ml of the solution placed in the cell.

Polarographic methods have been found quite useful for the direct assay of the naturally occurring nitro compounds. An assay procedure for chloramphenicol (Chloromycetin, d-threo-1-p-nitrophenyl-2-dichloroacetamido-1,3-propanediol) was developed by Hess (282) for use on samples providing as little as 30 $\mu$g/ml in the polarographic cell.

Another example of direct polarographic analysis of nitroaromatics is the determination of trinitrotoluene in the exudate of warheads (730); the exudates are dissolved in a suspension medium consisting of 1 part acetone with 3 parts of 0.1 $M$ LiCl. The TNT is determined at $-0.66$ V versus pool. Lewis (413) used an acetone suspension containing borax and sodium sulfite for the same purpose and for the determination of cyclotrimethylenetrinitramine (RDX, $-0.77$ V vs. SCE). Kolthoff and Bovey (370) used an alcoholic solution of ammonium chloride and ammonia for determination of m-dinitrobenzene. The determination of m-dinitrobenzene in the presence of 1,3,5-trinitrobenzene can be performed in 2 $M$ ammonium chloride adjusted to pH 12 (130). The analysis of dinitrotoluene, trinitrotoluene, and other propellants by polarographic methods has been described by Frey (202).

The direct determination of small amounts of nitrobenzene in aniline may be performed in a medium prepared by mixing 4 vols. of aniline with 1 vol. of hydrochloric acid (526), especially when nigrosine is used as maximum suppressor (262); $m$-dinitrobenzene (677) and nitrobenzene can be determined in the presence of dinitrobenzene and benzidine in a similar suspension medium (383). As expected, the determination of $\alpha$-nitronaphthalene in $\alpha$-naphthylamine can also be accomplished in hydrochloric acid medium (with methanol) (709).

Nitrobenzene can be determined in blood by polarographic methods (681); Roubal et al. (592) found that a suspension medium of 50% aqueous ethanol containing sodium sulfite, potassium carbonate, Metol, and bromthymol blue is suitable for determination of $m$-dinitrobenzene in blood. Chloramphenicol in blood and urine can be conveniently determined by polarographic methods (364). The determination of mono- and dinitroxylenes in alcoholic solutions of sodium acetate has been described by Hale (248).

When quantitative determination of nitroaromatic compounds cannot be performed directly on a sample because of interfering substances, separations are often accomplished by extraction or by adsorption on chromatographically active media. For example, small amounts of $m$-dinitrobenzene in crude nitrobenzene can be separated by chromatographic adsorption on a column packed with alumina; the eluent is a mixture of benzene and light petroleum. Adsorbed dinitrobenzene is extracted from dried alumina by ethanol, and an aliquot of the extract is added to a suspension medium containing 1 part of $M$ NaOH to 5 parts of $2 M$ NH$_4$Cl (419). Examples of the types of column chromatographic separations which may be used for polarographic determinations of nitro- and nitrosoaromatic compounds are afforded by the work of Schroeder et al. on derivatives of diphenylamine (615), and that of Landram et al. (396) with the dry-column technique of Loev and Goodman (439) for separation of nitrodiphenylamine from double-base propellant mixtures. The following are other examples of compounds that have been separated by column chromatography before polarographic analysis: dinitrobenzene and trinitrobenzene (129); polynitrostilbenes (680); monosubstituted derivatives of $o$- and $p$-nitroanilines (397); isomeric nitrophenols (88); 2,4-dinitrophenylhydrazones (63,443,557,585,589,670); and miscellaneous nitro- and nitrosoaromatics (152,153). Paper chromatography can also be used; for example, substituted trinitrobenzenes can be separated with the techniques described by Colman (119,120).

The determination of polarographically inactive compounds may be accomplished by nitration; for example, humic acids in soil were determined by Lindbeck and Young (419) after nitration, and toluene in blood and urine was determined in this way by Srbova (658).

It is of great interest to note that nitrate and nitrite ions react reproducibly with 2,6-xylenol to produce 4-nitro- and 4-nitrosoxylenol when the components are placed in a mixture consisting of 5 parts of sulfuric acid, 4 parts of water, and 1 part of acetic acid by volume (257); the products of reaction permit polarographic determination of the inorganic ions (258,259). The diffusion current of the nitroso derivative of xylenol is measured at $-0.15$ V versus a mercury–mercurous sulfate reference electrode prepared with the supporting electrolyte; the half-wave potential of the nitro compound occurs at $-0.32$ V.

Another route for rendering substances polarographically active involves attachment of electroreducible entities; for example, humic compounds can be determined after reaction with 2,4-dinitrophenyl-hydrazine (612), and antibodies can be determined similarly (746).

### 5. Coulometric Methods

A large variety of nitro and nitroso compounds may be determined by constant-potential coulometry with a mercury pool cathode, particularly since electrolytic reductions appear to be quantitative and more generally applicable than chemical reductions. However, because of instrumental limitations, the accuracy of coulometric analyses for nitro and nitroso compounds seldom is better than $\pm 1\%$ (standard deviation) at levels of concentration of $5 \times 10^{-4}$ $M$ or greater; concentrations as low as 20 ppm can be determined (390) with considerably less accuracy. Both nonaqueous solvents and mixtures of solvents and water are commonly used in coulometric procedures; conductivity is provided by salts such as lithium chloride or a tetraalkyl ammonium halide (156,390). Lithium chloride (0.1 $M$) in methanol is the most commonly used nonaqueous electrolyte for nitro and nitroso compounds, but the more expensive tetraalkyl ammonium halides are equally effective. Methanol, especially a 3 or 4:1 methanol–water mixture that is 0.1 $M$ in lithium chloride, has been found satisfactory for reduction of most nitro and nitroso compounds (672).

Generally, a silver–silver halide electrode is used as an anode; in some instances, the electrode is placed in a separate compartment and communication with the main body of the supporting electrolyte is provided by a glass frit (156,487) in one arm of some form of H-type cell (428,672). A variety of coulometers may be used, for example, copper

(155) or the classical silver coulometer, a coulometer based on the evolution of hydrogen and oxygen (421,422,425), or the titration coulometer described by Lingane and Small (427), that is, titration of hydroxide formed by the reaction

$$2\,Ag + 2\,Br^- + 2\,H_2O = 2\,AgBr + H_2 + 2\,OH^-$$

Colorimetric coulometers useful from 0.01 to 1.00 C, or even up to 10 C, have been described by Franklin and Roth (200); a current integrator has been used successfully by Meites (483). Highly stable electronic current integrators may also be used, especially when an electrolysis can be performed in less than 15 min.

Potentiostats that are used for coulometric analyses of nitroso and nitro compounds may be complex devices such as the apparatus described by Greenough et al. (240) and by Kaufman et al. (344), or simple manual instruments such as those described by Lingane (423) and by Kruse (390). Automatic circuitry must be properly designed in order to prevent severe oscillations of applied potentials because of time lags in the responses of the electrolytic elements (156); typical circuits for potentiostats which make use of solid-state operational amplifiers are given by Brown et al. (75,76) and by Bezman and McKinney (50). A completely transistorized coulometer capable of delivering 1 A at low voltages and small currents at 50 V has been described by Wood (742).

The major problem encountered in quantitative determinations of small amounts of nitro and nitroso (and other organic) compounds by constant-potential coulometry is the necessity for subtracting background corrections, which often are as large as the current consumed by the sample. The background current (i.e., the current observed at the desired potential when no sample is present) is very much greater with organic solvents than with aqueous suspension media. As can be anticipated, if the background correction is a significant portion of the total number of coulombs when small quantities of nitro or nitroso compounds are determined, even a small error in the correction leads to a significant error in the analysis of the sample. One of the most effective ways of reducing the background current is to electrolyze the supporting electrolyte before addition of the sample until a low, constant residual current is obtained; the final steady-state background current represents the lowest current to be achieved after complete reduction of a nitro or nitroso compound, although, as a rule, the steady-state current actually observed after complete reduction is somewhat higher. In these instances, it appears more appropriate to use the average background current (390). A discussion of background corrections and appropriate ways of application is presented by Meites and Moros (487). Another complicating factor is the asymptotic decay of current during the final stages of electrolysis.

As is customary in polarographic procedures, oxygen interference in coulometric procedures is avoided by deoxygenation of the supporting electrolyte by a current of an inert gas such as nitrogen, and a stream of the inert gas is passed continuously through or over the supporting electrolyte during electrolysis. The mercury cathode pool is continuously stirred during electrolysis, preferably by a magnetic bar stirrer, but since hydrogen discharge occurs at a lower voltage in stirred pools than in quiescent ones (590), an error may be introduced.

Although it is possible to perform constant-potential coulometric analyses of nitro and nitroso compounds at a variety of potentials, the pool potential ordinarily employed ranges from $-0.85$ to $-1.00$ V versus reference electrode (usually $-1.00$) for complete reduction of nitroaromatics (156,390) and is higher for nitroparaffins (e.g., $-0.90$ to $-1.30$ V). As can be anticipated, the results of preliminary experiments made with known amounts of pure compounds assure that conditions are appropriate for quantitative reductions and that interference from other electroreducible substances is either absent or accountable. In many instances, two electroreducible compounds can be determined if one can be reduced quantitatively at a more positive potential than the other. For example, Ehlers and Sease (156) determined nitrobenzene in a sample by preliminary reduction at $-0.85$ V and then determined nitromethane in the same sample by subsequent reduction at $-1.10$ V; Meites (482) has discussed the theoretical background of coulometric analyses in which two products are reduced successively.

Constant-potential coulometric methods are also widely used for preparation of sufficient products of electrolysis to permit isolation and analysis; this type of information suggests possible electrolysis mechanisms. For example, Laviron et al. (402) electrolyzed a solution of 31 mg of nitroguanidine in 150 ml of molar sodium hydroxide at $-1.10$ V (where the reduction presumably involves 2 F/mol); at the end of the electrolysis, the solution was brought to pH 7 with hydrochloric acid and evaporated to dryness under vacuum. The product exhibited the same infrared spectrum in a salt pellet as did a sample of nitrosoguanidine. After electrolysis of another sample of nitroguanidine at $-1.13$ V, acidification, and treatment with picric acid, the precipitated picrate was considered to be guanidine picrate because of its melting point and ir spectrum. These experimental results provide the basis for the reduction mechanisms discussed in Section VI.A.4.e.

The number of faradays involved in reductions at constant potential can, of course, be obtained directly from current–time observations or from coulometers operated in series with the experimental electrolysis systems (421,423,424). However, $n$ values can be determined more simply by a dip-type, thin-layer electrolysis cell in nonaqueous systems (478) or by the technique described by Rosie and Cooke (590), which is applicable to aqueous as well as nonaqueous systems.

A variety of systems have been devised for handling small quantities of material; a number of the so-called microcoulometric and milli-coulometric methods for determining the number of electrons involved in irreversible electrolytic processes at the dropping mercury electrode itself have been described by Reynolds and Shalgosky (580), as well as by others (166,159,223,300,484,722). A complete survey of the theory, experimental methods, and applications of coulometry to the study of electrode reactions has been given by Bard (33); Meites (485) has reviewed the applications of coulometry to determination of the rates of homogeneous reaction. Karp and Meites (339) and Rangarajan (568) have studied the effect on $n$-value determinations of reactions which occur in the diffusion layer.

It is necessary to bear in mind at all times that conditions at the electrodes of one type of electrolytic system are not necessarily the same as those in another type, even though there may be fair agreement between the current–voltage curves obtained with the dropping mercury electrode and the curves obtained with a mercury pool cathode, as Ehlers and Sease (156) found for nitromethane and 1-nitropropane. It is emphasized that electrolytic processes taking place at a constant-potential quiet-pool cathode are not exactly the same as those occurring at a dropping mercury electrode, and the results obtained from constant-potential pool electrolyses are therefore not necessarily indicative of processes at the dropping mercury electrode. Thus it cannot be assumed that coulometric methods which use a mercury pool cathode are merely large-scale adaptations of the polarographic method, for there are readily visible differences between results obtained when a polarized mercury pool is used as a device to obtain potential-current curves and ordinary polarographic results (671,672). For example, potential-current curves obtained with an unstirred pool of mercury as a cathode and a reference electrode (e.g., SCE) as the anode often yield graphs of potential versus current that have shapes highly reminiscent of polarographic waves with maxima; however, waves obtained with an unstirred pool do not have the plateau which is characteristic of dropping mercury electrode polarograms. Potential-current curves obtained with an unstirred mercury pool show a dependence of peak height on scanning rate; the peak height is linearly proportional to the square root of the scanning rate over a considerable range of scanning rates. Of particular interest is the observation that the potential versus current curve of the reduction of *m*-dinitrobenzene at a pool electrode (pH 3, 5% EtOH) shows two peaks, presumably corresponding to the separate reductions of the nitro groups; in contrast, the dropping mercury polarogram under identical conditions reveals only a single wave, and the half-wave potential is different from that observed with the pool electrode (about 0.06 V). Other reducible substances that cannot be resolved at a dropping mercury electrode often can be determined by quiet-pool polarography (673).

The potential-current curves obtained with unstirred mercury pools are theoretically predictable by equations previously derived for oscillographic reductions (567,630) for the case of ion-amalgam reductions with periodically applied triangular voltage. Similar curves are observed for reduction of organic compounds, but since reductions of organic compounds are generally irreversible, the actual shapes cannot be predicted at this time.

A mercury frit electrode is another form of quiet-pool electrode; contact between electrolyte and mercury is made in a glass frit, and the resulting immobilization of the electrolyte cuts convection to negligible levels and gives peak voltamograms even at very slow scanning rates (110). A logical and convenient form of the quiet mercury electrode is the hanging drop electrode formed at the end of a polarographic capillary; it has been found useful for interpretation of the electrolytic reduction and oxidation of nitro compounds by means of oscillopolarography (rapid and repetitive applications of scanning voltage to permit presentation of anodic and cathodic polarograms on an oscilloscope (346,358).

A mercury pool electrode with smooth agitation of the electrolyte in its vicinity provides high sensitivity for determination of nitro compounds; a pool contained in a 1.7-mm-i.d. tube that has been thrust about 5 mm into another rapidly rotating tube (600 rpm) yields half-peak potentials that are seldom 0.025 V more negative than the half-wave potentials obtained with conventional dropping mercury electrode systems. For all practical purposes, the results are very similar to potential-current curves obtained with quiet-pool electrolytic systems (17,549,590).

The mercury-film electrode described by Moros (505) and others (260) is still another type of electrode which has been found useful for quantitative determinations of nitro compounds; a small piece of platinum wetted by a thin film of mercury (about 10 $\mu$ thick) is used as the polarizable electrode in suspension media of the type ordinarily used for polarographic analysis with the dropping mercury electrode. Potential-current curves obtained with the mercury-film electrode show the characteristic peaks of quiet-pool electrodes. The mercury-film electrode often must be preconditioned before use; for example, in the determination of trace amounts of nitrobenzene in aniline, 25 ml of the sample is mixed with 5 ml of concentrated hydrochloric, the electrode is preconditioned at $-0.35$ V versus SCE to a stable reading, and then the peak current is obtained in the vicinity of $-0.50$ to $-0.60$ V. The sensitivity is such that 0.3–211 ppm of nitrobenzene can be determined ($2 \times 10^{-6}$ to $1.7 \times 10^{-3}$ M). Potential-current curves obtained with the mercury-film electrode of 3,5-dinitrobenzoic acid show two distinct peaks in acetate buffer at $-0.27$ and $-0.39$ V versus SCE.

Pyrolytic graphite electrodes, glassy carbon electrodes (751), and carbon paste electrodes (393,460) have also been used for analyses of nitro compounds. Pyrolytic graphite is useful for anodic oxidation at acidities as low as pH 1.6 (mercury is oxidized below pH 4 in 50% ethanolic buffer systems); the graphite electrode was used in a systematic study which confirmed that the nitrosobenzene–phenylhydroxylamine system is reversible (109,649,695).

### 6. Kjeldahl Digestion

Kjeldahl procedures which are also suitable for nitro compounds are described in Sections VI and VII of "Azo Group," Vol. 15. Several workers (142,178,463,464) have reported low results when a variety of Kjeldahl methods have been used directly on nitro group compounds. As a consequence, it is now considered necessary (25,107,448,749) to reduce nitro groups to amino groups before performing the Kjeldahl digestion, in spite of the fact that some workers (331) report quantitative formation of ammonia without preliminary reduction. When dealing with compounds of unknown structure, it is best to use a preliminary reduction step.

Among the early Kjeldahl procedures using a reduction is that of Flammand and Prager (193), which employs zinc dust and hydrochloric acid in alcohol for reduction of nitro groups; cupric sulfate serves as catalyst in the digestion. Margosches and Kristen (461,462) showed that the procedure was accurate for compounds such as p-nitrotoluene, dinitrobenzenes, dinitrophenols, and picric acid, but it yielded unacceptably low values for o- and m-nitrotoluene and o- and m-nitrobenzyl chlorides. The low values were attributed to loss of sample during digestion in hot sulfuric acid solution. A sealed tube digestion such as is described by White and Long (728) will prevent loss of nitrogen.

Of course, nearly any powerful reducing agent can be used to effect reduction of nitro nitrogen. Thus zinc dust in fuming sulfuric acid was employed as the reductant in a method described by Weizmann et al. (725), and zinc dust in formic acid or hydrochloric acid was used successfully by Dickinson (142) for a rather wide variety of mono- and polynitroaromatics. Phosphonium iodide has been used for reduction of trinitroaromatic compounds (331), and Friedrich (205) used red phosphorus and hydriodic acid in an involved microprocedure that included steps to remove all of the iodine before completion of the digestion. Reduction by chromous salts was investigated by Belcher and Bhatty (39); good results were obtained by Bezinger et al. (49) with lithium aluminum hydride. Titanous sulfate has been used by Callan et al. (284), but titanium chloride is recommended by Somers (653); the sample is quenched in 20% titanous solution and allowed to remain for 20 min, then the mixture is warmed for 2 min.

Reduction of nitro and azo nitrogen by sodium hydrosulfite, $Na_2S_2O_4$, was recommended by Simek (639); Shaefer and Becker found the reducing agent adequate for general purposes and especially appropriate for reduction of diazodinitrophenol before the Kjeldahl digestion (633a). A freshly prepared, concentrated, hot aqueous solution of the hydrosulfite is added to a boiling alcohol–water solution of the sample; diazo nitrogen is eliminated quantitatively as elemental nitrogen, and nitro groups are reduced to amino groups without formation of an intermediate hydrazine derivative. Alternatively, Grandmougin's procedure of adding the sample to a hot hydrosulfite solution may be used (237).

The presence of sufficient organic matter in the Kjeldahl digestion seemingly assures quantitative conversion to ammonia, and this may be the reason why some workers report good results when the Kjeldahl method is applied directly to nitro compounds. Thus Elek and Sobotka (158) found glucose effective in microprocedures, and Belcher et al. (38) found glucose to be a satisfactory reductant in submicroprocedures when digestion was performed at 420°C. Bradstreet (64) also reported that sucrose is better than salicylic acid in standard Kjeldahl procedures, especially when phosphoric acid is included. Bhat (51) used pure cotton cellulose as the reducing agent in micro Kjeldahl procedures. Harte (255) found dextrose more manageable in a semimicroprocedure of which the following is an example:

Place enough sample in a dry 100-ml Kjeldahl flask to yield 2–5 mg of nitrogen. In succession, add 300 mg of pure dextrose, 1–1.5 grams of $K_2SO_4$, 20 mg of $CuSO_4 \cdot 5\,H_2O$, and some bumping stones or No. 14 Alundum grains. Then wash down the neck of the flask with 4 ml of concentrated sulfuric acid; digest over a microburner while the flask is nearly horizontal. When the digestion mixture appears homogeneous, add 1 drop of selenium oxychloride.

Heat until the solution is clear and then 15 min longer. When the flask is cold, add 35 ml of water and 12 ml of 50% sodium hydroxide solution; then distil off the ammonia into 25 ml of 0.02 $N$ hydrochloric acid. Titrate the excess acid with 0.02 $N$ barium hydroxide, using mixed methyl red–methylene blue as indicator.

Thiosalicylic acid ($O$-mercaptobenzoic acid) is considered by McCutchan and Roth (479) to be a better reducing agent for nitro nitrogen in Kjeldahl digestions than the more widely used mixture of salicylic acid and sodium thiosulfate (22).

Hormann et al. (296) have determined dinitrophenyl derivatives colorimetrically with Nessler–Winkler reagent (mercuric iodide and sodium iodide in sodium hydroxide solution), after a Kjeldahl digestion with selenium catalyst.

## 7. Gasometric Methods

### a. NITROMETER METHODS

Methods that depend upon the evolution of a gas by reaction of the nitro group with other substances are seldom used except in laboratories specializing in the analysis of explosives, where determination of nitrato and nitrito (—$ONO_2$ and —ONO) groups is performed routinely with the aid of a nitrometer; in fact, many ingredients for munitions are assayed by archaic nitrometer methods in government laboratories (252) throughout the world. Nitrometer methods are predicated on the reaction of nitrogen-containing substances with sulfuric acid and mercury to produce nitric oxide. Nitric acid, nitrous acid, and organic compounds which produce these acids when treated with concentrated sulfuric acid form nitric oxide and therefore can be determined directly within the precision and accuracy indigenous to the method ($\pm 0.05\%$):

$$2\,HNO_3 + 3\,H_2SO_4 + 3\,Hg \rightarrow 3\,HgSO_4 + 4\,H_2O + 2\,NO$$
$$2\,HNO_2 + H_2SO_4 + Hg \rightarrow HgSO_4 + 2\,H_2O + 2\,NO$$

It is necessary to recognize that $C$-nitro compounds do not form nitric oxide in the nitrometer. However, all organic nitrogen-containing compounds can be transformed into nitrites or nitrates which then can be introduced into the nitrometer; for example, C—$NO_2$ compounds can be oxidized to nitric acid by heating for about 30 min in a mixture of concentrated sulfuric and chromic acids (333,527).

The nitramines, —N—$NO_2$, form nitric acid quantitatively by an undiscovered mechanism when treated with concentrated sulfuric acid (685); this fundamental difference in the reactivities of C—$NO_2$ and N—$NO_2$ groups (127) makes possible the nitrometer analysis of explosives containing nitroguanidine, $NH_2$—$C(NH)$—$NHNO_2$, or tetryl, $(NO_2)_3C_6H_2N(CH_3)NO_2$, even though trinitrotoluene or other aromatic nitro compounds may be present (128,499):

$$2 R_2N—NO_2 + 3 Hg + 3 H_2SO_4 \rightarrow 2 R_2NH + 3 HgSO_4 + 2 H_2O + 2 NO$$
$$2 R_2N—NO + Hg + H_2SO_4 \rightarrow 2 R_2NH + HgSO_4 + H_2O + 2 NO$$

However, estimation of tetryl and nitroguanidine by the nitrometer has its limitations (139,140,499), largely because of low solubility in sulfuric acid; decomposition of nitroguanidine to form products other than nitric acid may possibly be a competing reaction (135). As a rule, the rate of evolution of gas in the nitrometer method gives an insight as to the nature of the bonding force of the nitro group to nitrogen in the $N$-$NO_2$ bond. The $N$-nitro group also reacts with ferrous ion in acidic media (see Section VI.A.6). Additional details on nitrometer methods are given in the section on —ONO and —$ONO_2$ groups (Section VI.A.2 of "Nitrate and Nitrite Ester Group").

b. MISCELLANEOUS

One nitro group in tetranitromethane is more reactive than the others. In alkaline solutions, hydrazine very rapidly yields 0.5 mol of nitrogen for each mole of tetranitromethane. The aliphatic nitro compound can thus be determined in a nitrometer in the presence of nitrous acid, nitric acid, and the alkali-metal salt of nitroform (24).

Aromatic $C$-nitro groups can be reduced to amines and oxidized by potassium iodate in phosphoric acid at 300°C to nitrogen gas (530,675).

## B. NITROSO GROUP

### 1. Volumetric Methods

a. TITRATION WITH REDUCING AGENTS

#### (1) Reduction with Titanous Salts

Methods for reduction of nitroso compounds with titanous salts are very similar to the methods found suitable for nitro compounds.

The reduction of $C$-nitrosoaromatic compounds by titanous chloride has been described by Knecht and Hibbert (363) and Salvaterra (603); their methods require treatment of the compounds with an excess of titanous chloride at an elevated temperature. The necessity for strict observance of a temperature of 40–50° was noted, inasmuch as prolonged times at higher temperatures gave low results (363); presumably, nitric oxide escapes and the titanous salt is decomposed by reaction with water. Compounds such as *p*-nitrosodimethylaniline and *p*-nitrosophenol are titrated directly in a warm solution containing hydrochloric acid; methylene blue is used as indicator, and a stream of carbon dioxide serves to promote mixing and to exclude atmospheric oxygen. Four equivalents of titanous chloride are required for each nitroso group. Rathsburg (571) titrated nitroso compounds in strong hydrochloric acid solution (boiling). On the other hand, nitroso-$\beta$-naphthol is best determined by first reducing with excess titanous chloride at 60°C and then back-titrating with iron alum (363). Similarly, $\beta$-benzildioxime and benzophenonoxime must be reduced with an excess of hot titanous chloride and back-titrated with methylene blue to take advantage of a sharp color change at the end point; as would be anticipated, aliphatic oximes such as acetoxime or dimethylglyoxime cannot be analyzed in this manner. Callan and Strafford (100) have reported successful application of titanous chloride titrations to the nitroso compounds used as vulcanizing accelerators, and Dachselt (131) has also shown that it is possible to titrate *C*-nitroso compounds (aromatic) directly with titanous chloride at 50–80°C in solutions buffered with Rochelle salt (potentiometrically).

Salvaterra (603) suggested a modification of the above procedures for use with nitroso compounds. An excess of titanous chloride is added to the sample solution under an atmosphere of carbon dioxide; after reaction has proceeded for some time, a known volume of methylene blue solution is added until there is a sharp color change from green to yellowish brown.

The reduction of nitrosobenzene or phenylhydroxylamine by titanous chloride in acidic media is very many times more rapid than that of nitrobenzene, and the interaction of these two compounds to give azoxybenzene is also extremely fast. On the other hand, the rate of reduction of azoxybenzene is only about one-seventh as fast as that of nitrobenzene at the same molar concentration (515). On a submicroscale (about 50 $\mu$g), Belcher et al. (40) found that the same procedure used for the titanous reduction of nitro compounds (citrate buffer; excess of titanous sulfate and back titration) is successful with 1-nitroso-2-naphthol, sodium 2-nitroso-1-naphthol-4-sulfonate, and *N*-nitrosodiphenylamine (dissolved in acetic acid). No aliphatic *C*-nitroso compounds were tested.

Although most procedures for reducing aromatic *C*-nitroso as well as *C*-nitro compounds in titanous chloride solutions involve heat and even the refluxing of samples in strongly acidic solutions, Ma and Earley (447) found that reduction of both nitro and nitroso groups often can occur at room temperature in a sodium acetate buffer; however, in strong hydrochloric acid solution, the nitroso group is quantitatively reduced but the nitro group is not. Determination of the nitroso group in the presence of C-nitro groups by this procedure involves solution of the compound in water or ethanol, acidification with hydrochloric acid, addition of an excess of titanous solution, and back titration with ferric alum solution after a reaction time of 3 min. Details of the procedure are given in Section VII.F. Results accurate to ±0.3% were obtained for N-nitroso-substituted dimethylaniline, resorcinol, phenol, naphthol, diphenylamine, and R-salt; cupferron was not reduced quantitatively, but N-nitroso-N-phenylbenzylamine and N-methyl-N-nitrosoaniline had to be treated with an excess of 50–150% of titanous chloride and reduction allowed to proceed for 10 min before the hydrochloric acid was added.

Tiwari and Sharma (687) have described a semimicromethod for the determination of aromatic C-nitroso groups with titanous sulfate; the reduction is carried out with an excess of titanous sulfate in solutions buffered by potassium citrate, and the excess reductant is titrated with ferric sulfate, using potassium thiocyanate indicator [see Section VI.A.1.a(1)(a)].

Zimmerman and Lieber (747) found that the N-nitroso group in aliphatic nitrosamines such as nitrosoguanidine consumes 2 equiv. of titanous chloride when refluxed with excess titanous chloride in concentrated solutions of hydrochloric acid.

### (2) Reduction with Stannous Chloride

The reduction of aliphatic C-nitroso groups by stannous chloride is seldom quantitative, for a variety of products are formed, and the anticipated amine is usually obtained in low yield. There are a few instances, however, where yields of amines are quite high, although far from quantitative (113). In contrast, reduction of aromatic C-nitroso groups proceeds quite readily with formation of the amine; reaction rates tend to be somewhat slow, however, and a large excess of stannous ion is required to drive the reaction to completion. As a result, titrimetric methods using stannous ion are seldom employed for analysis of C-nitroso groups.

The reduction of N-nitroso groups can also be accomplished by stannous chloride, but yields are poor and reactions seldom proceed cleanly to amines; more frequently, stable hydrazines are formed, and the primary aromatic N—NO compounds often form oximes.

### (3) Reduction with Chromous Salts

Chromous salts reduce aromatic C-nitroso compounds to the cor-

responding amine (six electrons), and most N-nitrosoaromatic compounds to substituted hydrazines (four electrons). As a general rule, an
excess of chromous salt is added and the excess determined by back
titration with ferric ion; however, with many aromatic N-nitroso compounds, the excess reducing agent can be destroyed by aeration and the
substituted hydrazine determined oxidimetrically (e.g., ceric salts).

The reduction of aliphatic nitroso compounds by chromous salts has
not been studied sufficiently to permit the drawing of conclusions.

Bottei and Furman (59) reported that nitroso-R-salt is quantitatively
reduced by an excess of chromous chloride solution; the back titration
was accomplished with ferric alum solution, and the end point was
determined potentiometrically. Subsequently, modifications of this
method were used for the determination of several other nitroso compounds (211):

Aromatic C-nitroso compounds are dissolved in glacial acetic acid and then
acidified strongly with concentrated hydrochloric acid; carbon dioxide is bubbled
through the solution for 10 min, and then maintained over the solution for the
rest of the determination. An excess of chromous chloride solution (about 200%)
is added, and after about 1 min the excess chromous ion is titrated potentiometrically with ferric alum.

Values obtained with p-nitrosodimethylaniline were 0.4–2.6% low (16
determinations). In the case of cupferron (N-nitrosoaromatic), a mixture
of distilled water and hydrochloric acid was first deaerated with carbon
dioxide for about 10 min and then a 100% excess of chromous chloride
was added, followed by the sample. Cupferron is reductively cleaved to
phenylhydroxylamine and ammonia (6 equiv.); the error for 13 determinations ranged from +0.2 to −1.9%. N-Nitrosodiphenylamine and
N-nitroso-N-phenylbenzylamine were dissolved in ethanol and
deaerated; then chromous chloride reagent was added, followed by
concentrated hydrochloric acid. The error for 17 determinations of
N-nitrosodiphenylamine ranged from −1.5 to +2.2%, and for the substituted benzylamine, from −2.7 to +2.5% (29 determinations); the
nitroso group in these compounds was reduced to the corresponding
amino group (4 equiv.).

Tandon (679) found that 1-nitroso-2-naphthol could be determined
either by direct titration with chromous sulfate solution or by treatment
with excess chromous sulfate solution and back titration with ferric
alum. For the direct titration, Neutral Red, Phenosafranine, or p-
ethoxychrysoidine was used as indicator; tartrate was added when
methyl red or orange, Congo Red, Bismarck Brown, or tartrazine was
used.

The direct potentiometric titration of aromatic nitroso compounds in a variety of solvents with chromous sulfate solution was studied by Jucker (330); results within 1–2% are cited for nitrosophenol, nitrosodimethylanilines, and nitrosonaphthols. Most interestingly, it appears that the simultaneous determination of isomers is possible because of differences in redox potentials; values and redox curves are given for *o*- and *p*-nitrosodimethylaniline and 1-nitroso-2-naphthol in the presence of 3-nitroso-2-naphthol.

### (4) Reduction with Other Metal Salts

Trivalent molybdenum has been used by Gapchenko (215) as a reductant for aromatic *C*-nitroso compounds; pentavalent molybdenum is formed. A two- or threefold excess of the reductant is added to the sample in an atmosphere of inert gas; 2 or 3 min later, the excess reductant is titrated by standardized ferric alum in the presence of methylene blue as indicator. Cupferron is titrated in an alcoholic medium to prevent precipitation of the amine when the acidic solution of molybdenum is added.

Vanadous ion should be an excellent reducing agent for *C*-nitroso as well as *N*-nitroso compounds; the procedure of Gapchenko and Sheintis (216) appears to be adequate.

### b. REDUCTION BY HYDRIODIC ACID

Many nitroso compounds can be determined iodometrically because they oxidize iodides in acidic solutions. For example, Lobunets and Gortins'ka (437) determined nitrosobenzene with an accuracy of 0.16–0.33%; a 0.1-gram sample of nitrosobenzene in 25 ml of 60% methanol was allowed to react for a few minutes in a mixture of 30 ml of 6 $N$ hydrochloric acid and 15–20 ml of 20% potassium iodide solution before titration with thiosulfate. Becker and Shaefer (36) also found the iodometric procedure suitable for determination of the *N*-nitroso group of *N*-nitrosodiphenylamine in aged smokeless powder:

The material is dissolved in 25 ml of methanol and treated with 20 ml of a 5% solution of potassium iodide in 1:1 methanol–water. Carbon dioxide is passed through the solution for a few minutes, and then 5 ml of 1:1 hydrochloric acid solution is added. After 5 min, liberated iodine is titrated with sodium thiosulfate solution; 1 mol of the *N*-nitroso compound yields 1 equiv. of iodine.

Results of three determinations were 97.7, 98.4, and 99.0% of theoretical. A similar procedure for use on a semimicroscale is described by Tiwari and Sharma (688); it is of importance to note that C—NO$_2$ groups do not yield iodine in acidic solutions of alkali-metal iodides. Additional information on the reaction of hydriodic acid with *C*- and *N*-nitroso compounds is given in Section III.B.1.d and III.B.1.e.

Nitroso compounds can be determined iodometrically by the closed tube method of Aldrovandi and De Lorenzi (4), as described in Section VI.A.1.b.

### c. REDUCTION BY METALS AND AMALGAMS

The methods described in Section VI.A.1.c. on the determination of nitro groups can also be applied to aromatic C—NO groups; however, only the most vigorous of reducing couples can be used to convert aromatic nitroso groups quantitatively to amino or other groups for subsequent determinations. In view of the difficulty of obtaining a single product from reduction of nitroso compounds, this form of analysis is not recommended for general use. Aliphatic C—NO groups seldom undergo quantitative reduction to amino groups (21).

Nitroso compounds are reduced by metallic cadmium in a submicro method proposed by Budesinsky (87); the procedure is essentially the same as that described in Section VI.A.1.c.

### 2. Gravimetric Procedures

Although titrimetric procedures with strong reducing agents such as titanous and vanadous ions are in wide use, they are not convenient for a single or occasional analysis because two standard solutions are required; in particular, the reducing agent requires storage under an inert atmosphere and frequent standardization. The indirect gravimetric method for the determination of aromatic nitroso compounds proposed by Juvet et al. (335) can be conveniently used for occasional analyses. It involves reduction of the nitroso group by copper in sulfuric acid; the loss of weight incurred by the copper is used to compute the nitroso group content. It is very similar in principle to the method of Vanderzee and Edgell (713) discussed in Section VI.A.2; the use of copper, however, eliminates loss of metal by acid attack, and it appears that the concentration of acid and organic solvent has no appreciable effects on the analytical results. The method for reduction of nitroso groups is given in Section VII.C. With *p*-nitrosodiethylaniline, the authors found that reduction was complete within 1 hr in diethylene glycol dimethyl ether as solvent; 4 equiv. of copper/mol of nitroso compound were required, and the sample was found to correspond to a purity of $98.9 \pm 0.4\%$. No other compounds were tested.

### 3. Spectrometric Methods

#### a. INFRARED PROCEDURES

The direct spectrophotometric determination of nitroso compounds in the infrared region may be accomplished by absorptiometry at about 850 cm$^{-1}$ and at about 1520 or 1360 cm$^{-1}$; the latter two bands shift to lower wave numbers on dimerization, and it is often difficult to ensure that only absorption of the nitroso group is involved, for nitro compounds also absorb in these regions.

#### b. ULTRAVIOLET-VISIBLE SPECTROPHOTOMETRY

A procedure for the spectrophotometric quantitative estimation of nitroso compounds after chromatographic separation from analogous nitro compounds was developed by Edwards and Tate (153) and is described in Section V.C.2.d.

Nitrous acid, which is formed quantitatively from $N$-nitrosamines and $N$-nitrosamides by action of ultraviolet light, can be determined colorimetrically using Griess–Ilosvay reagent. A solution of the sample in aqueous or methanolic sodium carbonate solution is irradiated with an uv source (230 nm, mercury lamp); then Griess–Ilosvay reagent is added, and absorbance is read at 525 nm in a photometer. According to Daiber and Preussman (133), nitrosamines may be determined by this method in aqueous or methanolic solutions, but nitrosamides must be dissolved in methanol; a sensitivity of 1–2 $\mu$g/ml was found for seven nitrosamines and five nitrosamides.

Loo and Dion (440) have utilized the liberation of nitrous acid in hydrochloric acid solution for the colorimetric determination of 1,3-bis(2-chloroethyl)-1-nitrosourea (BCNU) in biological fluids; since the $N$-nitroso group is decomposed readily by strong acids, the procedure may find wide use.

The fluid is extracted with ether; the ether is separated and evaporated in a stream of nitrogen; the residue is taken up in dilute sulfanilamide reagent (0.5% in 2 $N$ HCl), maintained in a water bath (50°C) for 45 min, and then chilled. Bratton-Marshall reagent (66) is added, and absorbance at 540 nm is measured after 10 min. For samples containing 1–5 $\mu$g/ml, recovery of 78–95% has been demonstrated. The sulfanilamide reagent contains 1.67 grams of sulfanilamide/liter of 0.67 $N$ hydrochloric acid; Bratton-Marshall reagent is a solution of 0.3 gram of $N$-(1-naphthyl)-ethylenediamine dihydrochloride in 100 ml of water.

The $C$-nitroso and $N$-nitroso derivatives of aromatic compounds and corresponding derivatives of diphenylamine formed in the aging of double-base powders can be determined by uv spectroscopy after separation by chromatographic procedures (614,615). Similar analytical data and separation procedures are provided by the work of Edwards et al. (152).

### 4. Polarographic Methods

The polarographic reduction of nitroso compounds has not received much attention, inasmuch as they are not of great commercial importance. Moreover, nitroso compounds are relatively unstable and, under the conditions ordinarily employed in polarographic determinations, are reduced at positive potentials (especially in acidic media). The first examples of the polarograms of nitroso compounds were reported by Hertel and Lenz (280) in 1939.

### a. *C*-NITROSOPARAFFINS

Studies of the polarographic behavior of *C*-nitrosoparaffins are few. Results obtained thus far suggest that monomeric species operate through the general reversible reaction

$$R{-}NO + 2\,H^+ + 2\,e \rightleftharpoons RNHOH$$

with subsequent irreversible reduction to an amine:

$$RNHOH + 2\,H^+ + 2\,e \rightarrow RNH_2$$

Polarographic reduction of dimeric nitroso compounds involves a total of six electrons with formation of the corresponding substituted hydrazine:

$$\begin{array}{c}
R \\
\phantom{R}\diagdown \overset{+}{N}{=}\overset{+}{N} \diagup O^- \\
{}^-O \diagup \phantom{N} \diagdown R
\end{array}
\xrightarrow[6\,e]{6\,H^+}
\begin{array}{c}
R \phantom{aaa} H \\
\phantom{R}\diagdown N{-}N \diagup \\
H \diagup \phantom{NN} \diagdown R
\end{array}
+ 2\,H_2O$$

This behavior, in sharp contrast to the behavior of the aromatic nitroso compounds, is especially pronounced in the reduction of bis(nitroso-cyclohexane), which is dissociated to the monomer only to the extent of about 1%; the product is *sym*-dicyclohexylhydrazine (610).

### b. *N*-NITROSOPARAFFINS

The polarographic behavior of dimethyl-, diethyl-, di-*n*-propyl-, and diisopropylnitrosamine was investigated by Whitnack et al. (732). In 20% ethanolic solutions of dilute mineral or organic acids and in well-buffered acidic aqueous media, there is observed one well-defined but drawn-out wave with no maximum (gelatin); the waves are diffusion-controlled and can be used for quantitative work. Half-wave potentials in acidic media are given in Table XVI. Enough reduction products for study were prepared in a constant-potential coulometric apparatus (5 grams of $N$-nitrosamine) equipped with a mercury pool cathode and a rotating platinum electrode. The pH was controlled during the electrolysis, and the reduction of the $N$-nitrosamine was followed polarographically or spectrophotometrically. When the amount of $N$-nitrosamine in the catholyte became negligible, the catholyte was analyzed for 1,1-dialkylhydrazine and for free bases. Gases liberated during electrolysis were analyzed by mass spectroscopy.

Lund (444) also made a study of the polarographic and coulometric reduction of a group of $N$-nitrosamines; the isolated products prompted the suggestion that the reactions involved in the reduction are as follows:

*Acidic media, pH 1–5*

$$RRNNOH^+ + 4e + 4H^+ \rightarrow RRNNH_3^+ + 2H_2O$$

**Basic media**

$$2 RRNNO + 4e + 3H_2O \rightarrow 2 RRNH + N_2O + 4OH^-$$

On the other hand, Whitnack et al. (732) were not able to show definitively that only 4 F is involved in the reduction processes; for

TABLE XVI
Properties of $N$-Nitrosamines (720)

| $N$-Nitrosamine derivative | Potential (V)[a] | Distilled (%)[b] | Adsorbed by carbon (%) | Eluted by methanol (%) |
|---|---|---|---|---|
| Dimethylamine | −0.94 | 70 | 75 | 75 |
| Diethylamine | −0.88 | 80 | 96 | 94 |
| Dipropylamine | −0.84 | 85 | 100 | 72 |
| Dibutylamine | −0.79 | 100 | 100 | 26 |
| Dipentylamine | −0.78 | 78 | 93 | 0 |
| Dicyclohexylamine | −0.86 | 80 | 94 | 16 |
| Diethanolamine | −0.90 | 0 | 92 | 60 |
| Proline | −0.79 | 0 | 90 | 72 |
| $N$-Methylaniline | −0.69 | 96 | 100 | 0 |
| Dibenzylamine | −0.75 | 93 | 94 | 0 |

[a] Peak potential; derivative polarography in 0.2 $N$ HCl.
[b] Percentage of nitrosamine recovered on distillation to half volume from 20% sodium chloride.

example, a value of 4.72 F was found by polarography and 4.96 by microcoulometry for the reduction of dimethylnitrosamine in McIlvaine's buffer at pH 2.4 [also reported by Martin and Tashdjian (467) as 3.7–5.3 in acidic solution]. In contrast to the suggestions of Lund and others (as indicated above), Whitnack et al. prefer to consider the reduction of $N$-nitrosamines as a series of reactions:

$$R_2NNO + 4\,e^- + 4\,H^+ \rightarrow R_2NNH_2 + H_2O$$

$$R_2NNO + 6\,e^- + 6\,H^+ \rightarrow R_2NH + NH_3 + H_2O$$

$$2\,R_2NNO + 6\,e^- + 6\,H^+ \rightarrow 2\,R_2NH + N_2 + 2\,H_2O$$

$$2\,R_2NNO + 4\,e^- + 4\,H^+ \rightarrow 2\,R_2NH + N_2O + H_2O$$

All the above reactions take place during the reduction of $N$-nitrosodimethylamine at the mercury cathode in $3\,N$ $H_2SO_4$ because 1,1-dimethylhydrazine, dimethylamine, ammonia, nitrous oxide, and nitrogen have been identified as reduction products; however, the predominant reactions are the first and last. At higher pH values (e.g., 4–5), the first reaction is favored; above pH 5, the first reaction is subdued and the second and fourth reactions are favored.

Polarographic reductions of $N$-nitroso-$N$-isopropylaminoacetonitrile and $N$-nitroso-$N$-phenylaminoacetonitrile were studied by Kholodov et al. (357); only single four electron waves were found for both compounds below pH 6.5, but two-electron waves occurred in neutral or alkaline solutions.

The determination of small amounts of dimethylamine is an example of an analysis that is made possible by conversion to a polarographically active compound; the method was first described by Smales and Wilson (643) and later modified by English (170):

An aqueous solution containing 15–25% of aliphatic amines is treated for 10 min with a mixture of sodium nitrite and acetic acid at 25°C; then, after the excess nitrous acid has been removed by sulfamic acid and the solution made alkaline to phenolphthalein, sodium hydrosulfite is added to reduce nitromethane. The excess hydrosulfite is removed by titration with iodine; the solution is then rendered acidic with hydrochloric acid, and the polarogram is obtained. The polarogram is well defined, but the plateau is inclined at a much greater angle to the horizontal than is the condenser-current line; nevertheless, values are reproducible and accurately measurable.

Nitrosation of alkyl amines takes place readily in acetic acid but is prohibitively slow in mineral acid. According to English (170), monomethylamine reacts with nitrous acid to form methanol, water, and nitrogen, as is characteristic for a primary amine, but a side reaction forms nitromethane in amounts which interfere with the polarographic determination of $N$-nitrosodimethylamine. As a result, it is necessary to destroy nitromethane by reduction with sodium hydrosulfite in faintly alkaline solution; $N$-nitrosodimethylamine is not reduced under these conditions (even at 60°C). The sodium hydrosulfite must be free of thiosulfate; solutions containing hydrosulfite must not be exposed to atmospheric oxygen (thiosulfate is a product of oxidation of hydrosulfite) inasmuch as the tetrathionate formed by reaction with the iodine subsequently used to destroy hydrosulfite produces an interfering polarographic wave. Iodine oxidizes hydrosulfite to sulfate, not thiosulfate. In the procedure described above, ammonia in large amounts interferes owing to the consumption of nitrous acid; similarly, large amounts of trimethylamine appear to interfere because of the formation of a nitroso derivative. Of course, primary aromatic amines form diazonium compounds, and these or their coupling products are reducible at potentials in the vicinity of the $N$-nitroso derivatives of the secondary amines. Tertiary aromatic amines form $p$-nitrosoaromatic compounds which are reducible at potentials lower than those for $N$-nitrosamines. However, it is possible to determine secondary aromatic amines in the presence of secondary aliphatic amines inasmuch as there is a difference of at least 0.2 V between their half-wave potentials.

An alternative method for analysis of secondary amines in the presence of primary and tertiary amines has been given by Lund (444):

A solution of the amine is prepared so that it contains about 1 mg of material/ml; to 1 ml of the amine solution is added 1 ml of an acetate buffer which is made by mixing 10 ml of glacial acetic acid, 10 ml of $2 M$ sodium hydroxide, and 80 ml of water. Then 1 ml of a freshly prepared 20% sodium nitrite solution is added, and the mixture is heated for 15 min at 80°C. After cooling, there are added 1 ml of $4 M$ hydrochloric acid, 5 ml of a saturated aqueous solution of potassium chloride, 5 ml of ethanol, and 1 ml of a 5% aqueous solution of ammonium sulfamate. The polarogram of the resulting solution is obtained (may start at $-0.6$ V).

Lund claims that the above procedure leads to negligible side reactions with methylamine, ethylamine, and cyclohexylamine; trimethylamine is nitrosated to a small extent, but triethylamine and $N$-ethylpiperidine react only to a slight degree when present at concentrations less than $5 \times 10^{-3} M$ in the final solution.

The polarographic reduction of $N$-nitrosomethylurea in alkaline media produces ill-defined waves, and the results of coulometric studies suggest that this behavior can be attributed to cleavage of the N-NO bond. A single four-electron wave is observed in acidic media up to about pH 7; the reduction product is methyl-2-semicarbazide (402).

The products of reduction of nitrosoguanidine at the dropping mercury electrode are similar to those obtained by reduction of nitroguanidine, as indicated in an earlier section; thus aminoguanidine is the product of reduction in strongly acidic solution, and guanidine is formed in strongly basic solution. The two waves observed in polarograms of nitrosoguanidine at intermediate pH values are presumed to represent the formation of aminoguanidine and of guanidine. A pH range of 5–6 appears to be best for a separation of the first wave of nitrosoguanidine from the wave of nitroguanidine (see Table XIV); Whitnack and Gantz (731) have found that small amounts of nitrosoguanidine can be determined in nitroguanidine by polarography, but precise measurement of the first nitrosoguanidine wave is difficult when the nitrosoguanidine/nitroguanidine ratio exceeds 1:1.

The separation of N-nitrosamines from biological materials has recently been studied by Walters et al. (720); the determination of carcinogenic N-nitrosamines in foodstuffs containing sodium nitrite is of great importance. A wide variety of alkyl and aryl nitrosamines have been found to be strongly adsorbed (>90%) by carbon from aqueous solutions in which they are present at a 10-ppm level [with the exception of N-nitrosodimethylamine (75%) and the N-nitroso derivative of methyl-2-hydroxyethylamine (70%)]. The N-nitrosamines are so strongly adsorbed that the carbon can be washed repeatedly with water to remove interfering substances (e.g., sodium nitrite); the adsorbed compounds usually can be removed by elution with methanol. Subsequently, the methanolic eluate, made 0.2 N in hydrochloric acid, is examined by differential polarography. The disappearance of the wave without the appearance of a further wave at a more negative potential on addition of borax (pH 8.4) provides additional evidence of the presence of an N-nitrosamine. Similarly, the wave of an N-nitrosamine in acidic media should disappear on irradiation with ultraviolet light for 2 hr. Walters et al. used a differential cathode-ray polarograph; the voltage sweep is applied to two identical cells, the output currents of the two being in opposition to each other. Wave heights are observed on the cathode-ray tube. Addition of N-nitrosamine to one cell permits calibration *in situ*. Table XVI lists pertinent information on a selected series of N-nitrosamines.

### c. C-NITROSOAROMATICS

The C-nitrosoaromatic compounds in buffered alcoholic solutions are readily reducible over a wide range of pH values at the dropping mercury electrode. In general, below a pH of about 4 or 5, reduction of a C-nitrosoaromatic compound leads to the reversible formation of N-aryl hydroxylamine in a single two-electron reduction step; the first step (pH < 4) occurs at positive potentials (vs. SCE), but there is a poorly defined second step at more negative potentials (reduction to the amine):

$$C_6H_5NO + 2H^+ + 2e \rightarrow C_6H_5NHOH$$
$$C_6H_5NHOH + 2H^+ + 2e \rightarrow C_6H_5NH_2 + H_2O$$

In media of pH 4–10, only single waves are observed; the half-wave potentials are functions of pH (up to pH 10), but reduction to the amine is irreversible. Beyond pH 10, the dependence of half-wave potential on pH gradually decreases and disappears in strongly alkaline media. All substituted C-nitrosobenzenes appear to be reduced much as is nitrosobenzene; however, polarograms are quite complex whenever other substituent groups are also reduced (e.g., p-nitrosoiodobenzene and p-nitrosobenzaldehyde) (291). When substituents (other than the nitroso group) are —OH and —N(CH$_3$)$_2$, the compounds may behave abnormally in polarographic reductions. Suppression of maxima at high pH values is difficult with gelatin; agar appears to be satisfactory at a level of 0.02% (291).

The first reduction step of nitrosobenzene was studied in detail by Smith and Waller (649) and later by Holleck and Schindler (291); it was found that the N-phenylhydroxylamine–nitrosobenzene system is reversible (123) and that the half-wave potential of the system

$$2H^+ + 2e + C_6H_5NO \rightleftharpoons C_6H_5NHOH$$

can be expressed essentially as $E_{1/2}$ (vs. SCE) $= +0.33 - 0.061$ pH.

In contrast to the base itself, certain derivatives of phenylhydroxylamines are not polarographically reducible at all pH values; for example, N-(p-sulfonamido)phenylhydroxylamine yields a polarographic reduction wave in 0.1 N sodium hydroxide, but is not reducible at lower pH values (410). Since phenylhydroxylamine is converted to reducible azoxybenzene in alkaline solutions (469), it is suggested that a similar reaction may be the explanation for the behavior of the p-sulfonamido compound.

When the first reduction step of a C-nitrosoaromatic occurs at a positive potential, it is necessary (chloride interference) to replace the normally used calomel electrode with a saturated mercuric sulfate–mercury anode. Of course, at more negative potentials, chloride ion does not interfere, and it is possible to use the pool or a saturated calomel electrode for reference. Clark and Sorensen buffers prepared with sulfuric acid instead of hydrochloric acid are satisfactory suspension media (291); from 10 to 20% methanol concentrations may be required to dissolve nitrosoaromatics. Although the half-wave potentials of the first wave at pH 4–9 is linearly dependent on pH, in strongly acidic media the dependency of the half-wave potential on pH is even more pronounced; at very low pH values, it is suggested that the reduction reaction is (291)

$$C_6H_5NO + 3\,H^+ + 2\,e \rightarrow C_6H_5\overset{+}{N}H_2OH$$

Because the step height of the first reduction wave is very dependent on pH at low pH values, quantitative determinations of the nitroso group are best performed at higher pH values (acetate). In any event, the volatility of nitroso compounds must be given consideration when oxygen that is dissolved in suspension media is to be displaced by a stream of inert gas. Preferably, suspension media should be swept free of oxygen before the sample is introduced (small volume); of course, a small oxygen wave will be seen in the polarogram, but by standardizing the manipulations of sample introduction and running a series of blanks, it is possible to correct the polarogram of the sample for oxygen content (291).

Polarographic reduction of nitrosobenzene in dimethylformamide also involves an initial two-electron step (348). The oscillopolarographic reduction of nitrosobenzene in dimethylformamide (0.2 N NaNO₃) has been studied (350); two waves are observed, the first wave corresponding to formation of the free radical anions of nitrosobenzene. As can be anticipated, the radical anions quickly form azoxybenzene (the second wave).

Reduction of p-nitrosodimethylaniline takes place as one four-electron wave over the pH range 1–10 (444), even though, as noted in earlier paragraphs, reduction of C-nitrosoaromatics involves at least a preliminary two-electron step rapidly followed by another two-electron step yielding the amine. The pH dependency of the half-wave potential shows a transition zone between pH 4 and 5 which is characteristic (291) of the nitroso compounds, that is, formation of a salt of phenylhydroxylamine (or the phenylhydroxylammonium ion). Holleck and Schindler (291) suggest that the following reactions are involved in the reduction of p-nitrosodimethylaniline:

$$(CH_3)_2N-\!\!\!\bigcirc\!\!\!-NO + 2H^+ + 2e = (CH_3)_2N-\!\!\!\bigcirc\!\!\!-NHOH$$

$$(CH_3)_2N-\!\!\!\bigcirc\!\!\!-NHOH \rightleftharpoons (CH_3)_2\overset{+}{N}=\!\!\!\bigcirc\!\!\!=\bar{N} + H_2O$$

$$(CH_3)_2\overset{+}{N}=\!\!\!\bigcirc\!\!\!=\bar{N} + 2H^+ + 2e = (CH_3)_2N-\!\!\!\bigcirc\!\!\!-NH_2$$

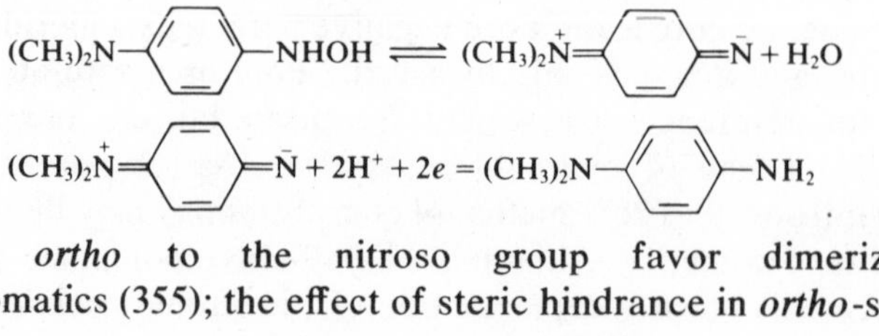

Groups *ortho* to the nitroso group favor dimerization of nitrosoaromatics (355); the effect of steric hindrance in *ortho*-substituted nitrosobenzenes is readily evident in the half-wave potentials of a series of substituted nitrosoaromatics. Polarograms of dimeric compounds show a first wave corresponding to reduction of the monomer and a second wave at a more negative potential corresponding to reduction of the dimer; the height of the second wave changes with temperature and is not linearly related to concentration of the sample (variable dimer content ) (295).

Holleck and Schindler (292) have shown that the longest wavelength band in the absorption spectra of substituted nitrosobenzenes (about 750 nm and interpreted as an $n - \pi^*$ transition) should be proportional to their reduction potentials inasmuch as only the energy of the $\pi^*$ levels are involved (assuming the $n$-level energy is independent of the nature of a substituent on the phenylnitroso radical).

As was indicated in the discussion of the polarographic behavior of $C$-nitro compounds, existence or formation of new resonating structures often complicates polarographic behavior. For example, the shape of the polarogram obtained with $p$-nitrosophenol as a function of pH has been shown to be governed by the relative amounts of $p$-quinoneoxime and $p$-nitrosophenol that are in equilibrium, by the effect of hydrogen ions on reduction products, and by the rates of the chemical reactions that reduction products can undergo (19,291). Thus the nitrosophenols, like the nitrophenols which can form quinoid structures, exhibit ECE reaction mechanisms:

$$ON{-}Ar{-}OH \xrightarrow[2\,H^+]{2\,e} HOHN{-}Ar{-}OH \xrightarrow{-H_2O} HN{=}Ar{=}O \xrightarrow[2\,H^+]{2\,e} H_2N{-}Ar{-}OH$$

In the reduction of $p$-nitrosophenol, as the pH of the suspension medium is increased beyond about 4, the first two-electron wave for reduction to hydroxylamine that is seen in more acidic media shifts to more negative values with increasing diffusion current levels until only a single four-electron reduction step (to amine) is seen in the polarogram. Clearly, in weakly acidic media, the first wave corresponds to the formation of $p$-hydroxyphenylhydroxylamine, which, within the lifetime of the mercury drop, loses water to form the quinoneimine; this is immediately reduced at more negative potentials to $p$-aminophenol. As was the case for nitrosobenzene, strongly acidic solutions are conducive to the formation of a salt of $p$-hydroxyphenylhydroxylamine, and the familiar nonlinear dependency of the half-wave potential on pH is observable at low pH values. $\alpha$-Nitroso-$\beta$-naphthol behaves like $p$-nitrosophenol; the first step occurs at $-0.01$ V versus SCE at pH 4.1 in acetate buffers (372). Additional studies of this compound have been reported by Wilson and Rhodes (737).

### d. N-Nitrosoaromatics

Lund (444) found that polarographic reduction of N-nitrosoaromatic compounds such as N-nitroso-N-methylaniline and N-nitrosodiphenylamine usually forms the corresponding hydrazine derivatives at pH 1–5:

$$\underset{R}{ArNNO} + 5\,H^+ + 4\,e \rightarrow \underset{R}{ArNNH_3^+} + H_2O$$

In basic solutions, the reduction products are the amines and nitrous acid:

$$2\,\underset{R}{ArNNO} + 4\,e + 3\,H_2O \rightarrow 2\,\underset{R}{ArNH} + N_2O + 4\,OH^-$$

The polarographic behavior of the ammonium salt of N-nitroso-phenylhydroxylamine (cupferron) was found by Kolthoff and Liberti (373) to be unlike the behavior of nitrosobenzene or phenylhydroxylamine. Some time later, Elving and Olson (160) made a more detailed study of the reduction of N-nitrosophenylhydroxylamine, N-nitroso-α-naphthylhydroxylamine (neocupferron), and N-nitroso-p-xenyl-hydroxylamine (p-phenylcupferron), used as the ammonium salts, and found that three well-defined polarographic waves can occur between pH values of −0.4 and 12. The single well-defined wave that appears in strongly acidic media splits into two waves at about pH 4–5; at higher pH values, three waves are observed. Coulometric studies, including isolation and identification of major reduction products formed at controlled potential, indicated that reduction in acidic solution (and up to about pH 8) involves a total of six electrons and yields the aromatic hydrazine. The reduction processes that form the third wave (appearing in alkaline media) involve only four electrons and yield the aromatic hydrocarbon.

Polarographic reduction of N-nitroso compounds can be the basis for the determination of amines after reaction with nitrous acid in an acetate buffer (pH 4) or in acetic acid (see Section VI.B.4.1). Typical aromatic amines that may be determined in this way are diphenylamine, centralites (after hydrolysis in 60% sulfuric acid to N-alkylaniline) (636), N-methyltoluenesulfonamide (293), and Terramycin (525). In a typical procedure, an alcoholic solution of diphenylamine and acetic acid is treated with sodium nitrite, the excess nitrous acid is destroyed by sulfamic acid, and the diffusion current is measured at −0.85 V (636).

### 5. Kjeldahl Digestion

The Kjeldahl procedures already described for the nitro group (Section VI.A.6) are also directly applicable to nitroso group determinations.

It is recommended that preliminary reduction of the nitroso group in acidic media be undertaken before the Kjeldahl digestion. The possibility of cleavage of $N$-nitroso compounds (release of nitrogen) must be taken into consideration when alkaline reducing systems are used.

## 6. Gasometric Methods

Nearly all aliphatic and aromatic compounds containing nitrosamine groups (N—NO) can be determined in the nitrometer (525); the N—NO₂ group may also respond similarly (see Section VI.A.7). In general, an $N$-nitroso compound yields nitric oxide in the nitrometer, but a $C$-nitroso compound like $p$-nitrosodimethylaniline does not. However, many $N$-nitroso compounds, especially the alkyl nitroso compounds, undergo secondary competitive reactions which lead to low results in nitrometer methods; for example, the nitrous acid released when an $N$-nitroso compound is treated with concentrated acid may be consumed by reaction with the sample or its decomposition products. Additionally, many aryl $N$-nitroso compounds undergo the Fischer-Hepp rearrangement in strong acids, whereby aryl $C$-nitroso compounds are formed which do not release nitric oxide in the nitrometer. For example, since $N$-nitroso-$N$-methylaniline (phenylmethylnitrosamine) has a free *para* position, it readily forms $p$-nitroso-$N$-methylaniline in acidic media; similarly, $N$-nitroso-$N$-methyl-$\beta$-naphthylamine forms $\alpha$-nitroso-$N$-methyl-$\beta$-naphthylamine.

Compounds containing an $N$-nitroso function that can be quantitatively reduced to hydrazine can be made to yield nitrogen on subsequent oxidation by mild agents such as ferricyanides (743):

$$R_2N—NO \xrightarrow{[H]} R_2N—NH_2 \xrightarrow{[Fe(CN)_6]^{3-}} N_2$$

It is to be recalled that most $N$-nitroso compounds form substituted hydrazines on reduction with zinc and acetic acid; however, quantitative reduction of $N$-nitroso groups to hydrazines may prove to be impossible in altogether too many instances. Lehmstedt (405) and Lehmstedt and Zumstein (406) have recommended ferrous chloride in hydrochloric acid as a convenient reagent for release of nitric oxide from aliphatic and aromatic $N$-nitroso compounds; most $N$-nitro compounds also respond:

$$R_2N—NO + FeCl_2 + HCl \rightarrow NO + R_2NH + FeCl_3$$
$$R_2N—NO_2 + 3\ FeCl_2 + 3\ HCl \rightarrow NO + R_2NH + 3\ FeCl_3 + H_2O$$

Determinations by this procedure generally tend to give low values for N—NO content, but the method is valuable in that —N—NO groups can often be determined in the presence of $C$-nitro and $C$-nitroso groups (488). Nitroso groups attached to heterocyclic nitrogen (such as piperidine) also liberate nitric oxide. Clearly, compounds that can release nitric acid on treatment with sulfuric acid form nitric oxide by reaction with ferrous ion (499):

$$NO_3^- + 3\,Fe^{2+} + 4\,H^+ \rightarrow NO + 3\,Fe^{3+} + 2\,H_2O$$

and those that form nitrous acid release nitric oxide, possibly through intermediate formation of nitrosylsulfuric acid:

$$HNO_2 + H_2SO_4 \rightarrow H(NO)SO_4 + H_2O$$
$$Fe^{2+} + H(NO)SO_4 \rightarrow Fe(NO)^{3+} + HSO_4^-$$
$$Fe(NO)^{3+} \rightarrow NO + Fe^{3+}$$

Jones and Kenner (320) found that cuprous chloride is as effective a reducing agent as ferrous chloride, but stannous chloride appeared to be unsatisfactory.

Phenylhydrazine in large excess reacts with many $C$-nitroso compounds to release nitrogen (111,488); the liberated nitrogen, which must be measured while saturated with benzene and water, is derived from the hydrazine and not the nitroso function, and the reaction may be considered simply as the oxidation of a bisubstituted hydrazine:

$$2\,R{-}NO + 2\,C_6H_5NHNH_2 = R{-}N{=}N{-}R + 2\,C_6H_6 + 2\,H_2O + 2\,N_2$$

The release of nitrogen is seldom quantitative, but as a rule yields exceed 95%. Aromatic $C$-nitro compounds generally do not react with phenylhydrazine at ordinary temperatures; however, the nitro group may be sufficiently activated in certain instances to cause release of nitrogen. At elevated temperatures (autoclave), the aromatic $C$-nitro compounds will release nitrogen (721). The aliphatic nitrosamines (N—NO), such as diethylamine and any type of oxime (C=NOH) do not release nitrogen; on the other hand, substituted aromatic nitroso compounds such as the nitrosoacids and nitrosoaldehydes will release nitrogen. Poly-$C$-nitrosoaromatic compounds release nitrogen when treated with phenylhydrazine; alkyl nitrites, $R_3C$—ONO, do not react with hydrazine to form nitrogen. It is of interest to note that aromatic nitrosamines which readily undergo rearrangement to $C$-nitroso compounds in other analytical methods may be determined by release of nitrogen from phenylhydrazine.

An indirect gasometric method for nitro or nitroso groups consists of determining the hydrogen released from the excess of sodium borohydride employed in reduction of the groups (78). Unfortunately, many other functional groups also consume hydrogen. In some instances, the direct measurement of hydrogen consumption is possible, especially when catalysts such as palladium on barium sulfate (296) are employed.

The reaction of $N$-nitrosodibenzylamines with sodium hydrosulfite at

60°C in basic ethanolic solutions yields nitrogen and hydrocarbon products (536); however, $N$-nitrosobenzylphenylamine does not release nitrogen.

## VII. LABORATORY PROCEDURES

The following procedures for determination of nitro and nitroso groups are given in great detail in order to facilitate their use by laboratory personnel. The procedures are representative of those in wide use today. Although each procedure has been checked in the author's laboratory with a wide variety of samples, it is recommended that at least two methods be used to perform analyses on samples of novel structure or on samples that are grossly contaminated with materials of unknown composition.

### A. DETERMINATION OF $C$-NITRO AND $C$-NITROSO COMPOUNDS BY REDUCTION WITH STANNOUS CHLORIDE

The basis of this method is the consumption of tin(II) chloride by nitro or nitroso groups (284); for example:

$$R\!-\!NO_2 + 3\,SnCl_2 + 6\,HCl \rightarrow RNH_2 + 3\,SnCl_4 + 2\,H_2O$$
$$R\!-\!NO + 2\,SnCl_2 + 4\,HCl \rightarrow RNH_2 + 2\,SnCl_4 + H_2O$$

The method is usually employed for determination of nitro compounds, but since stannous chloride is a powerful reducing agent, other functional groups will interfere (see Section III.B.1). The consumption of stannous ion is obtained iodimetrically as a difference between a blank and a sample; great care must be taken to ensure that the blank has received the same treatment as the sample. Moreover, since certain nitro compounds are more refractory than others, the length of time that the stannous chloride reagent is allowed to remain in contact with the sample must be determined by trial and error; the 1.5-hr treatment prescribed by the procedure will be found adequate for a large number of compounds.

#### Reagents

*Sulfuric acid*, 1:1 (v/v). Mix equal volumes of concentrated reagent-grade sulfuric acid and water; cool.

*Stannous chloride solution*, 2.5 N. Dissolve 283 grams of reagent-grade stannous chloride ($SnCl_2 \cdot 2\,H_2O$) in 245 ml of concentrated hydrochloric acid (d., 1.19), filter through glass wool, and dilute to 1 liter with water that has been deaerated and then saturated with carbon dioxide. The solution is oxidized by air; preferably, it is kept in an automatic buret under carbon dioxide. However, since the oxidation is slow and a blank is run concurrently, it may be kept in ordinary bottles if undue exposure to the atmosphere is prevented.

*Iodine, standardized*, 0.6 N. Prepare and standardize the solution by any convenient method.

*Starch solution*, 1%. Triturate 1 gram of soluble starch in a mortar with enough water to make a smooth, thick cream; pour into 100 ml of boiling distilled water, and continue the boiling until a nearly clear liquid is obtained. Cool; prepare daily.

*Alcohol.* Use aldehyde-free methanol or ethanol from a bottle that has not been unduly exposed to air; alternatively, purify the alcohol by refluxing it for several hours in a rapid stream of carbon dioxide.

**Apparatus**

*Reaction flask.* Fit a straight condenser about 400 mm long to a 300-ml round-bottom flask which has a ground-glass joint as an opening. Pass a narrow glass delivery tube (for carbon dioxide) through the condenser tube, and arrange for it to be held so that it can allow gas to pass over the solution in the flask or so that gas can be passed through the solution, if desired. At least four flasks and condensers should be provided so that two samples and two blanks can be run concurrently.

**Procedure 1**

Displace air in the reaction flask with a brisk stream of carbon dioxide, and then disconnect the flask from its condenser and introduce about 0.1–0.2 gram (weighed to ±0.0002 gram) of the nitro compound into the flask; continue the brisk flow of carbon dioxide. The sample weight should be sufficient to consume about two thirds of the stannous chloride. Through the top of the condenser, introduce 10 ml of 1:1 sulfuric acid solution into the flask and swirl to dissolve the sample.

If the compound is insoluble in the acid, add 5–6 ml (or more) of purified alcohol. Heat may be applied to accelerate dissolution. (Add the same amount of alcohol to the blank.)

When the sample is dissolved, lower the carbon dioxide inlet tube and deaerate the solution by a gentle stream of carbon dioxide (about 7 min). Then raise the gas delivery tube and allow it to drain; disconnect the flask from the condenser and quickly add (from a buret or an accurate pipet) exactly 10 ml of stannous chloride solution.

Reconnect the flask to its condenser, and let carbon dioxide displace any air which may have diffused into the flask. Now, set the flask in a water bath (start at about 50°C), and reduce the stream of carbon dioxide flowing over the solution so that a bubbler in series with the gas stream shows a flow rate of about two bubbles per second; maintain this flow rate for the period of time that the flask is in the water bath. Bring the water bath to a gentle boil, and continue heating the flask for 1.5 hr.

If the nitro compound is volatile, it may get up into the colder portions of the condenser tube; in this event, wash down the inside of the condenser tube with 2-ml portions of purified alcohol at 0.5-hr intervals. It is a good idea to wash down the condenser tube even though no condensed material can be seen.

The reduction is usually complete within 1.5 hr, but some samples may require prolonged heating; cool the contents of the flask, and then disconnect the flask from the condenser. Dilute the contents of the flask with about 150 ml of water that has been deaerated and saturated with carbon dioxide. Rapidly titrate the solution in the flask with 0.6 N iodine solution, adding starch solution near the end of the titration.

### Procedure 2

This procedure is especially useful for samples that dissolve readily in acetic acid.

Dissolve 1–2 grams of the sample in glacial acetic acid in a 250-ml volumetric flask, and bring to the mark with glacial acetic acid. Pipet 25 ml of this solution into a 100-ml volumetric flask (use 25 ml of glacial acetic acid for the blank), and transfer exactly 10 ml of stannous chloride reagent into the flask. Bring to the mark with deaerated water (saturated with carbon dioxide), and displace the air in the flask with carbon dioxide; immediately cap with a stopper carrying a bunsen valve. Put the capped flask in a boiling-water bath and allow to remain for at least 1.5 hr, or 0.5 hr after all initially insoluble matter has dissolved. Remove the flask, cool thoroughly, and fill to the mark (if necessary) with deaerated water; mix well.

By means of a pipet, transfer 25 ml of the solution to a flask containing about 125 ml of 1:10 hydrochloric acid (deaerated in the flask and kept under an atmosphere of carbon dioxide); titrate with 0.1 N iodine solution, using starch indicator. Repeat the titration on successive aliquots; each time rapidly run in the iodine solution closer and closer to the anticipated end-point volume before adding the starch indicator.

**Calculations**

Let  $B$ = iodine needed to titrate the blank (ml),
 $S$ = iodine needed to titrate the sample (ml),
 $W$ = weight of sample taken,
 $F$ = factor = $0.007667 \times$ Normality

**Procedure 1**

$$\frac{100(B - S)F}{W} = \quad \% - NO_2$$

**Procedure 2**

$$\frac{400(B - S)F}{W} = \quad \% - NO_2$$

### B. DETERMINATION OF PRIMARY AND SECONDARY
### $C$-NITROPARAFFINS BY CHLORINATION

The method is intended to be used as a criterion of purity or for samples containing only one nitroparaffin (497).

**Reagents**

*Sodium hypochlorite solution.* Use commercial bleach solution containing about 5.25 wt.% of sodium hypochlorite.

*Hypochlorite reagent A*, for primary nitroparaffin analysis. Dissolve 10 grams ($\pm 0.1$ gram) of reagent-grade sodium hydroxide pellets in about 100 ml of water contained in a flask. Add approximately 145 ml of commercial bleach solution, and dilute to 1 liter with water. Store in an amber glass bottle; the reagent is 0.25 M in sodium hydroxide.

*Hypochlorite reagent B*, for secondary nitroparaffin analysis. Dissolve 80 grams ($\pm 0.1$ gram) of reagent-grade sodium hydroxide pellets in a 1-liter flask containing about 200 ml of water. Cool, add approximately 145 ml of commercial bleach solution, and dilute to the mark with water. Store in an amber glass bottle; the reagent is 2.0 M in sodium hydroxide.

*Glacial acetic acid.* Use reagent-grade material.

*Potassium iodide solution.* Dissolve 20 grams of reagent-grade potassium iodide in 100 ml of distilled water.

*Sodium thiosulfate solution*, 0.1 *N*. Prepare and standardize the solution by any convenient method.

*Starch indicator.* Triturate 0.5 gram of soluble starch in a mortar with enough water to make a smooth, thick cream; pour into 100 ml of boiling water, and continue the boiling until a nearly clear liquid is obtained. Cool; prepare daily.

## Procedure for Primary Nitroparaffins

A sample weighing 0.6–0.7 gram for nitromethane or 1.2–1.4 grams for a higher nitroparaffin must be taken for analysis; determine the weight of the sample to the nearest 0.5 mg.

Weigh the sample into a tared 250-ml volumetric flask containing water, dilute to the mark with water, and mix thoroughly until the sample dissolves. Transfer a 10-ml aliquot of the sample solution into an iodine flask containing 25 ml of hypochlorite reagent A; concurrently, mix 25 ml of the hypochlorite reagent and 10 ml of water in another flask for a blank determination. Stopper the flasks, and allow to stand for 15 min at room temperature.

Cool the flasks in an ice bath, and to each add 15 ml of glacial acetic acid and 15 ml of 20% potassium iodide solution; mix well after each addition. Put a few milliliters of potassium iodide solution in the well surrounding the glass stoppers. Allow the flasks to stand in an ice bath for 10 min; then remove the stopper and wash the stopper and the well with water, allowing the washings to fall into the flask. Titrate the solution in the flasks with 0.1 $N$ sodium thiosulfate solution, using starch solution as the indicator near the end point.

## Procedure for Secondary Nitroparaffins.

Use the above procedure with hypochlorite reagent B.

## Calculations

The equivalent weights of nitroparaffins are determined by the equations

$$CH_3NO_2 + 3\ NaOCl \rightarrow CCl_3NO_2 + 3\ NaOH$$
$$RCH_2NO_2 + 2\ NaOCl \rightarrow RCCl_2NO_2 + 2\ NaOH$$
$$R_2CHNO_2 + NaOCl \rightarrow R_2CClNO_2 + NaOH$$

Now, let  $B$ = thiosulfate required by the blank (ml),
$S$ = thiosulfate required by the sample (ml),
$N$ = normality of thiosulfate,
$W$ = weight of sample (grams),
$V$ = volume aliquot (ml).

Then  $$\frac{25(B - S)N(\text{equiv. wt.})}{WV} = \%\ \text{nitroparaffin by weight}$$

## C. INDIRECT GRAVIMETRIC DETERMINATION OF THE NITRO- AND NITROSOAROMATIC GROUPS WITH COPPER

The procedure is especially useful for determination of pure nitro- and nitrosoaromatic compounds. Aliphatic nitro compounds react altogether too slowly, but the nitrosoparaffins react more rapidly (335). The method is also useful for determination of azo compounds. Most compounds are reduced to the corresponding amines, but side reactions may occur; other reducing agents, such as phenylhydrazine and azo compounds, consume copper in the procedure detailed below. Excessive amounts of chloride ion or other complexing anions and oxidizing agents such as nitrates and ferric ion interfere.

### Reagents

*Reagent-grade absolute ethanol.* Deaerate with carbon dioxide before use.

*Diethylene glycol dimethyl ether.* Distil from metallic copper; after long standing, it is advisable first to distil the solvent from dilute sulfuric acid saturated with ferrous sulfate to remove peroxides and then to distil it from copper. Deaerate the solvent with carbon dioxide before use in the following procedure.

*Electrolytic copper foil* (0.002-in. thick). Cut into 1-cm squares. Wash with dilute sulfuric acid, rinse with deaerated water and deaerated reagent-grade ethanol, dry under a stream of carbon dioxide, and store under vacuum. The metal surface must be shiny and have no spots of oxidation or corrosion products.

*Sulfuric acid*, 6 N. Mix 167 ml of reagent-grade sulfuric acid with 500 ml of distilled water, cool, and dilute to 1 liter with water. Saturate with carbon dioxide before use.

*Acetone.* Deaerate reagent-grade acetone with carbon dioxide.

### Apparatus

A 200-ml round-bottom flask with a ground-glass joint is fitted with a reflux condenser. Place a two-hole rubber stopper on top of the condenser, and insert a long glass tube in one hole for delivery of carbon dioxide into the reaction flask.

**Procedure**

Introduce 20 ml of ethanol or diethylene glycol dimethyl ether and 20 ml of 6 N sulfuric acid into the reaction flask, and deaerate the mixture with a few pieces of Dry Ice or by a stream of carbon dioxide. Dissolve in the solvent sufficient accurately weighed nitro compound to react with 0.5–1.0 gram of copper, and add 6–15 grams of clean, accurately weighed, dry copper. Run a blank concurrently. Reflux the sample and blank solutions for 1–2 hr while passing a slow stream of carbon dioxide into the condenser well above the condensation point in the condenser tube. After completion of the reaction (usually 1–2 hr), decant the liquid from the flask and rinse the copper rapidly six times with deaerated water, then three times with deaerated acetone. Dry the copper under vacuum, and weigh.

**Calculations**

The loss in weight of copper is related to the amount of nitro or nitroso compound according to the equations

$$3\,Cu + ArNO_2 + 7\,H^+ \rightarrow ArNH_3^+ + 2\,H_2O + 3\,Cu^{2+}$$
$$2\,Cu + ArNO + 5\,H^+ \rightarrow ArNH_3^+ + H_2O + 2\,Cu^{2+}$$

Hence let $L$ = loss (grams) in weight of copper,
$W$ = sample weight (grams).

Then

$$\% \text{ RNO}_2 \text{ by weight} = \frac{RNO_2 \times L \times 100}{190.62\ W}$$

and

$$\% \text{ RNO by weight} = \frac{RNO \times L \times 100}{127.08\ W}$$

## D. DETERMINATION OF EQUIVALENT WEIGHT OF AROMATIC C-NITRO AND C-NITROSO COMPOUNDS BY REDUCTION–DIAZOTIZATION (741)

This procedure is very easily performed on a wide variety of nitro and nitroso compounds. It is useful for establishing the identification of unknown nitro compounds, especially when other methods (e.g., infrared, ultraviolet, and mass spectrometry) can supply additional information.

**Reagents**

*Sodium nitrite solution*, 0.1 *N*. Dissolve about 7 grams of pure sodium nitrite in water, and then dilute to 1 liter. Dry pure sulfanilic acid monohydrate for at least 3 hr at 120°C to remove water of crystallization. Dissolve exactly 0.6928 gram (0.004 mol) of the anhydrous acid in 100 ml of water containing about 0.2 gram of sodium hydroxide. Then add 20 ml of concentrated hydrochloric acid, cool to 15°C, and titrate with the nitrite solution, using starch-iodide paper as an external indicator. Run a blank to the same end-point color intensity on the starch-iodide paper. Compute the normality of the sodium nitrite solution as follows:

$$N = \frac{4.0}{\text{ml of nitrite} - \text{ml of blank}}$$

Alternatively, the sodium nitrite solution can be standardized with 0.004 mol of *p*-nitroaniline. The solution is stable for at least 6 months.

*Glacial acetic acid.* Use reagent-grade material.

*Zinc dust.* Use technical grade or better.

*Sodium bromide.* Use U.S.P. or reagent-grade material.


**Procedure**

Transfer an accurately weighed sample (about 3 mequiv.—usually about 0.5 gram) to a 250-ml ground-joint Erlenmeyer flask, and add 25 ml of glacial acetic acid; dissolve the sample with the aid of heat if necessary, and then add 10 ml of concentrated hydrochloric acid and 25 ml of water and heat nearly to boiling. (If the sample cannot be maintained in solution, it is permissible to use more acetic acid and less water.) Add 5 grams of zinc dust in small portions and at such a rate that no material is lost by excessive frothing. Then attach a small reflux condenser, and reflux the mixture for 20–30 min to complete the reduction. If the color formed during the reaction fails to disappear at the end of the refluxing period, add a little more zinc dust and continue the refluxing a few minutes longer.

Remove unreacted zinc by filtering the hot solution through a small Buchner funnel; break up lumps of zinc with a glass rod, and rinse the reaction flask with small portions of hot water (pass the washings through the funnel to remove all amine from the residue). Quantitatively transfer the filtrate and washings (usually 200–250 ml) to a beaker, and cool the solution to 15–20°C. Make sure that the solution is distinctly acidic to Congo Red indicator (pH < 3); then add 1–2 grams of sodium bromide and stir to dissolve.

Titrate the solution with standardized sodium nitrite, using starch-iodide papers as an external indicator. At the end point, a faint blue spot forms immediately. Run a blank titration, using the same volume of water acidified with acetic acid. The blank should not be greater than 2–3 drops.

**Calculation**

$$\text{Equiv. wt.} = \frac{1000 \times \text{sample wt (grams)}}{\text{ml of NaNO}_2 \times N}$$

## E. DETERMINATION OF NITRO COMPOUNDS BY CONSTANT-POTENTIAL COULOMETRY

The method is predicated on the electrolytic reduction of the nitro group by coulometry at a potential of $-0.90$ V or less versus a silver–silver chloride reference electrode (390); other substances that can be reduced at this potential will interfere. By appropriate selection of the constant potential at which electrolysis is allowed to take place, aromaticmononitro compounds can be determined in the presence of aliphatic mononitro compounds by successive reductions (156).

**Reagents**

*Methanol.* Use reagent-grade methanol.

*Lithium chloride 0.5 M.* Prepare a 0.5 M solution of reagent lithium chloride in water.

*Boric acid.* Use reagent-grade material.

*Acetic anhydride.* Use reagent-grade or any "pure" variety.

*Sodium acetate.* Use reagent-grade trihydrate.

*Mercury.* Use triple-distilled or reagent-grade material; however, recovered mercury which has been washed with nitric acid and distilled is satisfactory.

*Hydrochloric acid, N.* Prepare an approximately normal solution of hydrochloric acid by mixing 9 ml of reagent-grade acid with 100 ml of water.

**Apparatus**

### a. Coulometric Cell

Almost any variety of coulometric cell can be used in which the potential of a mercury pool cathode can be maintained at a constant level with respect to a reference electrode that is in communication with the catholyte by a salt bridge; it must be possible to pass current between the pool cathode and a suitable anode. In a great many instances, it is best to isolate the anode compartment from the cathode compartment. An easily constructed apparatus for this purpose is shown in Fig. 7; it is essentially the arrangement suggested originally by Lingane et al. (428). Other, similar H-cell arrangements may be used; indeed, it is often possible to use open beakers in which the electrodes are in direct communication with each other. However, the author has found the cell of Fig. 7 adequate for most practical purposes even when the configuration is followed in principle rather than in exact embodiment.

The cathode compartment of the apparatus in Fig. 7 is made of a 250-ml Erlenmeyer flask or similar container, to which is fastened a side arm (with fine-porosity glass frit) communicating to an anode compartment. Solutions in the cathode and the anode compartments can be deaerated by passing an inert gas through the tubes indicated as N in Fig. 7; these tubes can be omitted, if desired, and deaeration effected by inserting a gas delivery tube into each compartment.

The interconnecting arm (about 50 mm long) should be made as wide as possible (e.g., 20 mm) to ensure a low-resistance path, and the glass frit should be sealed as closely as possible to the cathode compartment and about 10 mm above the bottom of the flask; the fine-porosity glass frit usually is a sufficient barrier to prevent undue mixing of catholyte

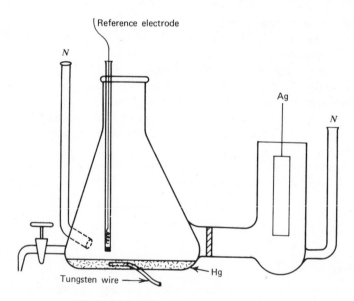

Fig. 7. Constant-potential coulometric cell.

and anolyte. A drain tube fitted with a stopcock is located about 5 mm above the bottom of the flask, and the stopcock is kept close to the flask in order to minimize the volume of trapped fluid in the drain tube. The drain tube permits removal of catholyte without disturbing the mercury pool.

A 5-mm layer of pure mercury is put into the bottom of the cathode compartment; electrical contact can be made via a tungsten wire sealed into the bottom of the Erlenmeyer flask or by other means. The anode is a silver–silver halide electrode formed by immersing in the anolyte a heavy silver plate about 50–60 cm$^2$ in exposed area or a coil made from a 30–40 cm length of 2-mm silver wire.

A silver–silver halide reference electrode is made from a small coil of silver wire enclosed in a 10-mm glass tube which has a fine-porosity glass frit at the end immersed in the catholyte (see Fig. 7). Conversion of potentials from one reference electrode to another has been discussed by Tsuji and Elving (691).

## B. Potentiostat

Any potentiostat may be used if it can deliver the high voltages (say 100 V) required for electrolysis of nonaqueous solutions. However, elaborate equipment is not necessary; for example, the circuit shown in Fig. 8 can be used effectively. The autotransformer, $T$, supplies ac voltage to the full-wave diode bridge ($D_{1-4}$); the capacitor-input filter section, $C_1$, $H$, $C_2$, provides relatively ripple-free dc power to potentiometer $R_1$, which acts as a voltage divider to supply the coulometer and electrolysis cell. The ammeter, $A$, provides indication of the current flowing through the electrolysis cell. The potential between the reference electrode and the pool is the potential indicated by the voltmeter when the galvonometer is at null. Of course, the combination $B$, $R_2$, $V$, and $G$ constitutes the familiar potentiometer setup used widely to measure electrode potentials; consequently, the combination can be replaced with a laboratory-type potentiometer. Alternatively, and more conveniently, the potential between the reference electrode and the pool can be measured with a vacuum tube voltmeter which has at least a 10–11 MΩ input resistance; in this instance, $V$, $G$, $R_2$, and $B$ are omitted.

## c. Coulometer

Although nearly any type of coulometer can be used, the one described below will provide accurate measurement of low coulomb values even when 100–150 C is required in an analysis. The coulometer (427)

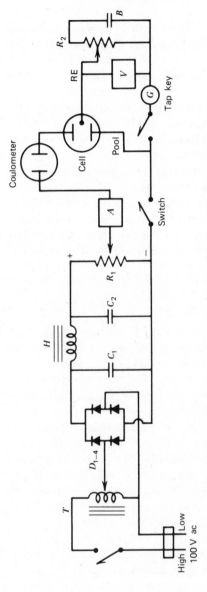

Fig. 8. Circuitry for constant-potential coulometry. $T$, autotransformer; $D_{1-4}$, diodes, IN3917; $C_1$ and $C_2$, 50-$\mu$F, 300-V electrolytic capacitor; $H$, 3–10 H, 100-mA filter choke; $B$, one or two 1.35-V mercury cells; $R_1$, 100-$\Omega$, 10-W potentiometer; $A$, 0–10, -100, -500 mA meter; $R_2$, 500-$\Omega$ potentiometer (3–10 turns); $V$, 0–1.5-V meter, 20,000 $\Omega$/V; $G$, galvanometer, 1 $\mu$A/div.; RE, reference electrode.

250

consists of a helical silver wire anode (area about 80 cm$^2$) and a platinum wire cathode suspended in an electrolyte which is 0.2 $M$ in potassium sulfate (to provide conductance) and 0.03 $M$ in potassium bromide. The net coulometer reaction is

$$2\,Ag + 2\,Br^- + 2\,H_2O = 2\,AgBr + H_2 + 2\,OH^-$$

The hydroxide ion that is produced is titrated with standard 0.01–0.03 $N$ hydrochloric acid solution; total coulombs = ml of acid × $N$ acid × 96.5 C/mequiv. The end point of the titration is best determined by a glass electrode, in every instance, at an indicated pH of 7; thus the pH of the electrolyte is brought to pH 7 before allowing current to pass through and then is returned to pH 7 after electrolysis. Of course, the electrolyte must be prevented from coming into contact with carbon dioxide; a stream of air free of carbon dioxide is bubbled through the electrolyte to provide agitation and to exclude atmospheric carbon dioxide. Owing to the sensitivity of silver bromide to light, the electrolyte must be protected at all times during an electrolysis, especially because hydrogen ion is formed:

$$AgBr + h\nu = Ag + \tfrac{1}{2}\,Br_2$$
$$Br_2 + H_2O = H^+ + Br^- + HOBr$$

The coulometer can readily determine 1 C with an accuracy of 1%, and when larger quantities (at least 10 C) are involved, the accuracy can be increased to ±0.1%.

### d. Preparation of Reference Electrode

Mix 400 ml of methanol and 100 ml of 0.5 $M$ lithium chloride solution, and adjust the pH to 2.0 with $N$ hydrochloric acid. Transfer about 50 ml of acidulated methanolic lithium chloride solution to a beaker; mix a drop of 0.1 $N$ silver nitrate solution with about 10 ml of the same solution, and put the mixture into the reference electrode tube. Connect the reference electrode to the positive pole of the potentiostat or other source of current, and use a platinum wire as a cathode. Put the reference electrode and the platinum wire in the acidulated methanolic lithium chloride solution, and by passage of current (a few milliamperes for 30–40 C) allow a film of silver chloride to form on the silver wire in the reference electrode tube.

**Procedure**

**a. Prereduction**

Take about 100 ml of electrolysis solution (4 parts of methanol mixed with 1 part of 0.5 $M$ aqueous lithium chloride) and adjust the pH to 2.0 with $N$ hydrochloric acid; put enough mercury to form a layer on the bottom of the electrolysis cell, and pour in the acidulated methanolic solution of lithium chloride. Place a stirring bar on the mercury pool, and then set the electrolysis apparatus on a magnetic stirrer. To remove dissolved oxygen, pass through the solution an oxygen-free stream of nitrogen gas (575) for at least 10 min while the magnetic bar is rotated at a very high speed. It is best to pass the gas first through a large volume of the electrolyte in order to saturate it with methanol. Now place the anode in its compartment and start the prereduction.

Position the reference electrode so that it is about 1 mm above the mercury pool. Then set the reference potential at $-0.90$ V for aromatic nitro compounds or $-0.95$ V for aliphatic nitro compounds (voltmeter of Fig. 8, set by $R_2$), turn on the electrolysis current, and increase the current through the electrolysis cell (ammeter $A$ in Fig. 8) until the galvanometer is nulled. Maintain the null point of the galvanometer until the current drops to a steady value of 3–4 mA (the actual value is best found by trial). Record the steady-state value as the initial background current. Disconnect the electrolysis cell, but maintain the flow of inert gas.

**b. Analysis of Sample**

Set up the coulometer and adjust the pH of the electrolyte to $7.0 \pm 0.2$; record the initial pH value. Place the coulometer in series with the electrolysis cell while the current is off (platinum cathode of coulometer connected to anode of electrolysis cell).

Add the sample of nitro or nitroso compound (an amount requiring 75–150 C is ideal, but lower values can be handled). Turn on the electrolysis current; at the same time, record the starting time and immediately increase the current flow through the electrolysis cell until the galvanometer is balanced or a current of not more than 75–80 mA flows through the electrolysis cell circuit. Maintain either a maximum current of 75–80 mA or a galvanometer null until the electrolysis current remains steady at a value below 5 mA for 10 min or decreases 0.1 mA over a 15-min interval; terminate the reduction by recording the final background current and noting the time at which current flow is interrupted. Also, disconnect the coulometer and remove the reference cell from the electrolysis solution.

Quickly titrate the coulometer solution to the pH value recorded initially. Calculate the total number of coulombs from

$$\text{Total coulombs} = \text{ml of acid} \times N \text{ acid} \times \frac{96.5 \text{ C}}{\text{mequiv.}}$$

Correct the computed value for the background current contribution:

Background correction

$$= \frac{\text{initial background (A) plus final background (A)} \times \text{total seconds}}{2}$$

Hence   Net coulombs = total coulombs − background correction

After each determination, regenerate the silver anodes by subjecting them to electrolysis as cathodes in water at about 50 mA (platinum anodes). When the electrodes appear clean or light gray, terminate the regeneration; wash the electrodes before use.

The reduction can also be performed in solutions that are more strongly buffered; for example, a 0.1 $M$ solution of boric acid in methanol is an excellent electrolyte for aromatic nitro compounds (− 1.00 V), and 0.1 $M$ acetic anhydride − 0.1 $M$ sodium acetate in methanol is suitable for aliphatic nitro compounds (− 1.20 V). Of course, the reference electrode must be prepared for use in the same electrolyte.

**Calculation**

A general formula is

$$\text{mequiv. of mononitro or mononitroso compound} = \frac{\text{net coulombs}}{n \times 96,500}$$

The value of $n$ is determined by the product formed by reduction; the following table indicates typical values of $n$ to be expected in acidic media:

| Functional group | Reduction product | $n$ |
|---|---|---|
| Arom. $CNO_2$ | $C—NH_2$ | 6 |
| Arom. CNO | $C—NH_2$ | 4 |
| Aliph. $CNO_2$ | C—NHOH | 4 |
| Arom. $NNO_2$ | $N—NH_2$ | 6 |
| Aliph. $NNO_2$ | $N—NH_2$ | 6 |
| Arom. NNO | $N—NH_2$ | 4 |
| Aliph. NNO | $N—NH_2$ | 4 |

A more general formula is

$$\% \text{ reducible compound} = \frac{(\text{net coulombs}) \text{ mol. wt. of compound}}{n \ (96.5) \ (\text{wt. of sample})}$$

## F. DETERMINATION OF AROMATIC NITRO AND NITROSO GROUPS BY REDUCTION WITH TITANOUS IONS

General volumetric methods suitable for determining nitro and nitroso groups by reduction with titanous salts are given in Section VII.C of "Diazonium Group," Vol. 15. Because titanous ion is a strong reductant, a variety of functional groups other than nitro and nitroso groups are also reduced. The methods given below are useful for reduction of aromatic C-nitro compounds; on the other hand, aliphatic nitro compounds and nitro groups on the side chains of aromatic compounds are not always quantitatively reduced.

One of the following procedures can often be used to differentiate between aromatic C-nitro and C-nitroso compounds (447). Macroprocedures are described, but semimicro procedures (one-fifth scale) can readily be adapted by appropriate reductions of all quantities and the use of correspondingly smaller apparatus.

**Reagents**

*Titanous chloride solution*, 0.2 N. Prepare as indicated in Section VII.C of "Diazonium Group," Vol. 15. Determine the titer of this solution in terms of standardized ferric alum solution (35,308,309,369,637). For this purpose, flush a titration flask with oxygen-free inert gas, and then add 30 ml of titanous chloride; add 20 ml of 12 N hydrochloric acid. Titrate rapidly with ferric alum until the pale blue color almost disappears, and then introduce 10 ml of 2.5 M ammonium thiocyanate solution; continue the titration until there is obtained a pink coloration which lasts at least 1 min.

*Ammonium thiocyanate solution*, 2.5 M. Dissolve 95 grams of the dry salt in 500 ml of water.

*Ferric ammonium sulfate solution*, 0.2 N. Prepare and standardize as indicated in Section VII.C of "Diazonium Group."

*Sodium acetate solution*, 2.5 M. Dissolve 170 grams of the trihydrate in 500 ml of water; filter if necessary.

**Special Apparatus**

Apparatus of the type indicated in Fig. 9 of Section VII.C of "Diazonium Group," Vol. 15, should be used; of course, the reservoir should be as large as possible (about 1 liter), and the buret should have a capacity of 50 ml (0.1-ml graduations). Alternatively, the titanous solution can be stored in a large bottle (painted black) under inert gas and delivered to a dispensing buret by pressure of the inert gas. Another form of dispensing apparatus has been described by Stone (668).

## Procedure

### a. Preparation of Sample Solution

Weigh 0.3750–1.000 gram of sample, transfer to a 100-ml volumetric flask, add about 80 ml of 95% ethanol, and mix until dissolution is complete; dilute to the mark with ethanol.

### b. Determination of $C$-Nitro Groups

Pipet 20 ml of the sample solution into the reaction flask, and add 25 ml of 2.5 $M$ sodium acetate. Flush the apparatus with a stream of oxygen-free inert gas. Titrate with 0.2 $N$ standardized titanous chloride solution until the solution in the reaction flask turns deep violet. Wait 3–5 min, and then add 20 ml of concentrated hydrochloric acid. While keeping a stream of inert gas flowing into the reaction flask, position the flask under a buret containing 0.2 $N$ ferric ammonium sulfate solution and titrate until the pale blue color almost disappears. Then add 10 ml of 2.5 $M$ ammonium thiocyanate solution, and continue the titration until a pink coloration is obtained. The end point is considered to be at hand when the pink color lasts for 1 min. Perform a blank titration.

Since the nitro group is ordinarily reduced to an amino group, $NO_2 \approx 6\ e$.

### c. Determination of $C$-Nitroso Groups

Follow the procedure given above for $C$-nitro groups, but use 25 ml of the sample solution. Perform a blank titration. For nitroso groups which are reduced to amino groups, $NO \approx 4\ e$.

### d. Determination of $N$-Nitroso Groups

Follow the procedure given above for $C$-nitro groups, but use 25 ml of the sample solution, add at least a 100% excess of the titanous solution, and let the reduction proceed for at least 15 min before adding hydrochloric acid. Perform blank titrations.

### e. Analysis of Mixtures of Aromatic *C*-Nitro and *C*-Nitroso Compounds

A large number of mononitro compounds and a few dinitro compounds are not reduced by titanous chloride under the conditions prescribed in the following procedure; in contrast, aromatic *C*-nitroso (but not *N*-nitroso) compounds are quantitatively reduced. The ease of reduction of aromatic *C*-nitro compounds increases as the electron-withdrawing properties of other substituents on the aromatic ring increase; thus 3,5-dinitrobenzoic acid, picric acid, and *p*-nitrobenzaldehyde are more or less reduced by titanous chloride in the following procedure.

Prepare a solution of the sample by dissolving 0.4–1.0 ($\pm 0.0005$) gram in 95% alcohol and diluting to exactly 100 ml. Transfer 25 ml of this solution to the reaction flask, and add 20 ml of 12 $M$ hydrochloric acid; flush the reaction flask with oxygen-free inert gas. Then add 30–35 ml of 0.2 $N$ titanous chloride solution and let stand for at least 5 min. Place the reaction flask under a buret containing ferric ammonium sulfate, and titrate until the blue color almost disappears. Add 10 ml of 2.5 $M$ ammonium thiocyanate solution, and titrate until a pink color is obtained which lasts for 1 min. Perform blank titrations. The accuracy of the method must be checked with samples prepared from the pure aromatic compounds.

## G. GENERAL PROCEDURE FOR THE QUANTITATIVE POLAROGRAPHIC DETERMINATION OF NITRO AND NITROSO COMPOUNDS

Many nitro and nitroso compounds are readily reducible in an acidic medium such as 0.05 $M$ sulfuric acid or 0.1 $M$ hydrochloric acid, and the waves often have plateaus which can be used for quantitative determination. Although these simple aqueous media are sufficient for very low concentrations of nitro or nitroso compounds, the inclusion of a solvent such as ethanol or methanol usually assures solubility of samples in relatively large amounts; moreover, solutions of samples or standards can be readily prepared in such solvents.

### Reagents

*Sulfuric acid suspension medium.* Dissolve 3 ml of concentrated sulfuric acid in 500 ml of water. Disperse 100 mg of gelatin in 100 ml of boiling water; cool. Mix the gelatin dispersion and the dilute sulfuric acid, and then dilute to 1000 ml.

*Ethanol, 95%.* Use reagent-grade material; methanol may be substituted.

## Apparatus

A dc polárograph and dropping mercury electrode system of nearly any variety can be used; the electrolysis cells can also be of any type, but they should permit use of a reference electrode (saturated calomel) for the best results. On the other hand, since it is assumed that the quantitative analysis is to be performed on samples containing a nitro or nitroso compound of known composition and that the pure compound at selected concentrations will be used to establish polarograms under defined conditions, a simple cell which permits application of potentials between the dropping mercury electrode and a pool of mercury will yield excellent results. Control of the temperature of the polarograph cell should be adequate to maintain contents to within $\pm 0.5°C$, and preferably to $\pm 0.2°C$.

## Procedure

Prepare a series of calibrant solutions by dissolving known weights of the pure compound in 95% ethanol; typically, these solutions should be from 0.005 to 1 M in mononitro compound. Pipet 10 ml of each of the calibrant solutions into 100-ml volumetric flasks, and then dilute to the mark with sulfuric acid suspension medium (containing gelatin).

Obtain a polarogram for each of the above solutions between 0 and 1.5 V versus SCE and at a constant head of mercury; keep the polarograph cell and its contents at some constant, known temperature which can also be maintained when samples are run at a later time. Of course, oxygen must be completely removed from solutions (and the polarograph cell) by an inert gas. Also obtain a polarogram of the suspension medium (10 ml of 95% ethanol diluted to 100 ml with sulfuric acid suspension medium). Find a voltage on the polarograms where the plateaus on the steps are sufficiently flat to permit accurate measurement of limiting currents. Subtract the limiting current of the suspension medium from the limiting currents obtained for the calibrating solutions, and plot the values obtained against the respective concentrations of nitro compound. Ideally, these points should fall on a straight line, but a curve is acceptable if it is reproducible.

Prepare a solution of the sample by dissolving a known weight in 100 ml of ethanol. Pipet 10 ml of the ethanolic sample solution into a 100-ml volumetric flask, and dilute to the mark with sulfuric acid suspension medium.

Rinse the polarograph cell with a mixture made by diluting 10 ml of ethanol to 100 ml with sulfuric acid suspension medium; then put enough of the solution in the cell to cover the dropping mercury electrode. Remove oxygen with a stream of oxygen-free nitrogen, and obtain a polarogram of the solution. Maintain the same temperature and the same head of mercury as were used in preparing the standard graph. Discard the solution in the polarographic cell, and rinse the cell with the sample solution. Put enough sample solution in the cell to cover the dropping mercury electrode. Remove oxygen by a stream of nitrogen; then obtain a polarogram at the same temperature used to obtain the calibration graph. Subtract the limiting current of the suspension medium from the limiting current of the sample solution. With this value of current, obtain the concentration of nitro compound in the sample solution from the calibration curve.

## H. DETERMINATION OF NITRO AND NITROSO COMPOUNDS BY ADSORPTION CHROMATOGRAPHY

This method is characteristic of chromatographic procedures used to separate mixtures of C-nitro, C-nitroso, N-nitro, and N-nitroso compounds. With minor variations, the procedures described below can be used to isolate nitrate esters.

### Reagents

*Silicic acid.* Use reagent-grade material, preferably Merck's silicic acid. Poorer grades of silicic acid are often improved by grinding (456).

*Celite.* Use No. 535 acid-washed grade (Johns-Manville). An alternative is calcined kieselguhr.

*Ethanol.* Use reagent-grade, anhydrous.

*Ligroin.* The fraction boiling at 60–70° (or a closer cut) is shaken several times with fuming sulfuric acid and then allowed to stand overnight with a portion of the acid. Then it is washed three times with aqueous sodium carbonate (5–10%) and several times with water, dried over sodium sulfate, and distilled from sodium metal.

*Acetone, ethyl ether, ethyl acetate, and benzene.* Reagent-grade materials are satisfactory.

### Apparatus

The chromatographic tubes preferably are about 9 mm in diameter (inside) with a column length of 130 mm, 10 mm in diameter and about 150 mm long, or about 19 mm in diameter and 250 mm long. The columns may be of almost any design; they are available from supply houses or can be improvised. The smaller column requires about 1 gram of adsorbent; the larger column can be packed with at least 15 grams. Arrangements should be provided to draw solvents through packed columns; for this purpose water-pump vacuum is ample, or, alter-

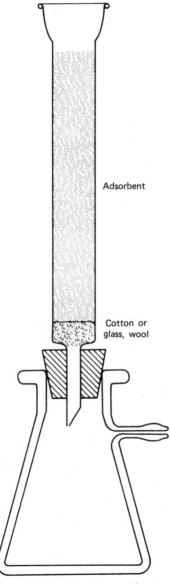

Adsorbent

Cotton or glass, wool

Fig. 9. Typical apparatus for chromatographic adsorption analysis.

natively, solvents can be driven through the columns by pressure or by combinations of pressure and vacuum. A simple form of chromatographic apparatus is shown in Fig. 9.

## Preparation of Columns

Prepare the adsorbent by mixing equal parts by weight of silicic acid (or alumina) with Celite; if the flow is too rapid, use a mixture of 2 parts of active material with Celite. The flow through a column should be of the order of 5–10 ml/min in a 19-mm tube. A plug of cotton is placed at the bottom of the chromatographic tube, and sufficient adsorbent is poured in to form a column 2–3 cm long. Gentle suction is applied and then released; the adsorbent in the tube is leveled by gentle tapping. Then, while gentle suction is applied, another 2–3 cm layer of powder is formed in the tube; suction is released and the tube tapped. This procedure is repeated until the tube is packed to the desired height. An alternative method involves use of a plunger-tamper to pack incremental volumes of the absorbent (669).

## Sample Solvents

All solvents should be purified by passage through columns of silica gel or alumina to remove impurities. When dealing with mixtures of substances of widely different compositions, it is important to use a solvent that can retain all the components in solution. For the series of compounds under consideration, the nitramines are sufficiently soluble only in such solvents as acetone, pyridine, ethyl acetate, and nitroparaffins. Unfortunately, these solvents are very strong developers or eluents for chromatographic adsorbents. However, since the nitramines are strongly adsorbed on silicic acid, they are retained in the upper regions of a column of adsorbent, and then they can be developed into compact zones by suitable mixtures of benzene and ligroin (see below). Other developers are 1:4 dioxane–ligroin, 1:9 ethyl acetate–ligroin, and mixtures of benzene and ethyl ether.

Samples containing insoluble substances such as nitrocellulose are customarily extracted with solvents such as methylene chloride. Thus, after such a sample has been extracted for about 3 hr in a Soxhlet extractor, the methylene chloride is evaporated at a low temperature and the residue is taken up in an appropriate solvent for placement on chromatographic columns. Since N-nitroso compounds are readily decomposed by light, extracts should be evaporated in subdued light.

## Procedure

### Separation of Nitroglycerin from N-Nitrosodiphenylamine, 2-Nitrodiphenylamine, and N-Nitroso-4-nitrodiphenylamine

Dissolve the sample (or the extraction residue) in 15 ml of a mixture of 2 parts by volume of benzene with 3 parts of ligroin. Transfer the sample solution to a $19 \times 250$ mm column packed with a mixture of 2 parts by weight of silicic acid and 1 part of Celite that has been prewashed by $V$ ml of ether and $2V$ ml of ligroin [$V$ ml is the volume of solvent required to wet the column completely (613)]. Develop with $2V$ ml of a 1:1 mixture of benzene and ligroin, and then follow with $V$ ml of ligroin. The nitroglycerine remains in the column; the other compounds are in the eluate. The position of the nitroglycerine in the column can be determined by pushing out the adsorbent column and then streaking with diphenylamine reagent (a 1% solution of diphenylamine in concentrated sulfuric acid).

### Separation of Nitroglycerin from N-Nitrosodiphenylamine and 2-Nitrodiphenylamine

Dissolve the sample in 10 ml of 1 vol. of benzene and 3 vols. of ligroin, and transfer to a $19 \times 250$ mm column packed with a mixture of 4 parts by weight of silicic acid with 1 part of Celite. Develop first with $1.8V$ ml of a mixture of 1 vol. of benzene and 4 vols. of ligroin, followed by $V$ ml of ligroin, and finally by $2V$ ml of a 2% solution of ether in ligroin. The nitroglycerin is retained in the upper part of the column, and N-nitrosodiphenylamine will be below it in the column; 2-nitrodiphenylamine will be in the eluate.

## REFERENCES

1. Adams, R., and F. L. Chen, in Gilman and Blatt, Eds., *Organic Syntheses*, Coll. Vol. I, John Wiley, New York, 1944, p. 240.

2. Albert, A., and B. Ritchie, *J. Chem. Soc.*, **1943**, 458.

3. Alberts, G. S., and I. Shain, *Anal. Chem.*, **35**, 1859 (1963).

4. Aldrovandi, R., and F. de Lorenzi, *Ann. Chim. (Rome)*, **42**, 298 (1952); through *Chem. Abstr.*, **46**, 10051 (1952).

5. Almstrom, G. K., *J. Prakt. Chem. (2)*, **95**, 257 (1917).

6. Altshuller, A. P., and I. R. Cohen, *Anal. Chem.*, **32**, 881 (1960).

7. Amarosa, M., and M. R. Cesaroni, *Gazz. Chim. Ital.*, **83**, 853 (1953).

8. Anderson, K. D., C. J. Crumpler, and D. L. Hammick, *J. Chem. Soc.*, **1935**, 1679.

9. Anderson, W., *J. Chem. Soc.*, **1952**, 1722.

10. Anding, C. E., Jr., B. Zieber, and W. M. Snalisoff, *Anal. Chem.*, **6**, 41 (1934).

11. Angell, F. G., *Analyst*, **78**, 603 (1953).

12. Anger, V., *Mikrochim. Acta*, **2**, 3 (1937).

13. Anger, V., *Mikrochim Acta*, **1960**, (1), 58.

14. Arai, T., M. Tsuchihashi, *Denki Kagaku*, **34**, 144 (1966); through *Chem. Abstr.*, **66**, 121519 (1967).

15. Armand, J., *Bull. Soc. Chim. Fr.*, **1965**, 1658.

16. Armand, J., *Bull. Soc. Chim. Fr.*, **1965**, 3246.

17. Arthur, P., et al., *Anal. Chem.*, **27**, 895 (1955).

18. Ashworth, M., *Collect. Czech. Chem. Commun.*, **13**, 229 (1948).

19. Astle, M. J., and W. V. McConnell, *J. Am. Chem. Soc.*, **65**, 35 (1943).

20. Aston, J. G., and D. E. Ailman, *J. Am. Chem. Soc.*, **60**, 1930 (1938).

21. Aston, J. G., D. F. Menard, and M. G. Mayberry, *J. Am. Chem. Soc.*, **54**, 1531 (1932).

22. Association of Official Agricultural Chemists, *Official and Tentative Methods of Analysis*, 6th ed., A.O.A.C., Washington, D.C., 1945.

23. Auerbach, M. E., *Anal. Chem.*, **22**, 1287 (1950).

23a. Bachman, W. E., *J. Am. Chem. Soc.*, **53**, 1524 (1931).

24. Baillie, A., A. K. Macbeth, and N. I. Maxwell, *J. Chem. Soc.*, **117**, 880 (1920); through *Chem. Abstr.*, **15**, 7501 (1921).

25. Baker, P. R. W., *Analyst*, **80**, 481 (1955).

26. Ballod, A. P., et al., *Zh. Anal. Khim.*, **14**, 188 (1959); through *Chem. Abstr.*, **53**, 12919 (1959).

27. Bamberger, E., *Berichte*, **30**, 1248 (1897).

28. Bamberger, E., and W. Ham, *Ann. Chem.*, **382**, 82 (1911).

29. Bamberger, E., and J. Meimberg, *Berichte*, **26**, 496 (1893).

30. Banerjee, P. C., *J. Indian Chem. Soc.*, **19**, 30 (1942); through *Chem. Abstr.*, **36**, 5724 (1942).

31. Banks, T., C. Vaughn, and L. N. Marshall, *Anal. Chem.*, **27**, 1348 (1955).

32. Bard, A. J., *Anal. Chem.*, **33**, 11 (1961).

33. Bard, A. J., Ed., *Electroanalytical Chemistry*, Marcel Dekker, New York, Vol. III, 1969, and Vol. IV, 1970.

34. deleted.

35. Becker, W. W., and W. E. Shaefer, in Mitchell, Kolthoff, Proskauer, and Weissberger, Eds., *Organic Analysis*, Vol. II, Interscience, New York, 1954, p. 76.

36. Becker, W. W., and W. E. Shaefer, in Mitchell, Kolthoff, Proskauer, and Weissberger, Eds., *Organic Analysis*, Vol. II, Interscience, New York, 1954, p. 95.

37. Bedard, M., J. L. Myers, and G. F. Wright, *Can. J. Chem.*, **40**, 2278 (1962).

38. Belcher, R., R. L. Bhasin, and T. S. West, *J. Chem. Soc.*, **1959**, 2585.

39. Belcher, R., and Bhatty, M. K., *Analyst*, **81**, 124 (1956).

40. Belcher, R., Y. A. Gawargious, and A. M. G. MacDonald, *J. Chem. Soc.*, Suppl. 1, **1964**, 5698.

41. Bell, J. A., and I. Dunstan, *J. Chromatogr.*, **24**, 253 (1966).

42. Bellamy, L. J., *The Infrared Spectra of Complex Molecules*, Methuen, London, 1958, Ch. 17.

43. Bennett, C. E., and P. J. Elving, *Collect. Czech. Chem. Commun.*, **25**, 3213 (1960); through *Chem. Abstr.*, **56**, 14016 (1962).

44. Bergman, I., and J. C. James, *Trans. Faraday Soc.*, **48**, 956 (1952).

45. Bethea, R. M., and F. S. Adams, Jr., *Anal. Chem.*, **33**, 832 (1961).

46. Bethea, R. M., and T. D. Wheelock, *Anal. Chem.*, **31**, 1834 (1959).

47. Beynon, J. H., *Mass Spectrometry and Its Applications to Organic Chemistry*, Elsevier, Amsterdam, 1960.

48. Beynon, J. H., R. A. Saunders, and A. E. Williams, *Ind. Chim. Belg*, **29**, 311 (1964).

49. Bezinger, N. N., T. I. Ovechkina, and G. D. Galpern, *Zh. Anal. Khim.*, **17**, 1027 (1962).

50. Bezman, R., and P. S. McKinney, *Anal. Chem.*, **41**, 1560 (1969).

51. Bhat, R. V., *Proc. Indian Acad. Sci.*, **A13**, 269 (1941); through *Chem. Abstr.*, **36**, 6211 (1941).

52. Birca-Galateanu, D., L. Arcan, and C. Lupu, *Rev. Phys. Acad. Pop. Roum.*, **4**(2), 177 (1959); through *Chem. Abstr.*, **54**, 23778 (1960).

53. Bitto, B. von, *Annalen*, **269**, 377 (1892).

54. Blanksma, J. J., *Rec. Trav. Chim.*, **28**, 105 (1910); through *Chem. Abstr.*, **3**, 1746 (1909).

55. Blom, J., *Biochem. Z.*, **194**, 385 (1928).

56. Bose, P. K., *Analyst*, **56**, 504 (1931).

57. Bose, P. K., *Z. Anal. Chem.*, **87**, 110 (1932).

58. Bost, R. W., and F. Nicholson, *Anal. Chem.*, **7**, 190 (1935).

59. Bottei, R. S., and N. H. Furman, *Anal. Chem.*, **27**, 1182 (1955).

60. Bowen, C. V., and F. I. Edwards, Jr., *Anal. Chem.*, **22**, 706 (1950).

61. Bowman, F. C., and W. W. Scott, *Ind. Eng. Chem.*, **7**, 766 (1915).

62. Boyer, J. H., and S. E. Ellzey, *J. Org. Chem.*, **24**, 2038 (1959).

63. Braddock, L. I., et al., *Anal. Chem.*, **25**, 301 (1953).

64. Bradstreet, R. B., *Anal. Chem.*, **32**, 114 (1960).

64a. Brand, K., and J. Steiner, *Berichte*, **55**, 875 (1922).

65. Brandt, W. W., J. E. DeVries, and E. St. C. Gantz, *Anal. Chem.*, **27**, 392 (1955).

66. Bratton, A. C., and E. K. Marshall, Jr., *J. Biol. Chem.*, **128**, 537 (1939).

67. Braude, E. A., and E. R. H. Jones, *J. Chem. Soc.*, **1945**, 498.

68. Braude, E. A., E. R. H. Jones, and G. G. Rose, *J. Chem. Soc.*, **1947**, 1104.

69. Breiter, M., M. Kleinerman, and P. Delahay, *J. Am. Chem. Soc.*, **80**, 5111 (1958).

70. Breyer, B., and H. H. Baver, *Austr. J. Chem.*, **8**, 425 (1955).

71. Brock, M. J., and M. J. Hannum, *Anal. Chem.*, **27**, 1374 (1955).

72. Brockman, F. J., D. C. Downing, and G. F. Wright, *Can. J. Res.*, **27B**, 469 (1949).

73. Brockman, H., and E. Meyer, *Naturwissenschaften*, **40**, 242 (1953).

74. Brown, E. L., and N. Campbell, *J. Chem. Soc.*, **1937**, 1698.

75. Brown, E. R., et al., *Anal. Chem.*, **40**, 1424 (1968).

76. Brown, E. R., D. E. Smith, and G. L. Booman, *Anal. Chem.*, **40**, 1411 (1968).

77. Brown, H. C., and K. Sivasankaran, *J. Am. Chem. Soc.*, **84**, 2828 (1962).

78. Brown, H. O., E. J. Mean, and B. C. S. Rao, *J. Am. Chem. Soc.*, **77**, 6209 (1955).

79. Brown, J. F., Jr., *J. Am. Chem. Soc.*, **77**, 6341 (1955).

80. Brown, S., *Chem. Trade J.*, **67**, 673 (1920): through *Chem. Abstr.*, **15**, 753 (1921).

81. Brownlie, I. A., *J. Chem. Soc.*, **1950**, 3062.

82. Bruss, D. B., and G. E. A. Wyld, *Anal. Chem.*, **29**, 232 (1957).

83. Buchman, E. R., M. J. Schlatter, and A. O. Reims, *J. Am. Chem. Soc.*, **64**, 2701 (1942).

84. Buck, J. S., and W. S. Ide, in Gilman and Blatt, Eds., *Organic Syntheses*, Coll. Vol. II, John Wiley, New York, 1944, p. 130.

85. Buckley, G. D., and C. W. Scaife, *J. Chem. Soc.*, **1947**, 1471.

86. Buckler, S. A., et al., *J. Org. Chem.*, **27**, 794 (1962).

87. Budesinsky, B., *Chem. Listy*, **50**, 1931 (1956).

88. Bunton, C. A., G. J. Minkoff, and R. I. Reed, *J. Chem. Soc.*, **1947**, 1416.

89. Burawoy, A., *J. Chem. Soc.*, **1939**, 1177.

90. Burkardt, L. A., *Anal. Chem.*, **28**, 323 (1956).

91. Burkardt, L. A., *Anal. Chem.*, **28**, 1271 (1956).

92. Burkardt, L. A., and J. H. Bryden, *Acta Crystallogr.* **7**, 135 (1954); through *Chem. Abstr.*, **48**, 4924 (1954).

93. Bush, M. T., O. Touster, and J. E. Brockman, *J. Biol. Chem.*, **188**, 685 (1951).

94. Butts, P. G., et al., *Anal. Chem.*, **20**, 947 (1948).

95. Bykhovskaya, M. S., *Org. Chem. Ind.* (*USSR*), **6**, 638 (1939); through *Chem. Abstr.*, **34**, 5325 (1940).

96. Cadle, S. H., P. R. Tice, and J. Q. Chambers, *J. Phys. Chem.*, **71**, 3517 (1967).

97. Caldin, E. F., and G. Long, *J. Chem. Soc.*, **1954**, 3737.

98. Callan, T., and J. A. R. Henderson, *J. Soc. Chem. Ind.*, **41**, 157T (1922).

99. Callan, T., J. A. R. Henderson, and N. Strafford, *J. Soc. Chem. Ind.*, **39**, 861 (1920).

100. Callan, T., and N. Strafford, *J. Soc. Chem. Ind.*, **43**, 1T (1924).

101. Camera, E., and D. Pravisani, *Anal. Chem.*, **39**, 1645 (1967).

102. Carter, C. L., and W. J. McChesney, *Nature*, **164**, 575 (1949).

103. deleted.

104. Cavett, J. W., and J. P. Heatis, *J.A.O.A.C.*, **42**, 239 (1959).

105. Chattaway, F. D., *J. Chem. Soc.*, **97**, 2100 (1910).

106. Cheronis, N. D., and J. B. Entrikin, *Semimicro Qualitative Organic Analysis*, Thomas Y. Crowell, New York, 1947.

107. Cheronis, N. D., and T. S. Ma, *Organic Functional Group Analysis by Micro- and Semimicro Methods*, Interscience, New York, 1964.

108. Chodkowski, J., and H. Czajka-Gluchowska, *Rocz. Chem.*, **31**, 1303 (1957); through *Chem. Abstr.*, **52**, 9816 (1958).

109. Chuang, L., L. Fried, and P. J. Elving, *Anal. Chem.*, **36**, 2426 (1964).

110. Clausen, J. G., G. B. Moss, and J. Jordan, *Anal. Chem.*, **38**, 1398 (1966).

111. Clauser, R., *Berichte*, **34**, 889 (1901).

112. Clifford, D. R., and D. A. M. Watkins, *J. Gas Chromatogr.*, **6**, 191 (1968).

113. Closs, G. L., and S. J. Brois, *J. Am. Chem. Soc.*, **82**, 6068 (1960).

114. Cohen, I. R., and A. P. Altshuller, *Anal. Chem.*, **31**, 1638 (1959).

115. Coldwell, B. B., *Analyst*, **84**, 665 (1959).

116. Collat, J. W., and J. J. Lingane, *J. Am. Chem. Soc.*, **76**, 4214 (1954).
117. Collin, J., *Roy. Soc. Sci. Liege*, **23**, 194 (1954).
118. Collin, J., *Roy. Soc. Sci. Liege*, **23**, 201 (1954).
119. Colman, D. M., *Anal. Chem.*, **35**, 652 (1963).
120. Colman, D. M., *J. Chromatogr.*, **8**, 399 (1962).
121. Colthup, N. B., L. H. Daly, and S. E. Wiberley, *Introduction to Infrared and Raman Spectroscopy*, Academic Press, New York, 1964, Ch. 11.
122. Conant, J. B., and B. B. Corson, in Gilman and Blatt, Eds., *Organic Syntheses*, Coll. Vol. II, John Wiley, New York, 1944, p. 33.
123. Conant, J. B., and R. E. Lutz, *J. Am. Chem. Soc.*, **45**, 1047 (1923).
124. Conduit, C. P., *J. Chem. Soc.*, **1959**, 3273.
125. Cooke, A. R., *Arch Biochem. Biophys.*, **55**, 114 (1955).
126. Cooper, P. J., and R. A. Hoodless, *Analyst*, **92**, 520 (1967).
127. Cope, W. C., and J. Barab, *J. Am. Chem. Soc.*, **38**, 2552 (1916).
128. Cottrell, T. L., C. A. Macinness, and E. M. Patterson, *Analyst*, **71**, 207 (1946).
129. Cruse, K., and R. Mittag, *Z. Anal. Chem.*, **131**, 273 (1950).
130. Cruse, K., and R. Harl, *Z. Elektrochem.*, **53**, 115 (1949).
131. Dachselt, E., *Z. Anal. Chem.*, **68**, 404 (1926); through *Chem. Abstr.*, **21**, 38 (1944).
132. Daftsios, A. C., and E. D. Schall, *J. Assoc. Agr. Chem.*, **45**, 228 (1962).
133. Daiber, D., and R. Preussman, *Z. Anal. Chem.*, **206**, 344 (1964).
134. D'Ans, J., and A. Kneip, *Berichte*, **48**, 1144 (1915).
135. Davis, T. L., *J. Am. Chem. Soc.*, **44**, 868 (1922).
136. Day, A. R., and W. T. Taggart, *Ind. Eng. Chem.*, **20**, 545 (1928).
137. Day, R. A., and R. M. Powers, *J. Am. Chem. Soc.*, **76**, 3085 (1954).
138. Dermer, O. C., and R. B. Smith, *J. Am. Chem. Soc.*, **61**, 748 (1939).
139. Desvergnes, L., *Anal. Chim. Appl.*, **8**, 353 (1926); through *Chem. Abstr.*, **21**, 823 (1927).
140. Desvergnes, L., *Chim. Ind.*, **28**, 1038 (1932).
141. DeVries, T., and R. Ivett, *Ind. Eng. Chem., Anal. Ed.*, **13**, 339 (1941).
142. Dickinson, W. E., *Anal. Chem.*, **30**, 992 (1958).
143. Dimofte, L., M. Starescu, and I. Rizescu, *Rev. Chim. (Bucharest)*, **1968**, 19(4), 228 (1968); through *Chem. Abstr.*, **69**, 49089 (1968).
144. Divers, E., *Phil. Trans.*, **163**, 359 (1873).
145. Dornow, A., and A. Miller, *Chem. Ber.*, **93**, 26, 32 (1960).
146. Duin, C. F. van, *Chem. Weekbl.*, **16**, 1071 (1919); through *Chem. Abstr.*, **13**, 3155 (1919).
147. Duin, D. F., van *Chem. Weekbl.*, **16**, 1111 (1919); through *Chem. Abstr.*, **13**, 1313 (1919).
148. Earl, J. R., R. J. W. Le Fevre, A. G. Pulford, and A. Walsh, *J. Chem. Soc.*, **1951**, 2207.
149. Earl, J. C., F. C. Ellsworth, F. C. Jones, and J. Kenner, *J. Chem. Soc.*, **1928**, 2697.
150. Earl, J. C., and J. Kenner, *J. Chem. Soc.*, **1927**, 2139.
151. Earley, J. V., and T. S. Ma, *Mikrochim. Acta*, **1960**, 685.
152. Edwards, W. R., Jr., O. S. Pascual, and C. W. Tate, *Anal. Chem.*, **28**, 1045 (1956).

153. Edwards, W. R., Jr., and C. W. Tate, *Anal. Chem.*, **23**, 826 (1951).

154. Eegriwe, E., *Z. Anal. Chem.*, **110**, 22 (1937).

155. Ehlers, V. B., and J. W. Sease, *Anal. Chem.*, **26**, 513 (1954).

156. Ehlers, V. B., and J. W. Sease, *Anal. Chem.*, **31**, 16 (1959).

157. Ehrlich, J., et al., *Science*, **106**, 417 (1947).

158. Elek, A., and Sobotka, H., *J. Am. Chem.*, **48**, 501 (1926).

159. Elofson, R. M., *Can. J. Chem.*, **36**, 1207 (1958).

160. Elving, P. J., and E. G. Olsen, *J. Am. Chem. Soc.*, **79**, 2697 (1957).

161. Elving, P. J., and J. M. Markowitz, *J. Org. Chem.*, **25**, 18 (1960).

162. Emery, T., and J. B. Nielands, *J. Am. Chem. Soc.*, **82**, 4903 (1960).

163. Emmons, W. D., *J. Am. Chem. Soc.*, **76**, 3468 (1954).

164. Emmons, W. D., *J. Am. Chem. Soc.*, **79**, 6522 (1957).

165. Emmons, W. D., and J. P. Freeman, *J. Am. Chem. Soc.*, **77**, 4387 (1955).

166. Emmons, W. D., and Pagano, and A. S., *J. Am. Chem. Soc.*, **77**, 4557 (1955).

167. Emmons, W. D., A. S. Pagano, and T. E. Stevens, *J. Org. Chem.*, **28**, 311 (1958).

168. Engelbertz, P., and E. Babel, *Zentralbl. Arbeitsmed. Arbeitsschutz*, **7**. 171 (1957); through *Chem. Abstr.*, **52**, 2664 (1958).

169. English, F. L., *Anal. Chem.*, **20**, 745 (1948).

170. English, F. L., *Anal. Chem.*, **23**, 344 (1951).

171. English, F. L., *Ind. Eng. Chem.*, **12**, 994 (1920).

172. Erp, H. van, *Rec. Trav. Chim.*, **14**, 1 (1895).

173. Essigmann, J. M., and P. Issenberg, *J. Food Sci.*, **37**, 684 (1972).

174. Evans, D. H., *Anal. Chem.*, **36**, 2027 (1964).

175. Evered, S., and F. H. Pollard, *J. Chromatogr.*, **4**, 451 (1960).

175a. Fainer, P., *Can. J. Chem.*, **29**, 46 (1951).

176. Fauth, M. I., and G. W. Roecker, *Anal. Chem.*, **33**, 894 (1961).

177. Fauth, M. I., and G. W. Roecker, *J. Chromatogr.*, **18**, 608 (1965).

178. Fauth, M. I., and H. Stalcup, *Anal. Chem.*, **30**, 1670 (1958).

179. Fear, D. L., and G. Burnet, *J. Chromatogr.*, **19**, 17 (1965).

180. Feigl, F., *Spot Tests, Organic Applications*, Vol. II, 4th ed., Elsevier, Amsterdam, 1954.

181. Feigl, F., and J. Amaral, *Mikrochim. Acta*, **1958**(3), 22.

182. Feigl, F., and V. Anger, *Mikrochim. Acta*, **15**, 183 (1934).

183. Feigl, F., and V. Gentil, *Anal. Chem.*, **27**, 432 (1955).

184. Feigl, F., and D. Goldstein, *Anal. Chem.*, **29**, 1521 (1957).

185. Feigl, F., and D. Goldstein, *Anal. Chem.*, **29**, 1522 (1957).

186. Feigl, F., and D. Goldstein, *Microchem. J.*, **1**, 177 (1957).

187. Feigl, F., and C. C. Neto, *Anal. Chem.*, **28**, 1311 (1956).

188. Feuer, H., and A. T. Nielsen, *J. Am. Chem. Soc.*, **84**, 688 (1962).

189. Feuer, H., C. Savides, and C. N. R. Rao, *Spectrochim. Acta*, **19**, 431 (1963).

190. Feuer, H., and B. F. Vincent, Jr., *Anal. Chem.*, **35**, 598 (1963).

191. Fischer, F. G., H. Eilingsfeld, and W. P. Neumann, *Annalen*, **651**, 49 (1962).

192. Fischer, W., and N. Steinbach, *Z. Anorg. Chem.*, **78**, 134 (1912).

193. Flammand, C., and B. Prager, *Berichte*, **38**, 559 (1905).

194. Flett, M., St. C., *Spectrochim. Acta*, **18**, 1537 (1962).

195. Florentin, D., and H. Vandenberg, *Bull Soc. Chim.*, **27**, 158 (1920).

196. Foley, R. L., W. M. Lee, and B. Musulin, *Anal. Chem.*, **36**, 1100 (1964).

197. Forster, O., *Chem. Z.*, **13**, 229 (1889); **14**, 1674 (1890).

198. Foster, R., and R. K. Mackie, *J. Chem. Soc.*, **1963**, 3796.

199. Franchimont, A. P. N., *Rec. Trav. Chim.*, **16**, 213 (1897).

200. Franklin, T. C., and C. C. Roth, *Anal. Chem.*, **27**, 1197 (1955).

201. Freeman, J., and K. S. McCallum, *J. Org. Chem.*, **21**, 472 (1956).

202. Frey, M., *Explosivestoffe*, **12**, (1964); through *Chem. Abstr.*, **64**, 19304 (1966).

203. Friedberg, M., and M. S. O'Dell, *Can. J. Chem.*, **37**, 1469 (1959).

204. Friedemann, F., *Z. Ges. Schiess-Sprengstoffw.*, **24**, 208 (1929).

205. Friedrich, A., *Z. Phys. Chem.*, **216**, 68 (1933).

206. Fritz, J. S., *Anal. Chem.*, **24**, 306 (1958).

207. Fritz, J. S., and N. M. Lisicki, *Anal. Chem.*, **23**, 589 (1951).

208. Fritz, J. S., A. J. Moye, and M. J. Richard, *Anal. Chem.*, **29**, 1685 (1957).

209. Fritz, J. S., and S. S. Yamamura, *Anal. Chem.*, **29**, 1079 (1957).

210. Frodyma, M. M., et al., *Anal. Chem.*, **35**, 1403 (1963).

211. Furman, N. H., and R. S. Bottei, *Anal. Chem.*, **29**, 121 (1957).

212. Gabriel, S., *Berichte*, **18**, 1251 (1885).

213. Gal, J., E. R. Stedronsky, and S. I. Miller, *Anal. Chem.*, **40**, 168 (1968).

214. Galbraith, H. W., E. F. Degering, and E. F. Hitch, *J. Am. Chem. Soc.*, **73**, 1323 (1951).

215. Gapchenko, M. V., *Zavod. Lab.*, **10**, 245 (1941); through *Chem. Abstr.*, **35**, 7312 (1941).

216. Gapchenko, M. V., and O. G. Sheintis, *Zavod. Lab.*, **9**, 562 (1940); through *Chem. Abstr.*, **34**, 7204 (1940).

217. Gast, J. H., and F. L. Estes, *Anal. Chem.*, **32**, 1712 (1960).

218. Gehring, D. G., and J. E. Shirk, *Anal. Chem.*, **39**, 1315 (1967).

219. Gergely, E., and T. Iredale, *J. Chem. Soc.*, **1951**, 3502.

220. Gergely, E., and T. Iredale, *J. Chem. Soc.*, **1953**, 3226.

221. German Pat. 643,058 (Mar. 31, 1937), E. Hoffa and H. Heyna; through *Chem. Abstr.*, **31**, 4508 (1937).

222. Geske, D. H., and A. H. Maki, *J. Am. Chem. Soc.*, **82**, 2671 (1960).

223. Gilbert, G. A., and E. K. Rideal, *Trans. Faraday. Soc.*, **47**, 396 (1951).

224. Gilman, H., and T. N. Goreau, *J. Am. Chem. Soc.*, **73**, 2939 (1951).

225. Gleu, K., and K. Pfannstiel, *J. Prakt. Chem.*, **146**, 129 (1936).

226. Glicksman, R., and C. K. Morehouse, *J. Electrochem. Soc.*, **106**, 288 (1959).

227. Glover, D. J., *Tetrahedron*, **19**, Suppl. 1, 219 (1963).

228. Glover, D. J., and E. G. Kayser, *Anal. Chem.*, **40**, 2055 (1968).

229. Glusker, D. L., and H. W. Thompson, *Spectrochim. Acta*, **6**, 434 (1954).

230. Goldschmidt, H., and M. Eckardt, *Z. Phys. Chem.*, **56**, 385 (1906).

231. Goldschmidt, H., and H. Larsen, *Z. Phys. Chem.*, **71**, 437 (1910).

232. Goldschmidt, H., and E. Sunde, *Z. Phys. Chem.*, **56**, 1 (1906).

233. Goward, G. W., C. E. Bricker, and W. C. Wildman, *J. Org. Chem.*, **20**, 378 (1955).

234. Gowenlock, B. G., and W. Luttke, *Quart. Rev.*, **12**, 321 (1958).

235. Gowenlock, B. G., and J. Trotman, *J. Chem. Soc.*, **1956**, 1670.

236. Grabiel, C. E., D. E. Bisgrove, and L. B. Clapp, *J. Am. Chem. Soc.*, **77**, 1293 (1955).

237. Grandmougin, E., *Berichte*, **40**, 422, 858 (1907).

238. Grebber, K., and J. V. Karabinos, *J. Res. Natl. Bur. Stand.*, **49**, 163 (1952).

239. Green, A. G., and A. R. Wahl, *Berichte*, **31**, 1078 (1898).

240. Greenough, M. L., W. E. Williams, Jr., and J. K. Taylor, *Rev. Sci. Instrum.*, **22**, 484 (1951).

241. Greenspan, F. P., *Ind. Eng. Chem.*, **39**, 847 (1947).

242. Grieg, W. N., and J. W. Rogers, *J. Am. Chem. Soc.*, **91**, 5495 (1969).

242a. Grindlay, J. W., *Anal. Chem.*, **44**, 1676 (1972).

243. Grundy, J., *Analyst*, **79**, 523 (1954).

244. Guilbault, G. G., and D. N. Kramer, *Anal. Chem.*, **38**, 834 (1966).

245. Gullstrom, D. K., H. P. Burchfield, and J. N. Judy, *Anal. Chem.*, **18**, 613 (1946).

246. Gutmann, A., *Berichte*, **45**, 821 (1912).

247. Gwatkin, R. B. L., *Nature*, **193**, 973 (1962).

248. Hale, C. H., *Anal. Chem.*, **23**, 572 (1951).

249. Halfter, G., *Z. Anal. Chem.*, **128**, 140 (1948).

250. Hammick, D. LL., and M. W. Lister, *J. Chem. Soc.*, **1937**, 489.

251. Hammond, G. S., and F. J. Modic, *J. Am. Chem. Soc.*, **75**, 1385 (1953).

252. *Handbook of the Joint Army–Navy–Air Force Panel on the Analytical Chemistry of Solid Propellants*, R. F. Muraca, Ed., Solid Propellant Information Agency, Silver Spring, Md., June 1962.

253. Hansen, R. L., and R. H. Young, *J. Phys. Chem.*, **70**, 1653 (1966).

254. Hantzsch, A., and O. W. Schultze, *Berichte*, **29**, 2251 (1896).

255. Harte, R. A., *Anal. Chem.*, **7**, 432 (1935).

256. Harthon, J. G. L., *Acta Chem. Scand.*, **15**, 1401 (1961).

257. Hartley, A. M., and R. I. Asai, *Anal. Chem.*, **35**, 1207 (1963).

258. Hartley, A. M., and R. M. Bly, *Anal. Chem.*, **35**, 2094 (1963).

259. Hartley, A. M., and D. J. Curran, *Anal. Chem.*, **35**, 686 (1963).

260. Hartley, A. M., A. G. Hiebert, and J. A. Cox, *J. Electroanal. Chem. Interfac. Electrochem.*, **17**, 81 (1968).

261. Hartley, A. M., and R. E. Visco, *Anal. Chem.*, **35**, 1871 (1963).

262. Haslam, J., and L. H. Cross, *J. Soc. Chem. Ind.*, **63**, 94 (1944).

263. Hass, H. B., and E. F. Riley, *Chem. Rev.*, **32**, 373 (1943).

264. Hass, H. B., A. G. Susie, and R. L. Heider, *J. Org. Chem.*, **15**, 8 (1950).

265. Haszeldine, R. N., *J. Chem. Soc.*, **1953**, 2075.

266. Haszeldine, R. N., *J. Chem. Soc.*, **1953**, 2525.

267. Haszeldine, R. N., and J. Jander, *J. Chem. Soc.*, **1954**, 691.

268. Haszeldine, R. N., and B. J. H. Mattinson, *J. Chem. Soc.*, **1955**, 4172.

269. Hatt, H. H., in Gilman and Blatt, Eds., *Organic Syntheses*, Coll. Vol. II, John Wiley, New York, 1944, p. 211.

270. Hawthorne, M. F., *J. Am. Chem. Soc.*, **79**, 2510 (1957).

271. Hearon, W. M., and R. G. Gustavson, *Anal. Chem.*, **9**, 352 (1937).

272. Heotis, J. P., and J. W. Cavett, *Anal. Chem.*, **31**, 1977 (1959).

273. Henderson, T., and A. K. Macbeth, *J. Chem. Soc.*, **121**, 892 (1922).

274. Henry, L., *Bull. Acad. Roy. Belg.*, **34**, 547 (1898); through *Chem. Zentralbl*, **69**, I, 193 (1898).

275. Herm, A. E., *Analyst*, **73**, 314 (1948).

276. Herman, H. B., and A. J. Bard, *Anal. Chem.*, **35**, 1121 (1963).

277. Herman, H. B., and A. J. Bard, *Anal. Chem.*, **36**, 510 (1964).

278. Herman, H. B., and A. J. Bard, *Anal. Chem.*, **36**, 971 (1964).

279. Herman, H. B., and A. J. Bard, *Anal. Chem.*, **37**, 590 (1965).

280. Hertel, E., and A. Lenz, *Z. Phys. Chem.*, **45B**, 395 (1939).

281. Hertwig, G., and W. Lipschitz, *Arch. Ges. Physiol.*, **183**, 275 (1920); through *Chem. Abstr.*, **15**, 708 (1921).

282. Hess, G. B., *Anal. Chem.*, **22**, 649 (1950).

283. Hickinbottom, W. J., *Reaction of Organic Compounds*, Longmans Green, London, 1936.

284. Hinkel, L. E., E. E. Ayling, and T. M. Walters, *J. Chem. Soc.*, **1939**, 403.

285. Hodgson, H. H., *J. Chem. Soc.*, **1937**, 520.

286. Hodgson, H. H., *J. Soc. Dyers Colour.*, **59**, 246 (1943); through *Chem. Abstr.*, **38**, 548 (1944).

287. Hodgson, H. H., and E. R. Ward, *J. Chem. Soc.*, **1947**, 242.

288. Hoffman, D., and G. Rathkamp, *Anal. Chem.*, **42**, 1643 (1970).

289. Holleck, L., and D. Becher, *J. Electroanal. Chem.*, **4**, 321 (1962).

290. Holleck, L., and H. Exner, *Naturwissenschaften*, **39**, 159 (1952).

291. Holleck, L., and R. Schindler, *Z. Elektrochem.*, **60**, 1138 (1956).

292. Holleck, L., and R. Schindler, *Z. Elektrochem.*, **60**, 1142 (1956).

293. Holleck, L., and R. Schindler, *Z. Phys. Chem.* (*Leipzig*), Sonderheft, July 1958, p. 197.

294. Holleck, L., R. Schindler, and O. Lohr, *Naturwissenschaften*, **46**, 625 (1959).

295. Holmes, R. R., *J. Org. Chem.*, **29**, 3076 (1964).

296. Hormann, H., J. Lamberts, and G. Fries, *Z. Physiol. Chem.*, **306**, 42 (1956).

297. Hornsby, S., and W. L. Peacock, *Chem. Ind.*, **1958**, 858.

298. Hubicki, W., and M. Dabkowska, *Anal. Chem.*, **33**, 90 (1961).

299. Huisgen, R. H., and H. Reimlinger, *Annalen*, **599**, 161 (1956).

300. Hume, D. N., *Anal. Chem.*, **28**, 629 (1956).

301. Idoux, J. P., *J. Chem. Soc.*, **1970**, 435.

302. Idoux, J. P., and W. Plain, *J. Chem. Educ.*, **49**, 133 (1972).

303. *Index of Mass Spectral Data*, ASTM Spec. Tech. Publ. 356, 1st ed., American Society for Testing and Materials, Philadelphia, Pa., 1963.

304. Ingersoll, H. W., L. J. Bircher, and M. M. Brubaker, in Gilman and Blatt, Eds., *Organic Syntheses*, Coll. Vol. I, John Wiley, New York, 1944, p. 485.

305. Ingold, C. K., *J. Chem. Soc.*, **125**, 93 (1924).

306. Ingold, C. K., *J. Chem. Soc.*, **127**, 513 (1925).

307. Ingold, C. K., and H. A. Piggott, *J. Chem. Soc.*, **125**, 173 (1924).

308. Ingram, G., in Wilson and Wilson, Eds., *Comprehensive Analytical Chemistry*, Vol. IB, Elsevier, p. 700.

309. Ingram, G., in Wilson and Wilson, Eds., *Comprehensive Analytical Chemistry*, Vol. IB, Elsevier, Amsterdam, 1960, p. 702.

310. Jacobs, W. A., and M. Heidelberger, *J. Am. Chem. Soc.*, **39**, 1435 (1917).

311. Jander, J., and R. N. Haszeldine, *J. Chem. Soc.*, **1954**, 912.

312. Jannakovdskis, D., and A. Wildeman, *Z. Naturforsch.*, **226**, 118 (1967).

313. Janovsky, J. V., *Berichte*, **24**, 971 (1891).

314. Jensen, H. J., and K. Gundersen, *Nature*, **175**, 341 (1955).

315. Johnson, C. L., *Anal. Chem.*, **25**, 1276 (1953).

316. Johnson, K., and E. F. Degering, *J. Org. Chem.*, **8**, 10 (1943).

317. Johnson, L. P., W. M. McNaab, and E. C. Wagner, *Anal. Chem.*, **27**, 1494 (1955).

318. Johnson, L. P., W. M. McNaab, and E. C. Wagner, *Anal. Chem.*, **28**, 392 (1956).

319. Jones, E. C. S., and J. Kenner, *J. Chem. Soc.*, **1930**, 919.

320. Jones, E. C. S., and J. Kenner, *J. Chem. Soc.*, **1932**, 711.

321. Jones, J. H., C. D. Ritchie, and K. S. Heine, Jr., *J. A.O.A.C.*, **41**, 749 (1958).

322. Jones, L. R., and J. A. Riddick, *Anal. Chem.*, **23**, 349 (1951).

323. Jones, L. R., and J. A. Riddick, *Anal. Chem.*, **24**, 1533 (1952).

324. Jones, L. R., and J. A. Riddick, *Anal. Chem.*, **28**, 254 (1956).

325. Jones, L. R., and J. A. Riddick, *Anal. Chem.*, **28**, 1137 (1956).

326. Jones, L. R., and J. A. Riddick, *Anal. Chem.*, **28**, 1493 (1956).

327. Jones, R. N., and Thorn, G. D., *Can. J. Chem.*, **27B**, 828 (1949).

328. Jones, W. H., *J. Am. Chem. Soc.*, **76**, 829 (1954).

329. Joshi, G. G., and N. M. Shah, *Curr. Sci. (India)*, **18**, 73 (1949); through *Chem. Abstr.*, **43**, 7438 (1949).

330. Jucker, H., *Anal. Chim. Acta*, **16**, 210 (1957).

331. Jullig, T., and J. Barbiere, *Mem. Poudres*, **27**, 127 (1937); through *Chem. Abstr.*, **31**, 7793 (1937).

332. Junell, R., *Z. Phys. Chem.*, **A141**, 71 (1929).

333. Jurecek, M., *Mikrochim. Acta*, **1962**, (5) 926.

334. Jurecek, M., V. Novak, and P. Kozak, *Talanta*, **9**, 72 (1962).

335. Juvet, R. S., Jr., M. C. Twickler, and L. C. Afremow, *Anal. Chim. Acta*, **22**, 87 (1960).

336. Kabasakalian, P., E. R. Townley, and M. D. Yudis, *J. Am. Chem. Soc.*, **84**, 2718 (1962).

336a. Kambert, M. J., L. A. Kaplan, and J. C. Dacons, *J. Org. Chem.*, **26**, 4371 (1961).

337. Kamlet, M. J., and D. J. Glover, *J. Org. Chem.*, **27**, 537 (1962).

338. Kamm, O., in Gilman & Blatt, Eds., *Organic Syntheses*, Coll. Vol. II, John Wiley, New York, 1944, p. 445.

339. Karp, S., and L. J. Meites, *J. Electroanal. Chem.*, **17**, 253 (1968).

340. Kastening, B., *Elektrochem. Acta*, **9**, 241 (1964).

341. Kastening, B., *Naturwissenschaften*, **47**, 443 (1960).

342. Kastening, B., and L. Holleck, *Z. Elektrochem.*, **63**, 166 (1959); **64**, 823 (1960).

343. Kastening, B., and L. Holleck, *Z. Elektrochem.*, **63**, 177 (1959).

344. Kauffman, F., E. Ossofsky, and H. J. Cook, *Anal. Chem.*, **26**, 516 (1954).

345. Kaye, S. M., *Anal. Chem.*, **27**, 292 (1955).

346. Kemula, W., and Z. Kublik, *Anal. Chim. Acta*, **18**, 104 (1958).

347. Kemula, W., and Z. Kublik, *Nature*, **182**, 793 (1958).

348. Kemula, W., and R. Sioda, *Bull. Acad. Polon. Sci., Ser. Sci. Chim.*, **10**, 107 (1962); through *Chem. Abstr.*, **57**, 7015 (1962).

349. Kemula, W., and R. Sioda, *Bull. Acad. Polon. Sci., Ser. Sci. Chim.*, **11**, 395 (1963); through *Chem. Abstr.*,

350. Kemula, W., and R. Sioda, *J. Electroanal. Chem.*, **6**, 183 (1963).

351. Kemula, W., and R. Sioda, *J. Electroanal. Chem.*, **7**, 233 (1964).

352. Kemula, W., and R. Sioda, *Nature*, **197**, 588 (1963).

353. Kemula, W., and R. Sioda, *Naturwissenschaften*, **50**, 70 (1963).

354. Kemula, W., and D. Sybilska, *Nature*, **185**, 237 (1960).

355. Keussler, J., and W. Luttke, *Z. Elektrochem.*, **63**, 614 (1959).

356. Khailov, V. S., B. B. Brandt, and G. N. Shcherbova, *Khim. Nauka. Prom.*, **2**, 806 (1957); through *Chem. Abstr.*, **52**, 9845 (1958).

357. Kholodov, L. E., V. V. Alekseev, and V. G. Yashunkii, *Zh. Fiz. Khim.*, **39**, 1566 (1965); through *Chem. Abstr.*, **63**, 11292 (1965); see also *Chem. Abstr.*, **65**, 3717 (1966).

358. Kitagawa, T., T. P. Layloff, and R. N. Adams, *Anal. Chem.*, **35**, 1086 (1963).

359. Klager, K., *Anal. Chem.*, **23**, 534 (1951).

360. Kleipool, R. J. C., and Heins, J. T., *Nature*, **203**, 1280 (1964).

361. Klemm, L. H., J. W. Sprague, Z. Ziffer, and B. I. MacGowan, *Anal. Chim. Acta*, **19**, 369 (1958).

362. Klimova, V. A., and K. S. Zabrodina, *Izv. Akad. Nauk. SSSR, Otdel. Khim. Nauk.*, **1961**, 160 (in English); through *Chem. Abstr.*, **27**, 964 (1933).

363. Knecht, E., and E. Hibbert, *New Reduction Methods in Volumetric Analysis*, 2nd ed., Longmans, Green, London, 1925.

364. Knobloch, E., and E. Svatek, *Collect. Czech. Chem. Commun.*, **20**, 1113 (1955).

365. Koch, L., and R. F. Milligan, *Anal. Chem.*, **22**, 1039 (1950).

366. Koelsch, F., *Z. Ges. Scheiss-Sprengstoffw.*, **12**, 109, 132, 155 (1918); through *Chem. Abstr.*, **12**, 427 (1918).

367. Kohler, E. P., and J. F. Stone, Jr., *J. Am. Chem. Soc.*, **52**, 761 (1930).

368. Kolbe, H., *J. Prakt. Chem.* [2], **5**, 429 (1872).

368a. Kolthoff, I. M., *Rec. Trav. Chim.*, **43**, 768 (1924).

369. Kolthoff, I. M., and R. Belcher, *Volumetric Analysis*, Vol. III, Interscience, New York, 1957, p. 610.

370. Kolthoff, I. M., and F. A. Bovey, *J. Am. Chem. Soc.*, **70**, 794 (1948).

371. Kolthoff, I. M., and N. H. Furman, *Potentiometric Titrations*, John Wiley, New York, 1926, p. 304.

372. Kolthoff, I. M., and A. Langer, *J. Am. Chem. Soc.*, **62**, 3172 (1940).

373. Kolthoff, I. M., and A. Liberti, *J. Am. Chem. Soc.*, **70**, 1884 (1948).

374. Kolthoff, I. M., and J. J. Lingane, *Polarography*, Vol. II, 2nd ed., Interscience, New York, 1952, Ch. XLII.

375. Kolthoff, I. M., and C. Robinson, *Rec. Trav. Chim.*, **45**, 169 (1926).

376. Koniecki, W. B., and A. L. Linch, *Anal. Chem.*, **30**, 1134 (1958).

377. Konovalov, M., *Berichte*, **28**, 1850 (1895).

378. Koppeschaar, W. F., *Z. Anal. Chem.*, **15**, 233 (1876).

379. Kornblum, N., et al., *J. Am. Chem. Soc.*, **77**, 5528, 6269 (1955); **78**, 1497 (1956).

380. Kornblum, N., and R. A. Brown, *J. Am. Chem. Soc.*, **87**, 1742 (1965).

381. Kornblum, N., B. Taube, and H. E. Ungnade, *J. Am. Chem. Soc.*, **76**, 3209 (1954).

382. Kornblum, N., H. E. Ungnade, and R. A. Smiley, *J. Org. Chem.*, **21**, 377 (1956).

383. Kornshunov, I. A., et al., *Zavod. Lab.*, **14**, 519 (1948); through *Chem. Abstr.*, **43**, 4976 (1949).

384. Kouba, D. L., R. C. Kicklighter, and W. W. Becker, *Anal. Chem.*, **20**, 948 (1948).

385. Kovac, J., and E. Sohler, *Z. Anal. Chem.*, **208**, 201 (1965).

386. Kraljic, I., M. Kopriva, and M. Pungersek, *Bull. Sci. Conseil Acad. RPF Yougosl.*, **3**, 104 (1957); *Chem. Abstr.* **52**, 8829 (1958).

387. Kratz, B., *Vom Wasser*, **17**, 83 (1949).

388. Kremera, E., N. Wakeman, and R. M. Hixon, in Gilman and Blatt, Eds., *Organic Syntheses*, Coll. Vol. I, John Wiley, New York, 1944, p. 511.

389. Kresze, G., and H. Manthey, *Chem. Ber.*, **89**, 1412 (1956).

390. Kruse, J. M., *Anal. Chem.*, **31**, 1854 (1959).

391. Kuhn, L. P., G. G. Kleinspan, and H. C. Duckworth, *J. Am. Chem. Soc.*, **89**, 3859 (1967).

392. Kuhn, W. E., in Gilman and Blatt, Eds., *Organic Syntheses*, Coll. Vol. II, John Wiley, New York, 1944, p. 447.

393. Kuwana, T. W., and W. G. French, *Anal. Chem.*, **36**, 243 (1964).

393a. Laccetti, M. A., S. Semel, and M. Roth, *Anal. Chem.*, **31**, 1049 (1959).

394. Lamberton, A. H., *Quart. Rev.*, **5**, 75 (1951).

395. Lamond, J., *Analyst*, **73**, 674 (1948).

396. Landram, G. K., A. A. Wickham, and R. F. DuBois, *Anal. Chem.*, **42**, 107 (1970).

397. Larson, J. E., and S. H. Harvey, *Chem. Ind.*, **1954**, 45.

398. Latimer, W. M., *The Oxidation States of the Elements and Their Potentials in Aqueous Solutions*, 2nd ed., Prentice-Hall, Englewood Cliffs, N. Y., 1952.

399. Lauer, W., *J. Am. Chem. Sec.*, **58**, 225 (1936).

400. Laviron, E., and P. Fournari, *Bull. Soc. Chim. Fr.*, **1966**, 518.

401. Laviron, E., P. Fournari, and J. Greusard, *Bull. Soc. Chim. Fr.*, **1967**, 1255.

402. Laviron, E., P. Fournari, and G. Refalo, *Bull. Soc. Chim. Fr.*, **1969**, 1024.

403. Lawford, D. J., and D. G. Harvey, *Analyst*, **78**, 63 (1953).

404. deleted.

405. Lehmstedt, K. D., *Berichte*, **60**, 1910 (1927).

406. Lehmstedt, K. D., and O. Zumstein, *Berichte*, **58**, 2024 (1925).

407. Leibmann, W., and J. T. Woods, *Anal. Chem.*, **29**, 1845 (1957).

408. LeRosen, A. L., R. T. Moravek, and J. K. Carlton, *Anal. Chem.*, **24**, 1335 (1952).

409. Lester, D., and L. A. Greenberg, *J. Am. Chem. Soc.*, **66**, 496 (1944).

410. Levitan, N. I., et al., *J. Am. Chem. Soc.*, **65**, 2265 (1943).

411. Levy, J. B., *Ind. Eng. Chem.*, **48**, 762 (1956).

412. Levy, W. J., and N. Campbell, *J. Chem. Soc.*, **1939**, 1442.

413. Lewis, D. T., *Analyst*, **79**, 644 (1954).

414. Lewis, G. N., and G. T. Seaborg, *J. Am. Chem. Soc.*, **62**, 2122 (1940).

415. Lieber, E., D. R. Levering, and L. J. Patterson, *Anal. Chem.*, **23**, 1594 (1951).

416. Liebermann, C. T., *Berichte*, **7**, 1098 (1874).

417. Liebermann, C., *Berichte*, **15**, 1529 (1882).

418. Limpricht, H., *Berichte*, **11**, 35 (1878).

419. Lindbeck, M. R., and J. L. Young, *Soil Sci.*, **101**, 366 (1966).

420. Lindenmeyer, P., and P. Harris, *J. Chem. Phys.*, **21**, 408 (1953).

421. Lingane, J. J., *Anal. Chim. Acta*, **2**, 584 (1948).

422. Lingane, J. J., *Faraday Soc. Disc.*, **1**, 203 (1947).

423. Lingane, J. J., *Ind. Eng. Chem., Anal. Ed.*, **16**, 150 (1944).

424. Lingane, J. J., *Ind. Eng. Chem., Anal. Ed.*, **18**, 429 (1945).

425. Lingane, J. J., *J. Am. Chem. Soc.*, **67**, 1916 (1945).

426. Lingane, J. J., and R. L. Pecsok, *Anal. Chem.*, **20**, 425 (1948).

427. Lingane, J. J., and L. A. Small, *Anal. Chem.*, **21**, 1119 (1949).

428. Lingane, J. J., C. G. Swain, and M. Fields, *J. Am. Chem. Soc.*, **65**, 1348 (1943).

429. Linnett, J. W., and R. M. Rosenberg, *Tetrahedron*, **20**, 53 (1964).

430. Lippincott, S. B., *J. Am. Chem. Soc.*, **62**, 2604 (1940).

431. Lippincott, S. B., and H. B. Hass, *Ind. Eng. Chem., Anal. Ed.*, **31**, 118 (1939).

432. Lipschitz, W., *Z. Physiol. Chem.*, **109**, 189 (1920); through *Chem. Abstr.*, **15**, 1932 (1921).

433. Lobunets, M. M., *Univ. Etat Kiev, Bull. Sci., Rec. Chim.*, No. 3, 71 (1937); through *Chem. Abstr.*, **33**, 2439 (1939).

434. Lobunets, M. M., *Univ. Etat Kiev, Bull. Sci. Rec. Chim.*, No. 4, 23 (1939); through *Chem. Abstr.*, **35**, 1356 (1941).

435. Lobunets, M. M., *Univ. Etat Kiev, Bull. Sci. Rec. Chim.*, No. 4, 41 (1939); through *Chem. Abstr.*, **35**, 1356 (1941).

436. Lobunets, M. M., *Zavod. Lab.*, **7**, 872 (1938); through *Chem. Abstr.*, **33**, 504 (1939).

437. Lobunets, M. M., and E. N. Gortins'ka *Univ. Etat Kiev, Bull. Sci. Rec. Chim.*, No. 4, 37 (1939); through *Chem. Abstr.*, **35**, 1356 (1941).

438. Lock, G., and F. Stitz, *Berichte*, **73B**, 1377 (1940); through *Chem. Abstr.*, **35**, 2867 (1941).

439. Loev, B., and M. M. Goodman, *Chem. Ind.*, **1967**, 2026.

440. Loo, T. L., and R. L. Dion, *J. Pharm. Sci.*, **54**, 809 (1965).

441. Lothrop, W. C., and G. R. Handrick, *Chem. Rev.*, **44**, 419 (1949).

442. Lothrop, W. C., G. R. Handrick, and R. M. Hainer, *J. Am. Chem. Soc.*, **73**, 3581 (1951).

443. Lucas, A. J., A. N. Prater, and R. E. Morris, *J. Am. Chem. Soc.*, **57**, 723 (1935).

444. Lund, H., *Acta Chem. Scand.*, **11**, 990 (1957).

445. Lutri, C., *Giorn. Chim. Ind. Appli.*, **2**, 557 (1920); through *Chem. Abstr.*, **17**, 450 (1923).

446. Luttke, W., *J. Phys. Radium*, **15**, 633 (1954); through *Chem. Abstr.*, **52**, 17960 (1958).

447. Ma, T. S., and J. V. Earley, *Mikrochim. Acta*, **1959**, 130.

448. Ma, T. S., R. E. Lang, and J. D. McKinley, Jr., *Mikrochim. Acta*, **1957**, 368.

449. MacBeth, A. K., and J. R. Price, *J. Chem. Soc.*, **1935**, 151.

450. Machle, W. F., E. W. Scott, and J. Treon, *J. Ind. Hyg. Toxicol.*, **22**, 315 (1940).

451. Mahood, S. A., and P. V. L. Schaffner, Gilman and Blatt, Eds., *Organic Syntheses*, Coll. Vol. II, John Wiley, New York, 1944. p. 160.

452. Mailhe, A., and M. Murat, *Bull. Soc. Chim.*, **7**, 952 (1910); through *Chem. Abstr.*, **5**, 470 (1911).

452a. Maine, P. H. de, et al., *J. Mol. Spectrosc.*, **4**, 398 (1960).

453. Mairanovskii, S. G., and O. M. Dolgaya, *Elektrokhimiya*, **5**, 893 (1969).

454. Maki, A. H., and D. H. Geske, *J. Am. Chem. Soc.*, **83**, 1852 (1961)).

455. Makovsky, A., and L. Lenji, *Chem. Rev.*, **58**, 627 (1958).

456. Malmberg, E. W., *Anal. Chem.*, **27**, 840 (1955).

457. Malmberg, E. W., K. N. Trueblood, and T. D. Waugh, *Anal. Chem.*, **25**, 901 (1953).

458. Manabe, O., and H. Hiyama, *J. Chem. Soc. Jap., Ind. Chem. Sect.*, **56**, 365 (1953); through *Chem. Abstr.*, **48**, 10412 (1954).

459. Manzoff, C. D., *Z. Nahr-Genussm.*, **27**, (1914); through *Chem. Abstr.*, **8**, 2248 (1914).

460. Marcoux, L. S., et al., *Anal. Chem.*, **37**, 1446 (1965).

461. Margosches, B. M., and W. Kristen, *Z. Ges. Schiess. Sprengstoffw.*, **18**, 39 (1923); through *Chem. Abstr.*, **17**, 3656 (1923).

462. Margosches, B. M., and W. Kristen, *Z. Ges. Schiess. Sprengstoffw.*, **18**, 73 (1923); through *Chem. Abstr.*, **17**, 3656 (1923).

463. Margosches, B. M., W. Kristen, and E. Scheinost, *Berichte*, **56B**, 1943 (1923).

464. Margosches, B. M., and E. Vogel, *Berichte*, **55B**, 1380 (1922).

465. Martin, A. J., *Anal. Chem.*, **29**, 79 (1957).

466. Martin, E. L., in Gilman and Blatt, Eds., *Organic Syntheses*, Coll. Vol. II, John Wiley, New York, 1944, p. 501.

467. Martin, R. B., and M. O. Tashdjian, *J. Phys. Chem.*, **60**, 1028 (1956).

468. Maruyama, S., *Sci. Papers Inst. Phys.-Chem. Res. (Tokyo)*, **16**, 196 (1931); through *Chem. Abstr.*, **26**, 396 (1932).

469. Marvel, C. S., in Gilman and Blatt, Eds., *Organic Syntheses*, Coll. Vol. I, John Wiley, New York, 1944, p. 177.

470. Marvel, C. S., F. D. Hager, and E. C. Caudle, in Gilman and Blatt, Eds., *Organic Syntheses*, Coll. Vol. I, John Wiley, New York, 1944, p. 224.

471. Marvel, C. S., and O. Kamm, *J. Am. Chem. Soc.*, **41**, 278 (1919).

472. Mason, J., and J. Dunderdale, *J. Chem. Soc.*, **1956**, 754.

473. Masui, M., and H. Sayo, *J. Chem. Soc.*, **1961**, 4773.

474. Masui, M., and H. Sayo, *J. Chem. Soc.*, **1961**, 5325.

475. Masui, M., and H. Sayo, *J. Chem. Soc.*, **1962**, 1733.

476. Matsumoto, H., et al., *Anal. Chem.*, **33**, 1442 (1961).

477. Mayer, A., and Vles, P. and F., *Compt. Rend.*, **171**, 1396 (1920); through *Chem. Abstr.*, **15**, 1578 (1921).

478. McClure, J. E., and D. L. Mariele, *Anal. Chem.*, **39**, 236 (1967).

479. McCutchan, P., and Roth, W. F., *Anal. Chem.*, **24**, 369 (1952).

480. McKay, A. F., *Chem. Rev.*, **51**, 301 (1952).

481. Meisenheimer, J., *Ann. Chem.*, **323**, 224 (1902).

482. Meites, L., *Anal. Chem.*, **27**, 1114 (1955).

483. Meites, L., *Anal. Chem.*, **27**, 1116 (1955).

484. Meites, L., *J. Am. Chem. Soc.*, **75**, 3809 (1953).

485. Meites, L., *Pure Appl. Chem.*, **18**, 35 (1969).

486. Meites, L., and T. Meites, *Anal. Chem.*, **28**, 103 (1956).

487. Meites, L., and S. A. Moros, *Anal. Chem.*, **31**, 23 (1959).

488. Meyer, H., *Analyse und Konstitutionsermittelung Organische Verbindungen*, 5th ed., Julius Springer, Berlin, 1931, p. 623.

489. Meyer, K. H., and P. Wertheimer, *Berichte*, **47**, 2374 (1914).

490. Meyer, V., *Chem. Berichte*, **9**, 384 (1876).

491. Meyer, V., and J. Locher, *Berichte*, **8**, 219 (1895).

492. Meyer, V., and O. Stuber, *Berichte*, **5**, 203 (1872).

493. Miguel, R., and A. Condylis, *Bull. Soc. Chim. Fr.*, **1955**, 236.

494. Miller, E. W., A. P. Arnold, and M. J. Astle, *J. Am. Chem. Soc.*, **70**, 3971 (1948).

495. Miller, J. M., and H. Pobiner, *Anal. Chem.*, **36**, 238 (1964).

496. Milner, G. W. C., *The Principles and Applications of Polarography*, Longmans, Green, London, 1957, Ch. 24.

497. Mirri, A. M., and P. Favero, *Ric. Sci.*, **29**, 106 (1959).

498. Mitchell, S., K. Schwarzwald, and G. K. Simpson, *J. Chem. Soc.*, **1941**, 602.

499. Mitra, B. N., and M. Srinivasan, *Analyst*, **70**, 418 (1945).

500. Momigny, J., *Roy. Soc. Sci. Liege*, **25**, 93 (1956).

501. Montagne, M., and P. Arnand, *Compt. Rend.*, **254**, 4001 (1962).

502. Moore, F. W., *J. Gen. Microbiol.*, **3**, 143 (1949).

503. Moore, H. C., *J. Ind. Eng. Chem.*, **12**, 669 (1920).

504. Moore, H. P., and C. R. Guertal, *J. Assoc. Agr. Chem.*, **43**, 308 (1960).

505. Moros, S. A., *Anal. Chem.*, **34**, 1584 (1962).

506. Morris, M. P., Pagan, C., and H. E. Warmke, *Science*, **119**, 322 (1954).

507. Mulliken, S. P., and E. R. Barker, *Am. Chem. J.*, **21**, 271 (1899).

508. Murray, M. J., and D. E. Waters, *J. Am. Chem. Soc.*, **60**, 2818 (1938).

509. Musha, S., *J. Chem. Soc. Jap.*, **66**, 38 (1945).

510. Musha, S., *J. Chem. Soc. Jap.*, **67**, 49 (1946).

511. Nagasawa, K., *Bunseki to Shiyaku*, **3**, 109 (1950); through *Chem. Abstr.*, **46**, 10042 (1952).

512. Nakamoto, K., and R. E. Rundle, *J. Am. Chem. Soc.*, **78**, 1113 (1956).

513. Neiman, M. B., et al., *Zavod. Lab.*, **15**, 1280 (1949); through *Chem. Abstr.*, **44**, 3845 (1950).

514. Nelson, L. S., and Laskowski, D. E., *Anal. Chem.*, **23**, 1495 (1951).

515. Newton, S. A., F. J. Stubbs, and C. Hinshelwood, *J. Chem. Soc.*, **1953**, 3384.

516. Nicholson, R. S., J. M. Wilson, and M. L. Omstead, *Anal. Chem.*, **38**, 542 (1966).

517. Nielsen, A. T., in H. Feuer, Ed., *The Chemistry of the Nitro and Nitroso Groups*, Interscience, New York, 1969.

518. Nielson, J. R., and D. C. Smith, *Ind. Eng. Chem., Anal. Ed.*, **15**, 609 (1943).

519. Nisida, S., *Bull. Inst. Phys.-Chem. Res. (Tokyo)*, **20**, 20 (1941); through *Chem. Abstr.*, **35**, 7320 (1941).

520. Noble, P., Jr., F. G. Borgardt, and W. L. Reed, *Chem. Rev.*, **64**, 19 (1964).

521. Noland, W. E., *Chem. Rev.*, **55**, 137 (1955).

522. Noland, W. E., and J. H. Sellsledt, *J. Org. Chem.*, **31**, 345 (1966).

523. Noller, C. R., *Chemistry of Organic Compounds*, W. B. Saunders, Philadelphia, 1965.

524. Norris, R. K., and S. Sternhell, *Austr. J. Chem.*, **19**, 841 (1966).

525. Noto, T., and G. Matsuoka, *Jap. Anal.*, **4**, 30 (1955); through *Chem. Abstr.*, **50**, 5244 (1956).

526. Novak, J. V. A., *Collect. Czech. Chem. Commun.*, **11**, 523 (1939).

527. Novak, V., et al., *Collect. Czech. Chem. Commun.*, **28**, 487 (1963).

528. Nystrom, R. F., and W. G. Brown, *J. Am. Chem. Soc.*, **70**, 3738 (1948).

529. Ogata, Y., M. Tsuchida, and Y. Takagi, *J. Am. Chem. Soc.*, **79**, 3397 (1957).

530. Ohashi, S., *Bull. Chem. Soc. Jap.*, **28**, 537 (1955).

531. Ohkuma, S., *J. Pharm. Soc. Jap.*, **75**, 1342 (1955); through *Chem. Abstr.*, **50**, 3956 (1956).

532. Ohkuma, S., *J. Pharm. Soc. Jap.*, **75**, 1430 (1955); through *Chem. Abstr.*, **50**, 3956 (1956).

533. Ohkuma, S., *Kagaku No Ryoiki (J. Jap. Chem.)*, **4**, 622 (1950).

534. O'Keefe, K., and P. R. Averell, *Anal. Chem.*, **23**, 1167 (1951).

535. Olivier, S. C. J., *Rec. Trav. Chim.*, **37**, 241 (1918); through *Chem. Abstr.*, **13**, 572 (1919).

536. Overberger, C. G., J. G. Lombardıno, and R. G. Hiskey, *J. Org. Chem.*, **22**, 858 (1957).

537. Page, J. E., J. W. Smith, and J. G. Waller, *J. Phys. Chem.*, **53**, 545 (1949).

538. Parsons, J. S., S. M. Tsang, M. P. DiGiaimo, R. Feinland, and R. A. L. Paylor, *Anal. Chem.*, **33**, 1858 (1961).

539. Pearson, D., and A. Morrissey, *Microchem. J.*, **6**, 175 (1960).

540. Pearson, J., *Trans. Faraday Soc.*, **44**, 683 (1948).

541. Pearson, J., *Trans. Faraday Soc.*, **44**, 692 (1948).

542. Pearson, J., *Trans. Faraday Soc.*, **45**, 199 (1949).

543. Per'e, M. I., *Univ. Etat Kiev, Bull. Sci. Rec. Chim.*, No. 3, 37 (1937); through *Chem. Abstr.*, **33**, 2439 (1939).

544. Per'e, M. I., and M. M. Lobunets, *Univ. Etat Kiev., Bull. Sci. Rec. Chim.*, **2**, No. 2, 45 (1936); through *Chem. Abstr.*, **31**, 2966 (1937).

545. Per'e, M. I., and M. M. Lobunets, *Univ. Etat Kiev, Bull. Sci. Rec. Chim.*, **2**, No. 2, 73 (1936); through *Chem. Abstr.*, **31**, 2969 (1937).

546. Per'e, M. I., and M. M. Lobunets, *Univ. Etat Kiev, Bull. Sci. Rec. Chim.*, No. 3, 43 (1937); through *Chem. Abstr.*, **33**, 2439 (1939).

547. Perret, J. G., and L. Holleck, *Z. Elektrochem.*, **60**, 463 (1956).

548. Petru, F., *Collect. Czech. Chem. Commun.*, **12**, 620 (1947).

549. Peurifoy, P. V., and W. G. Schrank, *Anal. Chem.*, **29**, 411 (1957).

550. Phillips, J. P., *J. Org. Chem.*, **27**, 1443 (1962).

551. Phillips, J. P., *Spectra-Structure Correlation*, Academic Press, New York, 1964, Ch. IV.

552. Pickard, R. H., and J. Kènyon, *J. Chem. Soc.*, **91**, 901 (1907).

553. Pickering, J. E., *Chem. Trade J.*, **70**, 144 (1922): through *Chem. Abstr.*, **17**, 450 (1923).

554. Pietrzyk, D. J., R. F. Breese, and L. B. Rogers, *J. Electrochem. Soc.*, **110**, 995 (1963).

555. Pietrzyk, D. J., and L. B. Rogers, *Anal. Chem.*, **34**, 936 (1962).

556. Pinnow, J., *Berichte*, **30**, 883 (1897).

557. Pippen, E. L., E. J. Eyring, and M. Nonaka, *Anal. Chem.*, **29**, 1305 (1957).

558. Popp, F. D., and H. P. Schultz, *Chem. Rev.*, **62**, 19 (1962).

559. Porter, C. C., *Anal. Chem.*, **27**, 805 (1955).

560. Prins, H. J., *Perfum. Essent. Oil Rec.*, **13**, 355 (1922); through *Chem. Abstr.*, **17**, 450 (1923).

561. Pristera, F., *Appl. Spectrosc.*, **7**, (3), 115 (1953).

562. Pristera, F., and M. Halik, *Anal. Chem.*, **27**, 217 (1955).

563. Radin, N., and T. DeVries, *Anal. Chem.*, **24**, 971 (1952).

564. Raikow, P. N., *Z. Angew. Chem.*, **29** (I), 196, 239 (1916); through *Chem. Abstr.*, **10**, 2786, 3139 (1910).

565. Ramachandran, L. K., *Anal. Chem.*, **33**, 1074 (1961).

566. Randle, R. R., and D. H. Whiffen, *J. Chem. Soc.*, **1952**, 4153.

567. Randles, J. E. B., *Trans. Faraday Soc.*, **44**, 327 (1948).

568. Rangarajan, S. K., *J. Electroanal. Chem.*, **21**, 257 (1969).

569. Rao, C. N. R., *Ultraviolet and Visible Spectroscopy—Chemical Applications*, Butterworths, London, 1961, p. 20.

570. Raschig, F., *Schwefel. Stickstoff.*, **1924**, 255; through *Chem. Abstr.*, **19**, 2484 (1925).

571. Rathsburg, H., *Berichte*, **54B**, 3183 (1921).

572. Rebstock, M. C., et al., *J. Am. Chem. Soc.*, **71**, 2458 (1949).

573. Reclaire, A., *Perfum. Essent. Oil Rec.*, **13**, 356 (1922); through *Chem. Abstr.*, **17**, 450 (1923).

574. Redemann, C. T., and C. E. Redemann, in Gilman and Blatt, Eds., *Organic Syntheses*, Coll. Vol. III, John Wiley, 1955, p. 69.

575. Reich, M., and H. Kapenekas, *Ind. Eng. Chem.*, **49**, 869 (1957).

576. Reihlen, H., and A. Hake, *Ann. Chem.*, **452**, 47 (1927); through *Chem. Abstr.*, **21**, 1418 (1927).

577. Reinmuth, W. H., *Anal. Chem.*, **32**, 1514 (1960).

578. Reitzenstein, F., and G. Stamm, *J. Prakt. Chem.*, **81**, 167 (1910); through *Chem. Abstr.*, **4**, 2280 (1910).

579. Reynolds, C. A., and D. C. Underwood, *Anal. Chem.*, **40**, 1983 (1968).

580. Reynolds, G. F., and H. I. Shalgosky, *Anal. Chim. Acta*, **10**, 386 (1954).

581. Rieger, P. H., and G. K. Fraenkel, *J. Chem. Phys.*, **39**, 609 (1963).

582. Rinkes, I. J., *Chem. Weekbl.*, **11**, 1062 (1914); through *Chem. Abstr.*, **9**, 625 (1915).

583. Riordan, J. F., M. Sokolovsky, and B. Vallee, *J. Am. Chem. Soc.*, **88**, 4104 (1966).

584. Roberts, J. D., and C. Green, *J. Am. Chem. Soc.*, **68**, 214 (1940).

585. Roberts, J. D., and C. Green, *Anal. Chem.*, **18**, 335 (1946).

586. Robertson, G. R., in Gilman & Blatt, Eds., *Organic Syntheses*, Coll. Vol. I, John Wiley, New York, 1944, p. 52.

587. Robertson, P. W., T. R. Hitchings, and G. M. Will, *J. Chem. Soc.*, **1950**, 808.

588. Robson, E., and J. M. Tedder, *Proc. Chem. Soc.*, **1963**, 13.

589. Rosen, A. A., K. V. Y. Sundstrom, and W. F. Vogel, *Anal. Chem.*, **24**, 412 (1952).

590. Rosie, D. J., and W. D. Cooke, *Anal. Chem.*, **27**, 1360 (1955).

591. Roth, M., and R. F. Wegman, *Anal. Chem.*, **30**, 2036 (1958).

592. Roubal, J., K. Tuhy, and F. Pokorny, *Cas. Lekavu Cesk.*, **85**, 1002 (1946); through *Chem. Abstr.*, **42**, 3454 (1948).

593. Rowe, M. L., *J. Gas Chromatogr.*, **4**, 420 (1966).

594. Rowe, M. L., *J. Gas Chromatogr.*, **5**, 531 (1967).

595. Rudolph, O., *Z. Anal. Chem.*, **60**, 239 (1921); through *Chem. Abstr.*, **15**, 3798 (1921).

596. Russell, J. H., *Rev. Pure Appl. Chem.*, **13**, 15 (1963).

597. Russian Pat. 56,179 (Dec. 31, 1939), B. I. Kissin and S. A. Vorobeichikov; through *Chem. Abstr.*, **38**, 4534 (1944).

598. Russian Pat. 58,088 (Oct. 31, 1940), B. I. Kissin; through *Chem. Abstr.*, **38**, 5756 (1944).

599. Ruzhentseva, A. K., and N. S. Goryacheva, *Dokl. Akad. Nauk. SSSR*, **81**, (1951).; through *Chem. Abstr.*, **46**, 3907 (1952).

600. Ruzhentseva, A. K., and V. V. Kolpakova, *Zh. Anal. Khim.*, **6**, 223 (1951); through *Chem. Abstr.*, **45**, 10138 (1951).

601. Sachs, F., and M. Craveri, *Berichte*, **38**, 3685 (1905).

602. Saito, T., T. Miyazaki, and K. Miyaki, *J. Pharm. Chem.*, **19**, 149 (1947); through *Chem. Abstr.*, **45**, 3277 (1951).

603. Salvaterra, H., *Chem. Ztg.*, **38**, 90 (1914); through *Chem. Abstr.*, **8**, 1554 (1914).

604. Sarson, R. D., *Anal. Chem.*, **30**, 933 (1958).

605. Sato, H., *Bull. Inst. Phys. Chem. Res.* (*Tokyo*), **16**, 804 (1937); through *Chem. Abstr.*, **32**, 4413 (1938).

606. Sawicki, E., and T. W. Stanley, *Anal. Chim. Acta*, **23**, 551 (1960).

607. Schall, R., *Compt. Rend.*, **239**, 1036 (1954).

608. Schechter, H., and H. L. Cates, Jr., *J. Org. Chem.*, **26**, 51 (1961).

609. Scheuler, F. W., and C. J. Hanna, *J. Am. Chem. Soc.*, **73**, 4996 (1951).

610. Schindler, R., W. Luttke, and L. Holleck, *Chem. Berichte*, **90**, 157 (1957).

611. Schmidt, E., A. Ashere, and L. Mayer, *Berichte*, **58B**, 2430 (1925).

612. Schnitzer, M., and S. I. Skinner, *Soil Sci.*, **101**, 120 (1966).

612a. Schoental, R., *Nature*, **188**, 420 (1960).

613. Schroeder, W. A., *N. Y. Acad. Sci. Ann.*, **49**, 204 (1948).

614. Schroeder, W. A., et al., *Anal. Chem.*, **23**, 1740 (1951).

615. Schroeder, W. A., et al., *Ind. Eng. Chem.*, **41**, 2818 (1949).

616. Schultz, G., G. Rohde, and E. Bosch, *Ann. Chem.*, **334**, 235 (1904).

616a. Schwartz, M., and E. A. Mark, *Anal. Chem.*, **38**, 610 (1966).

617. Scoggins, M. W., *Anal. Chem.*, **36**, 1152 (1964).

618. Scott, E. W., *J. Ind. Hyg. Toxicol.*, **25**, 20 (1943).

619. Scott, E. W., and J. F. Treon, *Anal. Chem.*, **12**, 189 (1940).

620. Seagers, W. J., and P. J. Elving, *J. Am. Chem. Soc.*, **72**, 3241 (1950).

621. Seagers, W. J., and P. J. Elving, *J. Am. Chem. Soc.*, **72**, 5183 (1950).

622. Seagers, W. J., and P. J. Elving, *Sb. Mezinar. Polarogr. Sjezdu Praze, 1st Congr. 1951*, Part I, 281; through *Chem. Abstr.*, **45**, 9449 (1952).

623. deleted.

624. Seely, G. R., *J. Phys. Chem.*, **73**, 117 (1969).

625. Sellers, D. E., and G. W. Leonard, Jr., *Anal. Chem.*, **33**, 334 (1961).

626. Sellers, D. E., and G. W. Leonard, Jr., *Anal. Chem.*, **34**, 1457 (1962).

627. Semel, S., M. A. Laccetti, and M. Roth, *Anal. Chem.*, **31**, 1050 (1959).

628. Semerano, G., *Ric. Sci.*, Suppl. 4, 11 (1959); through *Chem. Abstr.*, **54**, 108 (1960).

629. Sensabaugh, A., R. H. Cundiff, and P. C. Markunas, *Anal. Chem.*, **30**, 1445 (1958).

630. Sevcik, A., *Collect. Czech. Chem. Commun.*, **13**, 349 (1948); through *Chem. Abstr.*, **43**, 953 (1949).

631. Seyewitz, A., and Noel, *Bull. Soc. Chim.*, **3**, (4), 497 (1908); through *Chem. Abstr.*, **2**, 2229 (1908).

632. deleted.

633. Shea, F., and C. E. Watts, *Ind. Eng. Chem., Anal. Ed.*, **11**, 333 (1939).

633a. Shaefer, W. E., and W. W. Becker, *Anal. Chem.*, **19**, 307 (1947).

634. Shinozuka, F., and J. T. Stock, *Anal. Chem.*, **32**, 884 (1960).

635. Shriner, R. L., and R. C. Fuson, *Systematic Identification of Organic Compounds*, 3rd ed., John Wiley, New York, 1948.

636. Sifre, G., *Mem. Poudres*, **35**, 373 (1953); through *Chem. Abstr.*, **49**, 9448 (1955).

637. Siggia, S., *Quantitative Organic Analysis via Functional Groups*, 2nd ed., John Wiley, New York, 1954, p. 128.

638. Silverman, L., and W. G. Bradshaw, *Anal. Chem.*, **31**, 1672 (1959).

639. Simek, B. G., *Chem. Listy*, **25**, 322 (1931); through *Chem. Abstr.*, **25**, 5871 (1931).

640. Simpson, J. R., and W. C. Evans, *Biochem. J.*, **55**, (2), XXIV (1953).

641. Slough, W., *Trans. Faraday Soc.*, **57**, 366 (1961).

642. Slovetski, V. N., et al., *Izv. Akad. Nauk. SSSR, Otd. Khim. Nauk*, **1960**, 1709; through *Chem. Abstr.*, **51**, 7166 (1961).

643. Smales, A. A., and H. N. Wilson, *J. Soc. Chem. Ind.*, **67**, 210 (1948).

644. Smart, G. N. R., and G. F. Wright, *Can. J. Res.*, **26B**, 284 (1948).

645. Smith, B., and A. Haglund, *Acta Chem. Scand.*, **15**, 657 (1961).

646. Smith, G. B. L., and V. J. Sabette, *J. Am. Chem. Soc.*, **54**, 1034 (1932).

647. Smith, G. N., *Anal. Chem.*, **32**, 32 (1960).

648. Smith, G. N., and M. G. Swank, *Anal. Chem.*, **32**, 978 (1960).

649. Smith, J. W., and J. G. Waller, *Trans. Faraday Soc.*, **46**, 290 (1950).

650. Smyth, H. F., *J. Ind. Hyg.*, **13**, 87 (1931); through *Chem. Abstr.*, **25**, 3098 (1931).

651. Sokolova, N. P., Z. M. Skui'skaya, and A. A. Balandin, *Izv. Akad. Nauk SSSR, Ser. Khim*, **1966** (10) (1861); through *Chem. Abstr.*, **66**, 101411 (1967).

652. Soldate, A. M., and R. M. Noyes, *Anal. Chem.*, **19**, 442 (1947).

653. Somers, P. D., Jr., *Proc. Indian Acad. Sci.*, **54**, 117 (1944).

654. Someya, K., *Z. Anorg. Allgem. Chem.*, **169**, 293 (1928); through *Chem. Abstr.*, **22**, 2122 (1928).

655. Sommer, F., and H. Pincus, *Berichte*, **48**, 1963 (1915).

656. Spindler, P., *Annalen*, **224**, 288 (1884).

657. Spreter, V. Ch., and E. Briner, *Helv. Chim. Acta*, **32**, 215 (1949).

658. Srbova, J., *Prac. Lek.*, **4**, 47 (1952); through *Chem. Abstr.*, **49**, 4071 (1955).

659. Stafford, W. H., M. Los, and N. Thomson, *Chem. Ind.*, **1956**, 1277.

660. Stalcup, H., and R. W. Williams, *Anal. Chem.*, **27**, 543 (1955).

661. Stedronsky, E. R., et al., *J. Am. Chem. Soc.*, **90**, 993 (1968).

662. Stephen, M. J., and C. Hinshelwood, *J. Chem. Soc.*, **1955**, 1393.

663. Sternglanz, P. D., R. C. Thompson, and W. L. Savell, *Anal. Chem.*, **25**, 1111 (1953).

664. Sternglanz, P. D., R. C. Thompson, and W. L. Savell, *Anal. Chem.*, **27**, 392 (1955).

665. Stewart, P. E., and W. A. Bonner, *Anal. Chem.*, **22**, 793 (1950).

666. Stocesova, D., *Collect. Czech. Chem. Commun.*, **14**, 615 (1948).

667. Stock, J. T., *J. Chem. Soc.*, **1957**, 4532.

668. Stone, H. W., *Anal. Chem.*, **20**, 747 (1948).

669. Strain, H. H., *Ind. Eng. Chem., Anal. Ed.*, **14**, 245 (1942).

670. Strain, H. H., *J. Am. Chem. Soc.*, **57**, 758 (1935).

671. Streuli, C. A., and W. D. Cooke, *Anal. Chem.*, **25**, 1691 (1953).

672. Streuli, C. A., and W. D. Cooke, *Anal. Chem.*, **26**, 963 (1954).

673. Streuli, C. A., and W. D. Cooke, *Anal. Chem.*, **26**, 970 (1954).

674. Sweet, R. L., R. S. Spindt, and V. D. Meyer, *Am. Chem. Soc., Div. Petrol Chem., Gen. Pap.*, No. 34, 33 (1955).

675. Takagi, T., and N. Hyashi, *J. Chem. Soc. Jap.*, **78**, 445 (1957).

676. Takagi, S., and T. Ueda, *J. Pharm. Soc. Jap.*, **58**, 533 (1938).; through *Chem. Abstr.*, **32**, 8384 (1938).

677. Takeuchi, T., and M. Kasagi, *J. Chem. Soc. Jap., Ind. Chem. Sect.*, **52**, 64 (1949); through *Chem. Abstr.*, **45**, 1916 (1951).

678. Talsky, G., *Z. Anal. Chem.*, **191**, 191 (1962).

679. Tandon, J. P., *Z. Anal. Chem.*, **167**, 184 (1959).

680. Teague, A. F., W. A. Gey, and R. W. van Dolah, *Anal. Chem.*, **27**, 785 (1955).

681. Teisinger, J., *Mikrochem. Mikrochim. Acta*, **25**, 151 (1938); through *Chem. Abstr.*, **32**, 9136 (1938).

682. Terent'ev, A. P., and G. S. Goryachev, *Uch. Zap. (Wiss. Ber. Moskau Staats-Univ.)*, **3**, 277 (1934); through *Chem. Abstr.*, **30**, 8073 (1936).

683. Testa, A. C., and W. H. Reinmuth, *Anal. Chem.*, **32**, 1512 (1960).

684. Testa, A. C., and W. H. Reinmuth, *Anal. Chem.*, **32**, 1518 (1960).

685. Thiele, J., and A. Lachmann, *Annalen*, **288**, 267 (1895).

686. Tiwari, R. D., and J. P. Sharma, *Anal. Chem.*, **35**, 1307 (1963).

687. Tiwari, R. D., and J. P. Sharma, *Indian J. Chem.*, **2**, 173 (1964); through *Chem. Abstr.*, **61**, 10045 (1964).

688. Tiwari, R. D., and J. P. Sharma, *Proc. Natl. Acad. Sci., India Sect.*, **A33**, Pt. 3, 379 (1963); through *Chem. Abstr.*, **60**, 1111 (1964).

689. Trueblood, K. N., and E. W. Malmberg, *Anal. Chem.*, **21**, 1055 (1949).

690. Truffaut, G., and I. Pestac, *Compt. Rend. Acad. Agr. Fr.*, **29**, 381 (1943); through *Chem. Abstr.*, **40**, 5871 (1946).

691. Tsuji, K., and P. J. Elving, *Anal. Chem.*, **41**, 216 (1969).

692. Tsuji, K., and P. J. Elving, *Anal. Chem.*, **41**, 1571 (1969).

693. Turba, F., R. Haul, and G. Uhlen, *Angew. Chem.*, **61**, 74 (1949).

694. Turek, O., *Chim. Ind.*, **26**, 781 (1931).

695. Turner, W. R., and P. J. Elving, *Anal. Chem.*, **37**, 200 (1965).

696. Turney, T. A., and G. A. Wright, *Chem. Rev.*, **59**, 497 (1959).

697. Ubaldini, I., and F. Guerriere, *Ann. Chim. Appl.*, **38**, 702 (1948).

698. Ueno, S., and H. Sekiguchi, *J. Soc. Chem. Ind. Jap.*, **37B**, 235 (1934).

699. Ungnade, H. E., and L. W. Kissinger, *Tetrahedron*, **19**, Suppl. 1, 121 (1963).

700. Van Urk, H. W., *Chem. Weekbl.*, **21**, 169 (1924); through *Chem. Abstr.*, **18**, 2142 (1924).

701. U.S. Pat. 1,967,667 (July 24, 1934), H. B. Hass, E. B. Hodge, and B. M. Vanderbilt; through *Chem. Abstr.*, **28**, 5830 (1934).

702. U.S. Pat. 2,071,122 (Feb. 16, 1937), H. B. Hass and E. B. Hodge; through *Chem. Abstr.*, **31**, 2619 (1937).

703. U.S. Pat. 2,146,188 (Feb. 7, 1939), E. B. W. Kerone; through *Chem. Abstr.*, **33**, 3400 (1939).

704. U.S. Pat. 2,335,384 (Nov. 30, 1944), E. W. Bousquet, J. E. Kirby, and N. E. Searle; through *Chem. Abstr.*, **38**, 2834 (1944).

705. U.S. Pat. 2,365,981 (Dec. 26, 1944); J. B. Tindall; through *Chem. Abstr.*, **40**, 3126 (1946).

706. U.S. Pat. 2,392,611 (Jan. 8, 1946); E. M. Nygard, J. H. McCracken, and N. H. Hamilton; through *Chem. Abstr.*, **40**, 2960 (1946).

707. U.S. Pat. 2,649,361 (Aug. 18, 1953); R. Springer and W. R. Meyer; through *Chem. Abstr.*, **48**, 476 (1954).

708. U.S. Pat. 2,916,426 (Dec. 8, 1959), D. Horwitz and E. Cerwonka; through *Chem. Abstr.*, **54**, 6370 (1960).

709. Vainshtein, Y. I. *Zavod. Lab.*, **14**, 517 (1948); through *Chem. Abstr.*, **43**, 4978 (1949).

710. Vanags, G., *Z. Anal. Chem.*, **126**, 21 (1943); through *Chem. Abstr.*, **37**, 6593 (1943).

711. Vanags, G., and A. Lode, *Chem. Ber.*, **70**, 547 (1937).

712. Vanags, G., and E. Lukevic, *Zh. Obshch. Khim.*, **26**, 1400 (1956); through *Chem. Abstr.*, **50**, 14677 (1956).

713. Vanderzee, C. E., and W. F. Edgell, *Anal. Chem.*, **22**, 572 (1950).

714. Vanderzee, C. E., and W. F. Edgell, *J. Am. Chem. Soc.*, **72**, 2916 (1950).

715. Vanino, L., and F. Hartl, *Arch. Pharm.*, **244**, 216 (1906); through *Chem. Zentralbl.*, **77**, (II), 98 (1906).

716. Veibel, S., *Anal. Chem.*, **23**, 665 (1951).

717. Vertyulina, L. N., and N. I. Malyugina, *Zh. Obshch. Khim.*, **28**, 304 (1958); through *Chem. Abstr.*, **52**, 10765 (1958).

718. Wagner, C. D., R. H. Smith, and E. D. Peters, *Anal. Chem.*, **19**, 982 (1947).

719. Wagniere, G. H., in H. J. Feuer, Ed., *The Chemistry of the Nitroso and Nitro Groups*, Interscience, New York, 1969).

720. Walters, C. L., E. M. Johnson, and N. Roy, *Analyst*, **95**, 485 (1970).

721. Walther, R., *J. Prakt. Chemie*, **53**(2), 43 (1896).

722. Weaver, R. D., and G. C. Whitnack, *Anal. Chim. Acta*, **18**, 51 (1958).

723. Webster, M. S., *J. Chem. Soc.*, **1956**, 2841.

724. Weisburger, J. H., and E. K. Weisbarger, *Chem. Eng. News*, **44**(6), 124 (1966).

725. Weizmann, M., J. Yofe, and B. Kirzon, *Z. Physiol. Chem.*, **192**, 70 (1930); through *Chem. Abstr.*, **25**, 263 (1931).

726. White, E. H., and W. R. Feldman, *J. Am. Chem. Soc.*, **79**, 5832 (1957).

727. White, J. W., Jr., *Anal. Chem.*, **20**, 726 (1948).

728. White, L. M., and M. C. Long, *Anal. Chem.*, **23**, 363 (1951).

729. Whitmore, F. C., and M. G. Whitmore, in Gilman and Blatt, Eds., *Organic Syntheses*, Coll. Vol. I, John Wiley, New York, 1944, p. 401.

730. Whitnack, G. C., *Anal. Chem.*, **35**, 970 (1963).

731. Whitnack, G. C., and E. St. C. Ganz, *J. Electrochem. Soc.*, **106**, 422 (1959).

732. Whitnack, G. C., R. D. Weaver, and H. W. Kruse, U.S. Dept. of Commerce, National Technical Information Service, *AD 413 029*.

733. Wieland, H., and H. Jung, *Annalen*, **445**, 82 (1925).

734. Williams, F. T., Jr., P. W. K. Flanagan, W. J. Taylor, and H. Slochter, *J. Org. Chem.*, **30**, 2674 (1965).

735. Williams, R. L., R. J. Pace, and G. J. Jeacocke, *Spectrochim. Acta*, **20**, 225 (1964).

736. Wilson, H. N., and W. Hutchinson, *Analyst*, **72**, 432 (1947).

737. Wilson, R. F., and T. Rhodes, *Anal. Chem.*, **28**, 1199 (1956).

738. Wislicenus, W., and M. Waldmuller, *Berichte*, **41**, 3334 (1908).

739. Witry-Schwachtgen, G., *Inst. Grand-Ducal Luxemb., Sect. Sci., Nat. Phys. Math., Arch.*, **22**, 87 (1955); through *Chem. Abstr.*, **50**, 12728 (1956).

740. Wohl, A., *Berichte*, **27**, 1434 (1894).

741. Wolthuis, E., S. Kolk, and L. Schaap, *Anal. Chem.*, **26**, 1238 (1954).

742. Wood, K. I., *Anal. Chem.*, **37**, 442 (1965).

742a. Yasuda, S. K., and R. N. Rogers, *Anal. Chem.*, **32**, 910 (1960).

743. Yokoo, M., *Pharm. Bull. Jap.*, **6**, 64 (1958).

744. Young, S. W., and R. E. Swain, *J. Am. Chem. Soc.*, **19**, 812 (1897).

745. Zarembo, J. E., and I. Lysyj, *Anal. Chem.*, **31**, 1833 (1959).

746. Zikan, J., and J. Sterzl, *Nature*, **214**, 1225 (1967).

747. Zimmerman, R. P., and E. Lieber, *Anal. Chem.*, **22**, 1151 (1950).

748. Zincke, T., and E. Ellenberger, *Annalen*, **339**, 215 (1905).

749. Zinneke, F., *Angew. Chem.*, **64**, 220 (1952).

750. Zinsser, F. J., *Berichte*, **24**, 3556 (1891).

751. Zittel, H. E., and F. J. Miller, *Anal. Chem.*, **37**, 200 (1965).

# NITRATE AND
# NITRITE ESTER GROUPS

By R. F. Muraca, *College of Notre Dame,*
*Belmont, California*

**Contents**

## I. INTRODUCTION

Compounds that contain the nitrate and nitrite groups are esters of nitric and nitrous acids, respectively. These functional groups, —O—NO$_2$ and —O—NO, are structurally related to the nitro and nitroso groups discussed in the preceding section; however, their analytical reactions are sufficiently different to merit separate treatment. The nitrate group is often called the nitroxy group when it is an ester function.

There is great confusion in the older literature between the nomenclature of the nitrate and nitrite esters and the nitro compounds discussed in the preceding section. Nitrate esters are often called nitro compounds; for example, even to this day, glyceryl trinitrate is called nitroglycerin. This state of affairs comes about because the nitration of many alcohols often involves the mixtures of nitric and sulfuric acid that are used to produce nitro compounds from hydrocarbons. Fortunately, only a few compounds are involved in this confusion.

The characteristic feature of nitrate and nitrite esters is that the functional groups can be more or less readily split off by hydrolysis; thus the analytical reactions of nitrate and nitrite functional groups are in essence the reactions of the acids from which the esters are derived. As can be anticipated, there is a vast amount of literature concerned with the manufacture of glyceryl trinitrate (nitroglycerin) ethylene glycol dinitrate, and the cellulose nitrate esters (nitrocellulose); however, only a relatively small number of published articles deal with the analytical chemistry of these extraordinarily important compounds, and an even smaller number provide valuable insights into the reaction mechanisms involved in their analysis.

The nomenclature of compounds containing the nitrate and nitrite ester groups (as well as related groups) is straightforward; as a rule, the compounds are named as esters of the corresponding alcohols, for example, ethylene glycol dinitrate, glyceryl trinitrate, and isoamyl nitrite. Since the nitrate ester group is also called the nitroxy group, ethylene glycol dinitrate may be called 1,2-dinitroxyethane, and O$_2$NOCH$_2$—CH$_2$—CH$_2$ONO$_2$ 1,3-dinitroxypropane. Other esters of the inorganic oxygen-nitrogen acids and related derivatives included in this section are hyponitrites and thionitrites.

## II. SYNTHESIS AND OCCURRENCE

All the methods of forming esters can, in principle, be used to form compounds containing the nitrite and nitrate ester functional groups, but because many compounds of this type are readily decomposed or are sensitive, explosive materials, special methods have been developed to produce the commercially important members.

## A. NITRATE ESTERS

The preparation of nitrate esters generally involves one of two methods: esterification or reaction of an alkyl halide with silver nitrate, but extraordinary methods are often used for preparing special types of esters.

### 1. Esterification of Alcohols

The most common method of synthesizing mono- or polynitrate esters is the direct nitration of alcohols with nitric acid or with a mixture of nitric and sulfuric acids; in fact, the great majority of nitrate esters are formed by direct nitration with mixed acids.

#### a. MIXED ACIDS

As a rule, a mixture of nearly equal volumes of concentrated nitric acid (d., ~ 1.42) and sulfuric acid (d., ~ 1.84) is used together with a small amount of urea (or urea nitrate) to destroy nitrous acid which may be present initially or may be formed by attack of the alcohol. (When free nitrous acid is present, an alcohol may be oxidized so violently that explosion occurs.) Esterification usually is accomplished by slowly adding the alcohol (pure or in sulfuric acid) to a portion of mixed acids which is kept cold and mixed well (15). The amount of nitric acid is selected so that there is only a slight excess with respect to the amount of alcohol:

$$ROH + HONO_2 \rightarrow RONO_2 + H_2O$$

The ester forms rapidly, and it is separated by pouring the reaction mixture into cold water; occasionally, the ester separates (273), or it can be removed by careful distillation. In other instances, such as in the formation of diethylene glycol dinitrate, the ester forms an emulsion (240,241) which is difficult to break.

The sulfuric acid in the mixed-acid mixture described above serves to remove the water formed in the esterification, and the efficiency of a nitrating system under fixed conditions is largely dependent on the dehydrating value of the sulfuric acid (DVS) (241). Other dehydrating agents may be used, for example, acetic anhydride (189); esterification may be accomplished by simultaneous addition of 100% nitric acid (WFNA) and the alcohol to acetic anhydride (133). The reactive ingredient in mixtures of acetic anhydride and nitric acid may be acetyl nitrate (22) or dinitrogen pentoxide, $N_2O_5$ (19); after the acetic anhydride has removed all water as acetic acid:

$$Ac_2O + H_2O \rightarrow 2\,HOAc$$

both acetyl nitrate and dinitrogen pentoxide form in equilibrium amounts (19,22,90):

$$Ac_2O + HNO_3 \rightleftharpoons AcONO_2 + HOAc$$
$$AcONO_2 + HNO_3 \rightleftharpoons N_2O_5 + HOAc$$
$$Ac_2O + N_2O_5 \rightleftharpoons 2\,AcONO_2$$

Clearly, the concentration of acetic anhydride determines the preponderant nitrating species and may thus influence the rate of nitration (19). It has also been suggested that the protonated form of acetyl nitrate, $[CH_3COOHNO_2]^+$, is present in such a high concentration that it is the active substance in spite of its lower reactivity (22). Acetylation of hydroxyl groups is a competing reaction in nitrations performed with nitric acid–acetic anhydride mixtures (85) inasmuch as nitric acid serves as a catalyst. Schenk and Santiago (250) have made a detailed study of the nitration of alcoholic hydroxyl groups and have found that O-nitration of primary alcohols by nitric acid in acetic anhydride is quantitative at room temperature (20 min) in acetonitrile as a diluent. The optimum amount and rate of nitration was found to occur with a mixture of 1.2 parts of acetic anhydride and 1 part of nitric acid diluted with acetonitrile so that the nitric acid content is of the order of 0.42 $M$, and the hydroxyl compound is used in such quantity as to bring the final nitric acid concentration to about 0.2 $M$. As another example of nitrations performed under extremely anhydrous conditions, it is of interest to note that a mixture of nitric acid and phosphorus pentoxide is used to prepare di- and trinitroxycyclohexanes (43). Nitration of tertiary alcohols is complicated by accompanying hydrolysis; nitration in a water-immiscible solvent is a convenient route (23), but addition of nitric acid to an olefin is preferable (198).

Cellulose nitrates, commonly called nitrocelluloses, are prepared by soaking dry, pure cotton linters in mixed acid. The reaction is carried out rapidly and at as low a temperature as possible (30–40°C); the water content of the mixed acid is kept low to minimize degradation. After centrifuging and washing free of acid, the nitrated cotton is boiled in water to hydrolyze sulfate. It is usually stored wet and dehydrated by ethanol or butanol before use.

Three types of nitrated cellulose are produced:

1. Celluloid pyroxylin; 10.5–11% nitrogen; soluble in a mixture of ethanol and ether.
2. Soluble pyroxylin (dynamite cotton; collodion); 11.5–12.3% nitrogen; soluble in absolute ethanol.

3. Guncotton; 12.5–13.5% nitrogen; when over 13% nitrogen, insoluble in a mixture of 1 part of ethyl alcohol and 2 parts of ethyl ether (by volume) and insoluble in absolute ethyl alcohol. Standard military guncotton containing 12.8% nitrogen corresponds to the formula $C_{24}H_{30}O_{10}(NO_3)_{10}$. All commercial cellulose nitrates (<13.5% N) dissolve in acetone and low molecular weight esters. Cellulose dinitrate contains 11.11% nitrogen, and the trinitrate, 14.15%; obviously, commercial products are mixtures. Guncotton is the least degraded cellulosic structure; it has a chain length of about 3000 $C_6$ units and thus gives solutions of extremely high viscosity. For use in smokeless ammunition, guncotton is gelatinized to a dough by a solvent consisting of a mixture of approximately 1:2 ethanol-ethyl ether (plus acetone if nitroglycerin is present); then the dough is extruded and cut into cylindrical, perforated grains. The solvent is removed in a final operation.

Glyceryl trinitrate (nitroglycerin), ethylene glycol dinitrate, starch nitrate (nitrostarch), pentaerythritol tetranitrate (PETN), and n-propyl nitrate are produced commercially by esterification with mixed acids.

### b. NITRIC ACID ALONE

Absolute nitric acid (WFNA) may be used to prepare nitrate esters in the same way as reactions are carried out with nitric–sulfuric acid mixtures; urea is added to destroy nitrous acid. Occasionally, inert organic solvents are used as diluents to moderate the reaction; thus, by direct action of nitric acid, tert-amyl nitrate can be prepared in chloroform (198), and tert-butyl nitrate in methylene chloride at −30°C (182).

## 2. Alkyl Halides with Silver Nitrate

The metathetical reaction between silver nitrate and an alkyl halide is a useful method for preparing certain nitrate esters:

$$RX + AgNO_3 \rightarrow RONO_2 + AgX$$

The reaction may be made to take place heterogeneously or homogeneously.

### a. HETEROGENEOUS REACTIONS

Powdered silver nitrate is stirred into a solution of alkyl halide in an inert solvent such as benzene (147) or ether (7); however, nitromethane and nitrobenzene are very frequently used, and in some instances no solvent is necessary. Silver halide is slowly formed; after the reaction is complete, the solvent or the ester may be distilled off at a low temperature (vacuum).

### b. Homogeneous Reactions

A solution of silver nitrate in acetonitrile is used; alkyl iodides or bromides are needed for preparation of simple primary or secondary nitrates, but the chlorides of tertiary, allylic, or benzyl alcohols are sufficiently reactive (77).

### 3. Special Methods

There are a number of synthetic methods which are fairly limited in scope; for example, methyl nitrate can be obtained by the nitrolysis of methylnitramine:

$$CH_3NHNO_2 \xrightarrow{HNO_3} CH_3ONO_2 + N_2O$$

Transesterification often provides a convenient route for the preparation of nitrates:

$$ROH + R'ONO_2 \rightarrow RONO_2 + R'OH$$

Another route involves alcoholysis of an acyl nitrate (80), for example, preparation of ethyl nitrate by reaction of ethanol with benzoyl nitrate:

$$C_2H_5OH + C_6H_5COONO_2 \rightarrow C_6H_5COOH + C_2H_5ONO_2$$

The reaction of dinitrogen tetroxide with an olefin yields (among other products) a C-nitroso nitrate ester (199):

$$(CH_3)_2C{=}CH_2 + N_2O_4 \rightarrow (CH_3)_2\underset{\underset{NO}{|}}{C}CH_2ONO_2$$

Acyl nitrates such as acetyl nitrate, $CH_3COONO_2$, can be made in substantial quantities by action of dinitrogen pentoxide on the acid anhydride, but equilibrium quantities are produced on mixing acetic anhydride and absolute nitric acid. Benzoyl nitrate, $C_6H_5COONO_2$, is made by action of silver nitrate on benzoyl chloride at a low temperature ($-15°C$). Acetyl nitrate is rapidly becoming a common reagent for synthesizing nitrate esters; for example, with simple alkenes, acetyl nitrate forms (22,188) the following:

| Nitro derivative | Acetoxynitro derivative | cis-Nitronitroxy derivative |
|---|---|---|

Its reactions with alcohols, hydroxy esters, and vic-glycols were indicated in a preceding section.

## B. NITRITE ESTERS

Like the nitrate esters, the nitrite esters are usually prepared by either of two methods: direct esterification or reaction of an alkyl halide with silver nitrite.

### 1. Esterification of Alcohols

#### a. NITROUS ACID

The esterification of alcohols by nitrous acid in dilute aqueous solutions is remarkably rapid; this is in sharp contrast to esterifications of alcohols and acids, inasmuch as these are comparatively slow reactions which may take several hours even in the presence of a catalyst. For example, if an aqueous solution of sodium nitrite and benzyl alcohol is acidified with a little hydrochloric acid, it immediately becomes milky because of the separation of benzyl nitrite:

$$C_6H_5CH_2OH + HONO \rightarrow C_6H_5CH_2—ONO + H_2O$$

It is of interest to note that the —OH group in phenol does not form a nitrous ester; $C$-nitrosophenols are produced.

The release of nitrous acid by the addition of a strong acid to a mixture of alkali-metal nitrite and an alcohol often induces resinification reactions and destruction of the rather sensitive nitrite esters. Chrétien and Longi (42) have found that undesirable side reactions can be eliminated if the alcohol is mixed with a saturated solution of sodium nitrite and treated with a 40% solution of aluminum sulfate; dropwise addition of the aluminum sulfate solution over a 2-hr interval and thorough agitation of the mixture provides nearly quantitative conversions of most alcohols. The supernatant ester is drawn off and dried over anhydrous sodium sulfate. This nitrosation method should prove to be useful for analytical procedures.

#### b. NITROUS FUMES

Mixed oxides of nitrogen (NO, $NO_2$, $N_2O_4$, $N_2O_3$) are often used to prepare esters of nitrous acid; for this purpose, the fumes are prepared in the laboratory by reaction of nitric acid and copper. Alternatively, pure NO gas (commercially available in cylinders) is mixed with air. The nitrous fumes are passed into alcohols to form the esters.

#### c. NITROSYL CHLORIDE

Nitrous esters can be formed by passing vapors of nitrosyl chloride into a solution of an alcohol containing an equivalent of pyridine:

$$NOCl + ROH \rightarrow R—ONO + HCl$$

## 2. Alkyl Halides with Silver Nitrite

The metathetical reaction of an alkyl halide with silver nitrite at 80–110°C:

$$RX + AgNO_2 \rightarrow RONO + AgX$$

is not used very frequently for formation of nitrous esters because of poor yields. The reaction fails with methyl halide because the product is almost pure nitromethane, $C—NO_2$; with other alkyl halides, the amount of nitrite ester, $—O—NO$, that is formed varies with the nature of the alkyl radical, and a wide variety of products often results (alcohols, ketones, etc.) (147,149), especially when secondary halides are employed. In fact, nitrites may not be obtained at all; thus 1-bromo-heptane treated with silver nitrite at 85°C yields pure 1-heptyl nitrate (151). With straight-chain halides, the reaction appears to produce at least 70% of a $C$-nitroparaffin with (usually) at least 10% of a nitrite ester at low temperature. Branching in the vicinity of the reactive site hinders the reaction, but neopentyl iodide does not react (151). Secondary halides produce about 15% $C$-nitroparaffins and about 25–35% nitrite ester at low temperatures; sometimes, 10–30% olefin is also formed. The tertiary compounds produce only small amounts of $C$-nitroparaffins and about 50% nitrite esters at low temperature (150). The reaction of silver nitrite with $\alpha$-haloesters yields $\alpha$-nitroesters (146).

## 3. Transesterification

The alkyl group of a nitrous ester can be very readily exchanged for that of an alcohol in which it is dissolved. The exchange is common to all esters, but it occurs with exceptional rapidity with nitrous esters; for example, gaseous methyl nitrite is evolved when ethyl nitrite is dissolved in methanol. Nitrites can be removed quantitatively as methyl nitrite (78a,292a).

## C. HYPONITRITE ESTERS

These rather rare esters have been prepared by action of alkyl iodides on silver hyponitrite (220):

$$2\,RI + Ag—O—N{=}N—O—Ag \rightarrow RO—N{=}N—OR + 2\,AgI$$

As a rule, sodium hyponitrite is first prepared by reduction of 50% aqueous solution of sodium nitrite with sodium amalgam (62), although it can also be prepared by reaction of sodium with hydroxylamine hydrochloride (113,259). Sodium hyponitrite forms a pentahydrate and an octahydrate (219); silver hyponitrite is a yellow, insoluble salt readily precipitated on addition of the calculated amount of slightly acidic silver nitrate to a cold dilute solution of sodium hyponitrite (129). Magnesium, barium, cobalt, and lead salts are formed by metathesis in aqueous solution; other salts (e.g., barium) are obtained by metathesis in an ammoniacal solution of silver hyponitrite (167).

## D. THIONITRITE ESTERS

The preferred synthetic route for thionitrite esters is reaction of a mercaptan and an alkyl nitrite (165):

$$EtSH + EtONO \rightarrow EtOH + EtSNO$$

These rather unstable esters can also be prepared by action of nitrosyl chloride on a mercaptan at low temperatures (239,288):

$$EtSH + ClNO \rightarrow EtSNO + HCl$$

A more convenient procedure involves the reaction of $N_2O_3$ (equimolecular mixture of nitrogen dioxide and nitric oxide) with alkyl mercaptans at low temperature (92).

## E. OCCURRENCE

The nitrate, nitrite, and other esters discussed in this section are not found in nature; they are products of synthesis.

The nitrate esters of cellulose, starch, glycerin, ethylene glycol, and mannitol are used principally as high explosives and are often found in mixtures which may contain sodium nitrate, ammonium nitrate, wood pulp, kieselguhr, and diatomaceous earth. Double-base military smokeless powders such as Ballistite and cordite contain a mixture of about 60% cellulose nitrate with 40% nitroglycerin that is gelatinized with petrolatum. Rocket propellants often are double-base compositions containing aluminum powder, ammonium perchlorate, and small amounts of other substances (299). Gelatin dynamite (blasting gelatin) consists of nitroglycerin gelatinized with 8–12% of cellulose nitrate, but it is often mixed with wood pulp and sodium nitrate. The ordinary dynamites of today contain up to 55% of ammonium nitrate or sodium nitrate with calcium or magnesium carbonate (stabilizer), wood pulp, and glyceryl trinitrate; the dynamite invented by Alfred Nobel in 1867 consisted of nitroglycerin absorbed in kieselguhr (sodium carbonate stabilizer). Nonfreezing dynamites contain mixtures of ethylene glycol dinitrate with nitroglycerin, ammonium nitrate, sodium nitrate, woodpulp, and calcium carbonate. Nitrostarch dynamites are also extensively used for blasting.

Guncotton (cellulose nitrate) and many other nitrate esters used in explosives slowly decompose and release oxides of nitrogen, which catalyze further decomposition. Stabilizers which combine with nitrogen oxides are always included in explosive or propellant compositions in amounts ranging from 1 to 2% by weight; typical stabilizers are diphenylamine (which combines with nitric as well as nitrous acid) and the dialkylphenylureas, $OC(NR\phi)_2$, called centralites (hydrolyzed by traces of mineral acid to alkyl anilines that react with nitrogen oxides). A characteristic dialkylphenylurea in explosives is diethylphenylurea (ethyl centralite).

In addition to stabilizers, explosives often contain inert plasticizers such as dioctylphthalate (di-2-ethylhexylphthalate) (2-ethylhexyl-phthalate), dimethylphthalate, diethylphthalate, dibutylphthalate, tricresylphosphate, camphor, triacetin, and castor oil; active plasticizers are also used, for example, diethylene glycol dinitrate (DEGDN). Additionally, extruded grains of explosives contain additives such as nigrosine (C.I. 50420) or 0.01–0.2% of carbon black to prevent premature ignition of the internal sections of a powder grain by radiant energy.

A variety of ballistic modifiers such as potassium sulfate, lead compounds, and other burning-rate modifiers are also included in finished propellants. Dinitrotoluene is often incorporated in nitrocellulose propellant formulations to reduce the visibility of muzzle flash; nitroguanidine is used for the same purpose.

Whereas the cellulose nitrates are used extensively in explosives, the lower nitrogen-content products are incorporated in solutions as collodion or lacquers, and the intermediate nitrogen-content products are mixed with camphor to form highly inflammable Celluloid; however, these products are rapidly being replaced by newer synthetic materials.

Glyceryl nitrate is a short-duration vasodilator often used to relieve spasms of the coronary artery (angina pectoris); it is administered orally as a 1% alcoholic solution. If the alcoholic solution is allowed to come into contact with the tongue or with skin, violent headaches may result. Other nitrate esters commonly used in medicine are mannitol hexanitrate, pentaerythritol tetranitrate, and inositol hexanitrate; these substances are always diluted with lactose or mannitol and compressed into tablets.

Amyl nitrate is added to diesel fuel in order to increase the cetane numbers; a great number of alkyl nitrites and nitrates also act similarly to lower the critical compression ratio.

Nitrites relax the smooth muscles of the body and produce rapid lowering of the blood pressure; the active ingredient in ordinary "sweet spirit of niter" which exhibits diaphoretic and diuretic action, is ethyl nitrite at a concentration of 3.5–4.5% in alcohol.

Hyponitrites are rarely used in commerce; certain organic hyponitrites may find application as polymerization catalysts (288a).

## III. PROPERTIES

The molecular structure of the nitrate ester grouping appears to be a hybrid of equivalent resonance structures, as determined from electron diffraction (221) and infrared and Raman spectroscopy data (27), and the results of application of X-ray diffraction (20) and electric dipole moment methods (204):

$$R\diagdown_{O-\overset{+}{N}}\diagup^{O}_{O^-} \rightleftharpoons R\diagdown_{O-\overset{+}{N}}\diagup^{O^-}_{O}$$

The —ONO$_2$ group is planar, and there is free rotation about the O—N bond which is connected to R. In contrast, there is marked rigidity about the corresponding bond in the nitrous ester group; thus two planar *cis-trans* forms can exist:

$$R\diagdown_{O}\diagdown_{O\diagup N} \quad \text{and} \quad R\diagdown_{O}\diagdown_{N\diagup O}$$

The potential barrier to rotation is not high enough to prevent thermal interconversion of the forms. The primary alkyl nitrites ordinarily are mixtures of both *cis* and *trans* forms, with the *cis* in preponderance at low temperatures. The secondary alkyl nitrites exist mostly in the *trans* form, and the tertiary esters are essentially pure *trans* forms. Even nitrous acid itself is mostly *trans* at room temperature (127).

The structure of the hyponitrite esters suggests the possibility of a *cis* and a *trans* form:

$$\begin{array}{ccc} RO-N & & RO-N \\ \| & \text{and} & \| \\ N-OR & & RO-N \end{array}$$

However, available evidence indicates that only the *trans* form is capable of existence at ordinary temperatures (122).

### A. PHYSICAL PROPERTIES

The physical properties of a short series of nitrous and nitric esters are given in Table I; the compounds have been selected to demonstrate the range of properties which can be encountered.

The boiling points of the alkyl nitrites are lower than those of the corresponding alcohols because molecular association cannot take place. On the other hand, the alkyl nitrates contain a $\overset{+}{N}$-$\overset{-}{O}$ coordinate bond, and the electric dipole moment resulting therefrom leads to attraction between molecules and causes them to have lower vapor pressures than do

## TABLE I

### Physical Properties of Typical Nitrate and Nitrite Esters

| Compound | Molecular weight | Nitrogen (%) | Color and form | Melting point (°C) | Boiling point (°C) | Solubility | Ref.[a] |
|---|---|---|---|---|---|---|---|
| Methyl nitrate | 77.04 | 18.18 | Expl. vapor | -108 | 65 | $H_2O$, alc., ether | B1, 284 |
| Methyl nitrite | 61.04 | 22.95 | Gas | -17 | -12 | Alc., ether | B1, 284 |
| Ethyl nitrate | 91.07 | 15.38 | Inflam. | -112 | 88.7 | $H_2O$, alc. | B1[2], 328 |
| Ethyl nitrite | 75.07 | 18.66 | Colorless | — | 17 | $H_2O$, dec., alc. | B1[2], 328 |
| n-Propyl nitrate | 105.09 | 13.33 | Colorless | — | 110.5 | Sl. sol. $H_2O$; alc. | B1[2], 369 |
| n-Propyl nitrite | 89.09 | 15.72 | Colorless | + | 57 | Sl. sol. $H_2O$; alc. | B1[2], 369 |
| Isopropyl nitrate | 105.09 | 13.33 | Colorless | — | 101.4 | — | B1[3], 1465 |
| Isopropyl nitrite | 89.09 | 15.72 | Colorless | — | 45 | — | B1[3], 1464 |
| sec-Butyl nitrate | 119.12 | 11.76 | Colorless | — | 124 | Alc. | |
| sec-Butyl nitrite | 103.12 | 13.58 | Colorless | — | 68 | Sl. sol. $H_2O$; $CCl_4$ | B1[2], 402 |
| tert-Butyl nitrite | 103.12 | 13.58 | Colorless | — | 63 | Sl. sol. $H_2O$; $CHCl_3$ | B1[2], 415 |
| Isoamyl nitrate | 133.15 | 10.52 | Colorless | — | 148 | Sl. sol. $H_2O$; alc. | B1[3], 1642 |
| Isoamyl nitrite | 117.15 | 11.96 | Yellow; inflam. | — | — | Sl. sol. $H_2O$; alc. | B1[3], 1641 |
| n-Octyl nitrate | 175.23 | 7.993 | Colorless | — | $111^{20}$ | — | B1, 419 |
| n-Octyl nitrite | 159.23 | 8.796 | Yellow | — | 174 | Alc.; ether | — |
| Cyclohexyl nitrate | 145.16 | 9.649 | Colorless | — | 171 | $H_2O$; sl. sol. ether | — |
| Glycerol 1-mononitrate | 137.09 | 10.22 | Prisms | 58-9 | 155–160 | $H_2O$; sl. sol. ether | B1[2], 591 |
| Glycerol 2-mononitrate | 137.09 | 10.22 | Prisms | 58-9 | 155–160 | $H_2O$; ether | B1[2], 591 |
| Glycerol 1,3-dinitrate ($\frac{1}{2}$ $H_2O$) | 191.10 | 14.66 | Prisms | 26 | $148^{15}$ | $H_2O$; ether | |
| Ethylene glycol dinitrate | 152.06 | 18.42 | Yellow | -22.3 | 198 | Alc.; insol. $H_2O$ | B1[3], 2112 |
| Ethylene glycol dinitrite | 120.07 | 23.33 | — | < -15 | 98 | Alc.; insol. $H_2O$ | B1, 469 |
| d-Tartaric acid dinitrate | 240.08 | 11.67 | Needles | d. | — | Alc.: ether | B3[2], 328 |
| Glycerol trinitrate | 227.09 | 18.50 | Pale yellow | 13(2) | (256) | Insol. $H_2O$; ether | B1[2], 591 |
| Cellulose trinitrate | $(459.28)_x$ | 9.149 | Light yellow | — | — | Insol. $H_2O$; alc. | |
| Cellulose tetranitrate | $(504.28)_x$ | 11.11 | Light yellow | — | — | Alc.-ether | |
| Cellulose pentanitrate | $(549.27)_x$ | 12.75 | White amorph. | — | — | Alc.-ether | |
| Cellulose hexanitrate | $(594.27)_x$ | 14.14 | White amorph. | 160–170 | — | Nitrobenz. | |
| Erythritol tetranitrate | 302.11 | 18.54 | Leaves | 61 | — | Alc.; glyc. | B1[3], 2358 |
| Pentaerythritol tetranitrate | 316.14 | 17.72 | Tet. prisms | — | 140 | Acet., benz. | B9[2], 887 |
| Mannitol hexanitrate | 452.16 | 18.59 | $[\alpha]^{12}_{546} + 46.8$, ($C_2H_4Cl_2$)0.33% | 112 | (120) | Alc.; benz. | B1[3], 2404 |

[a] The number following the letter B indicates the volume of the main series of Beilstein, *Handbuch der organischen Chemie*; the superscript refers to the first, second, etc., supplement; and the following number is the page in the supplement.

the corresponding nitrites; as a result, the boiling points of the nitrate esters are not very much different from those of the parent alcohols.

Nitrite esters undergo photolytic reactions ($\lambda < 300$ m$\mu$); generally, a C-nitroso compound (dimer) is formed, for example, *tert*-butyl nitrite forms nitrosomethane:

$$(CH_3)_3CONO \rightarrow (CH_3)_2CO + MeNO$$

The photochemistry of primary, secondary, and tertiary alkyl nitrites in benzene has been studied by Kabasakalian et al. (130,131).

At elevated temperatures, of the order of 200°C, the gas-phase decomposition of nitrites may form nitric oxide, the alcohol, and the aldehyde (148):

$$2\,RCH_2\!-\!ONO \rightarrow 2\,NO + RCH_2OH + RCHO$$

However, recent studies by Gowenlock and Trotman (92) indicate that certain alkyl nitrite esters may produce ketones and C-nitrosoparaffins on pyrolysis:

$$R_1R_2R_3CONO \rightarrow R_1NO + R_2R_3CO$$

where $R_1$ is the alkyl group that contains more carbon atoms than the others ($R_2$ and $R_3$ may also be hydrogen). As a rule, the *cis* dimers of the C-nitrosoparaffins are formed. The thermal decomposition of ethyl nitrate has been studied by Levy (175), and a mechanism suggested; at 161–201°C and pressures of a few centimeters of mercury, ethyl nitrite is a major product, but methyl nitrite and nitromethane are also formed in minor amounts. In addition, small amounts of nitrogen dioxide are formed. Ultraviolet light also produces similar products.

Alkyl mononitrates, usually colorless, mobile liquids of pleasant odor, are insoluble in water and toxic. They burn quietly with a white flame, but will explode if heated above their boiling points; alkyl polynitrates are often sensitive to both heat and shock. The alkyl nitrites over $C_3$ are also colorless, toxic, and pleasant-smelling liquids; they decompose slowly at room temperature and evolve nitrous fumes.

The acyl nitrates show the same explosive tendencies as the alkyl nitrates; the nitrate group is mobile in these compounds, and they are often used as powerful nitrating agents. Moreover, the acyl nitrates are rapidly decomposed by water into the organic acid and nitric acid. With compounds such as acetanilide, anisole, and other monosubstituted aromatics, the acyl nitrates react to form predominantly *ortho*- instead of *para*-substituted products (135).

The simple esters of hyponitrous acid are colorless liquids that cannot be distilled without decomposition (even at reduced pressures) and that decompose merely on standing at room temperature. The higher molecular weight hyponitrites are unstable crystalline solids. The hyponitrite esters are rapidly hydrolyzed by water, especially at elevated temperatures; the decomposition may take place as follows:

$$RCH_2O-N \atop \| \atop N-OH_2CR \to RCH_2OH + RCHO + N_2$$

but Oza et al. (216) have found that the decomposition also may form nitrous oxide:

$$RN_2O_2 + H_2O \to R(OH)_2 + N_2O$$

The decomposition appears to involve formation of free radicals (as do the aliphatic azo compounds); in fact, the hyponitrites function as effective polymerization catalysts. The estimation of sodium hyponitrite (which can be obtained by alkaline hydrolysis of the esters) in the presence of sodium nitrite, sodium nitrate, and sodium carbonate has been described by Oza (217).

The thionitrites, RSNO, are rather unstable compounds; the lower molecular weight alkyl thionitrites are deep red liquids which, in contrast to the nitrites, are very resistant to hydrolysis by water, aqueous acids, and alkalies but are very readily oxidized by air (222).

## B. CHEMICAL PROPERTIES

The nitrite and hyponitrite esters are very readily hydrolyzed to the corresponding alcohols; in contrast, the nitrate and thionitrite esters are markedly resistant to hydrolysis, even more so than carboxylic acid esters. Of course, the presence of activating groups can markedly alter the resistance of a functional group to hydrolysis; for example, acyl nitrates are very readily hydrolyzed to the corresponding carboxylic acids and nitric acid, whereas the simple alkyl nitrates are very slowly hydrolyzed. Moreover, the hydrolysis of nitrate esters is not straightforward, and usually is accompanied by substitution and elimination reactions in which either $\alpha$- or $\beta$-hydrogen may be lost in elimination reactions. Thus hydrolysis of nitrate esters involves the following reactions (4,7,51,182,268):

*Nucleophilic attack on carbon*

$$RONO_2 + OH^- \to ROH + NO_3^- \qquad (S_N1 \text{ or } S_N2)$$

*Nucleophilic attack on $\beta$-hydrogen*

$$RCH_2CH_2ONO_2 + OH^- \to RCH=CH_2 + H_2O + NO_3^- \qquad (E1 \text{ or } E2)$$

*Nucleophilic attack on α-hydrogen*

$$RCH_2ONO_2 + OH^- \rightarrow RCHO + H_2O + NO_2^-  \quad (E2)$$

*Nucleophilic attack on nitrogen*

$$R_3CO^*NO_2 + OH^- \rightarrow R_3CO^*H + NO_3^-  \quad (S_N2)$$

Alkaline hydrolysis of nitrate esters is significantly slower than that of carboxylic esters; hydrolysis of primary and secondary nitrates is very slow in neutral solution (4,51), but in all instances nucleophilic attack on carbon predominates and nucleophilic attack on β-hydrogen is minor. For example, only 2% of olefin is formed in the alkaline hydrolysis of ethyl nitrate and 10% is produced by isopropyl nitrate (7). Benzyl nitrates are exceptions because they give high yields of benzaldehyde (182). In sharp contrast, tertiary alkyl nitrates readily undergo nucleophilic attack on β-hydrogen, and formation of olefin occurs extensively (as much as 45%) in the neutral or alkaline hydrolysis of *tert*-butyl nitrate (23).

It is emphasized that all of the above reactions occur during hydrolysis; thus, from the viewpoint of analytical chemistry, the possibility of formation of nitrite ion and a carbonyl compound by hydrolysis often proves to be an annoyance in dealing with nitrate esters, especially when it is necessary to identify nitrate and nitrite esters in admixture. Only traces of nitrite ion are formed in hydrolyses of methyl nitrate (7), but up to 35% is obtained from isobutyl nitrate (23).

Soluble inorganic hyponitrites are decomposed slowly by water; gases such as $N_2O$, NO, and $N_2$ are evolved (219).

## 1. Reduction

Mild reduction of nitrate esters yields the corresponding alcohol, but other products are formed by direct attack of the reducing agent on the nitrogen of the nitrate radical and the nucleophilic hydrolytic reactions discussed above. Reducing agents seldom appear to cleave the C-O bond that links the nitro group to the alkoxy group.

It is important to note that the alcohol is formed chiefly by direct attack of the reducing agent on the ester function; although the alcohol conceivably can be formed by hydrolysis, as a rule the rate of hydrolysis of nitrate esters is altogether too slow to account for the quantitative yields that are obtained within the relatively short periods of time involved in reductions.

In sharp contrast to the nitrate esters, the nitrite esters are readily hydrolyzed; accordingly, when these compounds are subjected to reduction by reagents such as tin and hydrochloric acid, they may first be hydrolyzed to the corresponding alcohols and then the nitrite radical is reduced to hydroxylamine or ammonia. In many instances, it is not clear whether hydrolysis or reduction takes place first or whether the two occur concurrently.

### a. NITRATE ESTERS

The reduction of nitrate esters by nearly all reducing systems proceeds without cleavage of the C-O bond that links the nitrate ester group to the remainder of the compound. However, reduction of a nitrate ester, for example, butyl nitrate (40), by strong hydriodic acid usually forms an alkyl iodide, nitrogen dioxide, and iodine. A similar cleavage of the C-O bond also takes place when the primary nitrate group of a sugar that has been nitrated in the 6-position is replaced by iodine on treatment with sodium iodide in acetone (58).

### (1) Reduction by Ferrous Salts

Nitrate esters are reduced by ferrous chloride in acetic acid, sulfuric acid, or hydrochloric acid. The reduction appears to be quantitative when the ester is refluxed with an excess of ferrous chloride; most nitrate esters are quite reactive, and so it appears that refluxing is usual merely to remove nitric oxide and thus complete the reaction:

$$R{-}ONO_2 + 3\,FeCl_2 + 3\,HCl \rightarrow ROH + 3\,FeCl_3 + NO + H_2O$$

Ordinarily a solution of ferrous chloride in hydrochloric acid is used (13,61,205); suggestions that hydrobromic acid improves quantitative reductions of nitrocellulose (264) have not been confirmed (226). In analytical procedures, the ferric iron formed by oxidation of ferrous iron is usually determined titrimetrically, and titanous ion is most frequently used for this purpose (264). An older method first advanced by Schulze (258) and Tiemann (289) depends on measurement of the nitric oxide released by refluxing a nitrate ester with a large excess of ferrous chloride solution; the method has been improved by Kamiike et al. (132), but gasometric procedures are not very popular. As has been implied above, reduction of nitrate ester by ferrous iron is complete because nitric oxide is removed by boiling. However, Mitra and Srinivasan (202) obtained fairly satisfactory results by slowly titrating with ferrous sulfate a cool solution of nitrocellulose dissolved in sulfuric acid (15°C); the end point was indicated by the permanent color of an $FeSO_4 \cdot NO$ complex.

Ferrous hydroxide in alkaline solution also reduces nitrates and nitrites; thus it is often possible to detect esters by the Hearon–Gustavson test described in Section V.B.1.a(1).

## (2) **Reduction by Titanous Salts**

Knecht and Hibbert (139) first employed titanous chloride for the direct reduction of potassium nitrate in alkaline solution; the ammonia formed by the reduction was removed by distillation and titrated with acid. The direct reduction of nitrate esters with titanous salts is also possible; however, Oldham (213) could not obtain stoichiometric consumption of titanous ion:

$$R—ONO_2 + 8\,Ti^{3+} + 8\,H^+ \rightarrow ROH + 8\,Ti^{4+} + NH_3 + 2\,H_2O$$

Each mole of the ester consumed only 7.5 mol ($\pm 1\%$) of titanous ion instead of 8. Even greater deviations have been observed in titanous reductions of nitroglycerin (72). It is emphasized, however, that a large excess of titanous salts in acidic media (even acetic acid) will completely reduce nitrate esters (95).

## (3) **Reduction by Metals**

Reduction of nitrate and nitrite esters by metal–acid couples generally yields the corresponding alcohol; typical couples are tin and hydrochloric acid, iron and acetic acid (58), and zinc and sulfuric acid. However, a magnesium–sulfuric acid couple is especially useful because the neutralized product solution can be evaporated to dryness without further degradation of the reduction products.

Sodium metal also reduces nitrate esters; for example, methyl, ethyl, and amyl nitrate form the corresponding alcohols and sodium nitrite (39), but the quantitative aspects of the reactions have not been explored.

Metal–base couples can also be used to reduce nitrate and nitrite esters.

## (4) **Reduction by Hydriodic Acid**

As already noted, strong hydriodic acid cleaves the C-O bond in nitrate esters and usually forms an alkyl iodide (40). The release of iodine is of analytical importance:

$$2\,RONO_2 + 8\,HI \rightarrow 2\,RI + 2\,NO + 3\,I_2 + 4\,H_2O$$

## (5) **Reduction by Sulfides**

The nitrate ester grouping exhibits characteristic oxidative powers when brought into contact with hydrogen sulfide or alkali sulfides. Thus hydrogen sulfide reacts with ethyl nitrate to form ethanol, ammonia, and sulfur (144). Ethanolic ammonium sulfide also reduces n-butyl nitrate to the corresponding alcohol; however, the products of sulfide reductions are dependent on pH and the molecular species of sulfide present (196). For example, when $n$-butyl nitrate is reduced with sodium hydrosulfide, $n$-butanol and sodium nitrite are the main products, but with ammonium hydrosulfide the main nitrogen-containing product is ammonia (along with 4 mol of sulfur/mol of ester). Above pH 13, sodium hydrosulfide or ammonium hydrosulfide reduces butyl nitrate to form nitrite ion, butanol, and 1 mol of sulfur/mol of ester. The reaction appears to be second order; it is catalyzed by hydroxyl ion and by polysulfide (23). It is of importance to note that in every instance the alcohol appears to be formed nearly quantitatively in sulfide reductions of nitrate esters (163).

### (6) Reduction by Hydrogen and Hydrides

Hydrogenolytic reactions are preferred when clean-cut denitrification is required and absence of aldehydes and other products is mandatory; moreover, reduction by other methods leaves inorganic salts, which may make difficult isolation of the desired product. Hydrogenolysis is especially useful for analytical procedures when it is desirable to destroy the nitrate functional group or to obtain the nearly pure alcohol corresponding to a nitrate ester.

Kuhn (156) has shown that nitrate groups may be smoothly reduced by catalytic hydrogenolysis with a palladium-on-calcium carbonate catalyst:

$$2\,RONO_2 + 5\,H_2 \rightarrow 2\,ROH + N_2 + 4\,H_2O$$

The usual copper-chromium catalysts are of little value inasmuch as they are active only at the elevated temperatures where considerable decomposition of the nitrate esters also occurs (e.g., 160–180°C). Nickel and platinum catalysts are effective at lower temperatures (60–70°C), but they are unsatisfactory because they reduce nitrate nitrogen to ammonia and unreduced nitrate groups are sensitive to ammonia (side reactions occur); unfortunately, the reduction of nitrate groups in the presence of nickel and platinum does not take place in acidic media, where the formation of ammonium ion would prevent the occurrence of the side reactions of nitrate esters.

The palladium-calcium carbonate catalyst is active under 1500 psi of hydrogen pressure at room temperature; however, Kuhn (156) also found that a palladium-on-charcoal catalyst is much more active at room temperature. It is of interest to note that the palladium-charcoal catalyst at 1500 psi of hydrogen pressure will form ammonia, but if the reaction is terminated when 2 mol of gas/mol of nitrate group has been consumed, no ammonia will be found in the reaction mixture. Stalcup et al. (276) have used palladium black catalyst with hydrogen at 30 psi to reduce nitroglycerin and *C*-nitro compounds at room temperature. The technique of Hormann et al. (120) may also be useful for hydrogenations at ambient pressures. The apparatus described by Southworth (275) can be used for this purpose with 1–500 mequiv. of material. Cheronis and Koeck (41) have presented a historical summary of low-pressure hydrogenations and also describe many useful contemporary techniques

Metal hydrides are useful reagents for reductive denitration of nitrate esters. For example, cyclohexyl, *n*-hexyl, and 2-octyl nitrates, as well as *cis*- and *trans*-cyclohexanediol dinitrates, are reduced essentially quantitatively to the alcohols by lithium aluminum hydride (274).

### (7) Electrolytic Reduction

The simple nitrate esters are readily reduced electrolytically. The reaction involves a two-electron change for each nitrate group and the production of the parent alcohol and nitrite ion (133). The two-electron process also takes place at the dropping mercury electrode in methanolic solutions containing lithium chloride (238), and Whitnack et al. (304) found that the same process occurs in the reduction of each nitrate group of ethylene glycol dinitrate, glyceryl trinitrate, and pentaerythritol tetranitrate in an ethanolic suspension medium containing tetramethylammonium chloride. The reduction process generally provides one wave which appears to be independent of hydrogen ion concentrations at pH values between 4 and 11; on the other hand, since glyceryl trinitrate, but not a primary ester such as *n*-butyl nitrate, shows two waves in alkaline media, it is suggested that the secondary nitrate group is more readily attacked by hydroxyl ion.

The coulometric reduction of simple nitrate esters with controlled potential at large mercury cathodes appears to provide a useful analytical method (134).

### (8) Miscellaneous Reducing Agents

Hydrazine and substituted hydrazines in the presence of platinum or palladium catalyst are effective reducing agents for nitrate esters. For example, *n*-hexyl nitrate is reduced by hydrazine to hexyl alcohol; however, nitrous oxide, nitrogen, and water are also formed, and there is a possibility that hydrazine can be alkylated by the ester. When methylhydrazine is used, methane and ethane are also formed (158). The rate of reduction is very much slower in the absence of the catalyst, and in many instances the reduction process is much more complex because hydrazoic acid and ammonia are formed (197).

Benzoin functions as an energetic hydrogen donor in the molten state; its reductive cleavage of some aliphatic compounds may be represented by the following reaction, in which X = halide, —SH, —NO$_2$, and so on (73,74):

$$RX + \phi CHOHCO\phi \rightarrow \phi COCO\phi + RH + HX$$

Accordingly, it has been observed that nitrate salts of organic bases and nitrate esters such as nitrocellulose form nitrous acid almost at once when fused with benzoin (75). The nitrous acid can be detected by paper moistened with Griess's reagent or by any other convenient method.

Reduction of nitrate ester groups by stannite ion has been reported by Sato (248). Murakami (207) claims that stannous chloride (and back titration with ferric ion) provides quantitative reduction of organic as well as inorganic nitrates and nitrate esters; further investigations are in order.

### b. NITRITE ESTERS

The reduction of nitrite esters by various reagents follows very similarly the reduction processes of the corresponding nitrate esters; consequently, reference should be made to the preceding discussion of the reduction of nitrate esters. It is also important to recognize that the reduction routes of nitrite esters may be influenced by the orientation of the nitrite ester group, that is, *cis* and *trans* forms may react differently. Another important difference between the processes involved in the reduction of nitrite esters and those of the corresponding nitrate esters arises from the relative ease of hydrolysis of the nitrite esters; in essence, when alkyl nitrites are reduced by reagents such as metal and acid, hydrolysis usually occurs before, or at least competitively with, reduction (formation of the corresponding alcohol); reduction products other than the alcohol are determined by the action of the reducing agent on the liberated nitrite ion.

Catalytic reduction of nitrite esters at ambient temperatures can form primary amines; at higher temperatures, secondary amines are formed, and nitriles are in predominance above 300°C.

### c. HYPONITRITE ESTERS

The inorganic salts of hyponitrous acid are not reduced by titanous chloride, stannous chloride, metal–acid couples, and amalgams (219).

### 2. Action of Acids and Bases

The esters discussed in this section are hydrolyzed more or less readily when brought into contact with acids or bases, but the products of hydrolysis are manifold; some note has already been made of the reductions which take place on hydrolysis and of the fact that detection of the esters is made possible by detection of nitrate and nitrite ions formed by hydrolysis. However, it is important to recognize that the presence of nitrite ion in the hydrolysis products of nitrate esters is not necessarily the result of a subsequent oxidation-reduction reaction between alcohol (or other organic products) and nitric acid; the formation of both nitrite and nitrate is the result of water performing as a nucleophilic agent (23).

The reaction for nucleophilic attack on $\alpha$-hydrogen:

$$HO^- + RCH_2ONO_2 \rightarrow RCHO + H_2O + NO_2^-$$

cannot occur with most tertiary nitrate esters, and therefore nitrite ion is not detected among the products of hydrolysis, although the nitrite ion may be subsequently formed from attack of organic residues by the nitric acid resulting from hydrolysis (in acidic media).

### a. ACTION OF ACIDS

The action of concentrated sulfuric acid on nitrate esters is complex, but it is of singular importance in analytical chemistry inasmuch as it provides the basis for the nitrometer method for assay of compounds such as glyceryl trinitrate and nitrocellulose. Cryoscopic and spectroscopic evidence suggests that nitrate esters in concentrated sulfuric acid form nitronium ion, $NO_2^+$; for example, the ultraviolet absorption spectrum of a sulfuric acid solution of ethyl nitrate is quite similar to that of a solution of nitric acid in sulfuric acid and altogether different from that of a chloroform solution of the ester (23). A van't Hoff $i$ factor approaching 5 has been found for ethyl nitrate (157). The overall reaction can be summarized as follows:

$$RONO_2 + 2 H_2SO_4 \rightarrow ROSO_3H + NO_2^+ + HSO_4^- + H_2O$$

The ester is not regenerated when the sulfuric acid solution is diluted with water; the presence of nitronium ion may be the explanation for the potency of sulfuric acid solutions of nitrate esters as nitration mixtures, and for their reactivity with mercury to form nitric oxide.

Solutions of certain highly alkylated benzyl nitrates in concentrated sulfuric acid do not form nitronium ion because it is consumed by the formation of a $C$-nitro group (272); for example, 5-bromo-2,3,4,6-tetramethylbenzyl nitrate is converted into 5-bromo-2,3,4,6-tetramethylnitrobenzene.

The inorganic hyponitrites are partially decomposed by dilute sulfuric or hydrochloric acid into $N_2O$, $NO$, and $N_2$; concentrated sulfuric acid causes complete decomposition, and the aforementioned gases, nitric acid, and nitrous acid are formed (219).

b. ACTION OF BASES

The reaction of alkyl nitrates with dry, alcoholic potassium hydroxide solution produces mixed ethers, instead of alcohols, and a variety of other products; sodium alkoxide reacts similarly. Of analytical importance is the fact that mixtures of strong bases and oxidizing agents such as hydrogen peroxide and sodium perborate quantitatively cleave nitrate esters (e.g., nitrocellulose) and form nitrites and nitrates as well as other products which can subsequently be reduced to ammonia. As early as 1868, it was noted that the reaction between potassium hydroxide and nitroglycerin is not a simple hydrolysis because nitrite, cyanide, oxalate, and formates are also formed; in spite of this observation and other, similar ones, saponification has been proposed periodically as a method of assay for nitroglycerin (260). A systematic study by Schulcek et al. (257) of the alkaline hydrolysis of pentaerythritol tetranitrate, glyceryl trinitrate, nitrostarch, and nitrocellulose in 1.5% sodium hydroxide has revealed that the nitrate/nitrite ratio is about 1:2 for nitroglycerin and nitrostarch, about 1:1.5 for nitrocellulose, and about 1:0.8 for pentaerythritol nitrate. Cyanides are not formed by pentaerythritol tetranitrate, and ammonia is not formed by nitrostarch, but nitrocellulose and pentaerythritol tetranitrate form both ammonia and cyanide.

### 3. Action of Oxidizing Agents

The nitrate ester group is quite resistant to oxidative attack; however, the nitrite ester group can be more or less readily oxidized to the nitrate ester group, although this reactivity is seldom used for analysis. Powerful oxidizing agents destroy organic structures and convert the nitrogen content of nitrite and nitrate esters to nitric acid.

The inorganic salts of hyponitrous acid are readily attacked by potassium permanganate; both nitric and nitrous acid are formed (219).

The alkyl thionitrites (especially methyl thionitrite) are very readily oxidized; even air will attack these substances (222).

### 4. Miscellaneous Reactions

The reaction of nitrate esters with ammonia and amines is worthy of note (23). When nitrate esters are heated with ammonia or primary or secondary aliphatic amines, N-alkylation occurs; thus methyl-, ethyl-, and n-propylamines have been prepared from the corresponding nitrates and ammonia in sealed tubes at 100°C, and piperidine and diethylamine have been alkylated by heating with primary, secondary, and tertiary alkyl nitrates. Primary aromatic amines react with alkyl nitrates to give the amine nitrate and the $N$-alkylated amine, but secondary and tertiary aromatic amines form complex mixtures of oxidation-reduction and condensation products, which are usually highly colored and difficult to characterize.

Pyridine and tertiary aliphatic amines react under reflux with primary nitrate esters to form quaternary ammonium salts (164), but the secondary and tertiary nitrate esters undergo elimination reactions to form olefins.

## C. GENERAL REFERENCES

Boschan, R., R. T., Merrow, and R. W. Van Dolah, "The Chemistry of Nitrate Esters," *Chem. Rev.*, **55**, 485 (1955). A review of literature through 1954.

Millar, I. T., and H. D. Springall, *A Shorter Sidgwick's Organic Chemistry of Nitrogen*, Clarendon Press, Oxford, 1969. A systematic survey of nitrogen compounds; a revision of *Sidgwick's Organic Chemistry of Nitrogen*. Chapter II describes the synthesis and reactions of the esters of hyponitrous, nitrous, and nitric acids.

Urbanski, T., *Chemistry and Technology of Explosives*, Vols. I and II, Macmillan, New York, 1964. A summary of synthetic methods, properties, and reactions of the nitrate esters used as explosives.

Davis, T. L., *Chemistry of Powder and Explosives*, John Wiley, New York, 1943.

## IV. TOXICITY AND INDUSTRIAL HYGIENE

The nitrate and nitrite esters should be treated as treacherously hazardous materials inasmuch as they are flammable, toxic, and explosive. Although nearly everyone treats glyceryl trinitrate with respect because it is commonly known to be highly sensitive to shock (with irrevocable consequences), the same courtesy is seldom extended to simple compounds such as amyl nitrate. To be sure, the simple compounds are in no way as sensitive to shock as are the polynitrate esters, but they are, nevertheless, explosive and toxic substances; the low molecular weight nitrates should not be heated much above 100°C because even traces of impurities may initiate an autocatalytic decomposition which possibly may lead to an explosion. The toxic qualities of nitrate and nitrite esters stem from their pronounced effects on the circulatory system and the blood; the simpler esters are extremely hazardous because their high vapor pressures increase the probability that lethal amounts can be inhaled. Constant vigilance must be maintained to ensure that only trivial amounts of relatively nonvolatile esters such as ethylene glycol dinitrate (EGDN) are present in the atmosphere surrounding industrial operations. Nitroglycerin is by far the least dangerous in industrial operations because its vapor pressure is at least 70 times lower than that of ethylene glycol dinitrate (28,193); generally, these dinitrate esters are present in workrooms in quantities ranging from 1 to 5 mg/m$^3$, while the concentration of nitroglycerin is usually lower than 0.1 mg/m$^3$ (34).

Binary solutions of ethylene glycol dinitrate and nitroglycerin are ideal with respect to vapor pressure and density relationships; it is of interest to note that the vapor pressure of nitroglycerin in the range of 10–50°C is lower than that of mercury (28).

Analysis of nitrate esters in air is often made on solutions obtained by passing air through solvents (154). However, absorption of the nitrogly- cerin content of air by bubbling through ordinary solvents such as water, alcohols, or acetone may be quite incomplete. Yagoda and Goldman (309) found olive and cottonseed oils to be good solvents but somewhat objectionable in subsequent analytical steps; however, the solubility of nitroglycerin in polyhydroxy solvents such as triethylene glycol or propylene glycol proved to be sufficient for trapping it in bubblers, and the absorbent (soluble in water) was not objectionable in subsequent reactions.

The nitrate and nitrite esters should be handled sparingly and in well-ventilated areas. At all times, contact of these substances with the person should be avoided inasmuch as they are absorbed through the skin as well as through the lungs and the digestive tract.

Nitrate (and nitrite) esters oxidize hemoglobin to methemoglobin, as do the nitro compounds. Moreover, by causing a depression of the muscles in the vascular walls, the esters cause a peripheral vasodilation which lowers systolic blood pressure and increases respiratory rates; vasodilator action, brought about by rigorously controlled adminis- trations of the esters, is the reason why amyl nitrate, glyceryl nitrate, and similar esters are widely used for relief of high blood pressure.

Severe headache and a feeling of pressure on the front and back of the head are the most characteristic symptoms of poisoning by nitrate esters. Dilation of the peripheral blood vessels leads to a shift of blood to the splanchnic area (the viscera, etc.) with a corresponding decrease in the blood supplied to the brain (in spite of vasodilation in the cerebral area); vertigo and fainting result quickly, and death may follow.

## V. QUALITATIVE DETERMINATIONS

### A. PREPARATION OF SAMPLES

Samples should be treated essentially as was described for the nitro and nitroso compounds (see Section V.A). It is necessary, however, to emphasize that nitrate and nitrite esters (especially the polynitrates) are sensitive, highly explosive substances which at all times must be handled with extreme caution; precautions are particularly in order when samples of novel structure are at hand. In contrast, mononitrate esters are rather insensitive compounds, although they can be detonated. The possibility that certain structures are sensitized to shock or will be induced to detonate by the presence of other substances must always be taken into consideration.

Substances such as pure nitroglycerin and mixtures containing esters are analyzed daily in many laboratories throughout the world. However, the above statement is not to be considered as an implication that the analysis of explosives is a perfectly safe operation; it is merely a recognition that the analytical and manipulative procedures used in these laboratories have withstood the test of time and that certain procedures are safe when used with known samples. There is no reason to assume that the very same procedures can be applied to new types of compounds with impunity. However, there is ample evidence that the time-tested analytical procedures may be applied to new compounds of closely similar structure.

The explosive nature of a nitrate ester is related to its oxygen balance (181); in fact, for any compound containing $X$ atoms of carbon, $Y$ atoms of hydrogen, and $Z$ atoms of oxygen,

$$\text{Oxygen balance} = \frac{1}{\text{mol. wt.}}\left[-1600\left(2X + \frac{Y}{2} - Z\right)\right]$$

Thus a compound having a composition which on explosion can be converted completely to carbon dioxide and water has zero balance, one lacking sufficient oxygen has a negative balance, and one containing excess oxygen has a positive balance. With few exceptions, explosives have a decidedly negative oxygen balance.

Although a term to describe explosive power quantitatively is difficult to devise, empirical tests are used for power and brisance measurements; a characteristic test is the so-called Trauzl lead block expansion test, in which the enlargement of a block of lead is determined after a standard weight of explosive is detonated in a specified cavity under controlled confinement. Trinitrotoluene is used as the standard and is arbitrarily assigned the value of 100.

The relationship of oxygen balance to power and brisance of some explosive substances is shown in Table II; oxygen balances close to zero (say, ±20) imply that compounds are high explosives. Power and brisance fall off rapidly for positive oxygen balances, and more slowly for negative values. Thus it is a simple matter to estimate the explosive hazard of N—O compounds.

TABLE II

Relationship of Oxygen Balance to Power and Brisance of Explosive Substances (181)

| Compound | Formula | Oxygen balance | Power and brisance |
|---|---|---|---|
| Tetranitromethane | $C(NO_2)_4$ | +49 | 55 |
| Nitroglycerine (NG) | $C_3H_5N_3O_9$ | +3.5 | 185 |
| Pentaerythritol tetranitrate (PETN) | $C_5H_8N_4O_{12}$ | −10.1 | 180 |
| Methyl nitrate | $CH_3NO_3$ | −10.4 | 182 |
| Ethylenedinitramine | $C_2H_6N_4O_4$ | −32 | 145 |
| Cyclotrimethylenetrinitramine (RDX) | $C_3H_6N_6O_6$ | −21.6 | 165 |
| Trinitrobenzene (TNB) | $C_6H_3N_3O_6$ | −56.4 | 120 |
| Ethyl nitrate | $C_2H_5NO_3$ | −61.5 | 135 |
| Trinitrotoluene (TNT) | $C_7H_5N_3O_6$ | −74 | 100[a] |
| Propyl nitrate | $C_3H_8NO_3$ | −106 | 68 |
| 2,4-Dinitrotoluene | $C_7H_6N_2O_4$ | −115 | 75 |

[a] Trinitrotoluene is the standard (= 100).

The brisance (shattering effect because of sudden release of energy) of many polynitrate esters is unparalleled, even in this age of nuclear "blasts." Consequently, samples of nitrate and nitrite esters should be handled with deliberate caution; for example, they should not be teased or subjected to attrition, tests for combustibility or thermal stability should be delegated to highly experienced personnel, and all operations should be performed at safe distances. As was indicated in the section entitled "Nitro and Nitroso Compounds," face shields and gloves should be worn at all times, and the sample should be kept as far away as possible from the person during manipulation.

Samples of nitrate and nitrite esters are seldom pure; for example, nitrate esters that are sensitive explosives usually are mixed with other materials as diluents. Mannitol hexanitrate is usually mixed with lactose or mannitol when it is used as a vasodilator, as are pentaerythritol tetranitrate, glyceryl trinitrate, and inositol hexanitrate (8). Even the commercial explosives of today are mixtures of esters and diluents; thus it is not uncommon to find tetranitrodiglycerin (TDNG), diethyleneglycol dinitrate (DEGDN), nitroglycerin (NG), and sucrose octanitrate (SON) in admixture with diluents and sundry stabilizers, as well as dinitrotoluene (DNT) and trinitrotoluene (TNT). As a rule, these mixtures are broken down into their component fractions by extraction with a variety of solvents (225). For example, TNDG and DEGDN are relatively insoluble in 60% acetic acid, whereas NG and ethylene glycol dinitrate (EGDN) are soluble. The mixture of NG and EGDN also can be separated chromatographically or determined by an infrared method.

It is not within the scope of this discussion to give specific directions on how to separate a wide variety of mixtures of nitrate and nitrite esters or how to remove the esters from diluents. It is appropriate to note, however, that many of the esters are soluble in organic solvents, whereas the diluents are not. A glance at Table I (properties) reveals that most esters are soluble in ether, chloroform, or alcohol; esters like nitrocellulose, however, are quite insoluble. In dealing with samples of unknown origin, the analyst should use a microscope at low magnification to attempt to detect components in simple admixture. A systematic study should be made to determine whether various components are extracted by a series of solvents. (**Caution:** Only small amounts of solvent extracts are to be evaporated since residues containing explosive esters often may detonate when removed from stabilizers.)

## B. DETECTION

### 1. Chemical Methods

#### a. COLOR REACTIONS

Many of the tests for detecting nitro and nitroso compounds listed in the preceding section on these substances can be used to detect nitrate and nitrite esters; it will be necessary to refer to that section to obtain full cognizance of the extent to which other nitrogen-oxygen functional groups can act as interferences.

Some of the following tests which depend on the release of nitrous acid or the nitrite ion from nitrite esters may suffer interference from nitrate esters because of the possible reduction of nitrate to nitrite. The possibility that nitrite ion can be generated from either nitrate or nitrite esters must always be taken into consideration.

#### (1) Test for Nitrogen- and Oxygen-Containing Groups

Feigl and Amaral (76) have described a test which depends on the finding that compounds with functional groups containing nitrogen and oxygen form nitrous acid ($N_2O_3$ or $N_2O_4$) on pyrolysis. The test is described in detail in Section V.B.1 of "Nitro and Nitroso Groups"; it may be difficult to apply the test to volatile alkyl nitrate and nitrite esters.

#### (2) Ferrous Hydroxide Oxidation

The general test given in Section V.B.1.a(1) of "Nitro and Nitroso Groups" will also provide a positive reaction with nitrate and nitrite esters. In essence, the test involves mixing a sample with freshly precipitated ferrous hydroxide; Hearon and Gustavson (110) have found that ethyl nitrite forms a brown color in 5 sec and that ethyl nitrate reacts after 30 sec. The rates of reaction are commensurate with the rates of hydrolysis of the esters, that is, the nitrite esters hydrolyze readily, and the nitrate esters are more difficult to hydrolyze. Inasmuch as the test is considered to be terminated after 5 min (for nitro compounds), it may be necessary to prolong the test for the more resistant esters.

### (3) Diphenylamine in Sulfuric Acid

The blue color that is obtained when $N$-nitrosamines and aliphatic $C$-nitro compounds (as well as other oxidizers) are mixed with diphenylamine and sulfuric acid is also given by nitrate and nitrite esters (93). The test, as given by Grebber and Karabinos for $C$-nitroparaffins, is described in detail in Section V.B.1.a(3)(a) of "Nitro and Nitroso Groups". A variation using diphenylbenzidine is described in Section V.B.1.a(4) of the same section for detecting $C$-nitroso compounds.

### (4) The Bose Test

Nitrous acid is released when $C$-nitro compounds are treated with hot, *concentrated* alkali solution; the Bose test (25) (see Section V.B.1 of "Nitro and Nitroso Groups") takes advantage of this fact and makes use of the Griess–Ilosvay reagent to detect liberated nitrite. However, since alkyl nitrites are often readily hydrolyzed, the Griess–Ilosvay test may also be used to detect nitrite and nitrate esters. It is to be noted that nitrite ion may not be generated by hydrolysis of all alkyl nitrates (see Section III.B) and that nitro- and nitroso-groups may form nitrite.

### (5) The Liebermann Test

This test is used widely for detection of $C$-nitroso compounds, as well as for nitrites; the details of an appropriate procedure are given in Section V.B.1.a(4) of "Nitro and Nitroso Groups."

### (6) Franchimont's Test

Nitrate esters usually form deep green colors in the Franchimont test, which is described in Section V.B.1a(6) of "Nitro and Nitroso Groups." However, the test is not very conclusive and should be used in conjunction with other tests.

### (7) Hydrolysis—Nitrate and Nitrite

In contrast to the great majority of $C$-nitro compounds, nitrate and nitrite esters can be hydrolyzed by treatment with *dilute* alkaline solutions [note difference from Bose test, (4) above]; the detection of nitrate and nitrite ion in the hydrolysate can be performed by any convenient colorimetric method. For example, the nitrite ion can be detected by diazotization of aniline in acetic acid and coupling with $\alpha$-naphthol in basic or neutral media. A colorimetric determination of nitrate when nitrite is present has been devised by Seaman et al. (261). Methods are available for determining nitrites in the presence of nitrates (78a,126a,292a).

### (8) Fusion with Benzoin

The nitrate group in esters such as nitrocellulose can be detected by release of nitrous acid on fusion with benzoin (mp, 137°C).

As described by Feigl (75), a tiny amount of the sample is placed in the bottom of a micro test tube, together with a few centigrams of benzoin; the open end of the tube is covered with a piece of filter paper moistened with Griess reagent (1 vol. of a solution of 1 gram of sulfanilic acid in 25 ml of acetic acid and 75 ml of water mixed with 1 vol. of a solution of 0.3 gram of $\alpha$-naphthylamine in 70 ml of water and 25 ml of acetic acid). The bottom of the tube is immersed in a glycerol bath which has been preheated to 140°C; if necessary, the temperature should be raised to 160°C. A positive result for the evolution of nitrous acid is indicated by the appearance of red areas on the reagent paper.

### (9) Miscellaneous

Alkyl nitrites (and tetranitromethane) form colorations with unsaturated compounds such as 1- and 2-pentene and $d$-limonene (100,111).

### 2. Physical Methods

Nitrate and nitrite esters can be detected with greater certainty by physical methods than by the chemical methods described above. However, in complex mixtures of nitro or nitroso compounds and inorganic nitrates (e.g., a propellant or explosive), both chemical and physical methods of identification by ultraviolet and infrared spectroscopy must be used, together with separations based on solubility, adsorption, boiling points, and so on. Brief summaries of ir and uv absorption bands useful for detecting the presence of nitrate and nitrite esters are given in Tables III, IV, and V.

The inorganic nitrate salts absorb strongly at 1380–1350 cm$^{-1}$ (7.24–7.41 $\mu$) because of asymmetric NO$_3$ stretch; additional bands of medium intensity are found at 833–813 cm$^{-1}$ (12.0–12.3 $\mu$) and weak bands at 741–725 cm$^{-1}$ (13.5–13.8 $\mu$). The inorganic nitrite salts show strong asymmetric NO$_2$ stretch absorptions at 1275–1235 cm$^{-1}$ (7.84–8.10 $\mu$) and medium-intensity bands at 833–820 cm$^{-1}$ (12.0–12.2 $\mu$).

TABLE III

Characteristic Bands in the Infrared Spectra of Typical Nitrate Esters

| Compound | Solvent or phase | Asymmetrical stretch | | Symmetrical stretch | | O—N stretch | | Out-of-plane | | NO$_2$ bend | | Ref. |
|---|---|---|---|---|---|---|---|---|---|---|---|---|
| | | ($\mu$) | (cm$^{-1}$) | ($\mu$) | (cm$^{-1}$) | ($\mu$) | (cm$^{-1}$) | ($\mu$) | (cm$^{-1}$) | ($\mu$) | (cm$^{-1}$) | |
| Methyl nitrate | Liq. | 6.12 | 1634 | 7.78 | 1285 | 11.63 | 860 | 13.16 | 760 | — | — | 30 |
| 1-Nitro-2-butyl nitrate | Liq. | 6.09 | 1642 | 7.84 | 1275 | 11.78 | 849 | 13.29 | 753 | 14.41 | 694 | 30 |
| 2-Nitrobutyl nitrate | Liq. | 6.06 | 1650 | 7.83 | 1277 | 11.85 | 844 | 13.30 | 752 | 14.50 | 690 | 30 |
| 2-Butyl nitrate | Liq. | 6.15 | 1626 | 7.86 | 1272 | 11.6 | 862 | — | — | — | — | 152 |
| t-Amyl nitrate | Liq. | 6.12 | 1634 | 7.82 | 1279 | 11.65 | 878 | 13.20 | 758 | 14.35 | 697 | 30 |
| 2-Pentyl nitrate | Liq. | 6.13 | 1631 | 7.82 | 1279 | 11.53 | 867 | 13.19 | 758 | 14.35 | 697 | 30 |
| 2-Nitrocyclohexyl nitrate (cis) | Liq. | 6.10 | 1639 | 7.86 | 1272 | 11.56 | 866 | 13.33 | 750 | 14.52 | 689 | 30 |
| 2-Nitrocyclohexyl nitrate (trans) | Liq. | 6.08 | 1645 | 7.84 | 1275 | 11.50 | 870 | 13.35 | 749 | 14.50 | 690 | 30 |
| 4-Nitrocyclohexyl nitrate (cis) | Liq. | 6.13 | 1631 | 7.83 | 1277 | 11.55 | 866 | 13.23 | 756 | 14.41 | 694 | 30 |
| 4-Nitrocyclohexyl nitrate (trans) | Liq. | 6.12 | 1634 | 7.83 | 1277 | 11.45 | 873 | 13.22 | 756 | 14.28 | 700 | 30 |
| Cyclohexyl nitrate | Liq. | 6.15 | 1626 | 7.83 | 1277 | 11.50 | 870 | 13.19 | 758 | 14.40 | 714 | 30 |
| Benzyl nitrate | Liq. | 6.10 | 1639 | 7.79 | 1284 | 11.62 | 861 | 13.19 | 758 | 14.34 | 697 | 30 |
| 1-Nitratooctane | Liq. | 6.12 | 1634 | 7.88 | 1269 | 11.6 | 862 | 13.2 | 758 | 14.4 | 694 | 189 |
| 1,10-Dinitratodecane | — | 6.13 | 1631 | 7.88 | 1269 | 11.6 | 862 | 13.2 | 758 | 14.3 | 699 | 189 |
| 1,2-Dinitratooctadecane | — | 6.10 | 1639 | 7.93 | 1261 | — | — | — | — | | | 189 |
| 18,19-Dinitratohexatricontane | Mull | 6.10 | 1639 | 7.89 | 1267 | 11.8 | 847 | 13.3 | 752 | 14.5 | 690 | 189 |
| Methyl 2-nitratooctadecanoate | — | 6.08 | 1645 | 7.89 | 1267 | 11.7 | 855 | 13.2 | 758 | 14.4 | 694 | 189 |
| Methyl 12-nitratooctadecanoate | — | 6.16 | 1623 | 7.91 | 1264 | 11.6 | 862 | 13.3 | 752 | 14.4 | 694 | 189 |
| 1,3-Didodecyl glyceryl nitrate | — | 6.12 | 1634 | 7.90 | 1266 | 11.9 | 840 | 13.3 | 752 | — | — | 189 |
| 1-Octadecyl glyceryl dinitrate | — | 6.02 | 1661 | 7.91 | 1264 | 11.9 | 840 | 13.3 | 752 | — | — | 189 |
| Cellulose nitrate | 2% Sol. | 6.05 | 1653 | 7.8 | 1280 | 11.9 | 840 | 13.3 | 752 | 14.4 | 694 | 232 |
| Glyceryl trinitrate | 2% Sol. | 6.1 | 1639 | 7.85 | 1274 | 11.9 | 840 | 13.3 | 752 | — | — | 232 |
| Ethylene glycol dinitrate | Film | 6.1 | 1639 | 7.85 | 1274 | 11.9 | 840 | 13.3 | 752 | 14.2 | 704 | 232 |
| Diethylene glycol dinitrate | Film | 6.13 | 1631 | 7.85 | 1274 | 11.8 | 847 | 13.2 | 752 | 14.2 | 704 | 232 |
| Mannitol hexanitrate | Mull | 6.13 | 1631 | 7.85 | 1274 | 11.8 | 847 | 13.3 | 752 | 14.3 | 694 | 232 |
| Pentaerythritol tetranitrate | Mull | 6.05 | 1653 | 7.80 | 1280 | 11.8 | 847 | 13.25 | 755 | 14.25 | 709 | 232 |
| | | | | 7.90 | 1266 | | | | | | | |

314

## TABLE IV
### Ultraviolet Absorbence Maxima for Typical Nitrate Esters

| Compound | Solvent | $\lambda_{max}$(nm) | $\epsilon_{max}$ | Ref. |
|---|---|---|---|---|
| Ethyl nitrate | EtOH | 265 | 14.8 | 128 |
| 1-Octyl nitrate | — | 270 | 15.0 | 237 |
| Cyclohexyl nitrate | — | 270 | 22.0 | 237 |
| $\alpha$-Methyl-D-glucose-6-nitrate | $H_2O$ | 265 | 19.0 | 237 |

## TABLE V
### Ultraviolet Spectra of Nitrite Esters

| Compound | Solvent | $\lambda_{max}$(nm) | $\epsilon_{max}$ | $\lambda_{min}$(nm) | $\epsilon_{min}$ | Ref. |
|---|---|---|---|---|---|---|
| Ethyl nitrite | Vapor | 384 | 27 | 382 | 26 | 108 |
| | | 369 | 56 | 364 | 41 | |
| | | 355 | 71 | 349 | 40 | |
| | | 343 | 67 | 336.5 | 31 | |
| | | 332 | 50 | 326 | 24 | |
| | | 321.5 | 33 | 317 | 21 | |
| | | 312.5 | 25 | 309 | 21 | |
| Butyl nitrite | $CCl_4$ | 370 | 74 | 364 | 56 | 108 |
| | | 357 | 89 | 350 | 54 | |
| | | 345 | 77 | 338 | 41 | |
| | | 334 | 55 | 327 | 30 | |
| | | 323.5 | 36 | 317 | 24 | |
| | | 314 | 26 | 310 | 24 | |
| $n$-Amyl nitrite | $CCl_4$ | 370.5 | 74 | 364 | 57 | 108 |
| | | 357 | 88 | 350 | 54 | |
| | | 345 | 76 | 337 | 40 | |
| | | 333.5 | 54 | 326 | 30 | |
| | | 323.5 | 35 | 318 | 24 | |
| | | 314 | 27 | 311 | 25 | |
| Isoamyl nitrite | $CCl_4$ | 370.5 | 68 | 364 | 52 | 108 |
| | | 357 | 82 | 350 | 49 | |
| | | 344.5 | 71 | 338 | 38 | |
| | | 333.5 | 51 | 338 | 28 | |
| | | 323.5 | 33 | 318 | 23 | |
| | | 314 | 25 | 311 | 24 | |

In the ultraviolet, nitrate salts such as ammonium nitrate, hexamine mononitrate, and hexamine dinitrate have absorption maxima at 301.5 nm with $\epsilon_{max}$ about 7.6 for each nitrate group (128).

### a. INFRARED SPECTROPHOTOMETRY

### (1) Nitrate Esters

The characteristic bands of the nitrate esters have their origin in the fundamental vibrations of the $XNO_2$ group. For example, the spectrum of nitromethane indicates that the $C—NO_2$ group fundamental frequencies are as follows:

| | |
|---|---|
| 6.37 $\mu$ (1570 cm$^{-1}$) | $NO_2$ asymmetrical stretch |
| 7.25 $\mu$ (1379 cm$^{-1}$) | $NO_2$ symmetrical stretch |
| 10.90 $\mu$ (917 cm$^{-1}$) | CN stretch |
| 15.24 $\mu$ (656 cm$^{-1}$) | $NO_2$ bend |
| 16.24 $\mu$ (616 cm$^{-1}$) | out-of-plane |
| 20.75 $\mu$ (482 cm$^{-1}$) | CNO bands |

As was indicated in Section V.B.2 of "Nitro and Nitroso Groups," the $NO_2$ asymmetrical and symmetrical stretching frequencies for a wide variety of nitroparaffins are well within the range of fundamental frequencies assigned to nitromethane; the most characteristic frequencies observed in nitroalkanes are the bands at 6.2–6.6 $\mu$ (1613–1515 cm$^{-1}$) and 7.2–7.6 $\mu$ (1389–1316 cm$^{-1}$). The asymmetric and symmetric $—NO_2$ frequencies are quite distinct in the nitrate esters, and they occur near 1640 cm$^{-1}$ (6.1 $\mu$) and 1270 cm$^{-1}$ (7.88 $\mu$), respectively. For a large series of compounds, the asymmetric band occurs well within the region 1660–1625 cm$^{-1}$ (6.02–6.15 $\mu$) and the symmetric band at 1285–1261 cm$^{-1}$ (7.78–7.93 $\mu$). Often, the bands are split, especially for vic-dinitrato compounds (71a). Thus it is evident that the nitrate esters are readily distinguished from the C-nitro and C-nitroso compounds even though the fundamental $XNO_2$ frequencies occur in the same general regions. The ir spectra of $\alpha,\omega$-dinitroxyalkanes (up to 1,10-dinitroxydecane) show an asymmetrical stretching band in the region 1626–1653 cm$^{-1}$ (6.15–6.05 $\mu$) which progressively shifts to lower frequencies as the carbon chain is lengthened. For example, the band located at 1635 cm$^{-1}$ in 1,4-dinitroxybutane shifts to 1627 cm$^{-1}$ in 1,7-dinitroxyheptane; as noted by Barefoot and Lawrence (11), the shift appears to be linear with the number of methylene groups in the chain.

### (2) Nitrite Esters

The frequencies assigned to the $NO_2$ asymmetrical stretch in nitrite esters appear as two extremely intense bands in the ranges 1681–1653 $cm^{-1}$ and 1626–1613$^{-1}$ (5.95–6.05 $\mu$ and 6.15–6.20 $\mu$, respectively) (285–287). The bands not only are distinctive but also can be readily distinguished from those of nitroso compounds [usually in the 1600–1500 $cm^{-1}$ (6.25–6.66 $\mu$) range] and nitrosamines (below 1500 $cm^{-1}$). The doubling also serves to distinguish the nitrite esters from the nitrate esters, but the possibility of coexistence must always be taken into consideration. The 1681–1653 $cm^{-1}$ band is attributed to the *trans* forms, and the 1626–1613 $cm^{-1}$ band to the *cis* forms of the nitrite esters (287). Methyl and ethyl nitrite show doublets in the vapor state, but as the molecular weight of the alkyl group increases, these bands shift to lower frequencies and each doublet becomes a single band; the bands are of approximately equal intensities for methyl nitrite, but the band at 5.77 $\mu$ (1675 $cm^{-1}$) predominates over that centered at 6.16 $\mu$ (1623 $cm^{-1}$) for compounds of higher molecular weights. Thus the ratio of intensity of the *trans* band to the *cis* band becomes about 3:1 in the nitrites of other primary alcohols; it is about 6:10 for secondary alcohols, and 35:50 in tertiary alcohol esters. Naturally, the amount of *cis* form that exists is less in secondary nitrites, and in tertiary nitrites the *cis* form is greatly reduced (287). Not only does the intensity ratio vary, but even the absolute intensities of the bands vary from primary to secondary to tertiary nitrites, the *trans* band becoming more and more intense and the *cis* band becoming weaker and weaker. Even more remarkable is the fact that the relative abundance of isomers is governed largely by the nature of the carbon adjacent to the nitrite group (i.e., whether it is primary, secondary, or tertiary) rather than by the length and shape of the carbon chain attached to the carbon.

The alkyl nitrites also exhibit strong N—O stretching absorptions in the 813–746 $cm^{-1}$ (12.3–13.4 $\mu$) range, and there are two bands attributable to —ONO bending deformations of *cis* and *trans* forms at 690–617 $cm^{-1}$ (14.5–16.2 $\mu$) and 625–565 $cm^{-1}$ (16.0–17.7 $\mu$)

Combinations and harmonics of the fundamental frequencies are also found in the spectra of alkyl nitrites, for example, at 2304–2252 $cm^{-1}$ (4.34–4.44 $\mu$), near 2500 $cm^{-1}$ (4.00 $\mu$), and at 3300–3195 $cm^{-1}$ (3.03–3.13 $\mu$).

### (3) Thionitrite Esters

The infrared spectra of methyl, ethyl, *n*-propyl, and isopropyl thionitrites (RSNO) in their gaseous state have been examined by Phillippe and Moore (222,223). All the thionitrites studied show a very strong N=O stretching vibration band at 1534 cm$^{-1}$ (6.52 $\mu$); in contrast to the corresponding nitrites, the band shows no indication of doubling, and its frequency location is constant throughout the series of thionitrites (perhaps rotational isomers cannot exist).

The C—S stretching frequencies occur in the range 685–730 cm$^{-1}$ (14.6–13.7 $\mu$) and are of medium intensity for the thionitrites.

### b. ULTRAVIOLET SPECTROPHOTOMETRY

### (1) Nitrate Esters

The nitrate esters exhibit weak absorptions attributed to $n \rightarrow \pi^*$ transitions (as do the nitrite esters, nitroso compounds, and azoxy compounds). As a rule, the nitrite esters give a spectrum in which there is no distinct maximum (128,255) and absorbance continuously increases as the wavelength is decreased, but there is an easily discerned inflection near 265–270 nm. (In contrast, in organic nitrate salt or the nitrate ion from an organic salt shows a clear, very low intensity maximum at $300 \pm 1$ nm.) Figure 1 shows typical nitrate and nitrite absorptions.

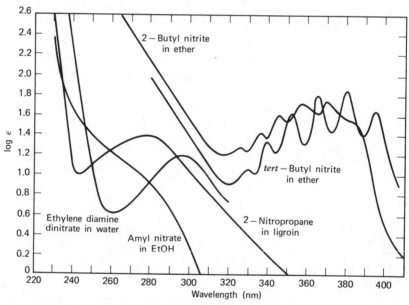

Fig. 1. Typical spectra of a nitroxy compound such as amyl nitrate and nitrite esters (*tert*- and 2-butyl nitrite). The spectra of a *C*-nitroparaffin and an organic nitrate salt are included for comparison. [Data from Ungnade and Smiley (295) combined with data from Jones and Thorn (128) for the ethylenediamine dinitrate in water.]

When a compound is both a nitrate ester and a nitrate (salt), for example, 1-nitroxy-3,6-diazahexane dinitrate:

$$O_2NO—CH_2—CH_2—\overset{+}{N}H_2—CH_2—CH_2—\overset{+}{N}H_3$$
$$\underset{NO_3^-}{\qquad\qquad\big|\qquad\qquad\qquad\qquad\quad\big|}$$

the uv absorbance curve is similar to the curve for a simple nitrate ester. However, Jones and Thorn (128) have shown that analysis of the shape and position of the absorption curve, along with comparisons with the spectra of known compounds, can yield the actual number of each type of group present.

### (2) Nitrite Esters

The nitrite esters exhibit a fine-structured band consisting of six or seven well-defined maxima roughly about 10 nm apart in the region 310–390 nm (see Fig. 1) and a high-intensity band at about 228 nm. The fine structure is little affected by polar solvents such as alcohol or water; as a rule, the uv spectrum of a nitrite ester is nearly the same for the vapor, the liquid, and its solutions in carbon tetrachloride, chloroform, and ligroin, but the relative heights of the various peaks in the fine-structured band may be altered in solvents such as acetonitrile and dimethylformamide. The change in heights is attributed to change in the *cis/trans* ratio (108,287).

Within each group of nitrites, the wavelengths of the fine-structure bands change little except when the electronic character of the $\alpha$-carbon is changed; an increase in the electron density of this carbon produces a bathochromic shift of all bands and an increase in absorption intensity (295). Molar absorptivities increase with the molecular weight of the alkyl group. The differences in the spectra of primary, secondary, and tertiary nitrites are so great that it is possible to differentiate between them; the bands are surprisingly sharp in tertiary nitrites (287).

As is evident from Fig. 1, the uv absorption spectra of nitroparaffins, alkyl nitrates, and alkyl nitrites differ sufficiently so that they can be identified when present as quite pure compounds; in some instances, it may be possible to detect alkyl nitrates in the presence of alkyl nitrites after treatment with methanol (trans esterification rapidly produces volatile methyl nitrite). Kornblum et al. (150,151) have utilized uv absorption in the region 300–400 nm (and higher) for determining alkyl nitrites in Victor Meyer reaction mixtures (see Section II.B.2 of "Nitro and Nitroso Groups,") which also may contain alkyl nitrates and nitroparaffins. An ether solution of the products (free of silver salts) is read directly, because C-nitroparaffins and alkyl nitrates do not absorb in this region. Nitrogen dioxide interferes because in ether solutions it shows five vibrational fine-structure bands at 337, 348, 359, 373, and 387 nm which may interdigitate with the fine-structure bands of alkyl nitrites or may interfere by superposition (295).

## C. IDENTIFICATION

Chemical methods for identifying specific nitrate and nitrite esters virtually always involve isolation and identification of the hydroxy compound from which the ester is derived; in isolated instances, specific tests have been developed to identify certain esters, and there are procedures whereby esters can be converted directly to derivatives of the parent hydroxy compounds.

Although alkaline hydrolysis provides the most convenient route for releasing the parent hydroxy compound from a nitrate or nitrite ester, the reaction is seldom straightforward and the by-products may prove annoying especially when derivatives are to be prepared for identification; the annoyance may quickly become an insurmountable difficulty with polynitrate esters of complex mixtures (e.g., nitrostarch). Reductive cleavage of esters is often more appropriate, but when reduction is effected by chemical reagents such as polysulfide or metals, isolation of the parent hydroxy compound from the reaction mixture may prove to be a challenge. On the other hand, catalytic hydrogenation nearly always provides the parent hydroxy compound in a medium from which it can readily be recovered; hydrogenation is, therefore, unquestionably the appropriate procedure for obtaining the parent hydroxy compound of a nitrate or nitrite ester.

Physical methods are also very useful for identifying products of hydrolysis of the esters or even the esters themselves; indeed, the use of physical methods should be investigated first, leaving chemical methods as the last resort. High-resolution mass spectroscopy is undoubtedly the most definitive and general physical method for identification of volatile esters; unfortunately, a large number of esters are not sufficiently volatile for analysis (e.g., nitrocellulose). Chromatographic procedures are readily applied to all types of esters (except insoluble ones such as nitrocellulose) and are sensitive as well as quite specific for identification of esters which have been prepared and characterized beforehand; although an unknown ester can be isolated by chromatographic procedures, it can be identified only by comparison with the chromatographic behavior of the identical species.

## 1. Chemical Methods

The parent hydroxy compound of a nitrate or nitrite ester can be regenerated as follows:

1. Hydrogenation.
2. Alkaline hydrolysis.
3. Reductive cleavage.
4. Transesterification.

The first method, catalytic hydrogenation, appears to be best because the hydroxy compound can often be obtained in pure form (from the reaction medium). The other methods (with the exception of transesterification) either yield variable amounts of the parent hydroxy compound, together with a variety of degradation products of the parent compound, or leave a reaction medium from which the hydroxy compound cannot readily be removed. The following discussion first describes procedures for isolating parent hydroxy compounds and then briefly indicates methods which can be used for identifying the hydroxy compounds of the types of esters most likely to be encountered in practical work.

### a. Isolation of Parent Hydroxyl Compound

### (1) Catalytic Hydrogenation

Palladium on calcium carbonate is an effective catalyst for the reduction of alkyl nitrate at room temperature and 1500 psi (156):

$$RONO_2 + 5 H_2 \rightarrow 2 ROH + N_2 + 4 H_2O$$

The conditions for hydrogenation appear to be quite critical because Stalcup et al. (276) report that the catalyst is ineffective at 30 psi in experiments on the hydrogenation of nitroglycerin; their work suggests that platinum oxide and palladium black are active catalysts at room temperature and 30 psi but that nitrate nitrogen is reduced to ammonia. Unfortunately, ammonia (but not nitrate ion) reacts with nitrate esters under the conditions of catalytic hydrogenation, and the resulting products may interfere with identification of the parent alcohol (see Section III.B.4 of "Nitro and Nitroso Groups"); for example, only 60% of the theoretical amount of glycerol (periodiate oxidation method) is formed when an alcoholic solution of nitroglycerin is reduced at room temperature with hydrogen at 30 psi in the presence of palladium black catalyst. Reductions of nitrate esters with palladium on charcoal, Raney nickel, or platinum also produce ammonia.

## (2) Hydrogenation by Hydrides

Lithium aluminum hydride reduces simple mono- and diesters quantitatively· to the corresponding alcohols, and the method is readily applicable to nitrate or nitrite esters which do not contain other functional groups that will be reduced or altered by alkalinity. The following procedure is typical for most nitrate esters:

A weighed portion of the ester is dissolved in dry ether (ethylene glycol diethyl ether), and an excess (about 10%) of a 20% solution of lithium aluminum hydride in ether is added dropwise (agitation). Ethylene glycol diethyl ether is also a suitable solvent for the hydride. The mixture is allowed to stand for at least 20 min and then is carefully acidified with 3 $N$ hydrochloric acid. The ether layer is removed, and the aqueous layer is further extracted with ether to effect complete removal of alcohol. Alternatively, the alcohols can be distilled from the aqueous layer after the forerun of ether. Higher molecular weight alcohols and polyhydroxy compounds remain in the aqueous layer and must be recovered by suitable methods.

## (3) Alkaline Hydrolysis

As was indicated in Section III.B, hydrolysis of nitrate esters is complicated by a variety of side reactions; on the other hand, nitrite esters are readily hydrolyzed. It is also necessary to be aware that certain esters, such as benzyl nitrate, yield aldehydes which on hydrolysis form condensation products rather than the expected alcohol.

In view of the surprising resistance of nitrate esters to hydrolytic attack, there is a tendency to use strenuous conditions with the result that degradation of the alcohol often occurs; in most instances, therefore, isolation of the parent hydroxy compound is best accomplished by hydrogenation. In any event, the amount of alkali used for hydrolysis should be kept at a minimum; the appropriate conditions should be established by a series of experiments. Fortunately, isolation of the parent hydroxy compound is not always necessary because derivatives of esters often can be formed in one step (see Section V.C.b).

Hydrolysis by alkali obviously cannot be used indiscriminately, for example, with esters made from hydroxy compounds that undergo condensation reactions or that are structurally altered by alkali (dehydrohalogenation).

The hydrolysis of small quantities of esters and the isolation of about 50 mg of an alcohol for preparation of derivatives are difficult. As a rule, the type of alcohol determines to a large extent the procedure for separation of derivatives, but the amount of sample available may be a major factor. An idea of the approximate structure of the parent hydroxy compound is of inestimable value in the charting of experimental procedures to be used for identification. Inasmuch as it is impossible to prescribe set procedures that are applicable to all samples requiring identification, only a few typical methods are given in the following paragraphs; it is hoped that they will serve as a guide to the selection of a method that will be applicable to the sample at hand.

### Method A

The ester is refluxed in an aqueous or alcoholic normal solution of sodium hydroxide. Generally, esters require only about 15 min for hydrolysis, but as much as 1 hr may be needed for certain types. Cool the hydrolysate and neutralize to phenolphthalein by cautious addition of 3 $M$ sulfuric acid; if the alcohol does not separate, add potassium carbonate to force it out of solution. In any event, repeatedly extract the alcohol from the saturated potassium carbonate medium with methanol, diethyl ether, or isopropyl ether and then evaporate the lighter solvent. Since many polyhydroxy compounds cannot be extracted from solution, the aqueous phase can be brought to dryness either on a steam bath or by evaporation under vacuum; the presence of sodium sulfate ordinarily does not interfere with the formation of derivatives of alcohols. In some instances, the neutral aqueous solution of the alcohol can be used to form derivatives.

### Method B

Warm the ester on a steam bath in aqueous 6 $M$ potassium hydroxide solution until decomposition is complete; cool, and then neutralize with 3 $M$ sulfuric acid (phenolphthalein). Salt out the alcohol with potassium carbonate, and follow, in general, the isolation procedure of Method A.

### Method C

Reflux the ester in a two-phase system consisting of 6 $M$ aqueous potassium hydroxide solution and a supernatant layer of diisopropyl ether until hydrolysis is complete; cool, and neutralize the aqueous layer with 6 $M$ sulfuric acid (phenolphthalein). If necessary, salt out the alcohol with potassium carbonate, and then continue as indicated in Method A.

### (4) Reductive Cleavage by Sulfides

Conversion of a nitrate or nitrite ester to its parent hydroxy compound can be achieved by reduction, but the products of cleavage often include a variety of other substances derived from the hydroxy compound. However, in reductions of nitrate esters with ammonium sulfide, the parent hydroxy compound is formed in sufficient preponderance to permit isolation without complications.

The alkaline sulfide reduction products of nitrate esters (especially the polynitrate esters) are complex; for example, Fischer (78) identified nitrate ions as well as ammonium, nitrite, and thiosulfate ions as the products of reaction of pentaerythritol tetranitrate with sodium sulfide (70% yield of the hydroxy compound). As the result of subsequent studies, Schmidt (251) suggests that the initial reaction in the reduction of pentaerythritol tetranitrate yields sodium hydroxide and hydroxylamine:

$$C(CH_2ONO_2)_4 + 6\,Na_2S + 11\,H_2O = 3\,Na_2S_2O_3 + 6\,NaOH + 4\,NH_2OH + C(CH_2OH)_4$$

and that the alkali-hydroxylamine couple (*in statu nascendi*) further reduces additional nitrate ester; experimental data suggest that, in the overall reaction, 1 mol of ester is reduced by 2 mol of sodium sulfide.

Esters that can be readily cleaved by metal–acid couples are best handled by treatment with magnesium and sulfuric acid, for after neutralization the resulting solution can often be used directly for formation of derivatives.

### Method A

Bubble hydrogen sulfide through cold, concentrated ammonium hydroxide solution (sp. gr. 0.88) until the resulting solution has a pH of $10 \pm 0.1$; cool the mixture while hydrogen sulfide is passed into it. Finally, dilute the reaction mixture with twice its volume of water. To a portion of the diluted ammonium sulfide (a large excess), cautiously add the nitrate ester dropwise and with agitation of the reaction mixture. Shake the mixture vigorously, and then let stand for a few minutes. Decompose the excess of ammonium sulfide by cautious addition of hydrochloric acid (to neutrality). Heat on a steam bath to coagulate sulfur; filter the resulting solution, and then shake with a globule of mercury to remove small amounts of sulfur compounds. Concentrate the solution by evaporation on a steam bath or under vacuum. If the alcohol is separable by salting out, saturate the solution (remaining after treatment with mercury) with potassium carbonate and extract with ether or alcohol.

### Method B

Prepare a solution of 6 N hydrochloric acid in methanol by mixing equal volumes of concentrated hydrochloric acid and methanol. Dissolve or suspend about 200 mg of the ester in the methanolic 6 N hydrochloric acid solution; cautiously add small amounts of magnesium metal, and agitate the mixture to ensure that the metal comes into contact with the solution of the ester. When reduction of the ester is considered to be complete, neutralize the resulting solution with alkali [until $Mg(OH)_2$ begins to precipitate] and then salt out the methanol and the product alcohol by saturation with potassium carbonate.

Alternatively, if the alcohol is nonvolatile, evaporate the neutral solution to dryness at low temperature; salts such as magnesium sulfate, sodium sulfate, and sodium nitrate seldom interfere with the formation of derivatives of hydroxy compounds.

### (5) Transesterification

This reaction may often be used when the sample can be obtained in essentially anhydrous condition; in the presence of small amounts (even traces) of alkali and a large excess of methanol, many high-boiling nitrate esters can be converted to their original alcohols by repeated distillation or refluxing at a temperature near the boiling of methanol and finally at a temperature that removes methanol:

$$MeOH + RONO_2 \xrightarrow{OH^-} ROH + MeONO_2$$

As was indicated in Section II.B.3, the nitrite esters undergo transesterification with remarkable ease and without catalysts. However, the nitrate esters are more resistant to transesterification and hence will often require prolonged treatment, but the investment in time is rewarded by purity of product, because side reactions seldom occur. The presence of original ester may prove to be an annoyance in subsequent operations.

### b. IDENTIFICATION OF PARENT HYDROXY COMPOUND

The following procedures are typical for the *direct* preparation of derivatives of the hydroxy compound from which esters have been derived:

#### 3,5-Dinitrobenzoate Esters of Volatile Alcohols

Dissolve 500 mg of 3,5-dinitrobenzoyl chloride in 3 ml of anhydrous pyridine contained in a test tube. Add about 200 mg of the ester and heat to boiling; a reflux condenser should be used (e.g., a smaller test tube full of cold water may be inserted in the larger test tube). Generally, the hydrolysis is complete within 15 min, but some esters may require as long as 1 hr. Cool the hydrolysate, and add 1–1.5 ml of 5% sulfuric acid to neutralize pyridine and 5 ml of water; shake well and filter off the 3,5-dinitrobenzoate. Dissolve the residue on the filter in ethyl or isopropyl ether. (Alternatively, extract the aqueous solution with ethyl or isopropyl ether). Wash the ethereal solution of the 3,5-dinitrobenzoate successively with 3-ml portions of water, 2% sodium hydroxide solution, and finally water. Evaporate the ether; dissolve the residue in less than 5 ml of methanol, and allow to crystallize. Determine the melting point; recrystallize and redetermine the melting point. If the 3,5-dinitrobenzoate is relatively insoluble in methanol, suspend the residue obtained from evaporation of the ether in about

5 ml of methanol, heat almost to boiling, and filter. The crystals on the filter are used for the melting point determination; crystals can also be obtained from the alcoholic filtrate.

### 3,5-Dinitrobenzoate Esters of High-Boiling Alcohols

The ester is hydrolyzed by a solution of potassium hydroxide. Add about 200 mg of the ester to 3 ml of aqueous 6 $M$ potassium hydroxide solution in a test tube. Heat in a boiling-water bath until the ester is decomposed. Cool the hydrolysate, and then neutralize the alkali by cautious addition of 6 ml of 3 $M$ sulfuric acid; to obtain neutrality, it is best to add about 5 ml of the acid in small increments (cooling!) and then to add a tiny drop of phenolphthalein indicator solution before continuing addition of acid (in drops) to a faint pink color. Now, if the alcohol has not separated, add enough solid potassium carbonate to saturate the solution and then extract the alcohol by ethyl or isopropyl ether; evaporate the ether. If the alcohol separates from the nearly neutralized hydrolysate, filter it off or remove it with the aid of ether; evaporate the ether. (The method is seldom applicable to sugars.) Prepare the 3,5-dinitrobenzoate derivative of the alcohol as follows. Transfer the alcohol obtained as indicated above to a dry test tube (the alcohol should be as water-free as possible). Prepare a solution of 500 mg of 3,5-dinitrobenzoyl chloride in 3 ml of pyridine, and pour it into the test tube containing the alcohol. Boil the mixture (use a reflux condenser) for at least 1 hr. Cool, and add 1–1.5 ml of 5% sulfuric acid to neutralize the pyridine; add 5 ml of water and shake well. The 3,5-dinitrobenzoate is treated as indicated in the preceding procedure.

The above procedures require that the 3,5-dinitrobenzoate be prepared under nearly anhydrous conditions in order to prevent hydrolysis of 3,5-dinitrobenzoyl chloride; only small amounts of water can be tolerated. However, Lipscomb and Baker (180) have shown that it is often possible to prepare the 3,5-dinitrobenzoate derivatives of some alcohols in the presence of much water. For this purpose, the acid chloride is dissolved in pure petroleum ether (ligroin) (112) and then shaken with an aqueous solution of the alcohol in the presence of sodium acetate. This method may be used on the neutralized hydrolysate obtained in the second procedure given above. Armstrong and Copenhaver (5) have shown that this method provides useful derivatives with alcohols as high as 1-eicosanol, $C_{20}H_{41}OH$.

### Derivatives of Polyhydric Alcohols

Many polyhydric alcohols (e.g., glycerol, 1,2-propanediol, and ethylene glycol) are best characterized as trityl ethers (94) or as urethanes; however, the diglycerols can be best identified by paper chromatography

(269). In any event, the alcohols must be first isolated as nearly anhydrous substances. The second procedure given above may be used for this purpose. After neutralization, the hydrolysate is saturated with potassium carbonate and extracted with ether; the ether is removed by evaporation, leaving essentially dry alcohol. For very soluble alcohols, an attempt should be made to cause separation by adding acetone or other solvents to the aqueous mixture. The procedure given below is typical for formation of urethanes from alcohols; it is also useful for Cellosolves and Carbitols (192).

Dry a test tube by heating it over a flame; loosely stopper it and allow to cool. Introduce about 30 mg of the alcohol and follow rapidly with 50–75 mg of p-nitrophenyl isocyanate; cork the tube and heat the contents in a steam bath for at least 5 min (at least 10 min longer for some alcohols). Then cool the mixture in an ice–salt bath for 20 min. Extract the solid with 1–3 ml of carbon tetrachloride (use more if the residue looks oily). Cool the carbon tetrachloride solution in ice; an oil or a solid should separate (if necessary, concentrate the solution by evaporation). Dissolve the solid by rewarming the solution; then place in an ice box and allow crystallization to proceed. The urethane can be recrystallized from carbon tetrachloride or from a 3:2 mixture of ligroin and carbon tetrachloride.

The above procedure can also be used to form phenylurethanes and $\alpha$-naphthylurethanes; in these instances, petroleum ether is recommended for extraction and crystallization. In many instances, formation of urethanes can be accelerated by addition of 1 drop of pyridine or a 10% solution of a trialkyl amine such as trimethylamine. Of interest is the finding by Dewey and Witt (59,60) that a mixture of equal parts of water and either methyl, ethyl, propyl, butyl, or amyl alcohol yields sufficient urethane for characterization; the mixture of urethane and diphenylurea is extracted with boiling petroleum ether. Of course, in these instances, a large excess of aryl isocyanate must be used.

The alcohols can also be converted to other derivatives, for example, trityl ethers [triphenyl methyl chloride in the presence of pyridine (94,262)], xanthates [aqueous potassium hydroxide, acetone, and carbon disulfide (194,302)], and O-alkyl saccharin derivatives [pseudo saccharin chloride (195)]. The identification of Cellosolves and Carbitols is accomplished best by conversion to xanthates as described by Whitmore and Lieber (302).

The esters of sugars and carbohydrates and their derivatives (e.g., sucrose octanitrate and mannitol hexanitrate) yield on hydrolysis the original carbohydrates or derivatives, but these do not form tractable 3,5-dinitrobenzoate and urethane derivatives, particularly since the starting material is seldom pure and contains small amounts of related compounds which are not readily removed by simple recrystallizations. Another complicating factor is that a given carbohydrate such as glucose seldom can be identified by a single physical constant (notable exceptions are levulose, starch, mannose, glycogen, and cellulose). On the other hand, derivatives of carbohydrates, such as mannitol and sorbitol, and polyols such as pentaerythritol form phenylurethanes or acetates. The following procedure is typical for production of completely esterified sugars or polyols from dry residues obtained by hydrolysis of nitrate or nitrite esters:

Heat about 500 mg of sodium acetate trihydrate in a test tube until water has been quite completely expelled (also heat the upper parts of the tube) and the salt is still liquid. Allow the salt to cool to room temperature; then add 7–8 ml of acetic anhydride, and heat to dissolve. Add about 200 mg of the sugar or polyol, and heat in a water bath (90–100°C) for at least 2 hr. Add 20–25 ml of water containing 4 ml of 10% sodium hydroxide solution, and let the tube stand in a cold-water bath for 12–18 hr. Scratch the tube to induce crystallization of any oil which separates. Filter the crystals; recrystallize from the minimum amount of boiling alcohol by addition of cold water until a permanent cloudiness is obtained. Several hours may be needed for complete recrystallization; repeat recrystallization until the melting point is constant.

Many sugars, carbohydrates, and derivatives are readily broken down, oxidized by air, or transformed more or less to related isomers when hydrolyzed in acidic or alkaline media (252). Consequently, hydrolyses of the esters of sugars and carbohydrates must be performed under controlled conditions; as a rule, hydrolysis is attempted by gentle heating of a slightly alkaline solution. The appropriate conditions are found largely by trial and error; some prior knowledge of the approximate structure involved is useful in selecting reasonable conditions. In any event, the products obtained from a series of hydrolyses performed under widely different conditions should be compared as a check on the possibility that side reactions have occurred in certain hydrolytic environments. For example, hydrolyses should be performed in sodium bicarbonate solutions, sodium acetate, sodium carbonate, 0.1 N sodium hydroxide, and progressively stronger solutions of sodium hydroxide; the increasing alkalinity represented by this series of solution provides a wide range of hydrolytic conditions. The time involved in the hydrolysis is another variable that can be controlled; ideally, the time should be as short as possible.

It must be recognized that hydrolysis may change the nature of the original carbohydrate and that derivatives formed in subsequent operations can only characterize the hydrolysis products. However, when the hydrolysis conditions are known quite precisely, it is generally possible to deduce the nature of the starting materials from the identity of the hydrolysis products; for corroboration, the effect of the hydrolysis conditions on the presumed starting material can be readily ascertained. Constant reference should be made to standard texts on sugars and carbohydates, especially for separation and identification of the individual carbohydrates in mixtures.

After hydrolysis is complete, the hydrolysate is neutralized to pH 7 to minimize further degradation of the products of hydrolysis. If the desired product separates, it should be filtered off, washed, and dried. If the product remains in solution or forms a colloidal system, it is best to evaporate the mixture of dryness, preferably at low temperatures (e.g., under vacuum). The reducing power of the hydrolysate should be checked with Fehling's solution (reducing sugars). A hydrolyzed product should be protected from bacterial attack from the time it is first obtained to the time it is characterized as a derivative.

In many instances, it is best to attempt to identify carbohydrates by special tests; a convenient system for testing for common carbohydrates such as cellulose, starch, soluble starch, dextrin, pectin, mannose, arabinose, lactose, sucrose, and glucose has been described by Dehn et al. (55). Many relatively pure sugars and polyhydric alcohols such as mannitol and sorbitol can be identified by their crystalline habits when thrown out of a saturated aqueous solution by acetone, alcohol, acetonitrile, or 1,4-dioxane; with the aid of a microscope, crystals prepared under similar conditions from known compounds are compared with those obtained from the sample or with photographs such as have been published by Quense and Dehn (234,235). Sugar and sugar derivatives may also be identified on thin-layer chromatographic plates by the colorations formed with perchloric acid (211); paper chromatography is suitable for many sugars and carbohydrates (126).

Carbohydrates may also be identified by infrared (155) and other instrumental methods. Of interest is the observation that $\beta$-D-glucose can be changed to $\alpha$-D-glucose monohydrate in improperly prepared potassium bromide pellets (12); the analyst must be alert to the need for appropriate pelleting procedures (201).

Some of the methods for forming derivatives of carbohydrates and other polyols may be used directly on their esters, inasmuch as conditions established for formation of derivatives also are optimum for hydrolysis of esters. For example, the following general procedure for formation of crystalline osazones of sugars often can be applied directly to nitrate and nitrite esters.

Place 2 ml of a neutral clear solution containing from 10 to 30 mg of the ester in a 15 × 150 mm test tube; add 0.4 gram of sodium acetate, 1 drop of a saturated solution of sodium bisulfite, and 0.2 gram of phenylhydrazine hydrochloride, swirling and shaking the mixture after each addition in order to hasten solution. Loosely stopper the test tube, and immerse it in a boiling-water bath for at least 30 min (longer if necessary); then remove from the bath and allow to cool slowly to room temperature. If enough sugar was taken as a sample, the osazones of the monosaccharides will crystallize while hot. Since the osazones of the disaccharides are soluble while the solution is hot, they can be separated from those of the monosaccharides by taking advantage of the difference in solubility.

Filter off the desired crop of crystals, and wash with 1 ml of water containing 1 drop of glacial acetic acid and then twice with 1 ml of water. Transfer the crystals to a small test tube, and add enough methanol to dissolve practically all of the osazone at the boiling point. Filter the hot solution and add water (at least one-tenth the volume of the methanolic filtrate); then cool in an ice-water bath. Filter off the crystals, and wash sparingly with a solution of 1 vol. of methanol and 3 vols. of water. Dry the crystals in a vacuum desiccator.

A neutral hydrolysate of an ester can be used in the above procedure; in some instances, it is best to triple the quantities specified in order to obtain manageable amounts for conversion to the osotriazole (see below). Although the melting point of the osazone can be used to identify its sugar progenitor, the osazones formed as indicated in the first paragraph above have characteristic crystalline habits which can be established under low magnification (×45) and compared with photomicrographs published by Hassid and McCready (105) or with the freshly prepared osazones of known sugars. It must be noted that there are certain limitations to the identification of sugars as their osazones, because a number of sugars have a common enolic form and so yield the same osazone. Thus the hexose sugars, *d*-glucose, *d*-mannose, and *d*-fructose, yield the same phenylosazone. Similarly, the same osazone is obtained from *d*-galactose, *d*-talose, and *d*-tagatose, from *d*-allose *d*-altrose, and *d*-pseudofructose, and from *d*-idose, *d*-gulose, and *d*-sorbose; *d*-arabinose and *d*-ribose yield one osazone, and *d*-xylose and *d*-lyxose give another. Moreover, the corresponding groups of sugars belonging to the *l*-series, having the same configuration beyond the second carbon atom, also yield the identical osazone, differing from the *d*-osazone only in the direction of optical rotation. Clearly, other tests must be made to identify a sugar belonging in a particular enolic series, for example, a study of the crystalline habit of the osazones with the aid of a microscope, as described by Hassid and McCready (105).

The *osotriazoles* have much sharper melting points than do the osazones; the osazones are first formed as noted above and then are converted to osotriazoles by oxidation with cupric ion (99,104):

$$
\begin{array}{ccc}
\underset{\substack{| \\ \text{(HCOH)}_3 \\ | \\ \text{CH}_2\text{OH}}}{\overset{\substack{\text{HC}=\text{NNHC}_6\text{H}_5 \\ | \\ \text{C}=\text{NNHC}_6\text{H}_5}}{\phantom{x}}} & \xrightarrow{\ \ \text{Cu}^{2+}\ \ } & \underset{\substack{| \\ \text{(HCOH)}_3 \\ | \\ \text{CH}_2\text{OH}}}{\overset{\substack{\text{HC}=\text{N} \\ | \qquad\ \ \diagdown \\ \qquad\qquad \text{N}-\text{C}_6\text{H}_5 + \text{C}_6\text{H}_5\text{NH}_2 \\ \text{C}=\text{N} \diagup}}{\phantom{x}}}
\end{array}
$$

Phenyl-D-glucosazone                Phenyl-D-glucosotriazole

A typical procedure for converting an osazone to an osotriazole is as follows. Place about 50 mg of the osazone in an 8-in. test tube; add 4 ml of water, 1 drop of 6 $N$ sulfuric acid, 150–200 mg of powdered cupric sulfate pentahydrate, 3 ml of isopropanol, and a bead or a boiling stone. Reflux the mixture for 1 hr. Transfer the yellow-green solution to a small evaporating dish, and reduce the volume to 1.5–2 ml on a steam bath. Cool the dish in ice water, and filter the crystals. Dissolve the crude product in 5–7 ml of boiling water, add charcoal, and filter. Allow the filtrate to remain in an icebox overnight. Filter the osotriazole, and wash the crystals with two 0.5-ml portions of water. Dry the crystals in air; then transfer to a small vessel, add 0.5 ml of 95% ethanol, and heat to boiling. Add 0.5 ml of water to the clear solution, and cool for 3–4 hr in an ice-salt mixture. Remove the crystals by filtration, wash with a few drops of water, and allow to dry.

The azoate derivatives of the sugars are useful not only for identification via melting points but also for chromatographic separation of mixtures (46,47). The reagent is $p$-phenylazobenzoyl chloride (azoyl chloride); in pyridine solutions, it reacts slowly with the hydroxyl groups of the sugar to form orange to dark-red azoates:

$$
\underset{\substack{\alpha\text{-D-glucose}}}{\overset{\substack{\text{HC}=\text{O} \\ | \\ \text{(HCOH)}_4 \\ | \\ \text{CH}_2\text{OH}}}{\phantom{x}}} + 5\ \underset{\text{Azoyl chloride}}{\text{C}_6\text{H}_5\text{N}=\text{NC}_6\text{H}_4\text{COCl}} \rightarrow \underset{\substack{\alpha\text{-Pentaazoyl-D-glucose} \\ (\alpha\text{-D-glucose azoate})}}{\overset{\substack{\text{HC}=\text{O} \\ | \\ [\text{HCOCOC}_6\text{H}_4\text{N}=\text{NC}_6\text{H}_5]_4 \\ | \\ \text{CH}_2\text{OCOC}_6\text{H}_4\text{N}=\text{NC}_6\text{H}_5}}{\phantom{x}}}
$$

The reaction requires from 8 to 20 days for completion at 0°C. Mutarotation occurs during the formation of azoates; thus a small amount of the other member of the anomeric pair is always obtained. The following procedure (47) usually provides over 90% conversion within 13 days:

Mix 50 mg of finely powdered sugar with 5 ml of anhydrous pyridine that has been cooled to 0°C, and then add 0.6 gram of azoyl chloride. Maintain the mixture in a closed container at 0°C for a total of 7 days; shake the mixture several times a day. Then remove the mixture from the cold box, and heat in an oven at 90°C for four days. Remove from the oven, add 0.25 ml of water, mix well, and allow to stand for 5 min; then pour the mixture slowly into 50–75 ml of cold water that is kept in constant agitation. Filter off the precipitate and wash it with a little alcohol (azoic acid can be recovered from the filtrates after acidification).

Dry the precipitate, grind it well, and reflux with 2–8 ml of chloroform; filter the mixture while it is hot. If there is too much residue, reflux it again and wash with hot chloroform until the filtrate is only golden in color. The residues are acid and acid decomposition products. Pour the combined chloroform extracts into six times their volume of methanol while stirring. Collect the coagulated precipitate by filtration. If the precipitate tends to be colloidal, boil the mixture and set aside overnight. Dry the crude azoate.

To ensure complete azoylation, dissolve the dry product in the minimum amount of hot pyridine and then add one-half its weight of azoyl chloride. Heat for 2 days at 90°C; then treat the mixture as before, and include two precipitations from chloroform to remove acids. Dry the azoate well before determining its melting point.

Certain sugars are decomposed by heat, for example, $\beta$-D-fructose and $\beta$-D-maltose. In these instances, maintain the azoylation mixture for 7 days at 0°C and then let stand at room temperature for 4 days. The aforementioned sugars (and others) form azoates which are quite soluble in methanol; accordingly, the chloroform extracts are drowned in aqueous methanol instead of pure methanol.

Esters of polysaccharides such as nitrostarch and nitrocellulose are most conveniently hydrolyzed by the reductimetric procedures described in an earlier section.

### c. Saponification Equivalent

The objective of this method of identification is to hydrolyze the ester and determine the amount of alkali which reacts with the acid produced. As a rule, the rate of saponification of an ester by alkali hydroxide is limited by the fact that most esters are sparingly soluble in aqueous media. Accordingly, the saponification media that are most frequently used are essentially anhydrous solvents containing alkali; potassium hydroxide in methanol, ethanol, propanol, and diethylene glycol (238) are often used as saponification media.

The saponification equivalent, usually expressed as milligrams of ester per milliequivalent of potassium hydroxide (or as the industrially important saponification number, i.e., milliequivalents of KOH × 56.1 per gram of ester) is a critical value for esters, and often can serve to narrow the field of search to only a small cluster of compounds. Unfortunately, a relatively large amount of ester must be available, and since the saponification of polynitrate esters does not proceed cleanly to a single product, saponification equivalents can be used only as guides and not as definitive values.

### d. Special Tests

Esters of cellulose and similar polysaccharides are difficult to identify because they are of high molecular weight and because the substances used to prepare them ordinarily exist as a multiplicity of polymeric compounds encompassing a wide range of molecular weights. Thus, not only is it difficult to obtain derivatives that can be crystallized, but also it is nearly impossible to obtain crystalline masses of fixed melting points. Additionally, materials of this type are seldom soluble, although they are dispersible as colloids in a limited number of solvents or solvent systems. For example, in Section II.A.1.a it was indicated that the percent nitrogen combined with cellulose (as nitrate) controls the dispersability of the ester in various solvents, and that guncotton is swelled into a doughy consistency by solvents in order to be cast into propellant grains.

A variety of special tests have been developed to identify some of the more intractable esters.

### (1) Nitrocellulose

The cellulosic component of nitrocellulose can be detected by the formation of an orange color when the sample is warmed with thiobarbituric acid and 85% phosphoric acid. The test is based on a more general color reaction for aromatic aldehydes, the condensation of aromatic aldehydes and thiobarbituric acid as first reported by Dox and Plaisance (64):

$$ArCHO + H_2CC(=O)NHC(=S)NHC=O \rightarrow ArCH=CC(=O)NHC(=S)NHC=O + H_2O$$

Feigl (75) noted that the Dox-Plaisance reaction could be applied to nitrocellulose since the latter is cleaved to hydroxymethylfurfural on warming with phosphoric acid:

Place a few milligrams of the sample in a small test tube, and then add about 50 mg of thiobarbituric acid and 2 drops of 85% phosphoric acid. Immerse the tube in a glycerol bath preheated to 130°C; a positive result is the development of an orange color in the mixture.

The test can be performed on as little as 0.5 mg of sample; nitrocellulose, celluloid, collodion, and pigmented nitrocellulose (pyroxylin) lacquer compositions respond to the test. The determination of cellulose resins in the *absence* of nitrocellulose has been investigated by Swann (282).

### (2) Polyglycols

Qualitative tests for certain important polyglycols can be used once the nitrate ester has been hydrolyzed; Orchin (214) describes qualitative tests involving identification of the products of periodate oxidation to distinguish between ethylene glycol, propylene glycol, glycerol, and diethylene glycol. Color reactions for hexoses and lactose, maltose, and sucrose have been described by Klein and Weissman (138).


## 2. Physical Methods

Physical methods are more generally useful for identification of the nitrate and nitrite esters, especially when data are on hand for all the esters which have been prepared. Paper or column chromatographic methods are very useful and more generally applicable than mass spectroscopy or gas chromatography; the sensitivity of certain esters to thermal degradation precludes their being vaporized without decomposition. Ultraviolet spectrophotometry yields data which cannot be considered conclusive; on the other hand, infrared spectrophotometric data can be compared point for point with the spectra of known compounds for definitive identification. In certain instances, data from uv spectrophotometry may be of assistance in identification of closely related materials.

Of course, when relatively large amounts of ester samples are available, and they can be purified reasonably well or can be readily hydrolyzed to their corresponding alcohols (as, e.g., the simpler liquid esters), density and refractive index often can be used for identification; these methods are obviously of sharply limited value.

### a. INFRARED SPECTROPHOTOMETRY

Identification of nitrate and nitrite esters depends on the existence of spectral features that betray the skeleton of the parent alcohol structure associated with the ester functional group; hence each ester has a characteristic spectrum that can serve to identify it. However, reference spectra for the nitrate and nitrite (and other) esters are not very common, and so it is often necessary to obtain the alcohol by hydrolysis and then to compare its spectrum with the more readily available spectra of alcohols.

Some of the characteristics of the infrared spectra of nitrate esters (and other oxidized nitrogen compounds) have been recorded by Brown (30), Jones and Thorn (128), Pristera and Fredericks (232), Pristera et al. (233), and Lieber et al. (177), and, of course, there are the usual voluminous compilations, including Sadtler Research Laboratories Catalog of Infrared Spectrograms, MCA Catalog of Infrared Spectral Data, and the ASTM-Wyandotte Index—Molecular Formula List of Compounds, Names, and References to Published Infrared Spectra. The infrared spectra of nitrite esters have been studied by Tarte (285–287).

The determination of nitrate esters in explosives is described by Pristera et al. (233).

Complete ir spectra for simple alkyl thionitrites have been recorded by Phillippe (222) and Phillippe and Moore (223). Complete ir spectra for propellant ingredients such as nitroglycerin, diethylene glycol dinitrate, triethylene glycol dinitrate, ethyl centralite, and a series of plasticizers have been recorded by Pristera (231). Data on $\alpha,\omega$-dinitroxyalkanes have been given by Barefoot and Lawrence (11). The ir spectra of nitrocellulose, cellulose derivatives, and over 75 carbohydrates have been recorded by Kuhn (155).

The ir absorption spectra of metal hyponitrites have been studied by LeFevre et al. (167); fragmentary information also has been reported by Kuhn and Lippincott (159). The spectra of metal hyponitrites that are free from impurities such as carbonates, nitrites, and nitrates show no characteristic features above $8.62\,\mu$ ($1160\,cm^{-1}$); for example, if carbonates are present, definite absorption appears in the $6.9\,\mu$ region ($1440\text{–}1450\,cm^{-1}$). The lack of absorption in the $5.75\,\mu$ ($1740\text{–}cm^{-1}$) region is considered to be evidence of the lack of the $-N{=}O$ group in hyponitrites (in comparison with alkyl nitrites), and absorption in the general region of $9\pm1\,\mu$ ($1000\text{–}1200\,cm^{-1}$) is considered to imply the existence of N—O double-single hybrid links ($-N{=}O$ to $-N-O$). Lead hyponitrite shows weak absorption at $8.87\,\mu$ ($1128\,cm^{-1}$), and its most intense absorption at $10.1\,\mu$ ($990\,cm^{-1}$); cobalt and magnesium salts exhibit weak absorptions at $8.70\,\mu$ ($1150\,cm^{-1}$) and strong absorptions at $9.52\,\mu$ ($1050\,cm^{-1}$). The barium salt shows weak absorption at $8.65\,\mu$ ($1157\,cm^{-1}$) and strong absorptions at $9.85\,\mu$ ($1015\,cm^{-1}$), $9.95\,\mu$ ($1006\,cm^{-1}$), $10.05\,\mu$ ($996\,cm^{-1}$), and (in contrast to the other salts) $19.42\,\mu$ ($515\,cm^{-1}$).

b. ULTRAVIOLET SPECTROPHOTOMETRY

The spectral characteristics of nitrite and nitrate esters in the ultraviolet region are seldom sufficiently distinctive to serve for definitive identification of individual compounds. However, the wavelengths of maximum absorption and the molar absorptivities (molar extinction coefficients) provide useful corroborative evidence for identities tentatively established by other methods (see Tables IV and V).

Spectral data for nitrate esters are given by Jones and Thorn (128), Kornblum et al. (152), Haszeldine (106), Boschan and Smith (24), Rao (237), Phillips (224), and Bellamy (14). A publication of Schroeder et al. (255) is especially useful because it displays curves over the region 220–500 nm for nitric esters and for a large series of related compounds.

Spectral data for nitrite esters are given by Tarte (287), Haszeldine

(106), Rao (237), Haszeldine and Mattinson (108), Ungnade and Smiley (295), and Haszeldine and Jander (107).

Standard collections of uv spectra often include a modest number of spectra of nitrate and nitrite esters.

### c. Mass Spectroscopy

Nitrate and nitrite esters that can be introduced into a mass spectrometer can be identified with certainty, especially with high-resolution instruments. In particular, the mass spectrometer can identify the individual esters in a mixture, and in many instances a quantitative estimate of the various ingredients can be had. Closely similar isomers can often be separated by paper or column chromatography (even gas chromatography) and then introduced into the mass spectrometer; even the alcohols obtained by hydrolysis of the esters can be identified.

The most prominent peak in the mass spectra of the nitrite esters, appearing at $m/e$ 30, is known to be the $NO^+$ ion; it is observed in all oxygenated nitrogen compounds and as a rule is larger than 10% of the base peak (the highest peak in the mass spectrum of a compound). The peak at $m/e$ 30 is nearly always absent in the spectra of alcohols (corresponding to $CH_2O^+$) and many other types of compounds; thus its appearance can be taken to indicate the presence of a nitro compound, nitrate ester, nitrite ester, nitramine, and so on. (This observation is helpful if the sample consists of a mixture of an alcohol and an oxygenated nitrogen compound.) The alkyl nitrites are readily identified because their spectra include many ions which comprise all or part of the —ONO group; all nitrites in which there is no substituent group on the carbon atom next to the —ONO group yield a large peak corresponding to $(CH_2ONO)^+$ at $m/e$ 60. Moreover, the essential lack of an ion at $m/e$ 46 $(NO_2^+)$ shows a preference for fragmentation at the bonds beta to the oxygen atom; a consequence of this behavior is a marked difference in the fragmentation pattern of isomeric butyl nitrites, shown in Fig. 2. The molecular ion (parent peak) of paraffin nitrite and nitrate esters is weak or virtually absent.

The $NO_2^+$ peak ($m/e$ 46) in the mass spectrum is a distinguishing characteristic of nitrate esters; this fragmentation peak is essentially absent from the spectra of alcohols, although it may appear as a peak of minor intensity in other types of compounds and is quite trivial in nitrite esters. Thus nitrate esters can be distinguished from other nitrogen compounds (only nitromethane gives a prominent 46 peak), but benzyl nitrate gives a moderately low 3% peak and, in addition, shows a parent peak (characteristic of aromatics), whereas the nitrite and nitrate esters seldom provide a parent peak of useful intensity.

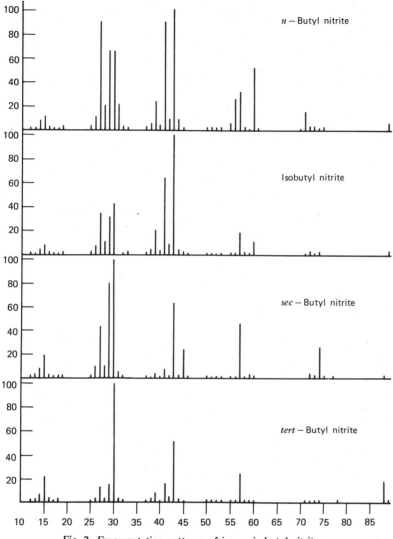

Fig. 2. Fragmentation patterns of isomeric butyl nitrites.

The mass spectra of the primary and secondary nitrate esters show large peaks at $m/e$ values corresponding to removal of the entire alkyl group attached to the $\alpha$-carbon atom:

$$R-\underset{\underset{R'}{|}}{CH}-ONO_2 \rightarrow \left[\underset{\underset{R'}{|}}{CH}-ONO_2\right]^+$$

A peak containing three oxygen atoms is definitive evidence for a nitrate ester, especially when accompanied by a large peak at $m/e$ 46 ($NO_2^+$). Fragmentation of this type occurs so readily in nitrate esters that peaks corresponding to the molecular ion (parent peaks) are very weak or essentially absent. The molecular weight of a nitrate ester can usually be determined by a study of the hydrocarbon fragments. As usual, molecular ion peaks can be stabilized by aromatic rings; thus, for example, benzyl nitrate provides a measurable parent peak. The spectra of primary nitrate esters such as ethyl-, $n$-butyl-, $n$-propyl-, and isoamyl nitrate show large peaks at $m/e$ 76, corresponding to $(CH_2ONO_2)^+$; secondary nitrate esters (e.g., isopropyl nitrate) show peaks corresponding to $[CH(CH_3)ONO_2]^+$ or, more generally, $[CH(R)ONO_2]^+$ at $m/e$ 90. Peaks at $m/e$ 76 or 90 seldom occur in most compounds (except as parent ions), and certainly very rarely in the nitrogen-oxygen compounds under discussion; thus the appearance of these peaks provides insights into the structure of the compounds under examination. Of extreme interest is the fact, noted above, that nitrate and nitrite esters usually show large peaks at $m/e$ values closely related to the alkyl group of the esterified alcohol; alcohols do not show prominent peaks at $m/e$ values corresponding to the R group in ROH, but $C$-nitroparaffins show hydrocarbon fragments up to the parent ion minus the —$NO_2$ group. The appearance of peaks such as 74, 88, 92, etc. (corresponding to $CH_2$ increments) in simple esters betrays branching at the $\alpha$-carbon.

The peak at $m/e$ 31 is a characteristic rearrangement peak that occurs in the spectra of nitrate and nitrite esters (ethyl nitrate is an exception); this peak is virtually absent in the mass spectra of most $C$-nitro compounds.

Reference spectra for nitrate and nitrite esters are given by Boschan and Smith (24); data may also be found in the *Index of Mass Spectral Data* (124). The mass spectra of a short series of simple alkyl nitrites have been recorded by D'Or and Collin (63), and Fraser and Paul (81,82) have reported on their extensive studies of mononitrate esters, polynitrate esters, and nitroalkyl nitrates.

### (d) Adsorption Chromatography

The separation of nitrate esters by column chromatography is used extensively in procedures for the analysis of explosives or mixtures of nitrate esters and nitro compounds. When working with a mixture of unknown substances, column chromatography ideally breaks down the mixture so that each component can be identified by its position on the substrate or by solution of the zones and application of infrared, ultraviolet (255), or mass spectrometric methods; with a mixture containing various amounts of a group of known substances (as is the case for explosives), systematic application of a series of chromatographic separations readily isolates the components, and the presence of a strange substance is easily discerned. Historically, column chromatographic procedures were eagerly pressed into service for analysis of nitrate esters and compositions containing them, followed by more recent application of thin-layer chromatographic techniques; unfortunately, the more discriminatory gas chromatographic procedures cannot be applied readily to sensitive, relatively nonvolatile substances.

A brief introduction to general chromatographic adsorption techniques has been given by Strain (280). The adsorbent found most useful for separation of nitrate esters is silica gel; for column chromatography the adsorbent is usually mixed with inert diluents such as kieselguhr or Celite-535 (Johns-Manville Co.) to increase filtration rates, and for thin-layer chromatography "silica gel G," that is, 200-mesh silica gel containing up to 15% of calcined calcium sulfate (plaster of Paris) as binder, is deposited in thin films on glass plates (191) or other supports. Silicic acid (2 parts plus 1 part diluent) has also been demonstrated to be generally useful in column chromatographic procedures for separation of nitro and nitroso compounds (292), as well as nitrate esters in admixture (254); common eluents such as alcohol, ether, and mixtures of alcohol and acetone in ether remove most compounds from silicic acid, but $N$-nitroso compounds need to be treated carefully in order to prevent decomposition. On silicic acid, ligroin or benzene solutions of ether, ethyl acetate, or acetone usually produce well-defined zones. Silicic acid columns usually require prewashing to produce clean-cut bands (254), but some separations can be effected without preliminary treatment (66); suitable prewashes are (1) successive additions of $V$ ml ether, $V$ ml acetone, and $2V$ ml petroleum ether; (2) $0.2V$ ml ether, $V$ ml of 1:1 acetone–ether, $0.8V$ ml of ether, and $V$ ml of ligroin ($V$ = milliliters required for complete wetting of column).

The adsorbents used in chromatographic separations preferably should be of the commercially available, standardized grades; however, adsorbents can be readily standardized by determination of $R_f$ values according to LeRosen (168–170) and Monaghan et al. (203) for a given substance such as $o$-nitroaniline. The best chromatograms are obtained when

$$R_f = \frac{\text{rate of movement of zone}}{\text{rate of movement of developing solvent}} = 0.1\text{–}0.3$$

but larger values often need to be tolerated when a relatively large number of components must be separated in a mixture, and some authors prefer values of 0.3–0.5. For full characterization of an adsorbent, other parameters must also be determined, and rigid control of the nature of the adsorbent must be exercised; the original literature should be consulted. Improvement of commercial silicic acid by grinding to more uniform, finer sizes has been reported by Malmberg (190). Fortunately, today there are a multitude of commercial adsorbents of rigidly controlled quality which can be expected to perform reproducibly and efficiently.

One of the major problems in the application of column or thin-layer chromatographic procedures to separate and identify nitrate esters is determination of the position of adsorbed materials, inasmuch as nitrate esters are usually colorless to the eye and do not fluoresce when illuminated with uv light. To reveal the presence of colorless substances, the adsorbent column is exposed and brushed, sprayed, or streaked with a variety of reagents; this technique was developed first by Zechmeister et al. (310). For nitrate esters, visualization reagents depend upon the oxidative powers of nitrogen functional groups or on the mere presence of any organic residue. Thus Ovenston (215) employed (1) a 0.5% solution of $K_2Cr_2O_7$ in 60% (w/w) $H_2SO_4$; (2), a 0.5% solution of $K_2Cr_2O_7$ in water, followed by a superimposed streak of 90% (w/w) $H_2SO_4$; (3) a 1% solution of diphenylamine in concentrated $H_2SO_4$ to determine nitrate esters; (4) 6$N$ NaOH; (5) a solution of bromocresol green. Schreoder has used a solution of sodium nitrite in sulfuric acid to detect diphenylamine (blue color) and a solution of diphenylamine in concentrated sulfuric acid to detect nitrate esters (254). Malins et al. (188,189) streaked thin-layer plates with 50% sulfuric acid to cause charring. In other applications, Malins et al. (189) localized nitrate esters on thin-layer chromatographic plates by staining with iodine vapors or by staining with 2,7'-dichlorofluorescein and inspecting under uv light (188); a solution of bromine or iodine in an appropriate solvent such as carbon tetrachloride is also useful. Convenient lists of streak reagents have been given by LeRosen et al. (171,172).

Kohlbeck (141) described thin-layer chromatographic separations of resorcinol and nitroglycerin on silica gel plates developed with 80:20 benzene–diethyl ether; a few years later, Macke (187) made use of a combination of thin-layer chromatography and spectrophotometry to isolate and determine components in aged double-base propellant. (Procedures for the determination of 2-nitrodiphenylamine, resorcinol, and nitroglycerin usually fail to distinguish between the compound of interest and many of the nitrated or dinitrated derivatives.) The propellant is extracted with ether, and the extract is allowed to evaporate; the residue (about 40 mg) is taken up in 5 ml of ethyl acetate and applied evenly across the bottom of the TLC plate (E. Merck's silica gel plates, Cat.

No. 5715). The chromatogram was developed by the ascending method with a mixture of 85 parts of benzene and 15 parts of ethyl acetate. After development, the section of the TLC plate containing the separated compounds of interest was removed and eluted from the support by ethyl acetate; the ethyl acetate was evaporated, and the individual components were dissolved in an appropriate solvent for quantitative analysis.

For the chromatographic system described above, Macke lists the following $R_f$ values: resazurin, resorufin, styphnic acid, 2,4-dinitroresorcinol, and 2,4-dinitrosoresorcinol, 0.00; 1-mononitroglycerin, 0.05; resorcinol, 0.26; 1,2-dinitroglycerin, 0.30; 1,3-dinitroglycerin, 0.47; triacetin, 0.52; 2,4'-dinitrodiphenylamine, 0.69; nitroglycerin, 0.74; 2,4-dinitrodiphenylamine, 0.79; 2-nitrodiphenylamine, 0.82; 2-nitroresorcinol, 0.84. Overlapping of zones occurs, especially with 1,3-dinitroglycerin and triacetin, and with 1,2-dinitroglycerin and resorcinol, but the lower nitrates do not interfere in the spectrophotometric determination of triacetin and resorcinol. Nitroglycerin is determined in 1,2-dichloroethane solution by its absorption at 6.03 $\mu$ (1659 cm$^{-1}$) with a baseline at 5.26 $\mu$ (1900 cm$^{-1}$); triacetin is determined in 1,2-dichloroethane solution at 5.73 $\mu$ (1745 cm$^{-1}$) with a baseline at 5.26 $\mu$ (1900 cm$^{-1}$). Resorcinol is determined in ethyl acetate solution at 272 nm (baseline at 300 nm), and 2-nitrodiphenylamine at 420 m$\mu$ (baseline at 700 nm).

Column chromatographic separations of 24 components commonly found in propellant explosives have been reported by Ovenston (215); Schroeder et al. (256), Schroeder (254), and Kohlbeck (141) give details of chromatographic separations applicable to propellants and explosives.

### e. GAS CHROMATOGRAPHY

Comparatively little work on the gas chromatography of nitrate and nitrite esters has been reported; unfortunately, this valuable method of separation is of limited use because the esters may pyrolyze. Even more irksome is the fact that pyrolysis products of the esters can nitrate or oxidize other substances. Although column temperatures are kept low, values as high as 160°C have been employed (vaporizer at 170°C). Nitroglycerin appears to decompose at temperatures greater than about 80°C (33).

It is of interest to note that Mikkelson and Richmond (200) separated $\alpha$-nitratoisobutyric acid as its methyl ester in a mixture of other methyl esters at a column temperature of 125°C (injection block limited to 145°C to prevent decomposition). To obtain symmetrical peaks, it was necessary to reduce the normal tendency of the support to act as an adsorbent; acceptable peak symmetry was obtained with 50–60 mesh Chromosorb pink treated with dichlorodimethylsilane vapor and then with 15 wt% of di-*n*-decylphthalate.

Evered and Pollard (71) found 60–80 mesh firebrick with 15–20% squalane or dinonylphthalate to be a suitable column packing for separation of mononitrate esters at column temperatures of 80–111°C when the carrier gas is a 3:1 mixture of nitrogen and hydrogen; a hydrogen flame detector was used. Dinonylphthalate was found to be particularly effective for separation of an alkyl nitrate from a nitroparaffin with the same number of carbon atoms; the following specific retention volumes have been obtained at 111°C: hexane, 16.7; heptane, 34; octane, 71; nonane, 138; ethyl nitrate, 42; *n*-propyl nitrate, 80.5; *n*-butyl nitrate, 166; *n*-amyl nitrate, 340; nitromethane, 42.5; nitroethane, 71.5; 1-nitropropane, 128; and 1-nitrobutane, 255. In earlier work, Pollard and Hardy (229) and Pollard et al. (230) also noted the efficacy of dinonylphthalate and dibutylphthalate (on kieselguhr) for separation of alkyl nitrates such as ethyl nitrate from ethyl alcohol. On a 4-mm column containing 2 grams of dibutylphthalate at 57°C and a flow rate of nitrogen of 30 ml/min, an injection of 200 $\mu$l of ethyl nitrate, alcohol, and water was separated into its components. Reaction products containing ethyl nitrate, ethyl nitrite, and water were also separated with the aid of similar columns. Fossel (79) has also reported satisfactory separation of glyceryl trinitrate and of chloroglyceryl dinitrate from ethanolic extracts of pharmaceutical tablets; a column temperature of 130°C (100°C for chloroglyceryl dinitrate) was adequate with a packing of 50–60 mesh Anakrom AB coated with 3% SE-30.

The gas chromatographic separation of alkyl polynitrates such as nitroglycerin, ethylene glycol dinitrate, and 1,2-dinitropropanediol was studied by Camera and Provisani (33). Decomposition of nitrate esters at elevated temperatures was minimized by use of very short columns [Trowell and Philpot (291) and Trowell (290) sought to avoid decomposition of nitrate esters by injecting samples directly on the column instead of through a vaporizer]; 35- and 50-cm stainless steel column lengths were found satisfactory when the shorter column (166 theor. plates) was operated at 145°C and the longer column (237 theor. plates) at 150°C. The evaporator was maintained at 160°C, and the helium flow rate was 250 ml/min. A hot-wire detector operating at 250 mA was used; chart speeds of the order of 0.33 in./min were adequate for sample sizes of 10 $\mu$l. The column packing was 40–60 mesh acid-washed Celite (C.Erba—C22ak) coated with 10% ethylene glycol succinate; partition liquids such as Carbowax-1500, silicone-E301, and Apiezon L are not as effective. As a safety precaution, the apparatus was washed with acetone after 50 analyses, inasmuch as complete evaporation does not occur (peaks become irregular). The retention times obtained for both columns are essentially the same, except for nitroglycerin (decomposition); at 145°C (35-cm column), the following relative retention times were observed: diethylene glycol dinitrate, 1.00; ethylene glycol dinitrate, 0.17; nitroglycerin, 1.95; 1,2-dinitropropanediol, 0.12; 1,5-dinitropentanediol, 0.66; triethylene glycol dinitrate, 3.98. The retention time of diethylene glycol dinitrate was of the order of 7 min.

A study of the application of electron capture detectors to gas chromatographic separations has shown that alkyl polynitrates can be determined at great sensitivity, for example, of the order of 1 $\mu$g (34); the detectors are much less sensitive to hydrocarbons and simple oxygenates than are flame ionization detectors. For the determination of vapors of alkyl polynitrates in air, it is necessary to ensure that the column and its packing will not retain nitroglycerin and other nitrate esters of low volatility. Excellent results were obtained with a 2-mm (i.d.) glass column, 25 cm long, packed with 10% Igepal CO-880 on siliconized 80–120 mesh Chromosorb P. Other operating parameters were as follows: column temperature, 120°C (160°C for nitroglycerin); vaporizing temperature, 170°C; nitrogen pressure, 2.5 kg/cm$^2$; nitrogen flow, 133 ml/min; chart speed, 0.33 in./min. The electron capture detector included a 300-mCi tritium source; it was operated at 20 V, that is, near the inflection point on the current–voltage curve. The steady-state curve was of the order of $3.5 \times 10^{-9}$ A, and the noise level was of the order of $2 \times 10^{-11}$ A.

The gas chromatographic system described above separates mixtures containing 1,2-propylene glycol dinitrate, ethylene glycol dinitrate, nitroglycerin, and mononitrotoluene (often present in the atmospheres of explosives establishments). Direct injection of polluted air by means of glass-Teflon syringes is not satisfactory because of marked absorption of nitroesters on syringe walls. Impinger sampling devices must be used; air can be bubbled at a rate of 1 liter/min through two impinger sampling glass tubes connected in series, each containing 10 ml of ethanol. A 5-$\mu$l sample of the combined alcohol volumes remaining in the impinger tubes is injected into the chromatograph; with a sampling of 10 liters of air, 0.02 mg/m$^3$ of 1,2-propylene glycol dinitrate, 0.08 mg/m$^3$ of mononitrotoluene, and 0.08 mg/m$^3$ of ethylene glycol dinitrate can be determined with a standard deviation of the order of 5%.

Trowell (290) devised a gas chromatographic system of analysis which permits determination of the nitrated derivatives of glycerin in aged double-base propellants. An instrument with dual flame ionization detectors was used; the columns were 6 ft long (0.125-in. tubing) packed with 2.5% OV-17 and 2.5% QF-1 (a trifluoropropyl-substituted silicone) on 60–80 mesh Gas Chrom Q. The column temperature was programmed from 70 to 230°C at 10°/min, and the detector temperature was held at 250°C. The injection port was placed inside the oven compartment to allow true on-column injection. Other operating parameters were as follows: helium flow, 65 ml/min; hydrogen flow, 28 ml/min; air flow, 500 ml/min; sample size, 4 μl. Samples of propellant extracts were obtained by treating 1–2 gram portions of propellant with methylene chloride or diethyl ether. After reduction of the volume of extract to about 2 ml, the residue was treated with 1 ml of N-O-bis(trimethyl-silyl)acetamide catalyzed with 1% of trimethylchlorosilane. The silanization reaction with resorcinol and nitroglycerin derivatives was essentially instantaneous. Dimethylphthalate was added to yield a final concentration of 1 mg/ml when the reaction mixture was diluted to 25 ml with dichloroethane. Chromatograms show clearly separated peaks for the denitrated derivatives of 1,3- and 1,2-dinitroglycerin (the 1,3-derivative is the only dinitro derivative present in aged propellants) and mononitroglycerin; resorcinol, triacetin, dimethylsebacate, and 2-nitrodiphenylamine are also separated. A small peak was obtained for the nitroglycerin derivative even though it was in preponderance in the sample (lack of response of the flame detector prevents quantitative measurement of nitroglycerin). Of interest is the observation that traces of 1-nitroglycerin and dinitroglycerin were present in freshly manufactured nitroglycerin.

The determination of peroxyacyl nitrates in polluted atmospheres has been reported by Darley et al. (54). An electron-capture detector was used because of its enhanced sensitivity to nitrato groups; however, the column used for analysis is of interest because it separates nitrate esters. The column consists of 3 ft of 1.5-mm-i.d. glass tubing packed with 5% Carbowax-400 on 100–120 mesh Chromosorb W operated at 35°C with a nitrogen carrier gas flow of 25 ml/min. It is possible to use 1–3 ml gas samples.

Paraskevopoulos and Cvetanovic (218) have studied the behavior of simple nitrate and nitrite esters (and corresponding nitro compounds) in capillary columns (300 ft, 0.015-in. i.d.) coated with dinonylphthalate and operated at 0°C with helium as a carrier and a hydrogen flame detector; the molar responses of the nitrate esters are generally somewhat lower than those of the corresponding nitrites and nitroalkanes (see Table VI). All esters can be readily separated from each other and from their corresponding nitro compounds. Harrison and Stevenson (101) also found that high column temperatures must be avoided when alkyl nitrites are analyzed; for example, methyl nitrite underwent pyrolysis at

TABLE VI

Relative Retention Times and Molar Responses of Alkyl Nitrates, Alkyl Nitrites, and Nitroalkanes (218).

Capillary column, 0.015-in. i.d., coated with dinonylphthalate; operating temperature 0°; helium carrier at 275 ml/min; hydrogen flame detector. Reference: benzene, relative retention time = 1.000, relative molar response = 100.

| Alkyl group | Nitrate ester | | Nitrite ester | | Nitroalkane | |
|---|---|---|---|---|---|---|
| | Retention | Response | Retention | Response | Retention | Response |
| Methyl | 0.48 | 6 | 0.016 | — | 1.29 | — |
| Ethyl | 1.42 | 20 | 0.050 | — | 3.06 | — |
| n-Propyl | 4.30 | 32 | 0.227 | (38) | 7.89 | 38 |
| 2-Propyl | 2.51 | (38) | 0.147 | 36 | 4.12 | 32 |
| n-Butyl | 14.98 | 38 | 0.747 | — | 26.99 | 49 |
| 2-Butyl | — | — | 0.401 | 58 | 11.23 | 53 |
| Isobutyl | 7.72 | — | 0.426 | 48 | 13.38 | — |
| tert-butyl | 4.65 | — | 0.341 | (62) | 5.50 | — |

column temperatures above ca. 100°C to yield NO, $N_2O$, and $CO_2$. However, even at temperatures below 35°C, recovery of methyl nitrite added to various inorganic gas mixtures was consistently 84–87%. Copper columns 10 and 30 ft (0.125-in. i.d.) were used; the packing was 60–80 mesh Diatoport S (diatomaceous earth treated with dimethyldichlorosilane) containing 20% $\beta,\beta'$-oxidipropionitrile. Thermal detectors with the longer columns at 35°C (helium carrier) showed partial resolution of $N_2$ and NO from $N_2O$ and $CO_2$ and excellent separation of methyl, ethyl, and propyl nitrites.

## f. OPTICAL ROTATION

A number of nitrate and nitrite esters are derivatives of alcohols or other polyhydroxy compounds that show optical activity. Thus, for example, the specific optical rotation of sucrose is +66.5°, but it drops to 56.05° on nitration (sucrose octanitrate). Unfortunately, by far the greater number of esters are optically inactive; on the other hand, optical rotation may be of use for the identification or corroboration of sugars and carbohydrates after deesterification.

## VI. QUANTITATIVE DETERMINATIONS

Quantitative procedures for nitrate and nitrite esters are quite similar to those discussed for nitro and nitroso groups in the preceding section of that name. However, from the standpoint of the number of nitrate ester samples analyzed by a given method of analysis, the nitrometer method, even though it is of limited applicability and is more hazardous than titrimetric or gravimetric methods, is used very frequently, inasmuch as it is often mandatory for the explosive nitrate esters used in the military propellant compositions of nearly every nation of the world. Indeed, the nitrometer method is far more accurate than many proposed methods, although recent developments have brought forth excellent methods which are not given a fair chance by those who do not relish change. Nevertheless, indirect determinations of nitrate esters such as ethylene glycol dinitrate and nitroglycerine with ferrous sulfate (back titration with titanous ion) are gaining popularity.

The bulk of available literature on the quantitative analysis of nitrate and nitrite esters relates to studies and comparisons of the accuracy and precision of the nitrometer method and titrimetric methods (notably ferrous and titanous titrations); such studies have been pursued from the dogmatic viewpoint that the accuracy of the nitrometer is unimpeachable and that other methods must provide results of the same magnitude and of equivalent precision. Since the Du Pont nitrometer handles volumes of gas of the order of 200 ml and has a buret calibrated to at least 0.1 ml, it is evident that individual values can be reported to at least 1 part in 2000 ($\pm0.05\%$); in contrast, ordinary titrimetric determinations of nitrate nitrogen require two titrations under conditions where endpoint determinations are seldom better than $\pm0.03$ ml for volumes of the order of 30 ml, and so titrimetric results are at a disadvantage by a factor of at least 2 and more probably 3 or 4. Thus, for a nitrate ester containing 13.50% nitrogen, nitrometer results are well within $\pm0.01\%$ absolute and titrimetric results will be within $\pm0.02$ or $\pm0.04\%$ absolute; statistical values reported by Pierson and Julian (226) support these statements.

## A. TITRIMETRIC PROCEDURES

### 1. Titration with Reducing Agents

Determinations of the organic nitrate ester group by titrations with reducing agents are patterned after analogous procedures for inorganic nitrates. Mild reducing agents convert primary nitrate esters to the corresponding nitrite esters, but the reaction is of little use for analytical purposes because reduction is seldom quantitative; on the other hand, moderately strong reducing agents form nitric oxide quantitatively under appropriate conditions:

$$RONO_2 + 3\,e + 3\,H^+ \rightarrow ROH + NO + H_2O$$

The products in the above equation could also be considered to be formed by a combination of hydrolysis of the primary nitrate ester followed by reduction of the resulting nitrate ion:

$$RONO_2 + HOH \rightarrow ROH + HNO_3$$

$$HNO_3 + 3\,e + 3\,H^+ \rightarrow NO + 2\,H_2O$$

Experimental evidence suggests that in some instances the nitrate ester group is attacked directly by the reducing agent and that primary nitrite esters may be formed as intermediates, but usually a reduction reaction appears to take place exclusively without suspicion of a preliminary hydrolysis. Strong reducing agents, such as titanous ion, form ammonia quantitatively:

$$RONO_2 + 8\,e + 9\,H^+ \rightarrow ROH + NH_4^+ + 2\,H_2O$$

### a. Titanous Reductions

Determination of nitrate esters, such as nitroglycerin, by reduction with titanous ion has proved troublesome. Oldham (213) found that only 7.5 mol of titanous ion was consumed for each nitrate group in aliphatic nitrate esters instead of the expected 8 mol, as indicated above for formation of the ammonium ion; Oldham suggested that the low value may stem from formation and escape of small amounts of nitric oxide. Shankster and Wilde (265) found that 18 mol of titanous chloride is consumed by 1 mol of nitroglycerin (three nitrate ester groups).

Fainer (72) made a detailed study of the titanous reduction of nitroglycerin and found that in strongly acidic solutions (sulfuric acid) the consumption of titanous salt corresponds to an equivalence of 9 mol/mol of the ester, suggesting hydrolysis to nitric acid and formation of nitric oxide. However, in solutions of moderate acidity where 18 equiv. of titanous salt are consumed per mole of nitroglycerin, it was demonstrated that secondary nitrate groups react differently than do primary ones:

*Primary*

$$RONO_2 + 8\,Ti^{3+} + 9\,H^+ \rightarrow ROH + 8\,Ti^{4+} + NH_4^+ + 2\,H_2O$$

*Secondary*

$$R_2CHONO_2 + 2\,Ti^{3+} + 2\,H^+ \rightarrow R_2CHONO + H_2O + 2\,Ti^{4+}$$

$$R_2CHONO + H_2O \rightarrow R_2CHOH + HNO_2$$

$$2\,HNO_2 \rightarrow H_2O + NO + NO_2$$

Hydrolysis of secondary nitrite esters in the presence of titanous sulfate was verified with secondary butyl nitrite; nitrogen trioxide was evolved, and no oxidation of titanous salt occurred. In a separate series of experiments, it was found that the reduction of nitroglycerin occurs in at least two steps, one corresponding to the partial reduction of the secondary nitrate group followed by hydrolysis, and the other to reduction of the two primary groups to ammonia; the first inflection in the potentiometric titration curve of nitroglycerin with titanous sulfate suggests that the secondary group is attacked first. Accordingly, the reduction, which involves 18 equiv. of titanous ion/mol of nitroglycerin, is formulated as follows:

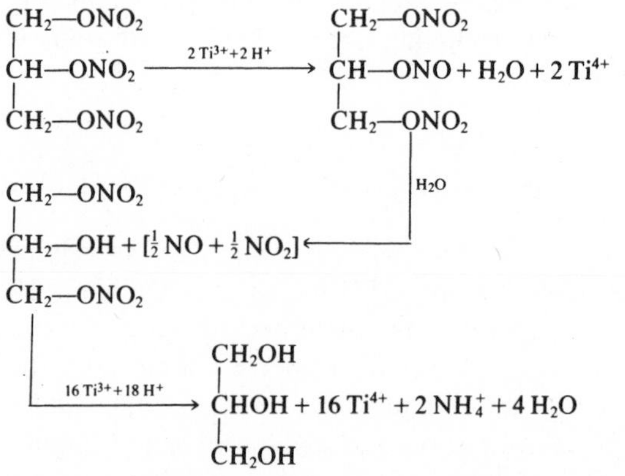

$$
\begin{array}{l}
CH_2-ONO_2 \\
| \\
CH-ONO_2 \\
| \\
CH_2-ONO_2
\end{array}
\xrightarrow{2\,Ti^{3+}+2\,H^+}
\begin{array}{l}
CH_2-ONO_2 \\
| \\
CH-ONO + H_2O + 2\,Ti^{4+} \\
| \\
CH_2-ONO_2
\end{array}
$$

$$
\begin{array}{l}
CH_2-ONO_2 \\
| \\
CH_2-OH + [\tfrac{1}{2}\,NO + \tfrac{1}{2}\,NO_2] \xleftarrow{\quad H_2O\quad} \\
| \\
CH_2-ONO_2
\end{array}
$$

$$
\xrightarrow{16\,Ti^{3+}+18\,H^+}
\begin{array}{l}
CH_2OH \\
| \\
CHOH + 16\,Ti^{4+} + 2\,NH_4^+ + 4\,H_2O \\
| \\
CH_2OH
\end{array}
$$

Fainer's (72) work also included a study of the reaction of primary nitrite esters with titanous sulfate under reflux; with $n$-butyl nitrite, the reaction appears to be quantitative with formation of ammonia:

$$RONO + 6\,Ti^{3+} + 7\,H^+ \rightarrow ROH + 6\,Ti^{4+} + NH_4^+ + H_2O$$

Thus, even if nitrites are intermediates in the reduction process for primary nitrate esters, the end products are the same.

Oldham (213) treated from 0.04 to 0.1 gram of sample dissolved in 10–15 ml of hot glacial acetic acid with an excess of 0.1 $N$ titanous sulfate; the solution was boiled for 20 min (no condenser), and the excess titanous titrated with ferric alum (methylene blue, C.I. 52015). Shankster and Wilde (265) added 25 ml of 0.05 $N$ titanous chloride to a solution of about 10 mg of nitroglycerin in 10 ml of methanol, and set the mixture aside for 10 min. Then the mixture was boiled for 10 min and cooled in ice for 10 min. The excess titanous ion was titrated with 0.05 $N$ ferric alum. Fainer's procedure is as follows (72):

Prepare a solution of about 0.2% nitroglycerin in acetic acid by transferring a known amount to a glass-stoppered volumetric flask and diluting to the mark with 70% acetic acid. Pipet a 10-ml aliquot of the nitroglycerin solution into a flask equipped with a side arm, and expell the air by a stream of carbon dioxide; maintain an atmosphere of carbon dioxide in the flask for the remainder of the determination. Add 20 ml of 0.2 $N$ titanous sulfate from a buret protected by carbon dioxide and then boil the contents of the flask under reflux for 10 min; cool to room temperature. Titrate the excess of titanous ion with 0.15 $N$ ferric alum; toward the end of the titration, add 5 ml of 5% ammonium thiocyanate as indicator.

Fainer's method differs from that of Shankster and Wilde in that 70% acetic acid is used instead of methanol to dissolve the sample.

The determination of nitrate ester groups in nitrocellulose by titanous chloride reduction methods is not satisfactory because of the insolubility of cellulose esters in general, especially when the nitrogen content is more than 13%. In fact, cellulose esters of high nitrogen content are not dispersed by concentrated sulfuric acid, and so the results of analyses by the nitrometer method (see below) are often too low. Stalcup and Williams (277) studied transnitration as a possible route for quantitative removal of the nitrate ester groups of nitrocellulose; in essence, transnitration involves strong-acid cleavage of the nitrate ion from the nitrate ester, and $C$-nitration of a sensitive compound by the liberated nitric acid. Salicylic acid was selected as the sensitive compound. Typically, low-nitrogen nitrocellulose samples (or other samples containing nitrate esters of nitramines, such as nitroguanidine, which release nitric acid in strong acids) are suspended in a mixture of sulfuric, salicylic, and acetic acids and allowed to remain at room temperature for 20 min. Nitrocellulose samples containing more than 13% nitrogen are suspended in concentrated sulfuric acid; after 20 min, a mixture of sulfuric, salicylic, and acetic acids is added and the whole allowed to stand for 10 min longer. Nitration of salicylic acid produces 5-nitrosalicylic acid (with some 3-isomer); the nitration residue is transferred to a reflux flask, and an excess of titanous chloride is added (inert gas blanket). After refluxing for about 1 min, the excess titanous ion is titrated with ferric alum, using ammonium thiocyanate as indicator.

Reductions of nitrate esters directly with titanous solutions do not appear to be stoichiometric; it is probable that small amounts of nitric acid are formed during the time that the reducing agent is allowed to remain in contact with the sample (at elevated temperatures), and that the nitric acid is reduced to a variety of products. For example, in dilute aqueous solution strongly acidulated with hydrochloric acid, reduction of nitric acid leads to a mixture of nitric oxide and nitrous oxide; in fact, it is claimed that, in acid concentrations greater than 0.1 $N$, nitrate ion is reduced via a four-electron change to nitrous oxide (31). If a primary aromatic amine is present while the titanous ion/nitric acid ratio is 2:1, nitrous acid is formed in the cold, and the amine is diazotized (139).

### b. FERROUS REDUCTIONS

Reductions with ferrous ions may be made to take place in dilute acids or in concentrated acids.

### (1) Dilute Acids

A large number of workers have found that reduction of nitrate esters by a moderate excess of ferrous ion proceeds smoothly and yields results which appear to be stoichiometric (±2 parts per thousand):

$$RONO_2 + 3\ FeCl_2 + 3\ HCl \rightarrow ROH + 3\ FeCl_3 + NO + H_2O$$

It is of interest to compare the above equation with one given in the preceding section for the reaction of titanous ion with a primary nitrate ester. The more powerful reducing agent (titanous) converts the nitrogen atom of the nitrate ester group to ammonia.

In general, samples are dissolved in acetic acid, rendered strongly acidic with hydrochloric acid, treated with an excess of ferrous chloride, and refluxed; the remaining ferrous ion (or the ferric ion produced) is titrated with a standard solution. Kolthoff et al. (143) studied the volumetric determination of nitrates (inorganic) with ferrous sulfate. Dickson and Easterbrook (61) and Huff and Leitch (121) were among the earlier workers who used to advantage the reducing action of ferrous salts on nitrate esters. Huff and Leitch back-titrated the excess ferrous salt with permanganate. Dickson and Easterbrook found that, when nitrate esters are treated with ferrous sulfate, any nitroaromatic compounds that are in the reaction mixture will be nitrated. Thus the authors report that mixtures of nitrotoluene and nitroglycerin give low values; on the other hand, a commercial mixture of dinitrotoluene, trinitrotoluene, and dinitrobenzene did not appear to be so sensitive to the nitrating action. Similarly, Becker (13) reported excellent results with mixtures of nitroglycerin and dinitrotoluene in a procedure essentially similar to one used by Knecht and Hibbert (139) for determination of nitric acid, that is, reduction with ferrous iron and titration of the resulting ferric iron with a standard solution of titanous chloride. Becker's finding that reductimetric titration of the resulting ferric iron is preferable to oxidimetric titration of the excess ferrous iron has been widely adopted for analysis of nitrate esters.

Ferrous–titanous titration procedures generally involve refluxing the sample with an excess of ferrous chloride and titration of the ferric iron by titanous chloride; the following reaction occurs:

$$R—ONO_2 + 3 FeCl_2 + 3 HCl \rightarrow ROH + 3 FeCl_3 + NO + H_2O$$

Details of the ferrous–titanous titration are given in Section VII; the procedure is useful when applied to nitrate esters that are soluble in the reaction mixture. Modifications of the original Becker procedure include use of ferrous ammonium sulfate as the reducing agent (e.g., Mil-Std 286B).

Many simple dinitroaromatic compounds are not reduced by ferrous chloride when some ferric chloride is present, and thus nitrate esters such as nitroglycerin and ethylene glycol dinitrate can be satisfactorily determined in admixture with 2,4-dinitrotoluene (13); moreover, nitro compounds can also be determined in the same sample. After nitrate ester groups have been reduced with ferrous chloride by boiling for 5 min and the resulting ferric ion has been titrated with titanous chloride, a large excess of titanous chloride is added and the mixture is refluxed to reduce the nitro compound; the remaining titanous chloride is back-titrated with ferric alum. Similarly, Stanek and Vacek (278) have reported that pentaerythritol tetranitrate (PETN) can be preferentially reduced by ferrous chloride in the presence of cyclo-trimethylenetrinitroamine (RDX); after reduction of the nitrate ester and titration of the resulting ferric iron with titanous chloride, the nitroamine is reduced by refluxing with an excess of titanous chloride in sodium citrate buffer, and the excess titanous is back-titrated with ferrous ammonium sulfate.

When the procedure is performed as outlined above and applied to the determination of nitroglycerin in propellants, stabilizers such as ethyl centralite and diphenylamine do not interfere; however, the ferrous–titanous procedure was found to give low and erratic results for the nitrate ester group content of nitrocellulose in lacquers. Under the most favorable conditions, results were concordant but about 0.05 unit lower than nitrometer value (lacquer-grade nitrocellulose contains about 12% N). Shaefer and Becker (264) claimed that the inclusion of hydrobromic acid in the ferrous reagent provided results concordant within a few hundredths of 1% with those obtained by nitrometer analysis. The reducing reagent is an approximately $2 N$ ferrous chloride solution containing 80 ml of 37% hydrochloric acid and 20 ml of 40% hydrobromic acid/liter.

Application of the Shaefer–Becker ferrous–titanous titration (HBr) procedure to the determination of the nitrogen content of propellants containing nitrocellulose often was found to give unsatisfactory results because of incomplete reduction of the nitrocellulose (high nitrogen content), which precipitated in the reaction mixture. Grodzinski (96) employed a mixture of equal volumes of acetic anhydride and acetic acid as a solvent; thus the greater part of the water introduced with the ferrous chloride reagent is fixed by the acetic anhydride, and precipitation of nitrocellulose from solution is prevented. It was also found advantageous to add hydrobromic acid to the reaction flask rather than to introduce it via the ferrous reagent solution (ferrous solutions that do not contain hydrobromic acid can be reduced by treatment with iron). In spite of the improvements in technique introduced by Grodzinski, when ferrous chloride solution containing hydrobromic acid is reduced, somewhat low and discordant results are obtained for the nitrogen contents of nitrocellulose propellants (14.6% N) containing stabilizers.

Pierson and Julian (226) modified and extended the ferrous–titanous method of Shaefer and Becker so that satisfactorily accurate and precise results can be obtained with a wide variety of nitrocellulose samples. Nitrocellulose or propellant residues that are not completely soluble in sulfuric or acetic acids can be analyzed; in the procedure, it is unnecessary for all nitrocellulose to be in solution at the beginning of reduction with ferrous solution. The procedure provides excellent results (±2 parts per thousand with nitrocellulose samples containing 13.4% nitrogen. The reduction of nitrocellulose is made to take place in a refluxing two-phase system; an upper layer of $n$-butyl acetate contains most of the nitrocellulose in solution, and a lower aqueous layer contains the ferrous reductant. Even though some of the nitrocellulose in the upper layer is not in solution, it is gradually reduced as refluxing continues. Early during the refluxing period, ferrous iron migrates from the lower layer into the upper one, and reduction occurs immediately. The upper layer rapidly turns dark green, and the lower layer becomes lighter in color. The color changes from light to dark green, followed by paling to yellow, are associated with the desirable type of redox reactions for the nitrate–ferrous systems. The concentration of hydrochloric acid in the two-phase system is critical and must be controlled; in high concentrations of hydrochloric acid, the reaction mixture very quickly turns yellow (signifying completion), but recovery values will be low and erratic. Pierson and Julian were not able to show that hydrobromic acid is· an adjuvant in the ferrous chloride solution; in fact, satatistical analysis of results obtained with and without hydrobromic acid suggested that its presence in the two-phase reagent system impairs accuracy and precision.

The solvent selected for the two-phase reduction system, $n$-butyl acetate, is highly soluble in glacial acetic acid, is relatively insoluble in cold water, and is an excellent solvent for nitrocellulose. In the two-phase system, the density of an acetic acid–$n$-butyl acetate solution is less than that of the ferrous solution; thus, since nitrocellulose is wetted by the solvent layer and tends to remain dissolved in it, undissolved particles of the nitrate ester stay in the solvent layer and so are kept away from the heated bottom of the reflux flask. Details of the Pierson–Julian method for analysis of nitrocellulose are given in Section VII.

Lambert and DuBois (162) have found that the disodium salt of ethylenediaminetetraacetic acid (EDTA) can be used as a titrant for determination of the ferric ion formed in ferrous reductions of nitrate esters. The advantages over the use of titanous titrants are manifold. For exa ple, EDTA solutions need not be stored in an inert atmosphere and standardized daily; they do not reduce or react with most organic residues and standard solutions may be prepared readily from accurately assayed material or by dilution of stronger solutions. Unfortunately, most colorimetric internal indicators such as Eriochrome Black T and Xylenol Orange cannot perform effectively because of interference from ferrous or ferric ions, as well as for a number of other reasons; thus it was necessary to use a potentiometric end-point procedure (Pt–SCE), but even this was found to be sensitive to pH. The recommended procedure is as follows:

Place a sample containing from 0.1 to 0.15 gram of nitroglycerin in a 500-ml reduction flask (standard taper top and a side inlet for gas). Add 25 ml of glacial acetic and 25 ml of 6 $N$ hydrochloric acid. Purge with nitrogen to remove air; add 25 ml of 0.7 $N$ ferrous ammonium sulfate solution, and then reflux the mixture until the sample changes from yellow-orange to reddish brown and then back to yellow-orange (5–10 min). Cool, and then transfer the mixture to a 250-ml tall-form beaker equipped with a rubber stopper carrying platinum and calomel electrodes, a glass electrode, nitrogen gas inlet and outlet tubes, and a buret tip. Adjust the pH of the mixture to 2.5 with 30% sodium hydroxide solution. Then titrate the ferric ion content of the mixture with 0.20 $M$ EDTA solution.

The authors claim that their procedure is at least as accurate as the Shaefer–Becker method procedure when potentiometric end points are used for both methods.

### (2) Concentrated Sulfuric Acid

Nitrate ester groupings are readily reduced by ferrous ion in cold, concentrated sulfuric acid because they are removed as nitric acid, which will react with ferrous sulfate as follows:

$$4\,FeSO_4 + 2\,HNO_3 + 2\,H_2SO_4 \rightarrow N_2O_3 + 2\,Fe_2(SO_4)_3 + 3\,H_2O$$

When a trace of ferrous sulfate is present in excess, there appears the characteristic pinkish brown coloration of the classical brown-ring test for nitrates. The method for titration of nitrates with ferrous sulfate in arsenic, phosphoric, or sulfuric acid was described by Bowman and Scott (26) and modified by Johnson (126a); some details of the titration method are given in Section VI.A.1.b of "Nitro and Nitroso Groups." Simple nitrate esters dissolve readily in cold sulfuric acid and thus can be titrated readily; however, esters such as nitrocellulose, of high nitrogen content, often are not dispersed, and erroneous results are obtained. Mitra and Srinivasan (202) were able to determine the nitrate nitrogen content of nitrocellulose by slow titration with ferrous sulfate at 15°C; Leclercq and Mathe (166) also were successful with nitrocellulose, which dissolves within 2 hr in 95% sulfuric acid at 0°C. When samples do not dissolve in sulfuric acid at 0°C, the method cannot be applied; moreover, low values for nitrate nitrogen will be obtained if the titration is performed at higher temperatures (e.g., 30–40°).

Satisfactory analysis of simple nitrate and nitrite esters can be achieved by the procedure of Murakami (208); the ester is reduced with ferrous iron, and the resulting ferric ion is titrated with stannous chloride.

c. STANNOUS REDUCTIONS

Nitrate ion is quantitatively reduced by stannous salts to hydroxylamine:

$$NO_3^- + 3\,Sn^{2+} + 7\,H^+ \rightarrow NH_2OH + 3\,Sn^{4+} + 2\,H_2O$$

The reduction is made to take place in an atmosphere of carbon dioxide or other inert gas. Samples are usually boiled in hydrochloric acid solutions containing an excess of stannous chloride, and the excess reducing agent is then back-titrated with ferric chloride, using a thiocyanate colorimetric end point (207), but potentiometric determinations of end points are also satisfactory. Following is a typical procedure:

Into a 300-ml conical flask fitted with a rubber stopper having gas inlet and outlet glass tubings, place a solution of the sample containing no more than 90 mg of $HNO_3$ (or its equivalent). Pass a rapid stream of inert gas into the flask to displace air, and then add 50.00 ml of 0.2 N $SnCl_2$ (in 3 N HCl); add enough hydrochloric acid to the flask to make the contents 7 N in HCl. Reduce the rate of flow of inert gas to 50 ml/min, and heat the contents of the flask to a gentle boil; boil the contents for 15 min to complete the reduction. Increase the rate of flow of the inert gas, and let the boiling subside. Titrate the solution in the flask with 0.2 N ferric chloride; keeping the temperature of the solution higher than 60°C. The first appearance of yellow ferric ion can be used as an indication of the end point, but thiocyanate is preferred as the indicator.

The procedure gives good results with nitrocellulose samples.

### d. Chromous Reductions

An excess of chromous ion reacts slowly with nitrate ion (and nitrite ion) (179); titanium (IV) sulfate appears to catalyze the reaction, and it is possible to titrate directly at 50°C with a bright platinum indicator electrode and a saturated calomel reference electrode (in an inert atmosphere).

### e. Reduction by Other Titrants

An excess of hypovanadous ammonium sulfate reacts with nitrate ions; thus it is anticipated that nitrate esters also should be reduced by this reagent, inasmuch as hypovanadous ion has a reducing potential that is between the values for chromous and titanous ions. For determination of nitrate ions, an excess of hypovanadous ammonium sulfate (9) is used in an acidic medium, and the excess hypovanadous ion is titrated with permanganate or ferric solution to a potentiometric end point (10,67). The reaction with nitrate ion is

$$NO_3^- + 8\, V^{2+} + 10\, H^+ \rightarrow NH_4^+\, 8\, V^{3+} + 3\, H_2O$$

Oxalic acid in the presence of manganese salts can reduce nitrate ion either to nitric oxide or to nitrous oxide, depending on the acidity of the reaction medium. Reduction to nitric oxide is favored at sulfuric acid concentrations greater than $1:5$:

$$2\, NO_3^- + 3\, C_2O_4^{2-} + 8\, H^+ \rightarrow 2\, NO + 6\, CO_2 + 4\, H_2O$$

Typically, procedures employ an excess of sodium oxalate in a solution containing about 5 grams of manganous sulfate and 12 ml of concentrated sulfuric acid in a volume of 100 ml; after refluxing the sample in the mixture, the excess of oxalic acid is determined by classical permanganometric methods.

Sulfamic acid can be made to react with nitrate ion in solutions of high acidity. Reaction conditions must be properly controlled to prevent interference of nitrites and ammonium salts, but it is also possible to adapt the method to the simultaneous determination of both nitrate and nitrite ion (91). The procedure has not been used for analysis of nitrate and nitrite esters.

### 2. Reduction with Hydriodic Acid

Nitrate esters are reducible by hydriodic acid; the method of Aldrovandi and De Lorenzi (1), in which the ester is sealed in a tube together with concentrated hydriodic acid (57%) and heated at 100–300° for a period of time, obviously will lead to a release of iodine equivalent to the number of nitrato (and nitrito) groups present. The liberated iodine is determined titrimetrically. However, the effect of free iodine (at elevated temperature) on the hydroxy compound formed by the reaction must be taken into consideration; in many instances, iodination of the hydroxy derivative may take place, and complex structures may be degraded. There is also a possibility that alkyl iodides may be formed (40). Details of the practical aspects of the method were presented in preceding sections.

Red phosphorus and hydriodic acid constitute a particularly powerful combination for reducing nitrate esters.

### 3. Reduction by Metals and Amalgams

According to Scoville (260), a Swedish chemist (Binz, 1905) first proposed to determine a small amount of nitroglycerin in pharmaceutical preparations by saponifying it with alcoholic potash, reducing the nitrate, cyanide, and so on thus formed to ammonia by a zinc-acid couple, and estimating the ammonia by distillation; the procedure was intended to circumvent difficulties with the methods for nitrate ester determinations in use at that time, namely, (1) extraction of the ester with chloroform or ether and evaporation of the solvent, and (2) saponification of the ester by alcoholic potash. Thus the earliest methods for the direct estimation of nitroglycerin in cordite (contains nitrocellulose) involved an ether extraction of nitroglycerin; however, it was soon found that ether cannot be removed without loss of the ester. To overcome this difficulty, Silberrad et al. (270) proposed determining the nitroglycerin content of the ethereal extracts by a mixture of 2 parts of zinc and 1 part of reduced iron in a 40% caustic soda solution to reduce nitrate groups to ammonia (passage of a slow stream of air through the mixture assisted transferral of ammonia). It is plausible to expect that reductive cleavage of the esters can take place whenever metal–acid or metal–base couples (and amalgams) come into contact with nitrate and nitrite esters, because hydrogenation has been shown to be an effective method for removal of the ester group [see Section III.B.1.a(6) of "Nitro and Nitroso Groups"]. The Silberrad procedure was considered a standard method for nearly 40 years, but it was noted that more refractory esters (e.g., nitrocellulose of high nitrogen content) were not cleaved quantitatively on reduction with metal couples and that even some of the simpler esters often were troublesome.

Analytical procedures that employ metals as reducing agents for nitrate and nitrite esters are of two general types; both depend on formation of ammonia and its titrimetric determination:

1. A metal-base couple (e.g., Devarda's alloy or "ferrum reductum") is used to convert ester nitrogen directly to ammonia.

2. Ester nitrogen is converted to nitrate ion by saponification in alkaline peroxide, and the inorganic ions formed by hydrolysis are converted to ammonia by Devarda's alloy.

### a. DEVARDA'S ALLOY

Methods for the direct reduction of nitrate and nitrite esters to ammonia by Devarda's alloy (50% Cu, 45% Al, 5% Zn) in alkaline solution are adaptations of Devarda's original method (56,57) for determination of nitrate ion (and nitrite ion). The ammonia is distilled into a given volume of standard acid, and the excess acid titrated with strong base (methyl red); an efficient scrubbing bulb is necessary to prevent carryover of sodium hydroxide by frothing and by the small droplets which are formed during decomposition of the alloy. Reduction by Devarda's method was one of the first analytical procedures used for nitrate esters; Koehler et al. (140) were among the early workers who employed the Devarda procedure for analysis of nitrocellulose and nitroglycerin, and a form of the Devarda reduction procedure was the official U.S.P. method for the determination of nitroglycerin in vasodilator tablets up to U.S.P. XVI. Cannon and Hewermann (35) recommend that nitroglycerin be extracted in ether and reduced with Devarda's alloy in alcoholic potassium hydroxide; the ammonia is collected in ice water and titrated directly with standard acid. It is claimed that alcoholic potash prevents foaming and frothing.

The Devarda reduction procedure may be used for the ultramicro-determination of nitrate esters in concentrates obtained by passing contaminated atmospheric air through solvents (154).

As noted earlier, saponification of nitrocellulose often forms a variety of products; however, Bush (32), Utz (297), and later Kruger (153) and Mutakami (209) have shown that saponification of nitrocellulose (and other esters) by strong alkali produces nitrate ion quantitatively when hydrogen peroxide or another powerful oxidant is present in excess. As a result, when the saponification mixture is subsequently treated with Devarda's alloy, formation of ammonia equivalent to the nitrogen content of the nitrocellulose is as complete as the Devarda method will allow. [Lacroix et al. (161) have recorded results of a statistical study of the Devarda method; nitrogen values of ±0.03% absolute are given as a limit.] Moreover, nitrocellulose samples are thoroughly disintegrated by the combined attack of alkali and oxidant; in fact, subsequent work has demonstrated that the success of an analytical method is in large measure determined by the solubility or dispersibility of the sample in the solvent system employed in the analytical procedure. [Frehden and Nicolescu (83) have confirmed that nitrocellulose is completely disintegrated by alkaline oxidation.]

In typical procedures, 50–100 mg of nitrocellulose is mixed with about 15 ml of 30% sodium hydroxide and 2 ml of 30% hydrogen peroxide and heated by steam before treatment with Devarda's alloy and additional caustic.

Mutakami (209) suggests dissolution of nitrocellulose in 50% potassium hydroxide solution and oxidation with 30% hydrogen peroxide at 60–70°C; when oxygen is no longer given off, nitrate ion is reduced by Devarda's alloy to ammonia, which is driven off into standard acid by passing purified air through the reaction mixture at 60–70°C and a pressure of 100–300 mm of Hg. It is claimed that spraying of caustic into the receiving flask is minimized when distillation is performed at pressures less than 1 atm; an improved form of scrubber is recommended.

Many comparisons have been made of the relative accuracies obtained by Devarda methods and by other methods for analysis of the nitrogen content of nitrocellulose; Verschragen (298) has reported characteristic values for Devarda reductions with and without peroxide oxidation and for nitrometer values and results obtained by ferrous chloride reductions on the same samples. Cope and Barab (49) and Cope and Taylor (50) have compared results obtained by the Devarda method on a variety of nitrate esters with those obtained with the nitrometer, the nitron gravimetric method, and the Dumas method. It is interesting to note that very few reports on such comparisons provide results obtained from statistical evaluations of well-designed interlaboratory testing programs, and it may be surmised that personal preferences are often reflected in conclusions drawn from data.

Muraour's method (210) for reduction of nitroglycerin is elaborate and operates under mild conditions; decomposition of nitroglycerin in acetone is begun with a mixture of 3.6% hydrogen peroxide in dilute alkali at 30–35°C and then completed with sodium perborate. Ammonia is formed with Devarda's alloy. The procedure is given in detail in Section VII.D; it is especially useful for analysis of all but the most refractory nitrate esters (e.g., high-nitrogen-content nitrocellulose). Becker (13) found that results obtained by Muraour's method closely confirmed theoretical values of nitroglycerin, but the time required for a determination (overnight) was considered too long for the needs of control laboratories.

b. OTHER METAL COUPLES

Reduction of nitrate esters has also been effected by gray impalpable aluminum power (263) (this is not the bright powder used in varnishes) and by a combination of zinc dust, ferrous sulfate, and 40% sodium hydroxide solution (109). As noted in the preceding section, Silberrad et al. (270) used a mixture of 2 parts of zinc and 1 part of reduced iron in a 40% caustic soda solution. Schraiber and Rubinshtein (267) have claimed satisfactory accuracy for the determination of nitroglycerin in pills (without prior extraction) by reduction with aluminum filings in alcoholic sodium hydroxide containing cupric hydroxide. Raney nickel in alkaline solution also appears to be effective for the determination of nitroglycerin in tablets (271).

The method of Ulsch (293,294) is based on the reduction of nitrate ion with "ferrum reductum" in acidic media; the iron powder must be quite pure (186). The procedure as set forth by Ulsch seldom gives precise results; however, Erdey et al. (70) have devised a new procedure in which nitrate ions are reduced by refluxing with a five- to ten-fold excess of iron powder in 2 N sulfuric acid containing 0.3 gram of nickel or cobalt sulfate as catalyst. The ammonia that is formed by reduction of nitrate ion is determined gravimetrically by precipitation with tetraphenylboron or titrimetrically after distillation in the usual manner.

## B. GRAVIMETRIC PROCEDURES

The classical gravimetric procedure for determination of nitrate ions by precipitation as the nitron salt can be applied to nitrate and nitrite esters after they have been saponified in the presence of an oxidant such as hydrogen peroxide.

A typical procedure for saponification of nitrite and nitrate esters is as follows:

Place a 0.2–0.3 gram sample of the ester in a 125-ml Erlenmeyer flask; add 5 ml of sodium hydroxide solution (40 grams/100 ml total volume), and follow with 10 ml of 3% hydrogen peroxide (preferably 1 ml of 30%). Heat the mixture carefully on a water bath; continue increasing the temperature of the mixture, and keep it on the water bath until frothing ceases. Then heat the mixture over a flame until decomposition is complete. Add 40–50 ml of water and 10 ml of 3% hydrogen peroxide; heat to 50°C. With a pipet, introduce 20 ml of 10% sulfuric acid at the bottom of the flask; carefully cause the solutions to mix. Be sure the final mixture is acidic.

The solution obtained from saponification of the ester contains nitrate ion, which may be precipitated by nitron (1,4-diphenyl-3,5-endanilo-4,5-dihydro-1,2,4-triazole, $C_{20}H_{16}N_4$). Any specific procedure for nitron precipitation may be used, but in general a 10% (w/v) solution of nitron in 5% (w/v) acetic acid is added in excess to a slightly acid, hot solution of nitrate; after cooling the mixture and allowing to stand overnight, it is filtered. The precipitate of nitron nitrate, $C_{20}H_{16}N_4 \cdot HNO_3$, is washed with ice water and dried to constant weight at 100–110°C (2–3 hr). Inasmuch as some other nitron salts are also quite insoluble, nitrite, bromide, iodide, chromate, chlorate, perchlorate, thiocyanate, ferrocyanide, ferricyanide, picrate, and oxalate ions must be absent or be removed before determination of nitrate; fortunately, nitrite and a number of the above ions are destroyed by attack from the hot alkaline peroxide solution used in saponification of the ester. Large amounts of chloride ion should not be present.

The nitron gravimetric procedure was studied by Cope and Barab (49) and by Cope and Taylor (50), and results obtained with it were compared with values obtained by the nitrometer method, Devarda's method, and the classical Dumas procedure. Results of the comparison suggest that in most instances the nitron method may give values that are lower by 0.1–0.2% absolute (nitrometer values as the standard); however, from a practical viewpoint, the nitron method is to be recommended for occasional analysis of nitrate (or nitrite) esters. Moreover, the nitron gravimetric method (alkaline peroxide saponification followed by nitron precipitation) does not require special apparatus or special skills for its execution.

Nitrates can also be precipitated by the dicyclohexylthallic ion according to Hartman and Bathge (103). The reagent is a solution of dicyclohexylthallic carbonate, $[(C_6H_{11})_2Tl]_2CO_3$; unfortunately, a wide variety of anions also form precipitates with the reagent.

### C. SPECTROPHOTOMETRIC PROCEDURES

Very few spectrophotometric methods have been developed for the direct determination of nitrate or nitrite esters; in the main, when spectrophotometric methods have been described, esters are decomposed to nitrate and nitrite ions and these ions are used as the basis for development of colored species that can be measured in the visible or ultraviolet. Infrared absorptiometric methods have only recently been improved to provide acceptable results for routine analysis where values must be in close agreement with those obtained by standard methods (e.g., the nitrometer method).

### 1. Infrared

The nitrite and nitrate esters usually have clear-cut absorptions in the regions about 6.1 $\mu$ (1639 $cm^{-1}$), 7.8 $\mu$ (1280 cm$^{-1}$), and 11.9 $\mu$ (840 cm$^{-1}$) which may be used for quantitative analysis. Thus simple esters and polyesters such as erythritol tetranitrate, pentaerythritol tetranitrate, mannitol hexanitrate, and glyceryl trinitrate are readily determined by measurement of their absorption in the 6.1-$\mu$ region. As a rule, the esters are seldom found neat, and so it is necessary to use separation procedures; for example, Carol (36) has described a group of typical procedures for extraction of mannitol hexanitrate, pentaerythritol tetranitrate, and erythritol tetranitrate from pills before the use of infrared spectrometry. When phenobarbital is absent, the esters are extracted by chloroform from an acidulated aqueous suspension of ground pills; if phenobarbital is present, it is removed from the chloroform extract by a chromatographic column packed with Celite which has been treated with tripotassium phosphate. Absorbance of the chloroform solutions in the 6.1-$\mu$ region is used to determine the content of esters from a calibration curve; potassium bromide pellets may also be prepared from residues obtained by evaporation of aliquots of the chloroform solutions. Woo et al. (307) have described a similar procedure for quantitative analysis of isosorbide dinitrate, (1,3,4,6-dianhydro-D-glucitol-2,5-dinitrate).

Rosenberger and Shoemaker (243) have found the 11.9-$\mu$ (840 cm$^{-1}$) region suitable for the determination of nitrocellulose in admixture with other cellulose resins. As is characteristic of many procedures for determining nitrocellulose in mixtures, the nitrate ester is extracted from other substances in a solvent such as acetone or methyl ethy ketone; obviously, the extraction procedure and the analysis are applicable only to nitrocellulose of about 12% nitrogen, because the determination of the nitrate ester is made in acetone solution (dispersion). In earlier work reported by Kuhn (155), attempts were made to measure the absorption of cellulose nitrate films obtained by evaporation of ethyl acetate solutions; absorbance in the 6-$\mu$ region was almost complete, and so it was suggested that the determination could be performed by measuring the ratio of the hydroxyl band at 3 $\mu$ to the C–H band at 3.5 $\mu$.

Levitsky and Norwitz (174) have reported the development of an accurate procedure for determination of the nitrogen content of nitrocellulose; absorbance of a solution of the ester is measured at one of the three strong bands of the nitrate ester group, the 6.1-$\mu$ (1639-cm$^{-1}$) band. The solvent must have low absorbance in the region of strong absorption by the nitrate ester group; of the solvents ordinarily used for nitrocellulose (esters, acetone, methyl ethyl ketone, cyclohexanone, dioxane, methanol, nitrobenzene, nitroethane, propylene oxide, pyridine, mixtures of ethanol and ether, and tetrahydrofuran), only acetone and tetrahydrofuran merit consideration. Acetone has a transmittance of 74% (0.2-mm cells) at 11.9 $\mu$ and cannot be used at 6.0 or 7.8 $\mu$; on the other hand, tetrahydrofuran has a transmittance of 87% (0.2-mm cell) at 6.0 $\mu$, and (of practical importance) the commercial high-purity product contains only 0.02% water. The procedure requires preparation of a 0.78% solution of nitrocellulose sample in tetrahydrofuran; the solution is allowed to stand overnight before its absorbance is measured in the region 5.80–6.05 $\mu$ (1724–1653 cm$^{-1}$) at a definite temperature. The calibration plot is a straight line between 12 and 13.6% nitrogen (the range of nitrogen found in the esters used for ammunition and explosives). Presumably, another calibration curve must be prepared for nitrocellulose of nitrogen contents between 10.5 and 12.0%; however, this is seldom necessary because esters of lower nitrogen content are of limited commercial importance (they are difficultly soluble in organic solvents). Over the range of 12.24–13.27% nitrogen, standard deviations averaged-about ±0.09%; results were not affected by the source of the cellulose used to prepare the ester. Clarkson and Robertson (45) later showed that more refined calculations than those indicated by Levitsky and Norwitz (1974) can provide results that approach those of the nitrometer in accuracy and reproducibility. With the improved method of computation, a standard deviation of 0.03% is obtainable; this compares favorably with the deviation obtained by experienced operators using the nitrometer method.

The methods discussed above have dealt only with the determination of a single nitrate ester. However, ir spectrophotometry often can be used for determination of one nitrate in the presence of another; for example, Pinchos (227) has reported the determination of diethylene glycol dinitrate and nitroglycerin in acetone solution. Both substances have a common absorption band in the 7.85-$\mu$ region [nitroglycerine at 7.86 $\mu$ (1273 cm$^{-1}$) and diethylene glycol dinitrate at 7.83 $\mu$ (1278 cm$^{-1}$), corresponding to the 7.80-$\mu$ (1283-cm$^{-1}$) absorption of nitrocellulose]. The dinitrate also shows an absorption peak at 8.77 $\mu$ (1140 cm$^{-1}$), characteristic of an ether group. Thus, with instruments of appropriate resolving power, it is possible to show that the nitrate band in the 7.85-$\mu$ (1274-cm$^{-1}$) region consists of two peaks when diethylene glycol dinitrate is present along with nitroglycerine; the amount of dinitrate can be estimated from its absorption at 8.77 $\mu$ (1140 cm$^{-1}$), and its contribution to absorption in the 7.85-$\mu$ region can be computed to obtain the residual absorption assignable to nitroglycerin.

## 2. Ultraviolet-Visible

The spectral characteristics of nitrate and nitrite esters in the ultraviolet and visible regions are seldom used as the basis for quantitative determinations. By and large, the esters are first hydrolyzed under controlled conditions to form nitrate or nitrite ions, and these inorganic species are then made to provide colorations which can be used for the indirect quantitative determination of the esters. Unfortunately, the majority of spectrophotometric methods for the analysis of nitrate and nitrite esters are useful only for the assay of small quantities of relatively pure esters; although most esters can usually be extracted from materials that interfere with the determination of nitrate and nitrite ions in hydrolysates, the extracts are often accompanied by traces of substances that interfere with colorimetric procedures.

It is emphasized that colorimetric procedures for nitrate and nitrite esters lack accuracy, generally providing values to ±5% absolute (at best, ±2%), and thus cannot be recommended for assay; the colorimetric methods are used for the determination of less than about 10 mg of nitrate ester. Methods for removal of nitrite from nitrate have been devised by Seaman et al. (261) and others (78a,126a,292a).

a. DIPHENYLAMINE REAGENT

Diphenylamine in sulfuric acid has long been known as a sensitive reagent for indicating the presence of nitrates by production of a blue coloration due to oxidation of the amine (145); naturally, other strong oxidizing agents such as nitrites, chromates, and ferric salts will interfere (142).

Roberts (242) has made a study of the sensitivity of the diphenylamine test as applied to the determination of the nitrate contents of lacquer films that contain nitrocellulose as the primary ingredient. The time required for a blue coloration of defined intensity to be produced by a single drop of diphenylamine reagent placed in contact with a polymer film is noted; the nitrate nitrogen content of the film is obtained from calibration curves. No single indicator solution is best for all nitrate contents, but a suitable reagent for general applicability consists of 0.1 gram of diphenylamine, 100 ml of sulfuric acid (96%), and 30 ml of water; comparison should be made with standard films.

b. PHENOLDISULFONIC ACID REAGENT

The reaction of nitrate ion obtained from hydrolysis of nitrate esters with an ammoniacal solution of 2,4-disulfonic acid produces an intense yellow color. The method was first used by Scoville (260) for the determination of nitroglycerin in pharmaceutical tablets. The reagent is prepared by treating phenol with fuming sulfuric acid (38); when properly prepared, it is stable for months. For nitrate analysis, the sample is evaporated to dryness, if possible, and the reagent is added; after warming to promote reaction, the mixture is diluted with water and concentrated ammonium hydroxide is added to develop the color (37). If the ester solution cannot be brought to dryness, a solution in glacial acetic acid is prepared and mixed with the reagent; Allert (2) found this technique suitable for determining nitroglycerin in pills, and it was the official method of the 1958 British Pharmacopeia.

The colored product appears to be the triammonium salt of the C-nitro derivative of 2,4-phenoldisulfonic acid (38); the greatest sensitivity of the method is realized when measurements are made at 410 nm (about 0.05 $\mu$g of nitrate nitrogen/ml). Because the absorption peak is broad, measurements can be made with reduced sensitivity at other wavelengths, for example, 70 $\mu$g/ml at 500 nm. Nitrites in excess of 0.2 $\mu$g/ml lead to erratic results and must be removed; alternatively, nitrite is oxidized to nitrate and the reported results corrected appropriately. Instead, sulfamate, urea, or thiourea may be used to destroy nitrite ion; since chloride ion interferes, treatment with silver nitrate is advisable.

The phenoldisulfonic acid method has also been applied directly to the determination of nitrate esters, for example, by Brooks and Badger (29) to cellulose nitrates; for a typical procedure where a solvent solution of nitrocellulose is evaporated to dryness and treated with the reagent, Gardon and Leopold (88) and Taras (284) have shown that reproducible results can be obtained when the pH of the final solution is maintained between 7.2 and 7.6, but it is necessary to use standard samples for preparation of calibration plots and to make correction for any variation in pH. The presence of phenolmonosulfonic acid in the reagent may give spurious results in the presence of very small or very large amounts of nitrate. Hohmann and Levine (118) have applied the phenoldisulfonic color reaction to eluates from a chromatographic column; the chromatographic separation removes nitroglycerin from inorganic nitrates and nitrites, as well as from mono- and dinitrate esters of glycerin.

The phenoldisulfonic acid method has received wide acceptance; thus, about 1950, the official method of the British Pharmacopeia for determination of pentaerythritol tetranitrate in tablets specified use of a phenoldisulfonic acid colorimetric method. Sarnoff (246) has shown that phenobarbital in tablets containing mannitol hexanitrate does not interfere in the direct spectrophotometric determination of the nitrate ester; in later work, it was demonstrated that alkali nitrates may be used as standards when absorbance is measured at 408 nm (247).

c. Xylenol Reagent

The method for determination of inorganic nitrates described by Blom and Treschow (17) is based on formation of 5-nitro-4-hydroxy-1,3-

dimethylbenzene. Of particular interest is the fact that the chromogenic compound can be isolated from other colored substances by means of steam distillation (183). Yagoda (308) demonstrated that the reagent, $m$-xylenol (2,4-dimethylphenol), can be nitrated to the chromogenic compound by the nitric acid which is released when certain nitrate esters are hydrolyzed by 62.5% sulfuric acid at room temperature [Lundgren and Cabback (183) used 65–70% sulfuric acid]; the procedure was checked with nitroglycerin, pentaerythritol tetranitrate, and erythritol tetranitrate, and it was found that an ester could be assayed by comparison with an inorganic nitrate standard (implying quantitative conversion of nitrate esters to nitrate ion under the hydrolytic conditions imposed by the method). Nitrite esters when hydrolyzed in acid solution give nitrous acid, which reacts with the xylenol to yield a variety of steam-distillable colored products that interfere with nitrate ester determinations.

Low concentrations of nitrate esters in air can be determined by the xylenol method (309); however, the reaction of nitrite esters leaves much to be desired, and thus their determination cannot be readily accomplished by the xylenol method.

Holler and Huch (119) determined that, of the six isomeric xylenols, 3,4-xylenol (3,4-dimethylphenol) is the best reagent; it forms 5-nitro-4-hydroxy-1,2-dimethylbenzene from the nitric acid liberated by action of 80% sulfuric acid on nitrate esters. (About 15% of the 3-nitro compound may also be formed.) The recommended concentration range is from 0.10 to 0.35 mg of nitrate nitrogen in 100 ml of solution (for a 1-cm photometric cell). The method was tested with decyl nitrate and $n$-amyl nitrate. When the method was applied to nitrocellulose, it gave low results because of incomplete hydrolysis in 80% sulfuric acid (even after 6 hr of reaction). Holler and Huch claim that stabilizers such as diphenylamine and diethyldiphenylurea (and their derivatives) do not interfere.

As can be anticipated, both nitro isomers formed with 3,4-xylenol (nitro group *ortho* to the hydroxyl group) can be distilled with steam, and so the chromogenic compounds can be effectively removed from many colored interferences. Chlorides and nitrites interfere but can be easily removed. Hydrogen peroxide must be absent, and although it can be readily removed by potassium permanganate, nitrites are oxidized (176); hence it is necessary to remove nitrites first. Hartley and Asai (102) have found 2,6-dimethylphenol to be more sensitive to nitrates, but since the 4-nitro derivative is not codistillable with water, the use of this reagent is seldom of value for analysis of nitrate esters.

## d. FERROUS SULFATE REAGENT

The red-purple color that develops when ferrous sulfate and small amounts of nitrates react in sulfuric acid has been used by Swann and Adams (283) as the basis of a rapid colorimetric procedure for nitrate esters. The chromogen is an addition compound of nitric oxide and ferrous sulfate, $(FeNO)SO_4$ (69). Following is the Swann–Adams procedure for determination of nitrate esters:

Dissolve 0.5 gram of ferrous sulfate heptahydrate in 25 ml of water, and mix with 75 ml of concentrated sulfuric acid; cool before use. Dissolve samples in water, acetone, or a similar solvent; use acetone for nitrocellulose. Take appropriate aliquots of the sample solution containing the equivalent of 0.3–1.5 mg of nitrate ($-NO_3$), place in 25-ml glass-stoppered flasks, and dry in an oven at 60°C (105°C for inorganic nitrates). Add exactly 20 ml of the ferrous sulfate reagent to the dried sample; stopper the flask, and allow to remain for 30 min (90 min for organic materials such as nitrocellulose). Determine absorbance at 520–525 nm; use the reagent as a blank. Use potassium nitrate as a standard for preparing calibration plots.

Bandelin and Pankratz (8) adapted the above procedure to the determination of nitrate esters in pharmaceutical products (vasodilator tablets); sodium sulfite is used to shorten the reaction time, and the absorptivity is obtained at 510 nm instead of the 520–525 nm range prescribed by Swann and Adams. Inasmuch as the nitrate esters (e.g., trinitroglycerin and pentaerythritol tetranitrate) in pharmaceutical products are always in admixture with desensitizing sugars or mannitol, a solvent extraction is employed to remove esters from inactive materials; chloroform is preferred, but acetone must be used for pentaerythritol esters.

Place the solvent extract in a 100-ml volumetric flask, and evaporate to dryness under reduced pressure (steam bath). Add 2 ml of glacial acetic acid, and swirl to dissolve the residue; dilute to the mark with ferrous sulfate reagent. (Prepare the reagent as follows: Dissolve 0.5 gram of ferrous sulfate heptahydrate in 250 ml of water; cautiously mix with 750 ml of concentrated sulfuric acid, and cool to room temperature.) Add 5 mg of sodium sulfite, stopper the flask, and agitate the contents with a magnetic stirrer. If a small quantity of white precipitate develops, remove it by centrifugation. Determine the absorbance of the solution at 510 nm.

An essentially similar procedure has been described by Fursov (86).

Laccetti et al. (160) have described still another modification of the original Swann–Adams colorimetric method; in essence, the concentration of ferrous sulfate is increased to enable determination of greater concentrations of the reacting constituents.

The reagent is prepared by dissolving 10.5 grams of anhydrous ferrous sulfate in 250 ml of boiling water, cooling, and then mixing the turbid solution with 600 ml of concentrated sulfuric acid; the cool solution is diluted to 1 liter with concentrated acid.

Transfer a 0.2500-gram sample to a 250-ml volumetric flask, and dilute to the mark with acetone. Pipet an aliquot containing 2 mequiv. of the sample (referring to the nitric oxide formed per mole of sample) into a 300-ml flask, and evaporate the solvent with a stream of dry air. Dissolve the residue in 50.00 ml of concentrated sulfuric acid; add 50.00 ml of cold (10–15°C) ferrous sulfate reagent, and cool the mixture in water or an ice bath, maintaining a temperature of 25–30°C. Allow the color to develop for 45 min, and determine absorbance at 525 nm, using an equivolume mixture of ferrous sulfate reagent and sulfuric acid as a blank.

In plotting absorbance versus milliequivalents of nitric oxide formed for potassium nitrate, nitroglycerin, pentaerythritol tetranitrate, triethyleneglycol dinitrate, $N$-methyl-$N$-nitro-2,4,6-trinitroaniline (tetryl), and nitroguanidine, it was found that all gave the same straight line, implying complete release of nitric oxide. On the other hand, hexahydro-1,3,5-trintro-$s$-triazine (RDX) and octahydro-1,3,5,7-tetranitro-$s$-tetrazine (HMX) apparently do not release nitric oxide completely, for the plots were straight lines of smaller slopes [see Section VI.A.3.1(i) of "Nitro and Nitroso Groups"]. Clearly, the method is applicable to nitrate esters and to nitroamines, N—NO$_2$. Moreover, the molecular weights of nitrate esters can be determined.

Although the ferrous sulfate–NO coloration is nearly specific for nitrate and nitrate esters, inorganic ions such as nitrites and thiosulfates interfere, whereas chromates, dichromates, sulfates, phosphates, chlorates, sulfites, acetates, and the halogens do not. On the other hand, organic materials such as oils and plasticizers produce spurious colors which often interfere, as do substances that char or discolor in contact with strong sulfuric acid solutions.

### e. GRIESS-ILOSVAY REAGENT

The Griess-Ilosvay reagent is used to determine nitrites by a series of reactions in which nitrous acid is allowed to diazotize an aryl amine and the resulting diazonium salt is converted to an azo dye by coupling with an appropriate aryl compound. Inasmuch as nitrite esters are readily hydrolyzed to nitrite ion [see Section V.B.I.a(7)], it is expected that a positive reaction will be obtained with the Griess-Ilosvay reagent and that, under standardized experimental conditions, the intensity of coloration can be correlated with the amount of ester.

A typical procedure for the colorimetric determination of nitrite esters has been described by Altschuller and Schwab (3):

Standard solutions are prepared by dissolving 1 ml of alkyl nitrite in 75 ml of glacial acetic acid and diluting to 250 ml with water. The standard solutions are stable for only about 0.5 hour at 0°C; in contrast, when the standard solutions are diluted 1–100, they are stable for 4 to 5 hr. (Neat nitrite esters are stable indefinitely when frozen.) An aliquot of a 1–100 dilution of the standard solution is added to 10 ml of a reagent prepared by dissolving 5 grams of sulfanilic acid in nearly 1 liter of water containing 140 ml of glacial acetic acid, adding 20 ml of 0.1% N-(1-naphthyl)ethylenediamine dihydrochloride solution, and diluting to 1 liter. After 10 mins, the absorption of the solution at 550 nm is obtained.

Nitrate esters usually do not interfere with the Griess test; for example, the Kaufman et al. (133) tests for nitrites (organic or inorganic) is as follows:

Dilute the solution so that the nitrite ester content is $1–5 \times 10^{-5}\ M$; to a 5-ml aliquot, add first 0.5 ml of 1% sulfanilic acid in glacial acetic acid and then 0.5 ml of a solution containing 0.1 gram of $\alpha$-naphthylamine in 75 ml of glacial acetic acid and 175 ml of water. Allow to stand at room temperature for 10 min, and measure the optical density at 470 nm.

### f. AZULENE REAGENT

A rapid colorimetric determination of small amounts of nitrite ion by diazotization of p-nitroaniline and coupling with an excess of azulene, $C_{10}H_8$, has been devised by Garcia (87); the method has also been applied by Sawicki et al. (249). The purplish pink azo dye, which absorbs at 515 nm, is 1-(p-nitrophenylazo)azulene:

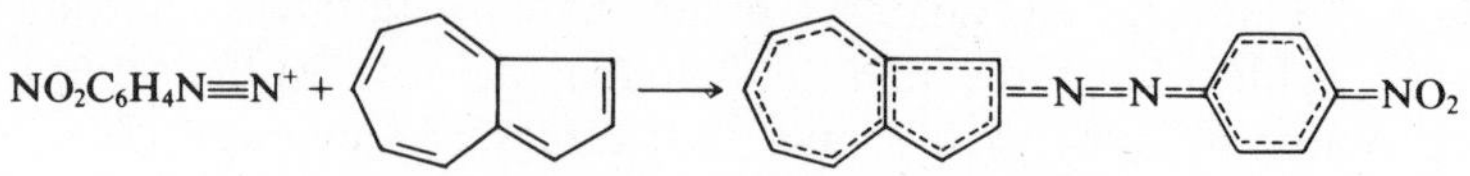

The color of the excess azulene is destroyed by strong perchloric acid; the azulenyl cation is formed:

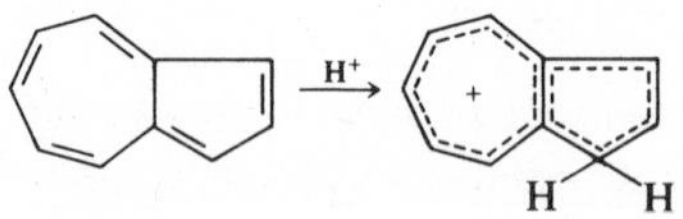

Ordinarily, coupling reactions in dilute solutions are relatively slow, often requiring 20–30 min for completion; coupling with azulene is so rapid that the entire analysis can be completed within 5 min. Moreover, 10,000 times more nitrate, sulfate, chloride, fluoride, and alkali metals than the amount of nitrate ion do not interfere; ions such as mercury(II), cobalt, nickel, aluminum, calcium, and magnesium do not interfere when present in concentrations 500 times the amount of nitrite ion, but iron(III) interferes seriously and copper(II) decreases the stability of the color.

Garcia's method has been applied by Hankonyi and Karas-Gasparec (98) to the alkaline hydrolysates of nitric esters; the procedure permits determination of esters in the concentration range of 5–50 $\mu$g in 10 ml, and is recommended for pentaerythritol tetranitrate (rel. std. dev., ±1.07 to ±1.96%) but not for nitroglycerin (rel. std. dev., ±2.32%).

Transfer an aliquot of solution (e.g., an acetone extract) containing between 5 and 50 $\mu$g of pentaerythritol tetranitrate to a 10-ml volumetric flask, and add 1 ml of $N$ NaOH. Keep the mixture in a boiling-water bath for 15 min, and then cool to room temperature. Add 1 ml of 2 $N$ HCl, and immediately follow with 1 ml of a solution of 0.4 gram of $p$-nitroaniline in 100 ml of glacial acetic acid, 1 ml of a solution of 0.08 gram of azulene in 100 ml of glacial acetic acid, and 4 ml of perchloric acid (mix well after each addition). Dilute to the mark with water, mix well, and measure the absorbance at 515 nm against a reagent blank. Establish calibration plots with a standard solution of the ester.

The procedure for other esters is the same except that alkaline hydrolysis can be performed at room temperature.

For every mole of ester, 1.63 mol of nitrite ion is released from pentaerythritol tetranitrate, 0.91 mol from erythritol tetranitrate, and 0.79 mol from glyceryl trinitrate; these data indicate that, when hydrolysis of the esters is not stoichiometric, quantitative results may be obtained (±2–3%) under rigidly controlled conditions.

### g. Miscellaneous Reagents

The hydrolysis of nitro esters in potassium hydroxide forms nitrites, among other products (see Section III.B.2 of "Nitro and Nitroso Groups"); under suitable conditions, nitrites will diazotize aromatic amines, and the resulting diazonium salts can be coupled with aromatic hydroxy compounds to form highly colored substances in solutions (52). (See the preceding section on azulene reagent.)

Swann (281) has reported that nitrocellulose reacts with alkali in acetone to form a yellow solution; absorption at 425 nm is related to the nitrocellulose content.

## D. POLAROGRAPHIC METHODS

Polarographic procedures are seldom used for the analysis of nitrate or nitrite esters because in these instances electroreduction processes are not sufficiently diagnostic; this comes about because the nitrate esters of commercial importance are nearly always found in admixture with a variety of $C$-nitroaromatic compounds or with nitrate esters of such similar structure that polarographic differentiation is not possible.

Among the earliest reports of the polarographic behavior of alkyl nitro compounds is a fragmentary note by Radin and DeVries (236) indicating that in nonaqueous solvents such as methanol (0.5 $M$ LiCl) $n$-butyl nitrate is reduced at about the same potential as are the nitroalkanes. No reduction was observed in glycerol (0.3 $M$ LiCl). A two-electron change was calculated for the reduction. Similarly, Kaufman et al. (133) found that cyclohexyl nitrate and ethyl nitrate are reduced by a two-electron process. (Cyclohexyl nitrate was chosen for study because it is less volatile than the lower members of the series and contains a ring structure and a secondary ester linkage—in close analogy to sugar and cellulose nitrates.) The alcohol and the nitrite ion were established as products of polarographic reduction:

$$RONO_2 + 2\,e \rightarrow RO^- + NO_2^-$$
$$RO^- + H_2O \rightarrow ROH + OH^-$$

A few experiments on the electroreduction of two dinitrates, $cis$- and $trans$-1,2-cyclohexanediol dinitrate, gave $n = 4$ with high accuracy, and indicated production of the corresponding diols in good yields. It is of special interest to note that cyclohexyl nitrite proved to be reducible at approximately the same potential as cyclohexyl nitrate, giving diffusion currents of comparable magnitudes. The reduction wave of the two esters was found to be pH independent, and the reduction was diffusion controlled between pH 3 and 13; in acidic media (pH 1 and 2), a large, smeared-out, double wave was obtained.

A few years later, Whitnack et al. (304) showed that the mechanism of reduction of polynitrate esters follows closely the reduction of the simple esters, but the reduction of polynitrate esters is dependent on pH; nevertheless, two electrons are involved in the reduction of each nitrate group in neutral or acidic media. The polynitrate esters studied were ethylene glycol dinitrate, glycerol trinitrate, and pentaerythritol tetranitrate. In alkaline solution, the secondary nitrate group is readily attacked by the large concentration of hydroxyl ion; thus a more complicated process is involved, and the nitroglycerin wave appears to break into two small waves which in combination are about 25% of the height of the one wave in neutral media (butyl nitrate does not produce a double wave in alkali). Excellent waves are obtained with all types of nitrate esters at pH 5. A procedure for the determination of nitroglycerin in double-base powder (i.e., also containing nitrocellulose) based on the above findings has been reported (303); an alcoholic extraction is used to separate nitroglycerin from the double base. Corrections must be applied if other reducible substances such as 2-nitrodiphenylamine. are present. The dc polarographic waves for polynitrate esters are long and drawn out; as a result, many electroreducible substances interfere.

As can be anticipated, dc polarography is not capable of distinguishing each of the nitrate groups in a polynitrate ester; however, Hetman (116) has shown that the derivative curve of the dc wave (obtained with a cathode-ray polarograph) clearly reveals the presence of three waves for nitroglycerin and two waves for ethylene glycol dinitrate. Of greater importance, however, is the finding that pyridine changes the reaction mechanism, so that nitroglycerin forms only a single peak; ethylene glycol dinitrate becomes irreducible. The addition of 5 ml of pyridine increases the pH of the base solution (10 ml $N$ KCl, 50 ml 2 $N$ $NH_4Cl$, and 40 ml water; initial pH = 6) to 7.3, but in aqueous buffers of the same pH, the reduction mechanism is not altered. In a more detailed publication, Hetman (115) described methods for the direct ac-polarographic determination of pentaerythritol tetranitrate (PETN) and cyclotrimethylenetrinitramine (RDX) in simultaneous admixture; however, even ac polarography suffers from the same interferences as dc polarography, and it is therefore indicated that $C$-nitro compounds such as dinitrotoluene, 2-nitrodiphenylamine, and trinitrotoluene interfere with the determination of one or both components. Nitroglycerin (NG) also interferes; thus the procedure for direct determination of PETN and RDX must be supplemented by a separate procedure for determination of NG and correction of the results of a direct analysis. Nitrocellulose does not interfere.

Ayres and Leonard (6) have described a rapid dc-polarographic method for the determination of pentaerythritol trinitrate in nitrocellulose propellants which may also contain nitroglycerin, nitrocellulose, 2-nitrodiphenylamine, and dibutylphthalate as major constituents. The wave height obtained from a solution of the sample in acetone is ascribed to total nitrate, that is, nitroglycerin, pentaerythritol trinitrate, and $C$-nitro compounds such as 2-nitrodphenylamine. An aliquot of the sample solution is treated with alcoholic alkali to decompose nitroglycerin. The resulting wave height is ascribed to pentaerythritol trinitrate and 2-nitrodiphenylamine; a spectrophotometric procedure is used to determine the 2-nitrodiphenylamine content of the sample, and the wave height is corrected for 2-nitrodiphenylamine.

The polarographic method of analysis for nitrate esters (dc) has been applied to the determination of additives in diesel fuels (266), small amounts of nitroglycerin in nitric and sulfuric acid (136), and nitrocellulose in Celluloid (53). Polarographic analyses of explosives and related products containing nitrates have been reported for nitroglycerin

(44,178); applications of cathode-ray polarography have been reported by Hetman (113,114,116) and by Williams and Kenyon (306). Ethyl nitrate and nitrite in aqueous solutions may be determined polarographically (18). Patsevich et al. (220a) also found that alkyl nitrites such as ethyl nitrite and butyl nitrite can be determined in an aqueous solution at pH > 7 (e.g., LiOH); half-wave potentials generally are of the order of −0.65 V.

## E. KJELDAHL METHODS

The Kjeldahl digestion procedure is not directly applicable to nitrate or nitrite esters; the ester groups must be brought into a form that can be converted to ammonia by digestion in hot sulfuric acid or reduced directly to ammonia. It is important to note that, when the functional groups are reduced directly to ammonia by Devarda's alloy or other metal couples in Kjeldahl flasks (coupled to scrubbers and condensers) so as to take advantage of the efficient distillation system used in the Kjeldahl method, the operations are often erroneously classified as Kjeldahl procedures; it is evident that they are reductions followed by Kjeldahl distillations.

The Kjeldahl-Gunning method should not be applied directly to nitrate and nitrite esters because of loss of nitrogen as nitric acid and nitrose gases. Methods involving addition of phenols followed by a reducing agent such as zinc dust (125) or of salicylic acid and sodium thiosulfate (48,70) are in wide use; phenols in strong sulfuric acid are nitrated by the esters, and the resulting C-nitrophenols are readily carburized in the Kjeldahl-Gunning digestion process. Nitrite esters may prove troublesome, however, because nitrous acid is momentarily released and the nitrogen oxides resulting from its decomposition usually escape. Reduction with phosphorus and hydrogen iodide is not satisfactory, although sealed tube procedures (1) may be found superior to a mere preliminary digestion with phosphorus and hydriodic acid (84).

Although the assay of pure esters (neat or in solution) may be possible by Kjeldahl methods, there is marked lack of specificity; results are of greatest value when the total nitrogen content of the esters can have its origin only in the ester function. Accordingly, determination of the ester functions by Kjeldahl procedures is not recommended.

## F. COMBUSTION METHODS

Conversion of combined nitrogen to elemental nitrogen by Dumas gasometric methods is an accepted analytical procedure for nitrate and nitrite esters. However, none of these methods is used extensively for ester analysis because elaborate apparatus is required and because personal skills need to be developed to a high degree if reliable results are to be obtained. More importantly, however, the esters often deflagrate suddenly rather than burn smoothly in the Dumas combustion train, and equipment is occasionally disintegrated by detonations. Needless to say, micro or semimicro Dumas procedures are preferable because small samples are less hazardous to the operator; Blais (16) has suggested modifications of the Dumas method to reduce hazards. Excellent discussions and reviews of the Dumas method have been published, for example, by Gustin (97), Steyermark (279), Schoniger (253), Kirsten (137), and Drew (65). Conventional Dumas techniques (Pregl) usually give low results. An adequate amount of metallic copper must be present in the Dumas combustion train to reduce the large amount of nitrogen oxides that are produced; when nitrate and nitrite esters deflagrate suddenly, the large volume of gas that is released passes through the train too quickly for complete reaction unless long, well-packed trains are used. Commercial automatic or semiautomatic nitrogen determinations give slightly better results for the nitrogen contents of explosives, probably because these devices have greater volumes than the ordinary semimicrotrains (301).

As a rule, volatile esters are weighed into capillaries or vessels which have a capillary to restrict the rate of release of the sample; release must be slow for accurate analysis. Samples that have essentially no volatility can be diluted by a liberal amount of powdered copper oxide and distributed over a section of the combustion tube so that only small amounts of the sample will be ignited at any given time. Dilution of samples of esters with nonnitrogenous matter is common practice in propellant laboratories; small amounts of glucose, sucrose, or starch seem to facilitate the decomposition of compounds with multiple nitrate ester groups. In spite of exercise of the above cautions, nitrogen values for polynitrate esters are low; for example, values typically between 17.70 and 18.00 are obtained for D-mannitol hexanitrate (18.59% theor.).

The inclusion of oxidants such as vanadium pentoxide, potassium dichromate, or potassium chlorate is recommended to assist decomposition (65,279); however, cobalt oxide has been found to be especially useful for nitrate esters by Borda and Hayward (21), and this oxidant is strongly recommended. The modified Dumas method using cobalt oxide is presented in detail in Section VII.E.

## G.  GASOMETRIC METHODS

### 1.  Nitrometer Method

The most common gasometric determination of nitrate and nitrite esters depends upon the release of nitric oxide when an ester (also inorganic nitrates, nitrites, and nitro- and nitrosoamines) reacts with mercury in concentrated sulfuric acid:

$$2HNO_3 + 6Hg + 3H_2SO_4 \rightarrow 3Hg_2SO_4 + 4H_2O + 2NO$$

$$2HNO_2 + 2Hg + \ H_2SO_4 \rightarrow \ Hg_2SO_4 + 2H_2O + 2NO$$

$$2R_2N\text{–}NO_2 + 6Hg + 3H_2SO_4 \rightarrow 3Hg_2SO_4 + 2R_2NH + 2H_2O + 2NO$$

$$2R_2N\text{–}NO \ + 2Hg + \ H_2SO_4 \rightarrow \ Hg_2SO_4 + 2R_2NH + 2NO$$

Although $C$-nitroaromatic compounds do not generate nitric oxide, there is some interference in the determination of nitrate esters such as nitroglycerin because the nitro compounds are converted to dinitro compounds and the loss of nitric acid invariably leads to low results (121). Moreover, since most aromatic compounds are capable of being nitrated, substances such as diphenylamine or symmetrical diethyl-diphenylurea which are compounded with propellants must be removed before analysis. Organic materials such as resin, camphor, paraffin, and petroleum jelly give low results; magnesium carbonate or other substances that release carbon dioxide produce high results, and sulfur also tends to give high results, probably because of release of some sulfur dioxide (212). Hyde (123) suggested that the nitrate esters be separated from nitroaromatics (and other nitratable substances) by extraction; however, the procedure was not used extensively because the recommended extraction involved carbon disulfide as one phase and a mixture of 3 parts of acetic acid to 1 of water as the other phase.

The determination has had a long history; Lunge (184,185) appears to have been the first to devise a simple gas buret which permitted introduction of reactive materials into a space immediately above mercury in the buret. Thus, for example, nitrate esters in sulfuric acid solution could be brought into contact with mercury in a gas buret, and the volume of nitric oxide could be measured. An excellent review of Lunge's nitrometer method, including a study of variables affecting its performance (especially with nitrocellulose), has been published by Newfield and Marx (212). Among the modifications of Lunge's "gas volumeter" made in subsequent years was F. I. Du Pont's improved model (a large Lunge buret), but of greatest importance is that at present the apparatus has been standardized; in fact, the most widely used apparatus for analysis of organic and inorganic nitrates is the Du Pont nitrometer described by Pitman (228) and subsequently included in the official procedures for analysis of the munitions of many governments throughout the world. The salient difference between the ordinary Lunge nitrometer and the Du Pont version is that a compensating tube containing a quantity of gas of known volume at standard conditions makes possible conversion of the volume of nitric oxide in the measuring buret to the volume that it would occupy under standard conditions of temperature and pressure. A semimicronitrometer which uses one-tenth the sample of the Du Pont nitrometer has been described by Elving and McElroy (68). The Du Pont nitrometer functions properly with 172–240 ml of nitric oxide; thus a deviation of 0.1 ml in readings of gas volumes represents less than ±1 part per thousand, and more frequently ±0.5 part per thousand. A detailed procedure for analysis of esters by the Du Pont nitrometer is given in Section VII.C.

## 2. Schulze–Tiemann Method

The reaction of ferrous chloride with nitrate and nitrite esters provides a convenient volumetric method, as was indicated in a preceding section; the Schulze–Tiemann method (258,289) is based on the same reaction, but the nitric oxide that is formed is measured in a gas buret:

$$HNO_3 + 3\,FeCl_2 + 3\,HCl \rightarrow 3\,FeCl_3 + 2\,H_2O + NO$$
$$RONO_2 + 3\,FeCl_2 + 3\,HCl \rightarrow 3\,FeCl_3 + ROH + NO + H_2O$$

Although the method was once considered more suitable than the nitrometer method for analysis of nitrocellulose samples that do not dissolve readily in sulfuric acid, it is seldom used at the present time even though improvements have been made (132) and the results compare favorably with those obtained with the nitrometer (173,244). The reagent is a saturated solution of ferrous chloride containing hydrochloric acid; a typical procedure involves use of a water-cooled reflux condenser and a stream of carbon dioxide to force nitric oxide out of the reaction vessel and into a gas buret filled with potassium hydroxide solution.

## H. EXTRACTION METHODS

Isolation of nitrate esters by selective extraction with suitable solvents is often employed as an analytical procedure when determination of the ester function by chemical procedures is precluded by the presence of other compounds with functional groups which interfere. Thus, for example, 65–85% acetic acid can be used to extract nitroglycerin (and ethyl centralite) from a propellant containing nitrocellulose (137) (N $\geqslant$

12.2%), and ether can serve to remove mannitol hexanitrate from inert matter in pills prepared for treatment of hypertension (296). The solvent is evaporated and the ester content determined by titration; alternatively, the insoluble residue remaining after solvent extraction is dried and weighed. Naturally, extraction methods are not specific and cannot be used in the presence of other solvent-soluble materials. For example, since propellants always contain additives that serve as stabilizers (e.g., diphenylamine and ethyl centralite) and the nitration derivatives of the stabilizers are also present in increasing amounts as the propellant ages, it follows that the residue may contain a variety of substances in addition to the desired component. Moreover, not all esters can be recovered quantitatively by evaporation of their solutions, and in many instances the "insoluble" fraction is extracted to a small but finite extent; for example, when a double-base propellant is extracted for 20 min at reflux with 84% acetic acid, the residue of nitrocellulose will be found to be too small. The solubility of nitrocellulose (N $\geqslant$ 12.2%) in 65–70% acetic acid is vanishingly small, but extraction time must be increased to 2–3 hr (117).

Although the accuracy of extraction methods can be questioned, they are widely used for the analysis of commercial products of relatively constant composition. Extraction methods provide excellent control procedures for rapid assay of a product, and since the composition of the product is more or less fixed, the solvent extraction system can be optimized; a short series of test runs in which the concentration of the extracted substance is deliberately varied will quickly provide a statistical evaluation of the accuracy of extraction.

Analysis of a propellant for nitrate esters by methods described in preceding sections very often involves a preliminary extraction designed to separate from interfering substances one or all of the various esters the propellant may contain; thus a double-base propellant containing nitroglycerin and nitrocellulose is usually extracted with acetic acid to separate nitroglycerin (and sundry substances including $C$-nitro, $C$-nitroso, $N$-nitro, and $N$-nitroso compounds) from nitrocellulose (the nitrocellulose is weighed, and the nitroglycerin content determined titrimetrically). The infrared spectra of 24 common ether-soluble ingredients found in propellants have been recorded by Pristera (231). Watts and Stalcup (300) have found that a low-boiling pentane-methylene chloride azeotrope (2–1) is best for removal of nitroglycerin and stabilizer components from cordite N, leaving the nitroguanidine content intact; the extract can be analyzed for its content of components, and nitroguanidine can be extracted from the initial propellant residue.

## VII.  LABORATORY PROCEDURES

### A.  FERROUS-TITANOUS TITRATION PROCEDURE FOR NITRATE ESTERS

The titration procedure described below is essentially similar to Becker's procedure for nitroglycerin (13); the nitrate ester in acetic acid is reduced by ferrous chloride (or ferrous ammonium sulfate):

$$C_3H_5(NO_3)_3 + 9\,FeCl_2 + 9\,HCl \rightarrow C_2H_5(OH)_3 + 9\,FeCl_3 + 3\,NO + 3\,H_2O$$

The amount of nitrate ester group is computed from the ferric ion that is produced; for this purpose, ferric ion is determined by titration with titanous chloride solution, using ammonium thiocyanate as the internal indicator. Substances such as 2,4-dinitrotoluene, diethyldiphenylurea, and diphenylamine do not interfere. The procedure is applicable to a wide variety of nitrate esters; however, the esters must be soluble in acetic acid. Nitrocellulose of high nitrogen content cannot be successfully analyzed by this procedure.

### Reagents

*Titanous chloride solution*, 0.2 $N$. Prepare and standardize with potassium dichromate as indicated in Section VII.C of "Diazonium Group," Vol. 15.

*Ferrous chloride solution*, 0.7 $N$. Dissolve 200 grams of $FeCl_2 \cdot 4\,H_2O$ in water containing 50 ml of concentrated hydrochloric acid; dilute to 1 liter with water. Store under inert gas as indicated in Section VII.C of "Diazonium Group."

*Ferric alum solution*, 0.15 $N$. Prepare and standardize as indicated in Section VII.C of "Diazonium Group"; however, use only 75 grams of the salt and 25 ml of sulfuric acid.

*Ferrous ammonium sulfate solution*, 0.7 $N$. This solution may be used in place of the ferrous chloride solution specified above. Dissolve 275 grams of $Fe(NH_4)_2(SO_4)_2 \cdot 6\,H_2O$ in 800 ml of water and 70 ml of concentrated sulfuric acid. Dilute to 1 liter with water. Reduce ferric iron by adding small portions of pure iron powder (reduced by hydrogen) until a drop of solution does not form an immediate pink color with ammonium thiocyanate solution. Filter, and store under inert gas.

*Ammonium thiocyanate solution*, 20%. Dissolve 20 grams of reagent-grade $NH_4CNS$ in 100 ml of water. The solution must be clear and colorless.

**Special Apparatus**

Use the apparatus suggested in Section VII.C of "Diazonium Group" for storing and dispensing titanous solution. A 250–300 ml Florence flask with a standard-taper ground-joint neck is required; in addition, the flask is to be fabricated with a short gas inlet tube on the shoulder. A reflux condenser with a mating ground joint is used in conjunction with the flask. For titration, the flask is fitted with a one-hole rubber stopper; a short piece of glass tubing is inserted into the stopper, and the tip of the buret is inserted into the tubing.

A supply of inert gas, preferably carbon dioxide, is required to displace air from the titanous storage vessel and from the Florence flask during reflux or titrations.

**Procedure**

Dissolve an accurately weighed sample of the ester in glacial acetic acid (sparged with inert gas to remove oxygen), and make up to such a volume in a volumetric flask that an aliquot of 25 ml contains nearly 8 mequiv. of the ester ($ONO_2/3$). Thus, for example, when nearly pure nitroglycerin is to be assayed, from 1.8 to 2.0 grams of sample is dissolved in 250 ml of glacial acetic acid.

Displace the air in the special Florence flask by passing through the side arm a current of carbon dioxide (3–5 min). Transfer to the Florence flask exactly 25 ml of the glacial acetic acid solution of sample; add 15 ml of ferrous chloride solution and 25 ml of 6 $M$ hydrochloric acid. Immediately connect the flask to the reflux condenser, and boil the contents gently for 5 min; be sure the boil keeps the contents mixed. Do not permit the flow of carbon dioxide to be interrupted at any time. (Glass beads may be used to prevent bumping.)

Increase the rate of flow of the carbon dioxide, and then immerse the flask (with condenser attached) into cold water; allow to cool to room temperature. The rate of flow of carbon dioxide must be rapid enough to prevent air from being drawn into the flask while it cools. Wash down the inside tube of the condenser with 10 ml of a solution made by mixing equal volumes of 6 $M$ hydrochloric acid and sparged glacial acetic acid. Allow the condenser to drain for about 1 min; then disconnect it from the flask. Place a one-hole stopper in the neck of the flask, and titrate with 0.2 $N$ titanous chloride solution until the reddish brown (or golden yellow) color of ferric ion is nearly discharged; then add 5 ml of 20% ammonium thiocyanate solution, and continue titration until the red color of ferric thiocyanate complex disappears. Record the amount of titanous chloride solution ($V$) used.

Conduct a blank determination (preferably concurrently with the sample); use 25.00 ml of glacial acetic acid, 25 ml of 6 $M$ hydrochloric acid, and 15.00 ml of ferrous chloride solution. Record the amount of titanous chloride solution required for the blank ($v$).

Calculate the percent of nitrate ester group (or its nitrogen equivalent) by means of the following:

$$\% \ ONO_2 = \frac{2.067(V - v)N}{W}$$

$$\% \ \text{nitrate N} = \frac{0.4669(V - v)N}{W}$$

Alternatively, if the composition of the ester is known, the purity may be obtained from the following:

$$\% \ \text{ester} = \frac{MN}{30 \ Wn}(V - v)$$

In the above equations,

$V$ = titanous chloride for sample (ml),
$v$ = titanous chloride for blank (ml),
$N$ = normality of titanous chloride,
$W$ = sample weight (grams),
$n$ = number of nitrate ester groups in molecule,
$M$ = molecular weight of ester.


## B. TITRIMETRIC DETERMINATION OF NITROGEN IN NITROCELLULOSE

This procedure is an adaptation of the ferrous reduction method described above; although it may be used with any nitrate ester, it is specifically designed to cope with nitrocellulose samples that are not completely soluble in acetic acid (as required by the ferrous–titanous titration procedure given above) or in sulfuric acid (as required by nitrometer methods). Many refractory nitrate esters may be analyzed by Grodzinski's method (acetic acid–acetic anhydride solvent) (96), but the method of Pierson and Julian (226) given below is more easily applied and is undoubtedly superior to other titrimetric methods for nitrocellulose. Moreover, the procedure can be used for assay of nitrocellulose because titration values obtained with it show no evidence of bias at the 5% level when compared with grand-average values for assay results by the nitrometer method (226).

## Reagents

*Titanous chloride solution,* 0.2 N. Prepare as indicated in Section VII.C of "Diazonium Group," Vol. 15; however, use 350 ml of concentrated hydrochloric acid/liter.

*Ferrous ammonium sulfate solution,* 0.7 N. Prepare as directed in the preceding procedure, but use 140 ml of concentrated sulfuric acid/liter.

*Acetic acid solution,* 68–70% W/W. Dilute 700 ml of glacial acetic acid to 1 liter with water; remove dissolved oxygen with carbon dioxide.

*n-Butyl acetate.* Use reagent-grade material.

*n-Pentane.* Use a technical-grade solvent.

*n-Hexane.* Use technical-grade solvent.

*Acetic acid, glacial.* Use reagent-grade material.

*Ammonium thiocyanate solution,* 20%. Dissolve 20 grams of reagent-grade $NH_4CNS$ in 100 ml of water. The solution must be clear and colorless.

## Special Apparatus

Solutions of titanous chloride and ferrous ammonium sulfate must be protected from air; they are stored and dispensed under carbon dioxide (see Section VII.C of "Diazonium Group").

The apparatus described above for the ferrous–titanous titration method will be adequate.

## Preparation of Sample

Nitrocellulose samples should be dried to constant weight in shallow weighing bottles in a vacuum oven at 60–70°C and a pressure of 2–5 cm of Hg.

Propellant samples (3–5 grams) which have been ground in a mill to pass through 20-mesh screens are extracted three times for 2 hr at steam-bath temperature with 68–70% acetic acid to remove nitroglycerin and other soluble nitrate esters. The mixtures are filtered through a medium frit, and the residues, consisting chiefly of nitrocellulose, are washed with hot water and dried in the filtering crucible in an oven at 100°C until the losses in weight between two weighings 1 hour apart do not exceed 2–3 mg. (Residues are preferably dried at a pressure of 2–5 cm of Hg at 60–70°C.)

Before weighing, all nitrocellulose samples should be cooled in a desiccator containing fresh Drierite, either with covers on the bottles or under vacuum.

**Procedure**

Transfer about 0.2–0.25 gram of nitrocellulose from a tared weighing bottle to the 300-ml reduction flask (with a side arm for gas inlet). Aliquots of propellant residues may be transferred directly from tared filter crucibles. Use a powder funnel to facilitate transfer of the sample to the reduction flask, and wash the sample into the flask with a stream of 3:1 n-pentane–n-hexane dispensed from a glass or polyethylene wash bottle; wash the funnel by pouring 45 ml of glacial acetic acid through it and into the reduction flask. Add a few beads or bumping stones, and then remove most of the pentane-hexane solvent by heating over a stream bath.

Add 25 ml of n-butyl acetate; connect the flask to the reflux condenser, pass a stream of carbon dioxide into the flask via the side arm, and gently reflux the solution for about 2 min. Now, without interrupting the flow of carbon dioxide, remove the reflux condenser, cool the contents of the flask nearly to room temperature (cold-water bath), and rapidly add 25 ml of 0.7 N ferrous ammonium sulfate solution, followed by 8–10 ml of concentrated hydrochloric acid. (The amount of acid is critical; the volume should be within the indicated limits and be constant in a given series of analyses.)

Replace the flask under the reflux condenser and heat at such a rate that boiling starts within 10 min; agitate the contents at 5-min intervals to bring down particles adhering to the sides of the flask. Two layers are evident in the early stages of heating; gradually, the upper layer will turn light green and then deepen to a very dark green. After some time, the two phases will appear to become one because both will have the same dark color. When the color of both phases changes to yellow, continue the boiling for about 10 min longer (total time: 30–40 min).

Increase the flow rate of carbon dioxide, and cool the flask and its contents in a water bath. When the contents are approximately at room temperature, loosen the condenser joint and wash down the condenser tube and joint with 30 ml of oxygen-free 68–70% acetic acid; connect the condenser to the flask and allow to drain for a few minutes.

Disconnect the flask and titrate its contents with 0.2 N titanous chloride solution; add 5 ml of ammonium thiocyanate solution near the end point, and titrate very slowly until the red color is discharged.

Blank determinations should be run concurrently; include all reagents in the same amounts, and do not omit any steps in the procedure.

Compute the percent nitrogen in nitrocellulose by means of the following equation:

$$\% \ N = \frac{0.46689(V - v)N}{W}$$

where  $V$ = titanous chloride for sample (ml),
       $v$ = titanous chloride for blank (ml),
       $N$ = normality of titanous chloride,
       $W$ = sample weight (grams).

If the nitrogen content of the nitrocellulose used in manufacturing a propellant is known or can be assigned a nominal value, the percent nitrocellulose in the propellant sample can be computed as follows:

$$\% \text{ NC} = \frac{46.689R(V - v)N}{PWF}$$

where  $P$ = propellant sample extracted (grams),
       $R$ = residue from propellant extraction (grams),
       $W$ = residue aliquot taken for analysis (grams),
       $V$ = titanous chloride for residue aliquot (ml)
       $v$ = titanous chloride for blank (ml),
       $N$ = normality of titanous chloride,
       $F$ = nitrogen (%) in NC used in propellant manufacture.

## C. DETERMINATION OF NITROGEN BY THE Du PONT NITROMETER

Determination of the nitrogen contents of nitrate and nitrite esters is frequently performed gasometrically by this method; the basic reaction is

$$2KNO_3 + 6Hg + 5H_2SO_4 \rightarrow 3Hg_2SO_4 + 2KHSO_4 + 4H_2O + 2NO$$

The volume of nitric oxide is measured under closely prescribed conditions. Any substance that releases nitric oxide in contact with mercury and concentrated sulfuric acid can be determined. Thus, in addition to nitrate and nitrite esters, the method may be used with inorganic nitrates and nitrites, nitramines, and nitrosamines (206).

### Reagents

*Potassium nitrate.* Recrystallize reagent-grade material three times from 95% (w/v) ethyl alcohol, grind to pass through a No. 100 U.S. Standard sieve, and dry for 2–3 hr at 135–150°C. Store in a glass-stoppered bottle in a desiccator

*Sulfuric acid.* Use nitrogen-free, reagent-grade, 94.5 ± 0.5% (w/w) acid.

### Special Apparatus

The Du Pont nitrometer is most frequently employed; it consists of five glass parts which are grouped into two divisions, a gas-measuring buret graduated to read directly in percent nitrogen and a generating bulb (see Fig. 3). The generating bulb (A) has a capacity of 300 ml and is in communication with leveling bulb B by means of heavy-walled rubber tubing. The upper part of the generating bulb is equipped with a two-way stopcock connecting either cup C or exit tube D with the bulb.

The gas-measuring buret (G) includes chamber F and stopcock E; the bottom of graduated tube G is connected to leveling bulb H by means of heavy-walled rubber tubing. A compensating tube I is also connected to

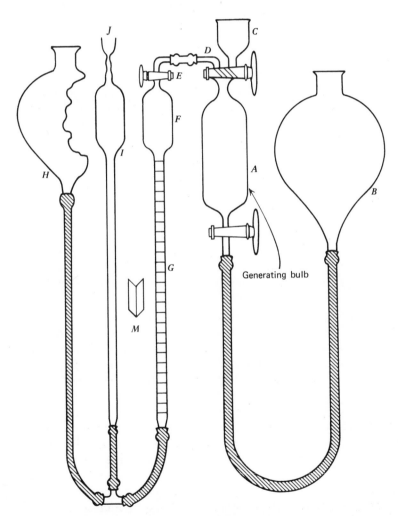

Fig. 3. Du Pont nitrometer.

the gas-measuring buret as shown in the diagram; the compensating tube has the same dimensions as the gas-measuring buret but is not graduated. Two mirrors ($M$) set at an appropriate angle permit simultaneous viewing of the level of mercury in graduated buret $G$ and in the compensating tube; thus it is a simple matter to bring mercury to the same level in both vessels. Tube $G$ is graduated to read from 10 to 14% nitrogen in 0.01% divisions. Between 171.8 and 240.4 ml of gas must be generated to obtain a reading.

## Standardization

Assemble and standardize the nitrometer as follows. Clean all the parts of the nitrometer with sulfuric acid–chromic acid solution, and then dry at 45–50°C for about 24 hr while drawing through a stream of dry air. Assemble the dry parts on a rack. Completely fill all the parts except the reservoirs with mercury. Close the stopcocks of the generating bulb to prevent loss of mercury and introduction of air; then disconnect the generating bulb from the measuring tube.

Place 25 ml of sulfuric acid in the cup of the generating bulb, and open both stopcocks on the bulb. Draw the sulfuric acid, together with air, into the generating bulb by lowering the mercury reservoir until the mercury level is at the lower shoulder of the bulb. Close both stopcocks on the generating bulb, and shake the bulb vigorously for 3 min to desiccate the air; replace the bulb on the rack, and let stand for 15 min.

Connect a piece of rubber tubing to the (unsealed) upper end of the compensating tube; raise reservoir $H$ about an inch above the end of the compensating tube, and permit mercury to fill the bore of the rubber tubing. Now connect the tubing to the generating bulb, making certain that no atmospheric air is trapped. Open the lower stopcock of the generating bulb, and turn the upper stopcock so that the generating bulb is connected to the compensating tube. Also, open the upper end of the measuring tube to the atmosphere. By regulating the heights of the two mercury reservoirs, transfer desiccated air from the generating bulb to the compensating tube until the level of the mercury in measuring tube $G$ is at 12.50 and the levels of mercury in the measuring tube and compensating tube are equal. Close the upper stopcock of the generating bulb, lower the level of the mercury in the compensating tube about 1 in., and then rapidly seal the upper glass and of the compensating tube with a flame.

Raise the level of the mercury in the measuring tube until the capillary tube $D$ is completely filled with mercury and then close the end of the capillary with a stopper made by inserting a glass rod plug into the rubber tubing connected to the capillary; be sure that no air is trapped.

Disconnect the rubber tubing from the generating bulb capillary ($D$). Hold a 50-ml beaker under the capillary, and then raise the reservoir ($B$) so as to eject the sulfuric acid into the beaker and fill the capillary completely with mercury; close the upper stopcock. Close the end of the capillary with a stopper made by inserting a glass rod plug into a short piece of rubber tubing; be sure no air is trapped.

Weigh (in a dry weighing bottle) exactly 1.0000 ($\pm 0.2$)mg of recrystallized and dried potassium nitrate. Transfer the salt to cup $C$ of the generating bulb.

With the upper stopcock of the generating bulb closed and the lower stopcock open, lower the mercury reservoir to reduce pressure in the generating bulb. Pour about 1 ml of sulfuric acid into the cup. By a series of quick openings of the upper stopcock, allow the potassium nitrate and acid to be drawn into the bulb; do not allow air to enter the bulb. Rinse the inside of the weighing bottle with two additional 1-ml portions of sulfuric acid, each time transferring the washings to the cup with the aid of a glass rod, rinsing the inside of the cup, and then drawing the acid into the generating bulb. Then rinse the cup with three portions of sulfuric acid (total 25 ml), and draw each rinsing into the generating bulb. Do not allow air to enter the bulb during the rinsing operations. Close the cup with a rubber stopper. With the mercury reservoir low enough to maintain a slightly reduced pressure in the generating bulb, and the lower stopcock of the generating bulb open, shake the bulb until most of the gas has been generated. Shake until the mercury in the bulb drops nearly to the level of the lower shoulder. Close the lower stopcock, and shake the bulb vigorously for an additional 3 min. Replace the bulb on the rack, and open the lower stopcock. Equalize the levels of the mercury in the generating bulb and the reservoir, and allow the bulb to remain undisturbed for 5 min. Close the lower stopcock, and shake the generating bulb for an additional minute. Replace the bulb on the rack.

Transfer gas from the generating bulb to the measuring tube on the rack as follows: Adjust the height of the measuring tube on the rack so that its capillary is aligned with the capillary of the measuring bulb. Remove stoppers; attach rubber tubing to the capillary of the measuring tube and fill its bore with mercury from the measuring tube. Connect the capillaries; be sure the connection is completely filled with mercury. Transfer the gas from the generating bulb to the measuring tube (allow some mercury to come over), and close the stopcock on the measuring tube. Sever the connection between the measuring tube and the generating bulb after closing the top stopcock of the bulb.

Continue the standardization as follows. Adjust the levels of the mercury in the measuring and compensating tubes to approximately the same height, and allow the gas in the measuring tube to stand for approximately 20 min to permit equilibration of temperature. **Note:** For standardization and analysis of samples, the room temperature must be 15–32°C. Check the mercury levels, and adjust the measuring tube and the mercury reservoir so that the mercury in the measuring tube is at the 13.855 mark (the percent nitrogen in potassium nitrate) and the mercury level in the compensating tube is in the same horizontal plane. Mark the compensating tube with a strip of gummed tape to show the level of the mercury. This is the standard reference mark used in determinations. Refill the measuring tube, generating bulb, and capillaries with mercury, and seal the ends of the capillaries with stoppers.

**Procedure**

Samples are handled in much the same way as the potassium nitrate standard used for standardization of the nitrometer. However, for specificity and as an example of the method by which nitrate esters are treated, the analysis of nitrocellulose will be described in detail.

Wash and dry the cup of the generating bulb, and then transfer about 1 gram of nitrocellulose to the cup. Measure into a small graduated cylinder 25 ml of sulfuric acid. Wash any residual sample from the weighing bottle into the nitrometer cup with 5 ml of sulfuric acid. Again wash the bottle with a 5-ml portion of sulfuric acid, and transfer the washings to the nitrometer cup. With a small stirring rod, stir the sulfuric acid in the cup until the nitrocellulose is emulsified. Lower the mercury reservoir sufficiently to maintain a reduced pressure in the generating bulb. Open wide the lower stopcock, and, with a series of quick openings of the upper stopcock, draw the acid mixture into the generating bulb. Make successive rinsings of the cup and the stirring rod with the rest of the sulfuric acid, and draw the acid into the generating bulb, admitting no air.

Close the upper stopcock of the generating bulb, leaving the lower one open, and adjust the mercury reservoir just low enough to maintain a slightly reduced pressure in the generating bulb. Shake the bulb gently until most of the gas has been generated. Adjust the mercury reservoir until the level of the mercury in the generating bulb is at the height of the lower shoulder; then close the lower stopcock and shake vigorously for 3 min. Replace the bulb on the rack, open the lower stopcock, and adjust the mercury reservoir until the mercury in the generating bulb is at approximately the same height as the mercury in the reservoir. This will bring the pressure inside the bulb to approximately 1 atm, and the solubility of nitric oxide in the sulfuric acid will be normal. Close the lower stopcock, and shake the bulb vigorously for an additional 1 min. Transfer the gas to the measuring tube as directed in the procedure for standardization.

Adjust the levels of the mercury in the compensating and measuring tubes to approximately the same height. Let the gas stand for 20 min (at 15–32°C) in order to permit equilibration of temperature. With the aid of a leveling device to check the levels of the mercury in the measuring and compensating tubes, adjust the level of mercury in the compensating tube to the standard volume mark while the level of the mercury in the measuring tube is in the same horizontal plane.

### Calculation

Obtain the percent nitrogen directly from the measuring tube if an exact 1-gram sample was used, or use the following equation:

$$\% \text{ N} = \frac{R}{W}$$

where  $R$ = reading of measuring tube (%),
 $W$ = sample weight (grams).

### Cleaning the Nitrometer

Clean the nitrometer before standardization and at least after every six or seven analyses of samples. For this purpose, transfer 25 ml of sulfuric acid to the cup of the generating bulb, and draw it into the bulb. Set the mercury level in the generating bulb at the lower shoulder of the bulb, and shake well so as to wet the inside surfaces with the sulfuric acid. Connect the capillary of the generating bulb to the capillary of the measuring tube with a piece of rubber tubing. Then transfer the acid from the generating bulb to the measuring tube, and fill the bulb and its capillary tube with mercury; close the upper stopcock of the bulb. Remove the rubber tubing from the capillary of the generating bulb, and stopper the end with glass-plugged rubber tubing; be sure no air is trapped. Lower the mercury level in the measuring tube, close the stopcock, and shake the tube so as to rinse its sides with acid. Then force the acid out of the measuring tube into a small beaker by means of the rubber tubing attached to the capillary of the measuring tube; fill the rubber tubing with mercury, and close the end with a glass plug. The nitrometer is now ready for another analysis or standardization. If the nitrometer is not to be used immediately, wash the generating bulb cup and dry with filter paper. Treatment of the nitrometer with sulfuric acid–chromic acid is necessary only when the glassware cannot be satisfactorily cleaned with rinsings of sulfuric acid carried out as described above.

### D. DETERMINATION OF NITROGEN IN NITROCELLULOSE BY PRELIMINARY ALKALINE HYDROLYSIS AND REDUCTION TO AMMONIA BY DEVARDA'S ALLOY

The following procedure follows closely Muraour's method for alkaline hydrolysis of nitrocellulose in the presence of hydrogen peroxide and sodium perborate. Although the method was developed specifically for nitrocellulose, it is suitable for nitroglycerin and is also applicable to a wide variety of nitrate and nitrite esters; however, loss of volatile esters must be prevented by use of efficient reflux condensers in the preliminary alkaline hydrolysis.

### Reagents

*Sodium hydroxide solution*, 5 N. Dissolve 210 grams of reagent-grade sodium hydroxide in distilled water, cool, and dilute to 1 liter.

*Sodium hydroxide solution*, 10 N. Dissolve 420 grams of reagent-grade sodium hydroxide in water; dilute to about 1 liter.

*Acetone.* Use reagent-grade material; alternatively, treat solvent acetone with permanganate and distil.

*Hydrogen peroxide*, 12-v. Dilute 30% reagent-grade hydrogen peroxide to a concentration of 3.6%. Verify the strength of the diluted solution by a permanganate titration.

*Devarda's alloy.* Use granular, reagent-grade metal (20 mesh or finer); the nitrogen content of the alloy must be negligible.

*Sodium perborate.* Use high-quality $NaBO_3 \cdot 4 H_2O$; the nitrogen content must be low.

*Sulfuric acid*, 0.33N. Mix 9.0 ml of concentrated sulfuric acid with recently boiled, cold, distilled water, and dilute to 1 liter. Determine the titer by titrating 25.00 ml of the mixture with sodium hydroxide, using methyl red as indicator.

*Sodium hydroxide*, 0.33N. Prepare carbonate-free sodium hydroxide, and standardize with potassium biphthalate in accordance with accepted procedures.

*Methyl red indicator*, 0.015%. Dissolve 0.015 gram of the indicator in absolute methanol or ethanol; use 2 ml.

## Procedure

Introduce about 0.84 gram of the nitrocellulose sample (ca. 12.5% N) into a 500-ml flask, and add 40 ml of acetone. Shake the flask intermittently over a period of 2 hr to dissolve the sample; alternatively, allow the mixture to stand overnight. When solution is complete, add at once a mixture of 50 ml of 12-vol. hydrogen peroxide with 4 ml of 5 N sodium hydroxide solution. Shake the contents of the flask from time to time until any solids are completely dissolved; heating to 30–35°C accelerates the dissolution of solids. When dissolution is complete, add 4 grams of sodium perborate, and allow to stand for at least 12 hr at room temperature. Then place the mixture in a water bath at a temperature of 60°C for 20 min; add 50 ml of water, and keep at 50°C until the perborate dissolves. Cool to room temperature.

Add 30 ml of 10 N sodium hydroxide solution and 80 ml of water; add 4 grams of Devarda's alloy, and immediately set the flask in place for distillation of ammonia into standard acid (25.00 ml of 0.33 N) in accordance with the usual Devarda procedure (for suitable apparatus, consult standard works). Foaming during distillation may prove troublesome; if the usual addition of a small amount of paraffin is ineffective, silicone antifoaming agents may be used. When distillation of ammonia is complete, the standard acid is back-titrated with sodium hydroxide solution to determine the amount of ammonia generated from the nitrate ester sample. Naturally, blanks should be run concurrently with samples.

For many purposes, it is best to standardize the method at the same time that nitrate ester samples are analyzed by using very pure potassium nitrate as the calibrant. Thus exactly 0.7500 gram of potassium nitrate is treated in precisely the manner indicated above, and the titer of the standard sodium hydroxide solution is computed from the following:

$$T = \text{grams of N/ml of base} = 0.13853 \frac{K}{A - B}$$

where  $K$ = weight of potassium nitrate taken (gram),
  $A$ = NaOH (ml) $\approx$ 25.00 ml $H_2SO_4$,
  $B$ = NaOH needed in back titration (ml),
  $C = (A - B)$ for a blank determination.

Hence the nitrogen content of a nitrate ester is:

$$\% \, N = \frac{100 \, T (A - B - C)}{W}$$

where  $W$ = sample weight of ester (gram)
  $T$ = titer of base,
and $A$, $B$, and $C$ are the same as before.

When the sodium hydroxide solution has been accurately standardized, and its normality $(N)$ is known, the nitrogen content of a nitrate ester is

$$\% \, N = \frac{1.40067(A - B - C)N}{W}$$

### E.  DETERMINATION OF TOTAL NITROGEN IN NITRITE AND NITRATE ESTERS BY MICRO DUMAS COMBUSTION

Analysis of organic compounds containing O—NO$_2$ or O—NO groups by the Dumas method as modified by Pregl and many others usually gives low and scattered values for nitrogen. The following modification (21) makes use of a cobalto-cobaltic oxide catalyst (89) in the permanent packing and also includes the catalyst (along with copper oxide) in the temporary filling. Satisfactory results can be obtained for mono- to octanitrate esters with minimum risk of explosive decomposition.

### Special Reagent

*Cobalto-cobaltic oxide,* $CO_3O_4$. Dissolve 145 grams of cobalt nitrate hexahydrate in 200 ml of water; add a solution of 88 grams of oxalic acid dihydrate in 300 ml of water. Filter the precipitate, and wash with water and ethanol. Remove the precipitate from the filter, and form a paste by mixing it with a slurry of equal parts of starch and water. Dry the paste at 110–120°C. Ignite the dry residue at 550–600°C for 2–3 hours, cool, grind, and sieve; collect the portion that passes U.S. screen No. 16 and is retained on No. 40 screen.

### Apparatus

Pregl-Dumas nitrogen microanalysis apparatus of any convenient design or manufacture may be used; preferably, the apparatus should be of the standard form described in detail in standard references on microanalytical procedures (279). The standard combustion tube is replaced with a silica combustion tube as shown in Fig. 4. A fixed electric furnace maintains the temperature of the permanent packing at 650°C, and a movable furnace, operating at 700–750°C, is used to pyrolyze the sample, which is contained in a silica capsule 9 cm long by 4 mm i.d.

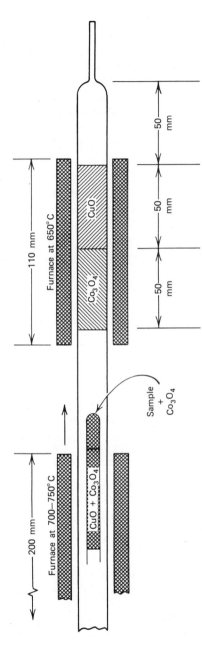

Fig. 4. Microcombustion tube for Dumas analysis of esters. [Adapted from Borda and Hayward (21).]

## Procedure

A sample weighing from 2 to 5 mg (depending on the nitrogen content) is placed in the capsule, together with 50 mg of finely powdered cobalto-cobaltic oxide. The capsule is stoppered and shaken so as to mix its contents. Then the capsule is filled with a mixture of 4 vols. of powdered cupric oxide and 1 vol. of powdered cobalto-cobaltic oxide and is placed in the combustion tube 5 cm away from the permanent packing, with the open end of the capsule facing the mouth of the combustion tube.

Air in the analysis system is displaced in the usual manner by carbon dioxide; then the stopcock between the combustion tube and the source of carbon dioxide is closed. The preheated movable furnace is moved slowly toward the sample, and the rate of motion is carefully controlled so that bubbles appear at the nitrometer at the rate of one per second, or less. When gas is no longer evolved, the movable furnace is stationed over the capsule, and carbon dioxide is allowed to enter the combustion tube so that nitrogen is swept into the gas-measuring buret at a rate of two to three bubbles per second. The remainder of the analysis is completed and the results are calculated in the usual manner.

## REFERENCES

1. Aldrovandi, R., and F. De Lorenzi, *Ann. Chim. (Rome)*, **42**, 298 (1952); through *Chem. Abstr.*, **46**, 10051 (1952).

2. Allert, J., *Dansk Tidsskr. Farm.*, **22**, 188 (1948); through *Chem. Abstr.*, **42**, 8714 (1948).

3. Altschuller, A. P., and C. A. Schwab, *Anal. Chem.*, **31**, 314 (1959).

4. Anbar, M., I. Dostrovsky, and D. Samuel, *J. Chem. Soc.*, **1954**, 3603.

5. Armstrong, M. D., and J. E. Copenhaver, *J. Am. Chem. Soc.*, **65**, 2252 (1943).

6. Ayres, W. M., and G. W. Leonard, *Anal. Chem.*, **31**, 1485 (1959).

7. Baker, J. W., and D. M. Easty, *J. Chem. Soc.*, **1952**, 1193.

8. Bandelin, F. J., and R. E. Pankratz, *Anal. Chem.*, **30**, 1435 (1958).

9. Banerjee, P. C., *J. Indian Chem. Soc.*, **12**, 198 (1935); through *Chem. Abstr.*, **29**, 5371 (1935).

10. Banerjee, P. C., *J. Indian Chem. Soc.*, **13**, 301 (1936); through *Chem. Abstr.*, **30**, 8072 (1936).

11. Barefoot, R. D., and A. R. Lawrence, *Appl. Spectrosc.*, **20**, 380 (1966).

12. Barker, S. A., E. J. Bourne, H. Weigel, and D. H. Whiffen, *Chem. Ind. (London)*, **1956**, 318.

13. Becker, W. W., *Ind. Eng. Chem., Anal. Ed.*, **5**, 152 (1933).

14. Bellamy, L. J., *The Infrared Spectra of Complex Molecules,* Methuen, London, 1958, Ch. 17.

15. Black, A. P., and F. H. Babers, in H. H. Blatt, Ed., *Organic Syntheses*, Coll. Vol. II, John Wiley, New York, 1943, p. 512.

16. Blais, M., *Microchem. J.*, **6**, 211 (1962).

17. Blom, J., and C. Treschow, *Z. Pflanzenernaehr. Dueng. Bodenk.*, **13A**, 159 (1929); through *Chem. Abstr.*, **23**, 4994 (1929).

18. Blyumberg, E. A., and V. L. Pikaeva, *Zh. Anal. Khim.*, **10**, 310 (1955); through *Chem. Abstr.*, **50**, 2369 (1956).

19. Bonner, T. G., *J. Chem. Soc.*, **1959**, 3908.

20. Booth, A. D., and F. J. Llewellyn, *J. Chem. Soc.*, **1947**, 837.

21. Borda, P., and L. D. Hayward, *Anal. Chem.*, **39**, 548 (1967).

22. Bordwell, F. G., and E. W. Garbisch, *J. Am. Chem. Soc.*, **82**, 3588 (1960).

23. Boschan, R., R. T. Merrow, and R. W. VanDolah, *Chem. Rev.*, **55**, 485 (1955).

24. Boschan, R., and R. Smith, U.S. Dept. of Commerce, Office of Technical Services, *PB*-151120. 1957.

25. Bose, P. K., *Analyst*, **56**, 504 (1931).

26. Bowman, F. C., and W. W. Scott, *Ind. Eng. Chem.*, **7**, 766 (1915).

27. Brand, J. C. D., and C. M. Cawthon, *J. Am. Chem. Soc.*, **77**, 319 (1955).

28. Brandner, J. D., *Ind. Eng. Chem.*, **30**, 681 (1938).

29. Brooks, C., and R. M. Badger, *J. Am. Chem. Soc.*, **72**, 1705 (1950).

30. Brown, J. F., Jr., *J. Am. Chem. Soc.*, **77**, 6341 (1955).

31. Burns, E. A., *Anal. Chim. Acta*, **26**, 143 (1962).

32. Busch, M., *J. Chem. Soc.*, **88**(I), 307 (1905); also *Z. Angew. Chem.*, **19**, 1329 (1906).

33. Camera, E., and D. Pravisani, *Anal. Chem.*, **36**, 2108 (1964).

34. Camera, E., and D. Pravisani, *Anal. Chem.*, **39**, 1645 (1967).

35. Cannon, J. H., and R. F. Heuermann, *J.A.O.A.C.*, **34**, 716 (1951).

36. Carol, J., *J.A.O.A.C.*, **43**, 259 (1960).

37. Chamot, E. M., and D. S. Pratt, *J. Am. Chem. Soc.*, **31**, 922 (1909).

38. Chamot, E. M., and D. S. Pratt, *J. Am. Chem. Soc.*, **32**, 630 (1910).

39. Chapman, E. T., and M. H. Smith, *J. Chem. Soc.*, **20**, 584 (1867).

40. Chapman, E. T., and M. H. Smith, *J. Chem. Soc.*, **22**, 158 (1869).

41. Cheronis, N. D., and M. Koeck, *J. Chem. Educ.*, **20**, 488 (1943).

42. Chrétien, A., and Y. Longi, *Compt. Rend.*, **220**, 746 (1945).

43. Christian, W. R., and C. B. Purves, *Can. J. Chem.*, **29**, 926 (1951).

44. Clarkson, A., and G. F. Reynolds, *J. Polarogr. Soc.*, **1960**, 4.

45. Clarkson, A., and C. M. Robertson, *Anal. Chem.*, **38**, 522 (1966).

46. Coleman, G. H., A. G. Farnham, and A. Miller, *J. Am. Chem. Soc.*, **64**, 1501 (1942).

47. Coleman, G. H., and C. M. McCloskey, *J. Am. Chem. Soc.*, **65**, 1588 (1943).

48. Cope, W. C., *Ind. Eng. Chem.*, **8**, 592 (1916).

49. Cope, W. C., and J. Barab, *J. Am. Chem. Soc.*, **39**, 504 (1917).

50. Cope, W. C., and G. B. Taylor, *U.S. Bur. Mines Tech. Pap.* 160, 1917; through *Chem. Abstr.*, **11**, 2277 (1917).

51. Cristol, S. J., B. Franzus, and A. Shadan, *J. Am. Chem. Soc.*, **77**, 2512 (1955).

52. Currah, J. E., paper presented at Analytical Subject Division, Chemical Institute of Canada, Quebec, February 1953.

53. Curti, R., and V. Riganti, *Rass. Chim.*, **16**, 176 (1964); through *Chem. Abstr.*, **66**, 56748 (1967).

54. Darley, E. F., K. A. Kettner, and E. R. Stephens, *Anal. Chem.*, **35**, 589, 1963.

55. Dehn, W. M., K. E. Jackson, and D. A. Bullard, *Ind. Eng. Chem., Anal. Ed.*, **4**, 413 (1932).

56. Devarda, A., *Chem. Ztg.*, **16**, 1952 (1892); through *Chem. Zentralbl.*, **1893**, I, 232.

57. Devarda, A., *Z. Anal. Chem.*, **33**, 113 (1894).

58. Dewar, J., and G. Fort, *J. Chem. Soc.*, **1944**, 492.

59. Dewey, B. F., and N. F. Witt, *Ind. Eng. Chem., Anal. Ed.*, **12**, 459 (1940).

60. Dewey, B. F., and N. F. Witt, *Ind. Eng. Chem., Anal. Ed.*, **14**, 648 (1942).

61. Dickson, W., and W. C. Easterbrook, *Analyst*, **47**, 112 (1922).

62. Divers, E., *J. Chem. Soc.*, **47**, 85 (1899).

63. D'Or, L., and J. Collin, *Bull. Soc. Roy. Sci. Liege*, **22**, 285 (1953).

64. Dox, A. W., and G. P. Plaisance, *J. Am. Chem. Soc.*, **38**, 2164 (1916).

65. Drew, H. D., in C. A. Streuli and P. R. Averell, Eds., *Analytical Chemistry of Nitrogen and Its Compounds*, Interscience, New York, 1970, Pt. I.

66. Edwards, W. R., Jr., and C. W. Tate, *Anal. Chem.*, **23**, 826 (1951).

67. Ellis, C. M., and A. I. Vogel, *Analyst*, **81**, 693 (1956).

68. Elving, P. J., and W. R. McElroy, *Ind. Eng. Chem., Anal. Ed.*, **14**, 84 (1942).

69. English, F. L., *Ind. Eng. Chem., Anal. Ed.*, **19**, 850 (1949).

70. Erdey, L., L. Polos, and Z. Gregorowicz, *Talanta*, **3**, 6 (1959).

71. Evered, S., and F. H. Pollard, *J. Chromatogr.*, **4**, 451 (1960).

71a. Fahr, E., *Annalen*, **638**, 1 (1960).

72. Fainer, P., *Can. J. Chem.*, **29**, 46 (1951).

73. Feigl, F., *Angew. Chem.*, **73**, 656 (1961).

74. Feigl, F., *Chemist-Analyst*, **51**, 4, 5 (1962).

75. Feigl, F., *Chemist-Analyst*, **52**, 47 (1963).

76. Feigl, F., and J. Amaral, *Mikrochim. Acta*, **1958**(3), 22.

77. Ferris, A. F., K. W. McLean, I. G. Marks, and W. D. Emmons, *J. Am. Chem. Soc.*, **75**, 4078 (1953).

78. Fischer, H., *Angew. Chem.*, **A60**, 334 (1948).

78a. Fischer, W., and N. Steinbach, *Z. Anorg. Chem.*, **78**, 134 (1912).

79. Fossel, E. T., *J. Gas Chromatogr.*, **3**, 179 (1965).

80. Francis, F. E., *J. Chem. Soc.*, **89**, 1 (1906).

81. Fraser, R. T. M., and N. C. Paul, *J. Chem. Soc.*, **1968(B)**, 659.

82. Fraser, R. T. M., and N. C. Paul, *J. Chem. Soc.*, **1968(B)**, 1407.

83. Frehden, O., and Z. Nicolescu, *Rev. Chim. (Bucharest)*, **9**, 688 (1958); through *Chem. Abstr.*, **54**, 21750 (1960).

84. Friedrich, A., E. Kuhaas, and R. Schnurch, *Z. Physiol. Chem.*, **216**, 68 (1933); through *Chem. Abstr.*, **27**, 2648 (1933).

85. Fritz, J. S., and G. H. Schenk, *Anal. Chem.*, **31**, 1808 (1959).

86. Fursov, A. F., *Aptechn. Delo*, **10**(1), 9 (1961); through *Chem. Abstr.*, **55**, 26370 (1961).

87. Garcia, E. E., *Anal. Chem.*, **39**, 1605 (1967).

88. Gardon, J. L., and B. Leopold, *Anal. Chem.*, **30**, 2057 (1958).

89. Gawargious, Y. A., and A. M. G. MacDonald, *Microchem. J.*, Symp. Ser. 2, 397 (1961).

90. Gold, V., E. P. Hughes, and C. K. Ingold, *J. Chem. Soc.*, **1950**, 2467.

91. Gottlieb, O. R., and M. T. Magalhaes, *Anal. Chem.*, **30**, 995 (1958).

92. Gowenlock, B. G., and J. Trotman, *J. Chem. Soc.*, **1956**, 1670.

93. Grebber, K., and Karabinos, J. V., *J. Res. Natl. Bur. Stand.*, **49**, 163 (1952).

94. Green, N., and M. W. Green, *J. Am. Chem. Soc.*, **66**, 1610 (1944).

95. Grodzinski, J., *Anal. Chem.*, **27**, 1765 (1955).

96. Grodzinski, J., *Anal. Chem.*, **29**, 150 (1957).

97. Gustin, G. M., in I. M. Kolthoff and P. J. Elving, Eds., *Treatise on Analytical Chemistry*, Part II, Vol. 11, Interscience, New York, 1965.

98. Hankonyi, V., and V. Karas-Gasparec, *Anal. Chem.*, **41**, 1849 (1969).

99. Hann, R. M., and C. S. Hudson, *J. Am. Chem. Soc.*, **66**, 735 (1944).

100. Harper, E. M., and A. K. Macbeth, *J. Chem. Soc.*, **107**, 87 (1915); through *Chem. Abstr.*, **9**, 1037 (1915).

101. Harrison, R. M., and F. J. Stevenson, *J. Gas Chromatogr.*, **3**, 240 (1965).

102. Hartley, A. M., and R. I. Asai, *Anal. Chem.*, **35**, 1207 (1963).

103. Hartman, H., and C. Bathge, *Angew. Chem.*, **65**, 107 (1953).

104. Haskins, W. T., R. M. Hann, and C. S. Hudson, *J. Am. Chem. Soc.*, **68**, 1766 (1946).

105. Hassid, W. Z., and R. M. McCready, *Ind. Eng. Chem., Anal. Ed.*, **14**, 683 (1942).

106. Haszeldine, R. N., *J. Chem. Soc.*, **1953**, 2525.

107. Haszeldine, R. N., and J. Jander, *J. Chem. Soc.*, **1954**, 691.

108. Haszeldine, R. N., and B. J. H. Mattinson, *J. Chem. Soc.*, **1955**, 4172.

109. Hayazu, R., *J. Pharm. Soc. Jap.*, **71**, 207 (1951); through *Chem. Abstr.*, **45**, 6537 (1951).

110. Hearon, W. M., and R. G. Gustavson, *Anal. Chem.*, **9**, 352 (1937).

111. Heilbrunner, E., *Helv. Chim. Acta*, **36**, 1121 (1953).

112. Henstock, H., *J. Chem. Soc.*, **1933**, 216.

113. Hetman, J. S., *Abh. Dtsch. Akad. Wiss. Berlin, Kl. Chem., Geol., Biol.*, **1964**(1), 169; through *Chem. Abstr.*, **64**, 4866 (1966).

114. Hetman, J. S., *Adv. Polarogr., Proc. Int. Congr., 2nd, Cambridge, 1960*, **2**, 640; through *Chem. Abstr.*, **57**, 8789 (1962).

115. Hetman, J. S., *Anal. Chem.*, **32**, 1699 (1960).

116. Hetman, J. S., *Talanta*, **5**, 267 (1960).

117. Hirschhorn, I. S., *Ind. Eng. Chem., Anal. Ed.*, **19**, 880 (1947).

118. Hohmann, J. R., and J. Levine, *J.A.O.A.C.*, **47**, 471 (1964).

119. Holler, A. C., and R. V. Huch, *Anal. Chem.*, **21**, 1385 (1949).

120. Hormann, H., J. Lamberts, and G. Fries, *Z. Physiol. Chem.*, **306**, 42 (1956).

121. Huff, W. J., and R. D. Leitch, *J. Am. Chem. Soc.*, **44**, 2643 (1922).

122. Hunter, E. C. E., and J. R. Partington, *J. Chem. Soc.*, **1933**, 309.

123. Hyde, A. L., *J. Am. Chem. Soc.*, **35**, 1173 (1913).

124. *Index of Mass Spectral Data*, ASTM STP-356, 1st ed., American Society for Testing and Materials, Philadelphia, Pa., 1963.

125. Jodlbauer, M., *Chem. Zentralbe.*, **57**, 433 (1886).

126. Jeanes, A., C. S. Wise, and R. J. Dimpler, *Anal. Chem.*, **23**, 415 (1951).

126a. Johnson, C. L., *Anal. Chem.*, **25**, 1276 (1953).

127. Jones, L. H., R. M. Badger, and C. E. Moore, *J. Chem. Phys.*, **19**, 1599 (1951).

128. Jones, R. N., and G. D. Thorn, *Can. J. Chem.*, **27B**, 828 (1949).

129. Jones, W. L., and A. W. Scott, *J. Am. Chem. Soc.*, **46**, 2172 (1924).

130. Kabasakalian, P., and E. R. Townley, *J. Org. Chem.*, **27**, 2918 (1962).

131. Kabasakalian, P., E. R. Townley, and M. D. Yudis, *J. Am. Chem. Soc.*, **84**, 2718 (1962).

132. Kamiike, O., T. Kodama, and G. Nakano, *J. Soc. Chem. Ind. Jap.*, **45**, 628 (1942); through *Chem. Abstr.*, **43**, 1966 (1949).

133. Kaufman, F., H. J. Cook, and S. M. Davis, *J. Am. Chem. Soc.*, **74**, 4997 (1952).

134. Kaufman, F., E. Ossofsky, and J. H. Cook, *Anal. Chem.*, **26**, 516 (1954).

135. Ketcham, R., et al., *J. Org. Chem.*, **28**, 2139 (1963).

136. Khamskii, E. V., and V. A. Il'ina, *Zavodsk. Lab.*, **29**, 799 (1963); through *Chem. Abstr.*, **59**, 10767 (1963).

137. Kirsten, W., in C. L. Wilson and D. W. Wilson, Eds., *Comprehensive Analytical Chemistry*, Vol. 1B, Elsevier, Amsterdam, 1960.

138. Klein, B., and M. Weissman, *Anal. Chem.*, **25**, 771 (1953).

139. Knecht, E., and E. Hibbert, *New Reduction Methods in Volumetric Analysis*, 2nd ed., Longmans, Green, London, 1925.

140. Koehler, A., M. Marqueyrol, and P. Jovinet, *Mem. Poudres*, **19**, 356 (1922); through *Chem. Abstr.*, **17**, 1775 (1923).

141. Kohlbeck, J. A., *Anal. Chem.*, **37**, 1282 (1965).

142. Kolthoff, I. M., and G. E. Noponen, *J. Am. Chem. Soc.*, **55**, 1448 (1933).

143. Kolthoff, I. M., E. B. Sandell, and B. Moskowitz, *J. Am. Chem. Soc.*, **55**, 1454 (1933).

144. Kopp, E., *Annalen*, **64**, 320 (1847).

145. Kopp, E., *Berichte*, **5**, 284 (1872).

146. Kornblum N., M. E. Chalmers, and R. Daniels, *J. Am. Chem. Soc.*, **77**, 6654 (1955).

147. Kornblum, N., N. N. Lichtin, J. T. Patton, and D. C. Iffland, *J. Am. Chem. Soc.*, **69**, 307 (1947).

148. Kornblum, N., and E. P. Oliveto, *J. Am. Chem. Soc.*, **71**, 226 (1949).

149. Kornblum, N., J. T. Patton, and J. B. Nordmann, *J. Am. Chem. Soc.*, **70**, 746 (1948).

150. Kornblum, N., R. A. Smiley, et al., *J. Am. Chem. Soc.*, **77**, 5528 (1955).

151. Kornblum, N., B. Taub, and H. E. Ungnade, *J. Am. Chem. Soc.*, **76**, 3209 (1954).

152. Kornblum, N., H. E. Ungnade, and R. A. Smiley, *J. Org. Chem.*, **21**, 377 (1956).

153. Kruger, D., *Z. Angew. Chem.*, **41**, 407 (1928).

154. Kuck, J. A., et al., *Anal. Chem.*, **22**, 604 (1950).

155. Kuhn, L. P., *Anal. Chem.*, **22**, 276 (1950).

156. Kuhn, L. P., *J. Am. Chem. Soc.*, **68**, 1761 (1946).

157. Kuhn, L. P., *J. Am. Chem. Soc.*, **69**, 1974 (1947).

158. Kuhn, L. P., *J. Am. Chem. Soc.*, **73**, 1510 (1951).

159. Kuhn, L., and E. R. Lippincott, *J. Am. Chem. Soc.*, **78**, 1820 (1956).

160. Laccetti, M. A., S. Semel, and M. Roth, *Anal. Chem.*, **31**, 1049 (1959).

161. Lacroix, Y., N. Bugat, and J. Mars, *Mem. Poudres*, **39**, 459 (1957); through *Chem. Abstr.*, **52**, 19686 (1958).

162. Lambert, R. S., and R. J. DuBois, *Anal. Chem.*, **39**, 427 (1965).

163. Lamond, J., *Analyst*, **73**, 674 (1948).

164. Lane, E. S., *J. Chem. Soc.*, **1953**, 1172.

165. Lecker, H., and W. Siefken, *Berichte*, **59**, 1314, 2594 (1926).

166. Leclercq, R., and J. Mathe, *Bull. Soc. Chim. Belg.*, **60**, 296 (1951); through *Chem. Abstr.*, **46**, 4232 (1952).

167. LeFevre, R. J. W., W. T. Oh, I. H. Reece, and R. L. Werner, *Aust. J. Chem.*, **10**, 361 (1957).

168. LeRosen, A. L., *J. Am. Chem. Soc.*, **64**, 1905 (1942).

169. LeRosen, A. L., *J. Am. Chem. Soc.*, **67**, 1683 (1945).

170. LeRosen, A. L., *J. Am. Chem. Soc.*, **69**, 87 (1947).

171. LeRosen, A. L., et al., *Anal. Chem.*, **22**, 809 (1950).

172. LeRosen, A. L., R. T. Moravek, and J. K. Carlton, *Anal. Chem.*, **24**, 1335 (1952).

173. Lesnicenko, K., *Chem. Obzor*, **10**, 140, 165, 192 (1935); through *Chem. Abstr.*, **30**, 3357 (1936).

174. Levitsky, H., and G. Norwitz, *Anal. Chem.*, **34**, 1167 (1962).

175. Levy, J. B., *J. Am. Chem. Soc.*, **76**, 3254 (1954).

176. Lewis, D. G., *Anal. Chem.*, **33**, 1127 (1961).

177. Lieber, E., D. R. Levering, and L. J. Patterson, *Anal. Chem.*, **23**, 1594 (1951).

178. Ligtenberg, H. L., *Mem. Poudres*, **44**, 391 (1962); through *Chem. Abstr.*, **62**, 8923 (1965).

179. Lingane, J. J., and R. L. Pecsok, *Anal. Chem.*, **21**, 622 (1949).

180. Lipscomb, W. N., and R. H. Baker, *J. Am. Chem. Soc.*, **64**, 179 (1942).

181. Lothrop, W. C., and G. R. Handrick, *Chem. Rev.*, **44**, 419 (1949).

182. Lucas, G. R., and L. P. Hammett, *J. Am. Chem. Soc.*, **64**, 1928 (1942).

183. Lundgren, P., and T. Canback, *Svensk Farm. Tid.*, **52**, 315 (1948); through *Chem. Abstr.*, **43**, 358 (1949).

184. Lunge, G., *J. Soc. Chem. Ind.*, **9**, 547 (1890).

185. Lunge, G., *J. Soc. Chem. Ind.*, **20**, 100 (1901).

186. Mach, F., and F. Sindlinger, *Z. Anal. Chem.*, **60**, 235 (1921).

187. Macke, G. F., *J. Chromatogr.*, **38**, H7 (1968).

188. Malins, D. C., and C. R. Houle, *J. Am. Oil Chem. Soc.*, **40**, 43 (1963).

189. Malins, D. C., J. C. Wekell, and C. R. Houle, *Anal. Chem.*, **36**, 658 (1964).

190. Malmberg, E. W., *Anal. Chem.*, **27**, 840 (1955).

191. Mangold, H. K., *J. Am. Oil Chem. Soc.*, **38**, 708 (1961).

192. Manning, J. F., and J. P. Mason, *J. Am. Chem. Soc.*, **62**, 3136 (1940).

193. Marshall, A., *J. Soc. Chem. Ind.*, **49**, 32 (1930).

194. Mason, J. P., and J. F. Manning, *J. Am. Chem. Soc.*, **62**, 1635 (1940).

195. Meadoe, J. R., and E. E. Reid, *J. Am. Chem. Soc.*, **65**, 457 (1943).

196. Merrow, R. T., S. J. Cristol, and R. W. VanDolah, *J. Am. Chem. Soc.*, **75**, 4259 (1953).

197. Merrow, R. T., and R. W. VanDolah, *J. Am. Chem. Soc.*, **76**, 4522 (1954).

198. Michael, A., and G. H. Carlson, *J. Am. Chem. Soc.*, **57**, 1268 (1935).

199. Michael, A., and G. H. Carlson, *J. Org. Chem.*, **5**, 1 (1940).

200. Mikkelson, L., and R. S. Richmond, *Anal. Chem.*, **34**, 74 (1962).

201. Milkey, R. G., *Anal. Chem.*, **30**, 1931 (1958).

202. Mitra, B. N., and M. Srinivasan, *J. Indian Chem. Soc.*, **21**, 397 (1944); through *Chem. Abstr.*, **39**, 4487 (1945).

203. Monaghan, P. H., H. A. Suter, and A. L. LeRosen, *Anal. Chem.*, **22**, 811 (1950).

204. Mortimer, C. T., H. Spedding, and H. D. Springall, *J. Chem. Soc.*, **1957**, 188.

205. Mullaly, M. A. C., *Analyst*, **80**, 237 (1955).

206. Muraca, R. F., Ed., *Handbook of JANAF Panel on Analytical Chemistry of Solid Propellants*, SPIA/H-2; No. 708.0, Recommended Specification for Nitrocellulose; see also JAN-N-244.

207. Murakami, T., *Bunseki Kagaku*, **7**, 766 (1958); through *Chem. Abstr.*, **53**, 19682 (1959).

208. Murakami, T., *Jap. Analy*, **4**, 630 (1955); through *Chem. Abstr.*, **50**, 16559 (1956).

209. Mutakami, K., *J. Soc. Org. Synth. Chem. (Jap.)*, **8**, 127 (1950); through *Chem. Abstr.*, **47**, 1385 (1953).

210. Muraour, H., *Bull. Soc. Chim.*, **45**, 1189 (1929).

211. Nagasawa, K., et al., *Anal. Chem.*, **42**, 1436 (1970).

212. Newfield, J., and J. S. Marx, *J. Am. Chem. Soc.*, **28**, 877 (1906).

213. Oldham, J. W. H., *J. Soc. Chem. Ind.*, **53**, 236T (1934).

214. Orchin, M., *J.A.O.A.C.*, **26**, 99 (1943).

215. Ovenston, T. C. J., *Analyst*, **74**, 344 (1949).

216. Oza, T. M., V. T. Oza, and N. L. Dipali, *J. Indian Chem. Soc.*, **28**, 15 (1951); through *Chem. Abstr.*, **45**, 8939 (1951).

217. Oza, V. T., *Anal. Chem.*, **29**, 453 (1957).

218. Paraskevopoulos, G., and R. J. Cvetanovic, *J. Chromatogr.*, **25**, 479 (1966).

219. Partington, J. R., and C. C. Shah, *J. Chem. Soc.*, **1931**, 2071; see also *Chem. Abstr.*, **25**, 5637 (1931).

220. Partington, J. R., and C. C. Shah, *J. Chem. Soc.*, **1932**, 2593.

220a. Patsevich, I. V., A. V. Topchiev, and V. Ya Shtern, *Zh. Anal. Khim.*, 13, 608 (1958), through *Chem. Abstr.*, **53**, 5976 (1959).

221. Pauling, L., and L. O. Brockway, *J. Am. Chem. Soc.*, **59**, 13 (1937).

222. Phillippe, R. J., *J. Mol. Spectrosc.*, **6**, 492 (1961).

223. Phillippe, R. J., and H. Moore, *Spectrochim. Acta*, **17**, 1004 (1961).

224. Phillips, J. P., *Spectra-Structure Correlation*, Academic Press, New York, 1964, Ch. IV.

225. Pierson, R. H., in F. J. Welcher, Ed., *Standard Methods of Chemical Analysis*, Vol. II-B; *Industrial and Natural Products and Noninstrumental Methods*, 6th ed., D. Van Nostrand, Princeton, N. J., 1963.

226. Pierson, R. H., and E. C. Julian, *Anal. Chem.*, **31**, 589 (1959).

227. Pinchas, S., *Anal. Chem.*, **23**, 201 (1951).

228. Pitman, J. R., *J. Soc. Chem. Ind.*, **19**, 983 (1900).

229. Pollard, F. H., and C. J. Hardy, *Chem. Ind. (London)*, **1956**, 527.

230. Pollard, F. H., A. E. Pedler, and C. J. Hardy, *Nature*, **174**, 979 (1954).

231. Pristera, F., *Anal. Chem.*, **25**, 844 (1953).

232. Pristera, F., and W. E. Fredericks, *Compilation of Infrared Spectra of Ingredients of Propellants and Explosives*, Picatinny Arsenal, October 1965; presented at 22nd Meeting of ICRPG (JANAF Panel).

233. Pristera, F., M. Halik, A. Castelli, and W. Fredericks, *Anal. Chem.*, **32**, 495 (1960).

234. Quense, J. A., and W. M. Dehn, *Ind. Eng. Chem., Anal. Ed.*, **11**, 555 (1939).

235. Quense, J. A., and W. M. Dehn, *Ind. Eng. Chem., Anal. Ed.*, **12**, 556 (1940).

236. Radin, N., and T. DeVries, *Anal. Chem.*, **24**, 971 (1952).

237. Rao, C. N. R., *Ultraviolet and Visible Spectroscopy—Chemical Applications*, Butterworths, London, 1961, p. 20.

238. Redemann, C. E., and H. J. Lucas, *Ind. Eng. Chem., Anal. Ed.*, **9**, 521 (1937).

239. Rheinboldt, H., *Berichte*, **59**, 1311 (1926).

240. Rinkenbach, W. H., *Ind. Eng. Chem.*, **19**, 925 (1927).

241. Rinkenbach, W. H., and H. A. Aaronson, *Ind. Eng. Chem.*, **23**, 160 (1931).

242. Roberts, A. G., *Anal. Chem.*, **21**, 813 (1949).

243. Rosenberger, H. M., and C. J. Shoemaker, *Anal. Chem.*, **31**, 1315 (1959).

244. Rubens, E., *Z. Ges. Schiess. Sprengstoffw.*, **28**, 172 (1933); through *Chem. Abstr.*, **27**, 4073 (1933).

245. deleted.

246. Sarnoff, E., *J.A.O.A.C.*, **38**, 637 (1955).

247. Sarnoff, E., *J.A.O.A.C.*, **39**, 630 (1956).

248. Sato, H., *Bull. Inst. Phys. Chem. Res. (Tokyo)*, **16**, 804 (1937); through *Chem. Abstr.*, **32**, 4413 (1938).

249. Sawicki, E., T. W. Stanley, J. Pfaff, and A. D'Amico, *Talanta*, **10**, 641 (1963).

250. Schenk, G. H., and M. Santiago, *Anal. Chem.*, **39**, 1795 (1967).

251. Schmidt, O., *Angew. Chem.*, **62**, 22 (1950).

252. Schoch, T. J., and C. C. Jensen, *Ind. Eng. Chem., Anal. Ed.*, **12**, 531 (1940).

253. Schoniger, W., in C. N. Reilly, Ed., *Advances in Analytical Chemistry and Instrumentation*, Vol. 1, Interscience, New York, 1960.

254. Schroeder, W. A., *Ann. N. Y. Acad. Sci.*, **49**, 204 (1948).

255. Schroeder, W. A., et al., *Anal. Chem.*, **23**, 1740 (1951).

256. Schroeder, W. A., et al., *Ind. Eng. Chem.*, **41**, 2818 (1949).

257. Schulek, E., K. Burger, and M. Feher, *Z. Anal. Chem.*, **177**, 81 (1960).

258. Schulze, N., *Rep. Pharm.*, **20**, 800 (1871); through *Bull. Soc. Chim. Belg.*, **60**, 296 (1951).

259. Scott, A. W., *J. Am. Chem. Soc.*, **49**, 986 (1927).

260. Scoville, W. L., *Am. J. Pharm.*, **83**, 359 (1911).

261. Seaman, W., et al., *Ind. Eng. Chem., Anal. Ed.*, **14**, 420 (1942).

262. Seikel, M. K., and E. H. Huntress, *J. Am. Chem. Soc.*, **63**, 593 (1941).

263. Seyewetz, A., *Bull. Soc. Chim.*, **45**, 463 (1929); through *Chem. Abstr.*, **23**, 4421 (1929).

264. Shaefer, W. E., and W. W. Becker, *Anal. Chem.*, **25**, 1226 (1953).

265. Shankster, H., and T. H. Wilde, *J. Soc. Chem. Ind.*, **57**, 91 (1938).

266. Shimonaev, G. S., and L. S. Stepanova, *Khim. Tekhnol. Topliv Masel*, **7**(9), 67 (1962); through *Chem. Abstr.*, **58**, 1282 (1963).

267. Shraiber, M. S., and B. A. Rubinshtein, *Aptechn. Delo*, **3**(5), 46 (1954). through *Chem. Abstr.*, **49**, 2674 (1955).

268. Shriner, R., and E. Parker, *J. Am. Chem. Soc.*, **55**, 766 (1933).

269. Siegel, H., A. B. Bullock, and G. B. Carter, *Anal. Chem.*, **36**, 502 (1964).

270. Silberrad, O., H. A. Phillips, and H. J. Merriman, *J. Soc. Chem. Ind.*, **25**, 628 (1906).

271. Simonyi, I., and G. Tokar, *Acta Pharm. Hung.*, **27**, 17 (1957); through *Chem. Abstr.*, **52**, 20888 (1959).

272. Smith, L. I., and J. W. Horner, *J. Am. Chem. Soc.*, **62**, 1349 (1940).

273. Snyder, H. R., R. G. Handrick, and L. A. Brooks, in L. I. Smith, Ed., *Organic Syntheses*, Vol. 22, John Wiley, New York, 1942, p. 5.

274. Soffer, L. M., E. W. Parrotta, and J. DiDomenico, *J. Am. Chem. Soc.*, **74**, 5301 (1952).

275. Southworth, B. C., *Anal. Chem.*, **28**, 1611 (1956).

276. Stalcup, H., M. I. Fauth, J. O. Watts, and R. W. Williams, *Anal. Chem.*, **29**, 1482 (1957).

277. Stalcup, H., and R. W. Williams, *Anal. Chem.*, **27**, 543 (1955).

278. Stanek, J., and J. Vacek, *Chem. Prumysl*, **8**(33), 361 (1958); through *Chem. Abstr.*, **52**, 19715 (1958).

279. Steyermark, A., *Quantitative Organic Microanalysis*, 2nd ed., Academic Press, New York, 1961.

280. Strain, H. H., *Ind. Eng. Chem.*, *Anal. Ed.*, **14**, 245 (1942).

281. Swann, M. H., *Anal. Chem.*, **29**, 1504 (1957).

282. Swann, M. H., *Anal. Chem.*, **29**, 1505 (1957).

283. Swann, M. H., and M. L. Adams, *Anal. Chem.*, **28**, 1630 (1956).

284. Taras, M. J., *Anal. Chem.*, **22**, 1020 (1950).

285. Tarte, P., *Bull. Soc. Chim. Belg.*, **60**, 227 (1951).

286. Tarte, P., *Bull. Soc. Chim. Belg.*, **60**, 240 (1951).

287. Tarte, P., *J. Chem. Phys.*, **20**, 1570 (1952).

288. Tasker, H. S., and H. O. Jones, *J. Chem. Soc.*, **95**, 1910 (1909).

288a. U.S. Pat. 3,042,663 (July 3, 1962), G. Scott; through *Chem. Abstr.*, **57**, 11396 (1962).

289. Tiemann, F., *Berichte*, **6**, 1034 (1873).

290. Trowell, J. M., *Anal. Chem.*, **42**, 1440 (1970).

291. Trowell, J. M., and M. C. Philpot, *Anal. Chem.*, **41**, 166 (1969).

292. Trueblood, K. N., and E. W. Malmberg, *J. Am. Chem. Soc.*, **72**, 4112 (1950).

292a. Ubaldini, I., and F. Guerriere, *Ann. Chim. Appl.*, **38**, 702 (1948).

293. Ulsch, K., *Angew. Chem.*, **4**, 241 (1891).

294. Ulsch, K., *Z. Anal. Chem.*, **30**, 175 (1891).

295. Ungnade, H. E., and R. A. Smiley, *J. Org. Chem.*, **21**, 993 (1956).

296. U.S. Pharmacopoeia, 14th revision, Mack, Easton, Pa., 1950.

297. Utz, F., *Z. Anal. Chem.*, **47**, 142 (1908).

298. Verschragen, P., *Anal. Chim. Acta*, **12**, 227 (1955).

299. Warren, F. A., *Rocket Propellants*, Reinhold, New York, 1958.

300. Watts, J. O., and H. Stalcup, *Anal. Chem.*, **29**, 253 (1957).

301. Wheeler, P. P., and M. I. Fauth, *Microchem. J.*, **9**, 309 (1965).

302. Whitmore, W. F., and E. Lieber, *Ind. Eng. Chem.*, *Anal. Ed.*, **7**, 127 (1935).

303. Whitnack, G. C., M. M. Mayfield, and E. StC. Gantz, *Anal. Chem.*, **27**, 899 (1955).

304. Whitnack, G. C., J. M. Nielsen, and E. StC. Gantz, *J. Am. Chem. Soc.*, **76**, 4711 (1954).

305. Wight, T. W., Naval Ordnance Inspection Laboratory, Caerwent, *N.O.I.L. Rep. CR* 2/48, May 1948.

306. Williams, A. F., and D. Kenyon, *Adv. Polarogr., Proc. Int. Congr., 2nd, Cambridge, 1959*, **2**, 565; through *Chem. Abstr.*, **57**, 8790 (1962).

307. Woo, D., J. K. C. Yen, and P. Sofronas, *Anal. Chem.*, **45**, 2144 (1973).

308. Yagoda, H., *Ind. Eng. Chem., Anal. Ed.*, **15**, 27 (1943).

309. Yagoda, H., and F. H. Goldman, *J. Ind. Hyg. Toxical.*, **25**, 440 (1943).

310. Zechmeister, L., L. von Cholnoky, and E. Ujhelyi, *Bull. Soc. Chim. Biol.*, **18**, 1885 (1936).

# NITRILE, ISOCYANIDE, CYANAMIDE, AND CARBODIIMIDE GROUPS

By R. F. Muraca, *College of Notre Dame Belmont, California*

**Contents**

# I. INTRODUCTION

Organic compounds with the functional group —C≡N connected directly to a carbon atom are known as nitriles, and those with the group represented in the ground state by the resonance forms $R—\overset{+}{N}\overset{\rightleftharpoons}{\equiv}\overset{-}{C}:$ ↔R—N̈=C: are called isocyanides (the name adopted by IUPAC). Existence of these functional groups implies that hydrogen cyanide should have tautomers of the forms H—C≡N and H—N̈=C:, but the latter form has not been observed.

Before about 1958, the classical method of Hofmann (131) was the principal route for the synthesis of isocyanides; today, the method serves as "the carbylamine test" for amines, that is, when a primary amine, chloroform, and alcoholic potassium hydroxide are heated, there results the characteristic, repulsive odor of an isocyanide. The isocyanides are often called isonitriles and are termed carbylamines in older literature, but these names are to be abandoned.

The nomenclature of the nitriles and isocyanides is simple and straightforward. The nitriles are named (one word) for the acids into which they can be hydrolyzed; for example, $CH_3CN$ is acetonitrile and $CH_3CH_2CN$ is propionitrile. For compounds such as "Cyclohexyl-nitrile," $C_6H_{11}CN$, where the name does not suggest the acid into which the nitrile can be hydrolyzed, the term "Carbonitrile" is used, and the compound under discussion is named cyclohexyl carbonitrile. The isocyanides are named (two words) to indicate the residue to which the —NC group is attached; thus $CH_3NC$ is methyl isocyanide and $C_6H_5NC$ is phenyl isocyanide.

The cyanohydrins, or $\alpha$-hydroxynitriles, are also nitriles, $\diagdown$C(OH)CN, but as they are in equilibrium with a carbonyl compound and cyanide;

$$\diagdown C(OH)CN \rightleftharpoons \diagdown C{=}O + HCN$$

analysis of the functional group is best considered from the viewpoint of the reactions of the carbonyl group and the inorganic cyanide ion.

The acyl cyanides, RCOCN, constitute an important group of nitriles and are often named in accordance with the acyl radical; for example, $CH_3COCN$ is acetyl cyanide or pyruvonitrile, and $C_6H_5COCN$ is benzoyl cyanide or phenylglyoxylonitrile.

Cyanamide, $H_2N{-}C{\equiv}N$, also is a nitrile group compound, but it differs from the "nitriles" in that the cyanide function is linked to a nitrogen atom instead of a carbon atom. The hydrogens of the amido group of cyanamide are slightly acidic in character and can form derivatives. The dimer of cyanamide, dicyandiamide or cyanoguanidine, $H_2NC({=}NH)NHCN$, also forms derivatives by replacement of hydrogen atoms. Carbodiimides are structures that contain the functional group $-N{=}C{=}N-$; they are named as disubstituted carbodiimides.

Nitrile $N$-oxides, $RC{\equiv}N^+{-}{}^-O$ are unstable compounds, seldom isolated, and more often existing transitorily as intermediates in compounds having C—N bonds; on the other hand, some nitrile oxides are sufficiently stable to be isolated. The nitrile oxides will be discussed in a following section as fulminides.

## II. SYNTHESIS AND OCCURRENCE

### A. NITRILES

Nitriles can be synthesized by a wide variety of methods, but only a few are of interest to analytical chemists; the small amounts of isonitriles that are also formed usually are removed by a wash with dilute hydrochloric acid.

### 1. Metallic Cyanide and Organic Halide

The reaction of a metallic cyanide and an organic halide (bromide or chloride) often is used in laboratory work because starting materials are readily available:

$$R—Cl + NaCN \rightarrow RCN + NaCl$$

The reaction is slow and must take place in a solvent such as ethylene glycol or alcohol (173), but some nitriles can be prepared in aqueous solutions; for example, adiponitrile is formed by reaction of aqueous sodium cyanide and 1,4-dichlorobutane (292). The metallic cyanides usually employed are those of the alkali metals or the alkaline earths such as calcium; different results can be obtained with sodium and potassium cyanides. Cuprous cyanide is used for the preparation of unsaturated nitriles such as allyl cyanide (273); *isocyanides* are formed by reaction of silver cyanide with alkyl iodides (see below). Acyl cyanides,

$$R—C\overset{\displaystyle O}{\underset{\displaystyle CN}{\big<}}$$

are readily prepared from acyl bromides and cuprous cyanide.

## 2. An Aromatic Sulfonate and a Metallic Cyanide

The reagents are mixed and heated to distil the nitrile (306):

$$ArSO_2ONa + NaCN \rightarrow ArCN + Na_2SO_3$$

Unfortunately, yields are often low.

## 3. Dehydration of an Amide

$$RCONH_2 \rightarrow RCN + H_2O$$

Thionyl chloride (247), phorphorus pentoxide, and phosphorus oxychloride (157) are often used as dehydrating agents but even acetic anhydride suffices (293).

## 4. Fatty Acids and Ammonia at High Temperature

Fatty acids such as stearic, oleic, and palmitic can be converted to nitriles in one step by forming the ammonium salt and then heating; the process is also used on a large scale for commercial preparation of amides. Ammonia is passed into fatty acids to form a mixture of salts and amides; the amides are vaporized and passed over an alumina catalyst at 200–300°C (158). In other processes, fatty acids are treated with ammonia in the presence of a dehydration catalyst and activated $Al_2O_3$ or $SiO_2$ at 200–300°C (294); adiponitrile (for manufacture of nylon) can be made from adipic acid in this way.

## 5. Aromatic Hydrocarbons and Ammonia

Aromatic hydrocarbons with a methyl group react with ammonia in the presence of a catalyst such as molybdenum oxide on alumina at temperatures in excess of 500°C:

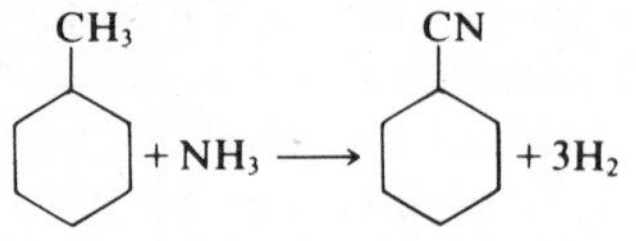

A catalyst such as bismuth, tin, or antimony phosphomolybdate may be employed when air is used to oxidize hydrogen.

## 6. Addition of Hydrogen Cyanide to an Unsaturated Compound

A typical example of the addition of hydrogen cyanide to a double bond is the preparation of acrylonitrile (intermediate for nylon and rubber) from acetylene (106):

$$HCN + HC\equiv CH \rightarrow CH_2=CHCN$$

Cuprous ammonium chloride in dilute hydrochloric acid at 70–90°C is a catalyst for the reaction. However, this process is no longer used, and more recent synthetic routes for preparation of acrylonitrile are based on the catalytic ammonoxidation (oxyamination) of propylene with ammonia and air (114); bismuth molybdate is a typical catalyst:

$$2\,CH_3CN=CH_2 + 2\,NH_3 + 3\,O_2 \rightarrow 2\,CH_2=CHCN + 3\,H_2O$$

## 7. Diazonium Salt Reaction with Cuprous Cyanide

Aromatic nitriles may be prepared by the Sandmeyer reaction (232) [see also Section III.B.6.b(2) of "Diazonium Group," Vol. 15].

$$ArN_2Cl \rightarrow ArCN + N_2$$

## 8. Addition of Hydrogen Cyanide to Carbonyl Compounds

Hydrogen cyanide adds to carbonyl compounds to give cyanohydrins:

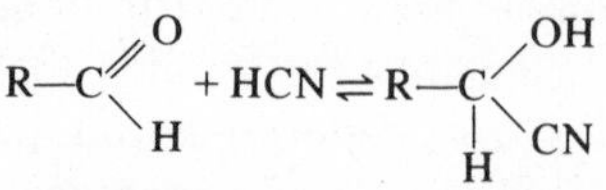

The reaction is reversible; favorable equilibrium is established with aliphatic and alicylic aldehydes and ketones, but alkyl aryl ketones such as acetophenone barely react, and diaryl ketones do not react at all

(169). A more convenient synthetic route is the dehydration of an oxime, for example, by acetic anhydride.

Acyl cyanides can be made by reaction of hydrogen cyanide with anhydrides at about 250°C (291):

$$(CH_3CO)_2O + HCN \xrightarrow{\text{catalyst}} CH_3COCN + CH_2COOH$$

## B. ISOCYANIDES

The first isocyanides (carbylamines) were prepared by Gautier in 1866 (105) by action of silver cyanide on alkyl iodides (cf. synthesis of nitriles). In fact, isocyanides are always formed as by-products when nitriles are prepared by reaction of a metallic cyanide with an alkyl, aryl, or alkyl aryl halide (especially at elevated temperatures); since isocyanides have lower boiling points than the corresponding nitriles, they can usually be removed by distillation.

### 1. Silver Cyanide and an Alkyl Iodide

Reaction of alkyl iodide with stoichiometric quantities of silver cyanide forms a complex:

$$RI + AgCN \rightarrow (AgCNR)^+I^-$$

which is decomposed by treatment with alkali cyanide:

$$(AgCNR)^+I^- \xrightarrow{\text{KCN}} RNC + KAg(CN)_2$$

Yields are disappointingly low; loss of volatile isocyanides must be prevented.

### 2. Primary Amines, Chloroform, and Strong Base

The Hofmann carbylamine reaction is useful for preparation of all types of isocyanides, but yields often may be very low:

$$RNH_2 + CHCl_3 + 3 NaOH \rightarrow RNC + 3 NaCl + 3 H_2O$$

### 3. Dehydration of N-Monosubstituted Formamides

Combinations of phosgene and trialkyl amines such as trimethylamine (284) or pyridine act on N-substituted alkyl or aryl formamides to form isocyanides in good yields. Typically, phosgene is led into a vigorously agitated solution or suspension of the formamide in trimethylamine until reaction is complete. Then ammonia is led into the mixture, ammonium chloride is filtered off, and the filtrate is evaporated *in vacuo* to obtain the crude isocyanide (287). The dehydration may also be accomplished by phosphorus oxychloride in the presence of pyridine or potassium *tert*-butoxide, by aryl sulfochlorides such as benzenesulfonyl chloride in pyridine (129), or by toluene-*p*-sulfonyl chloride in quinoline (48):

$$RNHCHO + ArSO_2Cl + C_5H_5N \rightarrow RNC + C_5H_5NSO_3H + C_5H_5N \cdot HCl$$

### 4. *N*-Alkylation of Hydrogen Cyanide with Olefins

The synthesis yields isocyanides remarkably free of isomeric nitriles; in general, 4 parts of an unsymmetrically disubstituted olefins, 4 parts of hydrogen cyanide, and 1 part of cuprous bromide are heated for more than 5 hr at 100°C (pressure vessel). Complexes of the type $(RNC)_3CuBr$ and $(RNC)CuCN$ are formed; isocyanides are liberated by aqueous alkali cyanide (206).

## C. CYANAMIDES

The most important derivative is calcium cyanamide, manufactured by passing nitrogen through a mixture of calcium carbide with about 10% calcium oxide at a temperature of about 1100°C:

$$CaC_2 + N_2 \rightarrow CaNCN + C$$

The reaction is exothermic; the cyanamide process was the first commercially important method for the fixation of nitrogen.

In the laboratory, cyanamide is obtained readily by treatment of a cold (not saturated) solution of thiourea with stoichiometric amounts of freshly precipitated lead hydroxide or mercuric oxide; incidentally, thiourea can be made by treating calcium cyanamide with hydrogen sulfide:

$$CaNCN + 2 H_2S \xrightarrow{165°C} CaS + H_2NCSNH_2$$
$$\xrightarrow{HgO} H_2NCN + HgS + H_2O$$

Cyanamide can also be synthesized by reaction of cyanogen chloride with ammonia; its sodium salt, $Na_2NCN$, is made by heating sodium amide with carbon or sodium cyanide.

*N*-Substituted cyanamides can be synthesized directly because both hydrogens on the amine group of cyanamide are replaceable. For example, diethylcyanamide, $(C_2H_5)_2NCN$, and its homologs are obtained by action of alkyl iodides or sulfates on calcium cyanamide. The reaction of cyanogen bromide with amines (Von Braun) is frequently used to synthesize alkyl- as well as aryl-substituted cyanamides:

$$R_3N + BrCN \rightarrow RBr + R_2NCN$$

For example, benzylcyanamide, $C_6H_5CH_2NHCN$, may be prepared by mixing a solution of benzylamine in ethyl acetate with a concentrated warm solution of potassium cyanide and then adding a solution of bromine in ethyl acetate.

Aryl cyanamides also can be prepared from aryl thioureas by oxidation with strong solutions of sodium iodate and sodium hydroxide or with substances such as lead oxide or mercuric oxide.

Acetylcyanamide is readily prepared by reaction of cyanamide with acetyl chloride.

### D. CARBODIIMIDES

A commonly used procedure for synthesis of mixed carbodiimides involves reaction of the dihalogen derivative of the isocyanide group with an amine hydrochloride:

$$RN\!\!=\!\!CCl_2 + R'NH_3Cl \xrightarrow{-3\,HCl} R\!-\!N\!\!=\!\!C\!\!=\!\!N\!-\!R'$$

Usually, the reaction is allowed to take place at 175° in a solvent such as dichlorobenzene. Mixed carbodiimides are also prepared by a two-step reaction of an isocyanate and a dialkyl phosphoramidate (which is readily made from a dialkyl phosphite, carbon tetrachloride, and an amine):

$$(RO)_2P(O)NHR' \xrightarrow{NaH} (RO)_2P(ONa)NR \xrightarrow{R'N=C=O} R\!-\!N\!\!=\!\!C\!\!=\!\!N\!-\!R'$$

Symmetrical carbodiimides can be prepared from two molecules of the corresponding isocyanate; phosphine oxide catalyzes loss of carbon dioxide:

$$2\,RN\!\!=\!\!C\!\!=\!\!O \xrightarrow{R_3PO} R\!-\!N\!\!=\!\!C\!\!=\!\!N\!-\!R + CO_2$$

For laboratory work, carbodiimides are prepared from solutions of N,N'-dialkyl thioureas in ether by removal of hydrogen sulfide with freshly prepared mercuric oxide or (for large-scale industrial purposes) sodium hypochlorite:

$$RNCH(S)NHR' + HgO \rightarrow RN\!\!=\!\!C\!\!=\!\!NR' + HgS + H_2O$$

Isocyanates can be converted to carbodiimides by special catalysts such as phospholene oxides or sulfides:

$$2\,RNCO \rightarrow RN\!\!=\!\!C\!\!=\!\!NR + CO_2$$

The products may be contaminated by derivatives of isocyanuric acid and polymers of carbodiimides. A typical catalyst is 3-methyl-1-ethyl-3-phospholene-1-oxide:

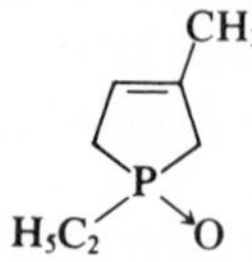

but manganese, iron, or copper naphthenates are also catalysts, even though less active. (The tertiary phosphines polymerize isocyanates to dimers and trimers.)

Frequently used carbodiimides are dicyclohexylcarbodiimide, diisopropylcarbodiimide, and ethyl-(3-dimethylaminopropyl)carbodiimide.

### E.  OCCURRENCE

Nitriles have been detected in minor amounts in plants, usually as glycosides of mandelonitrile (such as amygdalin), but glycosides of the cyanohydrins of other aldehydes and ketones have also been reported. Glycosides of the cyanohydrins generally are not found in plants that also produce alkaloids or terpenes (229). Ricinine [1-methyl-3-cyano-4-methoxy-2(1H)-pyridone] is an alkaloid in castor beans. [The glycoside in cassava (*Manihot,* manioc) must be removed to prepare edible tapioca.] The defensive secretion of millipedes may be benzaldehyde cyanohydrin (79) or a readily hydrolyzable derivative.

Phenylacetonitrile and $\beta$-phenylpropionitrile have been isolated from essential oils (nasturtium and watercress, respectively); nitriles such as geranonitrile and cinnamonitrile appear to be useful as stabilizers for the color and scent of toilet preparations containing unsaturated aldehydes (e.g., citral) (20).

A wide variety of nitriles are used in commerce for synthesis of dyes (168,295), as solvents [e.g., acetonitrile (55)], for modifying cellulose, for elastomers (notably acrylonitrile), as intermediates for innumerable syntheses, for refining of petroleum, as fungicides (35,99), as flotation agents, and for the manufacture of polymers (e.g., nylon via adiponitrile). Derivatives of tetracyanoethylene are used in scintillation counters, as phosphors for cathode-ray tubes, as fumigants, and as mildew preventatives; of great interest is the possibility of producing polymers from tetracyanoethylene.

The isocyanides are seldom found in natural products, but the biosynthesis of Xanthocillin suggests that living organisms are capable of producing isocyanide functional groups (230). Isocyanides are still under investigation for applications of commercial importance, for example, as pesticides (98).

Calcium cyanamide is an important nitrogen fertilizer; it can be used as a soil fumigant and as a defoliant. It is an important intermediate in the manufacture of sodium cyanide, guanidine, thiourea, and ammonia. Calcium cyanamide has been investigated as an agent for treatment of alcoholism.

Carbodiimides are usually too reactive to be found in nature, but as products of synthesis they are finding widespread application as stabilizers for foam polymers and for modifying sodium cellulose. Some carbodiimides are used for treatment of protein and polyamide fibers to increase the washfastness of dyes.

## III.  PROPERTIES

The physical properties of typical nitrile, isocyanide, and cyanamide compounds are given in Table I.

### A.  PHYSICAL PROPERTIES

#### 1.  Nitriles

The triple linkage in the nitriles, —C≡N, is the root of the reactivity that characterizes the functional group. The C-N triple bond in the nitrile group is considerably different from the acetylenic C-C bond in that it is stronger and much more polar, as indicated by the high dipole moment (4.0 D) of the simple nitriles; in fact, the dipole moment approaches (to about 70%) the value expected if one of the triple bonds were fully ionized. In view of their high dielectric constants, it is not surprising to find that the low molecular weight nitriles are somewhat soluble in water (the higher nitriles are only slightly soluble). Because of the dipole, nitriles are subject both to electrophilic attack on nitrogen and to nucleophilic attack on carbon; thus addition reactions are catalyzed both by acids and by bases.

The hydrogens on $\alpha$-carbons of nitriles are nearly as acidic as those alpha to carbonyl groups; thus it is possible to form salts with strong bases and to alkylate the alpha positions of nitriles with alkyl halides by reactions similar to those used with the esters of malonic and acetoacetic acids. The reactivity comes about because the nitrile group is strongly electronegative ($s,p$ hybridization); the electronegative nitrogen makes $\alpha$-hydrogens readily removable.

The saturated aliphatic nitriles up to about palmitonitrile (i.e., up to about $C_{14}$) are colorless liquids of pleasant odor (in contrast to isocy-

TABLE I

Physical Properties of Typical Nitriles, Isocyanides, Cyanamides, and Carbodiimides

| Compound | Formula | Molecular weight | CN (%) | Color and form | Melting point (°C) | Boiling point[a] (°C) | Solubility | Ref.[b] |
|---|---|---|---|---|---|---|---|---|
| Nitriles | | | | | | | | |
| Acetonitrile | $CH_3CN$ | 41.05 | 63.38 | Colorless liq | -45 | 81.6 | $H_2O$, alc., ether | B2², 181 |
| Propionitrile | $CH_3CH_2CN$ | 55.08 | 47.24 | Colorless liq | -91.9 | 97.2 | $H_2O$, alc., ether | B2², 225 |
| Butyronitrile | $CH_3(CH_2)_2CN$ | 69.11 | 37.65 | Colorless liq. | -112 | 118 | Alc., ether | B2², 252 |
| Valeronitrile (n-butyl) | $CH_3(CH_2)_3CN$ | 83.13 | 31.30 | Colorless liq. | -96 | 141 | Alc., ether | B2², 267 |
| Pivalonitrile (t-butyl) | $(CH_3)_3CCN$ | 83.13 | 31.30 | Colorless liq. | 15 | 105 | Alc., ether | B2², 181 |
| Capronitrile (n-amyl) | $CH_3(CH_2)_4CN$ | 97.16 | 26.78 | Colorless liq. | -45 | 160 | Alc. | B2³, 286 |
| 3-Cyanopentane | $(C_2H_5)_2CCN$ | 97.16 | 26.78 | Colorless liq. | — | 150 | Alc., ether | B2², 292 |
| Pelargonitrile (n-octyl) | $CH_3(CH_2)_7CN$ | 139.24 | 18.69 | Colorless liq. | -34 | 224 | Ether | B2³, 823 |
| Lauronitrile (n-undecyl) | $CH_3(CH_2)_{10}CN$ | 181.32 | 14.35 | Colorless liq. | 4 | 253 | Alc., ether, chl | B2, 363 |
| Stearonitrile (n-heptadecyl) | $CH_3(CH_2)_{16}CN$ | 265.49 | 9.80 | Waxy | 40.9 | 357 | Alc., ether, chl | B2, 384 |
| Cyanoacetic acid | $NCCH_2COOH$ | 85.06 | 30.59 | Dimorphic plates | 66 | — | Alc., ether | B2², 530 |
| Cyanoacetamide | $NCCH_2CONH_2$ | 84.08 | 30.94 | Needles | 121 | — | $H_2O$, alc. | B2², 534 |
| Allyl cyanide | $CH_2{=}CHCH_2CN$ | 67.09 | 38.78 | Liq. | -87 | 119 | Alc. | B2³, 389 |
| Acrylonitrile | $H_2C{=}CHCN$ | 53.06 | 49.03 | Colorless | -82 | 77.5 | Alc., ether, $M_2O$ | B2³, 1234 |
| Allylacetonitrile | $CH_2{=}CHCH_2CH_2CN$ | 81.12 | 32.07 | Liq. | — | 140 | Alc., ether | B2, 426 |
| Pyruvonitrile | $CH_3COCN$ | 69.06 | 37.67 | Rhomb | — | 92.3 | Ether | B3³, 402 |
| Succinonitrile | $NCCH_2CH_2CN$ | 80.09 | 64.97 | Colorless wax | 54.5 | 266 | $H_2O$, alc., benz. | B2, 615 |
| Adiponitrile | $NC(CH_2)_4CN$ | 108.14 | 48.12 | Colorless liq. | 1 | 295 | Alc., chl. | B2², 576 |
| Tetracyanoethylene | $(CN)_2C{=}C(CN)_2$ | 128.09 | 81.25 | Colorless | 199 | 223 s. | Chlorobenz. | 73 |
| Benzonitrile | $C_6H_5CN$ | 103.12 | 25.23 | Colorless liq. | -13 | 191 | Alc., ether | B9², 196 |
| Phenylacetonitrile | $C_6H_5CH_2CN$ | 117.15 | 22.21 | Needles | 24 | 233 | Alc., ether | B9², 330 |
| Phthalonitrile | $1,2{-}(CN)_2C_6H_4$ | 128.13 | 40.61 | Colorless needles | 141 | — | $H_2O$, alc., ether | B4², 602 |
| Terephthalonitrile | $1,4{-}(CN)_2C_6H_4$ | 128.13 | 40.61 | Needles | 224 | s. | Sl. sol. benz. | B9², 613 |

| | Formula | Mol. wt. | Appearance | M.p. | B.p. | Solubility | Beilstein[b] |
|---|---|---|---|---|---|---|---|
| **Cyanohydrins** | | | | | | | |
| Glycolonitrile | $CH_2(OH)CN$ | 57.05 | Colorless liq. | $-70$ | 183 d. | $H_2O$, alc., ether | B3³, 174 |
| Ethylene cyanohydrin | $CH_2(OH)CH_2CN$ | 71.08 | Colorless liq. | $-46$ | 228 | $H_2O$, alc. | B3², 213 |
| Lactonitrile | $CH_3CH(OH)CN$ | 71.08 | Yellow liq. | $-40$ | 183 | $H_2O$, alc., ether | B3³, 209 |
| Mandelonitrile | $C_6H_5CH(OH)CN$ | 133.15 | Yellow liq. | 29 | 170 | Alc., ether | B10², 60 |
| **Isocyanides** | | | | | | | |
| Methyl isocyanide | $CH_3NC$ | 41.05 | Colorless liq. | $-45$ | 59.6 | Alc., ether | B4¹, 119 |
| Ethyl isocyanide | $CH_3CH_2NC$ | 55.08 | Colorless liq. | $-66$ | 78.5 | Alc., ether | B4³, 205 |
| Phenyl isocyanide | $C_6H_5NC$ | 103.12 | Unstable green liq. | — | 78 | Benz., ether | B12³, 313 |
| **Cyanamides** | | | | | | | |
| Cyanamide | $H_2NCN$ | 42.04 | Needles | 46 | — | $H_2O$, alc., ether | B3, 74 |
| Dicyandiamide | $H_2NCN(NH)HCN$ | 84.08 | Monocl. | 208 | d. | $H_2O$, alc. | B4², 474 |
| Dimethylcyanamide | $(CH_3)_2NCN$ | 70.09 | Liq. | — | 162 | $H_2O$, alc., ether | B4³, 145 |
| Diethylcyanamide | $(CH_3CH_2)_2NCN$ | 98.15 | Liq. | — | 190 | Alc., ether | B4, 121 |
| Diphenylcyanamide | $(C_6H_5)_2NCN$ | 194.24 | Prisms | 73 | 235⁶⁰ | Alc. | B7, 430 |
| **Carbodiimides** | | | | | | | |
| Dimethylcarbodiimide | $CH_3NCNCH_3$ | 70.10 | Colorless | — | 76 | Ether | 224 |
| Di-n-propylcarbodiimide | $C_3H_7NCNC_3H_7$ | 126.20 | Colorless | — | 53¹⁰ | Ether | B4², 627 |
| Di-n-butylcarbodiimide | $C_4H_9NCNC_4H_9$ | 154.26 | Colorless | — | 84¹⁰ | Ether | 239 |
| Diisobutylcarbodiimide | $C_4H_9NCNC_4H_9$ | 154.26 | Colorless | — | 72¹⁰ | Ether | 239 |
| Dicyclohexylcarbodiimide | $C_6H_{11}NCNC_6H_{11}$ | 206.33 | Colorless | 35 | 123⁶ | Ether | 239 |
| Diphenylcarbodiimide | $C_6H_5NCNC_6H_5$ | 194.24 | Colorless | 169 | 331 | Ether | B12², 246 |

[a] A superscript refers to the pressure (mm of Hg); s. = sublimes; d. = decomposes.
[b] The number following the letter B indicates the volume of the main series of Beilstein, *Handbuch der organischen Chemie*; the superscript refers to the first, second, etc., supplement; and the following number is the page in the supplement.

417

anides), but they are often contaminated with malodorous isocyanides. The higher homologs are odorless, crystalline substances. Most nitriles boil without decomposition at temperatures lower than those for the corresponding acids; they are not soluble in aqueous acids and are too weakly basic to form salts in spite of the presence of a pair of unshared electrons on the nitrogen atom of the functional group (some protonation occurs with dry hydrochloric acid).

## 2. Isocyanides

The structure currently assigned to the isocyanide group is in accord with the reactivity of an unshared pair of electrons and the reactivity expected of a triple bond; moreover, the absorption of the isocyanide group in the infrared at about 2125 cm$^{-1}$ (4.71 $\mu$) is in the triple-bond stretching region [nitriles absorb at about 2245 cm$^{-1}$ (4.45 $\mu$), monosubstituted acetylenes at 2140–2100 cm$^{-1}$ (4.67–4.76 $\mu$), and a C-N double bond at 1690–1640 cm$^{-1}$ (5.92–6.10 $\mu$)].

There is probably a real difference in the structure of the isocyanide group when attached to an alkyl radical and when attached to an aryl radical because $Ar^-=\overset{+}{N}=C$: may be an important resonance form; thus lower infrared absorption (> 20 cm$^{-1}$) of the aryl isocyanide group may be attributable to the resonance form, but since the isocyanide stretching frequency increases in alkyl as well as aryl compounds with increasing polarity of the solvent system, other factors are obviously involved. Polar solvents appear to enhance the existence of the $R-\overset{+}{N}{\equiv}C$: resonance form (192). The results of extensive microwave studies confirm early postulates that the C-N-C bond system in the isocyanide group is linear. Moreover, isocyanides are the only stable organic compounds with divalent carbon; alpha addition and multicomponent reactions of the isonitriles are commonplace.

The isocyanides are colorless liquids with characteristically repulsive, powerful odors. Dipole moments are somewhat lower than those of corresponding nitriles (3.44 D) (302); accordingly, boiling points are about 20° lower than those of the isomeric nitriles. Isocyanides are not stable; they readily polymerize on storage or when in contact with ground glass, and are hydrolyzed at elevated temperatures to formic acid and a primary amine:

$$2\,H_2O + RNC \rightarrow RNH_2 + HCOOH$$

The products of hydrolytic reaction suggest that R is connected to the nitrogen atom of the isocyanide group (in contrast, nitriles are hydrolyzed to RCOOH). The isocyanides are readily hydrolyzed in aqueous acids (sometimes with explosive violence), but are essentially resistant to alkalies. In spite of the differences in reactivity between nitriles and isocyanides, the strengths of C-N bonds in the groups must be quite similar since their ir absorption bands are only $100 \text{ cm}^{-1}$ apart. Heats of formation and other thermal data also confirm the similarity of bond strengths.

Isomerism in the isocyanide–nitrile system occurs smoothly as a first-order process in the gas phase, and production of more stable nitriles occurs readily at 100°C in the presence of metallic cyanides. Unfortunately, the thermal rearrangement of isocyanides involves often a complex series of reactions; for example, at 100°C ethyl isocyanide forms a trimer which rearranges at 160°C to the nitrile. In spite of the multiplicity of reaction routes, the thermal rearrangement of isocyanides to nitriles at temperatures usually above 180°C is considered to be a *general* reaction (49).

Isocyanides polymerize readily when in contact with anhydrous acid or salts of nickel, cobalt, and palladium. Polymerization also takes place at room temperature during storage; for example, pure phenyl isocyanide rapidly becomes discolored, but a small amount of aniline or hydrazine acts as preservative. The blue nonvolatile product that phenyl isocyanide forms readily is the tetramer [the dianil of indigo (49)].

Polymers of the isocyanides are detected by absorption in the ir region arising from the presence of the imine group, $=C=N—$; generally, absorption is in the $1645–1615 \text{ cm}^{-1}$ (6.08–6.19 $\mu$) region, but two or three prominent, sharp absorption bands can also be found in the $910–825 \text{ cm}^{-1}$ (10.99–12.12 $\mu$) region (192).

### 3. Cyanamides

Cyanamide and its salts with the alkali metals and alkaline earths form an ion that is isostructural and isoelectronic with carbon dioxide. Cyanamide is quite stable in aqueous solutions below pH 5; however, it readily polymerizes on standing to the dimer dicyandiamide, $H_2NC(=NH)NHCN$. In alkaline solution or when its solutions are boiled, cyanamide dimerizes as well as hydrolyzes; the dimer is also formed when calcium cyanamide is heated with water. Thus the cyanamide of commerce usually contains much of the dimer and the tautomer, $H_2N—C(=NH)—N=C=NH$, which contains the carbodiimide group, $—N=C=N—$.

Cyanamide forms colorless, hygroscopic crystals that melt at about 45–46°C; it is very soluble in a number of solvents (e.g., ether, alcohol, and water) but is only slightly soluble in carbon disulfide, benzene, and carbon tetrachloride. Cyanamide attacks the skin much like caustic alkalies; above its melting point, it readily dimerizes. The basic nature of cyanamide is exhibited by formation of its dihydrochloride when dry hydrogen chloride and cyanamide are brought together in ether; the hydrochloride can be crystallized from water. Many substituted compounds of cyanamide or dicyandiamide are liquids or low-melting solids which can be distilled at low pressures. The solubility differences between cyanamide, dicyandiamide, calcium cyanamide, and related compounds such as guanidine and biguanidine are often sufficient to permit good separation; for example, dicyandiamide can be extracted from calcium cyanamide by 95% ethanol, and cyanamide can be extracted from dicyandiamide by ether.

## 4. Carbodiimides

Monomeric carbodiimides are very reactive and also tend to polymerize readily; in fact, dimethylcarbodiimide polymerizes so fast at room temperature that it must be prepared and stored at low temperature (224). The liquid aromatic carbodiimides polymerize readily, but the solid ones are quite stable; electron-attracting groups increase the tendency to polymerize. Polymers of the carbodiimides may have structures such as the following:

$$
\begin{bmatrix}
 & R & & R & & R & \\
-C & -N & -C & -N & -C & -N- \\
N & & N & & N & \\
R & & R & & R &
\end{bmatrix}_x
$$

In spite of great activity, the carbodiimides are structurally quite stable and can be purified by vacuum distillation. However, the liquid carbodiimides, and especially the lighter molecular weight aliphatic homologs, decompose slowly on standing; for example, freshly prepared carbodiimides are neutral to moist litmus, but after a time they become yellow, deposit a sediment, and turn litmus blue (238). The alicyclic carbodiimides, such as dicyclohexylcarbodiimide, are very stable.

## B. CHEMICAL PROPERTIES

The reactions of the nitriles and isocyanides are governed in large measure by the reactivity of the triple bond in the functional groups; thus, since the triple bond in nitriles is strongly polarized (high dipole moment), the functional group is subject to electrophilic attack on nitrogen as well as nucleophilic attack on carbon. The reactions of isocyanides also reflect the high polarity of the functional group and the unshared pair of electrons on the carbon of the functional group; reactions usually involve nucleophilic attack on the reagent followed by nucleophilic attack on carbon, with the result that both portions of the reagent add to the isocyanide carbon atom:

$$\overset{+}{R}\overset{-}{N}{\equiv}\overset{-}{C}: + A:B \rightarrow [R\overset{+}{N}{\equiv}C:A + {}^-:B] \rightarrow RN = C \underset{B}{\overset{A}{\diagup}}$$

It is important to note that nitriles and isocyanides (with few exceptions) do not undergo hydrolysis to form hydrogen cyanide; the cyanohydrins are obvious exceptions, as are tetracyanoethylene and acyl cyanides, which react with water as follows:

$$(CN)_2C{=}C(CN)_2 + H_2O \rightarrow (CN)_2C{=}C(CN)(OH) + HCN$$
$$RCO(CN) + H_2O \rightarrow R{-}COOH + HCN$$

The following discussion summarizes the reactions of the nitrile, isocyanide, cyanamide, and carbodiimide groups that are of importance in analytical procedures.

## 1. Reduction

Unlike the functional groups discussed in earlier sections, nitriles and isocyanides seldom can be straightforwardly and quantitatively reduced to a single product. As a rule, all alkyl nitriles behave irregularly when reduced with couples such as zinc and acid, but the reactions of the aromatic nitriles are sharply dependent on the nature of other substituent groups. It is important to recognize that reduction of nitriles or isocyanides does *not* yield cyanide, although there are instances where hydrogen cyanide can be released (see Section V.B.I.b) by other reactions.

### a. NITRILES

The reduction of nitriles may lead to a multiplicity of products, and only the most powerful reducing agents (also, catalytic hydrogenation) are in any way useful for analytical purposes. In principle, reduction of a nitrile should provide an amine, and as long ago as 1862 Mendius (188) found that tin and hydrochloric acid add four atoms of hydrogen to the nitrile group:

$$RCN + 2 H_2 \rightarrow RCH_2NH_2$$

However, it is known that secondary and tertiary amines are also formed because the reduction route involves formation of an imine as an intermediate, and, depending on the relative stability of the imine and the conditions established in the reaction medium, the primary amine and the imine interact:

$$RCH{=}NH + H_2NCH_2R \rightarrow R{-}\underset{\underset{NH_2}{|}}{CH}{-}NHCH_2R \rightarrow RCH{=}NCH_2R + NH_3$$

The product undergoes further hydrogenation to form a secondary amine:

$$RCH{=}NCH_2R + H_2 \rightarrow RCH_2NHCH_2R$$

The secondary amine thus produced then undergoes a similar set of reactions to form a tertiary amine. To a certain extent, the formation of secondary and tertiary amines can be sharply limited by reducing the nitriles in the presence of a large amount of ammonia; unfortunately, although ammonia is valuable in practical processes for industrial preparation of amines by hydrogenation, it is of little use for quantitative analytical procedures.

Aldimines can be readily hydrolyzed to aldehydes. In fact, when nitriles are treated with an ethereal solution of anhydrous stannous chloride in the presence of hydrogen chloride, compounds of aldimines and stannic chloride are formed (iminochlorostannates) from which aldehydes can be recovered by steam distillation; this useful synthetic route was explored by Stephen (262,263), Hantzsch (122), Turner (282), and Pyryalova and Zil'berman (221):

$$RCN + SnCl_2 + 4\,HCl \rightarrow [RCH{=}\overset{+}{N}H_2][HSnCl_6]^- \xrightarrow{H_2O} RCHO + (NH_4)HSnCl_6$$

The applicability of the Stephen method for preparation of aldehydes has been reviewed by Mosettig (195).

### (1) Sodium and Alcohol

Reduction of nitriles by the sodium–alcohol couple yields primary amines; dispersions of sodium in hydrocarbons are useful:

$$RCN + 4\,Na + 4\,C_2H_5OH \rightarrow RCH_2NH_2 + 4\,C_2H_5ONa$$

Reductions are seldom quantitative because many nitriles undergo competing condensation reactions. For example, adiponitrile forms 2-imino-1-cyanocyclopentane when treated with a trace of sodium ethoxide in ethanolic solution, and benzyl cyanide with sodium forms 1,3-diamino-2-phenylnaphthalene      or      4-amino-2,6-dibenzyl-5-phenyl-pyrimidine. Nitriles without an $\alpha$-hydrogen (to the nitrile group) do not undergo condensations of the type noted above; however, $n$-merization is possible with formation of nitrogen-ring compounds.

## (2) Lithium Aluminum Hydride

A number of nitriles can be reduced to the corresponding amines by solutions of lithium aluminum hydride in ether [see Section V.D.1.c(2)]; however, the great majority of nitriles are not reduced quantitatively, and condensation reactions often compete successfully to form ring structures containing nitrogen. A large excess of hydride must be used to ensure adequate reduction of the nitrile; when the amount of hydride is limited (or reaction has not been allowed to go to completion), aldehydes will be formed instead of amines (36):

$$RC{\equiv}N \xrightarrow{MH} RCH{=}NM \xrightarrow{MH} RCH_2NM_2$$

$$\downarrow H_2O \qquad\qquad \downarrow H_2O$$

$$RCHO \qquad\qquad RCH_2NH_2$$

where $M = LiAl/4$.

The cyanohydrins form 2-hydroxyamines; other compounds, including amides and imides, are also reduced to aldehydes by lithium aluminum hydride (36):

$$RCONH_2 \xrightarrow{MH} RCH(OM)NH_2 \xrightarrow{H_2O} RCHO$$

The reducing power of the hydride may be decreased by replacement of some of its hydrogen with alkoxyl groups; reduction can be stopped at the aldimine stage (which on subsequent hydrolysis will yield aldehydes). Alkoxylation of lithium aluminum hydride is accomplished simply by mixing 3 mol of ethanol with 1 mol of the hydride:

$$LiAlH_4 + 3\,C_2H_5OH \rightarrow LiAl(OC_2H_5)_3H + 3\,H_2$$

The lithium aluminum triethoxide hydride reaction with nitriles and subsequent hydrolysis is represented as follows:

$$RCN + LiAl(OEt)_3H \rightarrow [RCHN]^-[LiAl(OEt)_3]^+ \xrightarrow{5\,H_2O} RCHO + NH_3 + LiOH + Al(OH)_3$$
$$+ 3\,EtOH$$

Sodium aluminum hydride also is an effective reducing agent (324).

## (3) Hydrogenation

Catalytic hydrogenation is a preferred procedure for reduction of nitriles because it does not leave extraneous materials in the reaction mixture; unfortunately, hydrogenation has not been found useful for analytical work because both primary and secondary amines are produced. Inasmuch as the mechanism of hydrogenation involves the aldimine intermediary indicated above, it follows that formation of secondary (and tertiary) amines is repressed by high partial pressures of ammonia. Raney nickel catalysts are widely used in industrial processes (315), but cobalt oxide and oxide or sulfide catalysts of tungsten or molybdenum may also be employed. Active catalysts such as platinum oxide (47), palladium on charcoal, and palladium-Norite (125) are ordinarily used in laboratory work. Typically, hydrogenation with Raney nickel is made to take place in industrial processes at pressures of 70–270 atm in the presence of ammonia, but relatively large amounts of the catalyst are required (e.g., 100–300% of the weight of nitrile). In the absence of ammonia, good yields of secondary amines are obtained along with primary amine (41). A catalyst consisting of 5% rhodium on alumina is far more effective than nickel catalysts, for it permits hydrogenation to take place at 2–3 atm of hydrogen pressure (in ammoniacal methanol) with a catalyst loading of 10–20% of the weight of nitrile (97). The rhodium catalyst, however, does not significantly improve yields of amines (average = 70%); it does not hydrogenate aromatic nuclei. Alumina impregnated with copper and zinc chromites preferably is used for low-temperature reduction of aliphatic amines with hydrogen, but yields are seldom better than 70% (34).

Hydrogenation of acyloyl nitriles with nickel catalysts at 200–225°C yields a mixture of cyanohydrins and nitriles (276):

$$RCOCN \xrightarrow{H_2} R-CHOHCN + RCH_2CN$$

### (4) Electrolytic Methods

The electrolytic reduction of a few nitriles has been studied (218), and from the review that has been published by Tomilov et al. (278) it appears that, although a wide variety of nitrile compounds can be reduced under proper conditions, the nitrile group itself is very resistant. Lead, mercury, or nickel-lead cathodes have been used for reduction of the C-N triple bond in classical constant-current procedures, that is, in procedures where the electrolytic current is maintained more or less constant by manual adjustment of the applied potential; in view of the marked effect of potential on the route of an electrochemical reduction, such procedures are not selective or efficient. Moreover, since hydrogen may be given off at the cathode, reduction products may be derived from hydrogenation rather than from the direct addition of electrons.

The nitrile group (e.g., acetonitrile) is not reducible at the dropping mercury electrode in suspension media of the type ordinarily employed in polarographic procedures (31). However, when the nitrile group is conjugated with a double bond, reduction of the *double bond* can take place at very negative potentials ($E_{1/2} > 2.0$); for this purpose, the suspension medium must contain tetraalkylammonium halide salts or other cations which are discharged at very negative potentials (30). The nitrile group itself is reduced electrolytically only under conditions which suggest that catalytic hydrogenation is involved (see below).

### (a) Aromatic Nitriles

Ahrens (2) was among the earliest workers to report that the nitrile group of aromatic nitriles can be reduced electrolytically to amines; benzylamine was identified as the main reduction product of benzonitrile formed at a lead electrode in an acidic solution. Later, Ogura (203) was able to show that benzylamine was formed not only in dilute sulfuric acid but also in solutions of sodium hydroxide and solutions of ammonium sulfate. However, yields of the amine were poor, ranging from 24% in 2.5% $H_2SO_4$ to 49.5% in 7.5% aqueous ammonium sulfate solution. Yields were also poor with amalgamated lead or zinc electrodes, mercury electrodes, and platinum. A yield of 57% amine was obtained on reduction of p-methylbenzonitrile with lead electrodes in ammonium sulfate solutions.

Ohta (204) reported good yields for the reduction of benzonitrile in hydrochloric acid solution with palladium-plated platinum electrodes. The high yields are in conformity with the reduction mechanisms noted in earlier sections; aldimines are formed, and their reaction with primary amines to form secondary amine side products is suppressed by hydrochloric acid ($-NH_2 \rightarrow -NH_3Cl$).

The polarographic reduction of aromatic nitriles is discussed in detail in Section VI.D.1.

### (b) Aliphatic Nitriles

Saturated aliphatic nitriles may be reduced to amines during classical constant-current electrolyses when catalytic substances are present. Ohta (204) found that $CH_3CN$, $NCCH_2CN$, $NC(CH_2)_2CN$, $NC(CH_2)_4CN$, $NCCH_2CH_2OH$, and $NCCH_2COOH$ in aqueous hydrochloric acid are reduced at palladium-plated platinum electrodes; about 30% of ethylenecyanohydrin is reduced, but the other nitriles are reduced more completely, and the surprising 100% reduction of acetonitrile prompts suspicion that the palladium electrode provides a convenient site for hydrogenation rather than true electrolytic reduction. In fact, Janardhan (141) prepared dodecyl-, hexadecyl-, and octadecylamines by reduction of the corresponding nitriles at a copper electrode in methanol saturated with ammonium chloride and in the presence of a catalyst such as Raney nickel (85% yields at c.d. = 5 A/dm$^2$); similarly, Kawamura and Suzuki

(153) found that benzyl cyanide is reduced at nickel-palladium cathodes ($2\ A/dm^2$) in an electrolyte consisting of an aqueous solution of acetic acid and hydrochloric acids.

In the absence of catalysts, the aliphatic nitrile group often is remarkably resistant to reduction by electrolysis. Thus Baizer (10,11) and Baizer and Anderson (12,13) found that reduction of concentrated solutions of acrylonitrile in aqueous tetramethylammonium *p*-toluenesulfonate at lead and mercury cathodes provides 100% yields of adiponitrile at current efficiencies of nearly 100%, provided that pH is maintained constant. When the acrylonitrile concentrations were below 10%, or when the supporting electrolyte contained alkali-metal cations, increasing amounts of propionitrile appeared in the reduction product. These observations are in keeping with the results of a more extensive study of the reduction of a number of $\alpha,\beta$-unsaturated nitriles by Sevast'yonova and Tomilov (244–246); in general, reduction of $\alpha,\beta$-unsaturated nitriles takes place with a dianion as intermediary:

$$CH_2{=}CHCN + 2\ e^- \rightarrow CH_2CHCN^{2-}$$

At high concentrations of acrylonitrile, hydrodimerization takes place, that is, coupling at the beta position and attachment of hydrogen at the alpha position:

$$CH_2CHCN^{2-} + CH_2CHCN + 2\ H^+ \rightarrow \begin{array}{l} CH_2CH_2CN \\ | \\ CH_2CH_2CN \end{array}$$

At low concentrations of acrylonitrile, the dianion is protonated to propionitrile:

$$CH_2CHCN^{2-} + 2\ H^+ \rightarrow CH_3CH_2CN$$

Wiemann and Bouguerra (318) also have observed the hydrodimerization of crotononitrile to $\beta,\beta'$-dimethyladiponitrile; additional evidence of the presence of free radical ions is found in the work of Bargain (16) on $\alpha$-ethylenic nitriles.

The formation of monomeric and dimeric products by reduction of $\alpha,\beta$-unsaturated nitriles may be strongly influenced by adsorption of the nitriles on the electrode surfaces. The work of Franklin and Sothern (95) has demonstrated that adsorption of propionitrile, acetonitrile, butyronitrile, capronitrile, isocapronitrile, and phenylacetonitrile on platinized platinum electrodes is expressed by Freundlich's equation. Kaabak et al. (149) have proposed a mechanism for the reduction of vinylacrylonitrile which requires, as a first step, adsorption of the nitrile on the electrode surface; the ratio of products obtained (3- and 4-pentenonitriles and dimers) in classical constant-current reductions with lead-, tin-, zinc-, and mercury-plated electrodes did not depend on the nature of the cathode or the current density. Similarly, Tomilov et al. (277) found that electrolytic reduction of the double bonds in vinylacetonitrile, methacrylonitrile, and crotononitrile occurred simultaneously with formation of dimeric products in 0.7 $N$ sodium hydroxide solution; copper, zinc, tin, and graphite cathodes were used. With copper electrodes, some primary and secondary amines were formed. Smirnov et al. (255) found adiponitrile to be the main reduction product of 1,4-dicyano-1-butane in neutral solutions of potassium phosphate, but the yield was a sharp function of the concentration of the starting material (some dimer formed). At tin cathodes, in alkaline electrolytes, $\alpha,\beta$-unsaturated nitriles produce organotin compounds; for example, acrylonitrile yields tetra($\beta$-cyanoethyl)tin, and methacrylonitrile yields tetra($\beta$-cyanopropyl)tin.

Polarographic reductions of aliphatic nitriles are discussed in Section VI.D.2.

### (5) Miscellaneous Reductions

Reduction of nitriles with Raney nickel alloy in alkaline solution yields amines (58,207); however, double bonds are also reduced, even when the reaction is performed at room temperature to improve the yield of amine (average, about 70%) (261).

Raney nickel in acidic solutions (75% v/v aqueous formic acid) produces aldehydes directly from nitriles; refluxing time can be shortened by use of large amounts of the alloy, for example, five times the weight of nitrile (81). Yields of aldehydes are generally high, usually over 85%, but many nitriles perform poorly, for example, 1-naphthonitrile (60%).

The carbonyl group in acyl nitriles is preferentially reduced, and it is possible to form (in poor yields) cyanohydrins; for example, the zinc-hydrochloric acid couple reduces benzoyl cyanide to the cyanohydrin of benzaldehyde (162):

$$C_6H_5COCN \xrightarrow{[H]} C_6H_5CHOCN \leftrightarrow C_6H_5CHO + HCN$$

Hydriodic acid reduces nitriles and is especially effective at high temperatures and pressures in sealed tubes; reductions are alleged to be quantitative (4,100,228).

Ferrous hydroxide catalyzed by silver (and other metals such as nickel, copper, and mercury) slowly reduces nitriles in strongly alkaline solution, and the reaction has been used for the quantitative estimation of nitriles (177,297). (See Section VI.A.2.b)

### b. Isocyanides

The isocyanides as a class are, per se, strong reducers. Vigorous reduction of alkyl or aryl isocyanides produces secondary amines with one methyl group:

$$RNC + 2 H_2 \rightarrow RNHCH_3$$

The reduction can be effected by sodium in alcohol (especially the aromatic isonitriles), by metal hydrides, or by catalytic hydrogenation (nickel at 180°C). The aromatic isocyanides are preferably reduced by sodium and pentanol; the aliphatic isocyanides are best reduced by hydrogen and catalyst. On the other hand, Ugi and Bodesheim (285) claim that isocyanides are quantitatively reduced by alkali metals or alkaline earth metals in liquid ammonia at −40 to −35°C. The alkyl isocyanides form alkanes and metal cyanides, for example,

$$RNC + 2 Na + NH_3 \rightarrow RH + NaCN + NaNH_2$$

No quantitative data are available for this potentially useful reaction.

### c. Cyanamides

Cyanamide and dicyandiamide are cleaved by a metal-acid couple such as zinc and hydrochloric acid to hydrocyanic acid, but the final product is ammonia in the case of cyanamide; guanidine is formed from the dimer. In general, on reduction with zinc and hydrochloric acid, $N$-cyano compounds form some hydrocyanic acid, which can be detected by sensitive tests (87,250).

Catalytic reductions of substituted $N$-cyano compounds with palladium at room temperature in the presence of ammonium salts or mineral acids appear to take place readily and provide high yields, but cyanamide itself is a catalyst poison (78).

### d. Carbodiimides

Hydrogenation of the carbodiimide group yields the corresponding formamidine:

$$R-N=C=N-R \xrightarrow{H_2} HC{\overset{\displaystyle NR}{\underset{\displaystyle NHR}{\Big\langle}}}$$

If the hydrogenation takes place in dilute acid or alkalies, the formamidine can be hydrolyzed to an amide and amine, and the amide can be hydrolyzed still further:

$$RN=C=NR \xrightarrow[H_2O]{H_2} RNH_2 + HC{\overset{\displaystyle O}{\underset{\displaystyle NHR}{\Big\langle}}}$$

However, if sodium and alcohol are used, only amines are formed:

$$R-N=C=N-R \xrightarrow[\text{EtOH}]{\text{Na}} RNH_2 + RNHCH_3$$

The reduction by sodium and alcohol is quite complex because alkali catalyzes hydrolysis, and many sensitive carbodiimides rapidly polymerize in the presence of alkali (see Section III.B.4.d). Conditions for quantitative reduction of the carbodiimide group to defined products have not been established.

## 2. Oxidation

### a. NITRILES

Strong oxidants react with the nitrile group to form the carboxylic acid group:

$$RCN \rightarrow RCOOH$$

The direct oxidation of nitriles by powerful oxidants is not useful for quantitative purposes because of the inevitable attack on the radical to which the nitrile group is attached. Moreover, as a rule, oxidation takes place in strongly acidic or alkaline solutions where hydrolysis may also play an important part. Thus, in some instances, the oxidation reactions which are used for analysis of the nitrile group may be regarded as hydrolyses in which the intermediate amide (often surprisingly resistant to further hydrolysis) is broken down by oxidative attack; in other instances, reactions may be regarded purely as oxidative conversions of nitriles to amides (or cyanates) followed by hydrolysis.

Nitriles are converted to amides by action of hydrogen peroxide in alkaline solution at a rate that is about 10,000 times faster than simple alkaline hydrolysis (222); the reaction has been studied by McMaster and Langreck (185), and in greater detail by Wiberg (316,317):

$$RCN + 2 H_2O_2 \rightarrow RCONH_2 + O_2 + H_2O$$

The rate of reaction shows a first-order dependence on the concentrations of nitrile, hydrogen peroxide, and hydroxyl ion; for benzonitrile, it was found that a nitrile $N$-oxide ($RC\overset{+}{N}-\overset{-}{O}$) is not an intermediate. The initial attack at the nitrile carbon is by peroxide ion, $OOH^-$, and the intermediate product is a peroxyimidic acid:

$$\phi\text{-CN} \xrightarrow{\text{(OOH)}^-} \underset{\text{OOH}}{\phi-C}=N^- \xrightarrow{H^+} \underset{\text{OOH}}{\phi-C}=NH \xrightarrow{H_2O_2} \underset{\text{O}}{\phi-C}-NH_2 + O_2 + H_2O$$

Unfortunately, the strongly oxidizing medium often attacks the radical attached to the nitrile group; as a result, the oxidative hydrolysis of nitriles rarely is quantitative with respect to formation of their corresponding carboxylic acids (314).

### b. ISOCYANIDES

The isocyanide group is readily oxidized by mercuric oxide to an isocyanate group; the formation of mercury is readily evident, and the reaction can be used to distinguish nitriles from isocyanides:

$$RNC + HgO \rightarrow RNCO + Hg$$

but the quantitative aspects of the reaction have not been studied in detail. Travagli (279) has reported that mercuric salts catalyze the hydrolysis of nitriles, but have no effect on amides. Ozone, peracids, and silver oxide also oxidize the isocyanide group, and even atmospheric oxygen can form the isocyanate group when certain catalysts are present. Under special conditions, dimethyl sulfoxide also oxidizes the isocyanide group (148).

### 3. Action of Acids

### a. NITRILES

The nitrile group is hydrolyzed in stages by aqueous solutions of mineral acids; the amide group is an intermediate stage:

$$RCN \xrightarrow{H_2O} R-C\underset{NH_2}{\overset{O}{\diagup}} \xrightarrow{H_2O} RCOOH + NH_3$$

The overall reaction is

$$RC{\equiv}N + 2 H_2O + H^+ \rightarrow RCOOH + NH_4^+$$

The amides of some nitriles cannot be isolated because they are readily hydrolyzed, but when an amide is resistant to hydrolysis (e.g., one obtained from the *ortho*-substituted aromatic nitriles, or tertiary alkyl cyanides), hydrolysis of a nitrile to an amide often can be effected nearly completely by cold concentrated sulfuric acid, concentrated hydrochloric acid, glacial phosphoric acid, or a mixture of sulfuric and glacial acetic acids; the acid mixtures are cautiously poured into water to isolate the amides. Hantzsch (123) and Liler and Kosanovic (174) have studied the hydrolysis of simple nitriles in concentrated acids and found that reaction in anhydrous sulfuric acid takes place rapidly to form a compound of the form $[RC(OSO_2OH)NH_2]^+[SO_4H]^-$, which is rapidly hydrolyzed to the amide when the mixture is poured into water. The primary reaction appears to be the formation of a salt of the nitrile and sulfuric acid, $[RCNH]^+[SO_4H]^-$, which reacts with more sulfuric acid under anhydrous conditions to give the compound indicated above, or with the 0.2% water ordinarily in concentrated sulfuric acid to form the salt of the amide, $[RC(OH)NH_2]^+[SO_4H]^-$. Janz and Danyluk (142) have studied the formation of salts of nitriles.

Krieble and Noll (165) studied the rates of hydrolysis of acetonitrile, propionitrile, cyanoacetic acid, and $\alpha$- and $\beta$-hydroxypropionitrile, and found that the rate increases rapidly with increasing concentration of hydrochloric acid; more importantly, however, aqueous solutions of sulfuric acid proved much less effective than hydrochloric acid.

It is necessary to emphasize that, although many nitriles generally can be hydrolyzed in strongly acidic media, some compounds show remarkable resistance. Thus some nitriles need to be treated for several hours with 90% sulfuric acid at 120–130°C merely to convert them to the corresponding amide; concentrated (10 $N$) hydrochloric acid at 100°C (under pressure) is a more effective medium.

Condensation reactions can also take place in strong acids; for example, benzonitrile and other aromatic nitriles can be trimerized to form a variety of triazine derivatives, thus effectively blocking the quantitative conversion of the nitrile group to an amide group. A typical triazine formed from benzonitrile is 2,4,6-triphenyl-1,3,5-triazine (cyaphenine), which resists reduction (yields lophine and HCN).

Nitriles can react with carboxylic acids at elevated temperatures (200–280°C), but in the presence of mineral acids it is possible to have reaction take place at lower temperatures (71). In fact, the nitrile and carboxyl groups of cyano acids react slowly in the cold; for example, $o$-cyanobenzoic acid forms phthalimide:

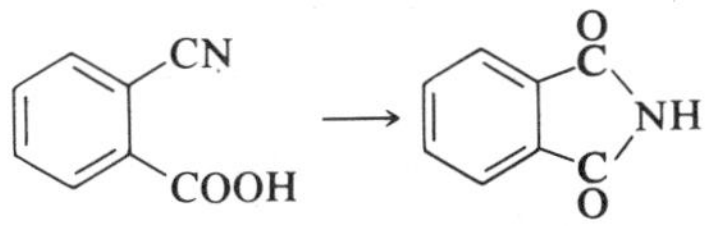

However, reaction of nitriles with carboxylic acids does not follow the same pattern (328) because a secondary amide is formed:

$$RCN + R'COOH \rightarrow RCONHCOR'$$

but exchange can also take place:

$$RCN + R'COOH \rightarrow RCOOH + R'CN$$

Further complications arise if hydrogen halides are present:

$$RCN + 2\,R'COOH + HCl \rightarrow RCONH_2 \cdot HCl + (R'CO)_2O$$
$$RCN + R'COOH + 2\,HCl \rightarrow RCONH_2 \cdot HCl + R'COCl$$

Reactions of the types noted above undoubtedly give rise to complications in analytical procedures for hydrolysis of nitriles by acids.

b. Isocyanides

Hydrolysis of isocyanides by aqueous solutions of acids yields a primary amine and formic acid:

$$RNC \rightarrow RNHCHO \rightarrow RNH_2 + HCOOH$$

Hydrolysis can take place quite violently in acidic solutions, but even hot water (especially at 180°) will cause the hydrolysis to occur.

Carboxylic acids react with isocyanides to produce acid anhydrides and formamides:

$$RNC + 2 R'COOH \rightarrow \overset{\displaystyle O}{\overset{\displaystyle \|}{HC}}-NHR + (R'CO)_2O$$

Reaction of isocyanides with glacial acetic acid may be used to form characterizable derivatives. However, the reaction of isocyanides and acids is not always simple, for example, the three-component system which includes carboxylic acids and a keto group compound [the Passerini reaction (209)]:

$$RCOOH + R'COR'' + R'''NC \rightarrow \underset{\overset{\displaystyle \|}{O} \ \ \overset{\displaystyle |}{R''}}{\overset{\overset{\displaystyle R' \ \ O}{\overset{\displaystyle | \ \ \|}{}}}{RCOC}} - CNHR'''$$

The reaction of isocyanides with oxalic acid is different from the more general reaction of dicarboxylic acids and is of interest for purposes of analysis (see Section V.C.1.c).

Anhydrous acids add to the isocyanide group:

$$RNC + HX \rightarrow RN{=}CHX$$

The products are formidoyl derivatives.

c. Cyanamides

Cyanamide is fairly stable in aqueous acidic solutions of pH less than about 5. Dicyandiamide is also quite stable in cold acidic solutions (it is a weak acid); when calcium cyanamide or cyanamide itself is heated with dilute sulfuric acid, the dicyandiamide which is first formed is hydrolyzed to guanylurea (dicyandiamidine):

$$H_2NCH(NH)NHCN + H_2O \rightarrow \underset{\overset{\displaystyle \|}{NH} \ \ \overset{\displaystyle \|}{O}}{H_2NCNHCNH_2}$$

Guanylurea also can be hydrolyzed to guanidine:

$$H_2NCNHCONH_2 + H_2O \rightarrow HN{=}C(NH_2)_2 + CO_2 + NH_3$$
$$\underset{NH}{\overset{\|}{\phantom{.}}}$$

Concentrated sulfuric acid decomposes dicyandiamide to guanidine, but N-substituted alkyl cyanamides are hydrolyzed in acidic media:

$$R_2NCN + 2\,H_2O + 2\,HCl \rightarrow NH_4Cl + CO_2 + R_2NH{\cdot}HCl$$

### d. CARBODIIMIDES

The hydrolysis of carbodiimides to disubstituted ureas is catalyzed by acids:

$$RN{=}C{=}NR + H_2O \rightarrow RNHCONHR$$

The sensitivity of carbodiimide compounds to aqueous acid varies considerably, some being more readily hydrolyzed than others; in particular, the monomeric carbodiimides are far more readily hydrolyzed than the polymeric. Zetsche and Fredrich (327) have noted that polymeric forms in ether suspension are converted by trichloroacetic acid to the corresponding ureides after a few days, but boiling alcohol containing hydrochloric acid was essentially without action. Weith (311) has found that, although an aromatic carbodiimide such as diphenylcarbodiimide may be readily hydrolyzed, others require drastic conditions such as repeated evaporations with hydrogen chloride in alcohol, to force adequate conversion to the substituted urea.

Hydrogen sulfide or hydrogen selenide often reacts readily with certain types of carbodiimides and may give rise to mixtures of ureas and thioureas or selenoureas, but the reactions are of little value for analytical work; moreover, the carbodiimide compounds show wide ranges of reactivity with these substances. Hydrogen cyanide (as aqueous cyanide ion) reacts to form $\alpha$-cyano-$N,N'$-disubstituted formamidines:

$$RN{=}C{=}NR + HCN \rightarrow RNHC{=}NR$$
$$\underset{CN}{\overset{|}{\phantom{.}}}$$

Acids seldom catalyze alcoholysis of the carbodiimide group.

The reaction of monocarboxylic acids with carbodiimide depends largely on what conditions prevail and whether the acid/carbodiimide ratio is mole to mole or in excess:

$$R'COOH + NR{=}C{=}NR \rightarrow R(R'CO)NCONHR$$
$$2\,R'COOH + RN{=}C{=}NR \rightarrow R'COOCOR' + RNHCONHR$$

An indicated by the above reactions, an equimolar mixture forms an N-acyl urea, but an excess leads to an acid anhydride and a urea.

The reaction of carbodiimides with oxalic acid is quantitative (see Sections V.E.1.a and V.I.1.a). Other dicarboxylic acids add directly to carbodiimides to form ring structures.

### 4. Action of Alkalies

#### d. NITRILES

The action of aqueous solutions of alkalies on the nitrile group parallels the action of aqueous acids, that is, hydrolysis proceeds via an amide as an intermediate, but the final products usually are the salt of a carboxylic acid and ammonia:

$$RCN + H_2O + NaOH \rightarrow RCOONa + NH_3$$

Solutions of potassium hydroxide in glycerine or ethylene glycol are recommended because they can be heated to higher temperatures than aqueous solutions (231). The simple nitriles are much less reactive than their corresponding amides, and as the complexity of the nitriles increases, the rate-governing step is the hydrolytic attack on the amide; some amides are remarkably resistant to hydrolysis. However, as was indicated in a preceding section, formation of amide often can be markedly accelerated by use of alkaline peroxide; the reaction mechanism involves nucleophilic attack by hydroperoxide ion (see Section III.B.2.a).

Though alkaline hydrolytic reactions of the nitrile group are considered to be broadly applicable, few are sufficiently quantitative to permit determination of nitrile compounds (as ammonia); steric hindrance and condensation reactions play important roles in determining the outcome of hydrolytic reactions. Galat (102) found that aromatic nitriles are very readily hydrolyzed to amides (but not to acids) by refluxing in alcohol-water solution while in contact with an anionic exchange resin (Amberlite IRA-400).

#### b. ISOCYANIDES

The isocyanides are remarkably stable in the presence of dilute aqueous alkalies; however, at elevated temperatures, a primary amine and a salt of formic acid are formed by hydrolysis. The isocyanide group is somewhat reactive with more concentrated alkalies and with sodium ethoxide in ethanol, but methyl isocyanide is quite resistant.

## c. Cyanamides

Cyanamide dimerizes readily to dicyandiamide at pH 7–12:

$$H_2NCN + H_2NCN \rightarrow H_2NCNHCN$$
$$\underset{NH}{\|}$$

In strongly alkaline solutions, dicyandiamide is hydrolyzed to urea.

Ammonia (and its salts) reacts with cyanamide to form guanidine, but dicyandiamide forms biguanidine:

$$H_2NCN + NH_3 \rightarrow \begin{matrix} H_2N \\ \\ H_2N \end{matrix} \!\!\! C\!\!=\!\!NH$$

$$H_2NCNHCN \rightarrow H_2N\!-\!C\!-\!N\!-\!C\!-\!NH_2$$
$$\underset{NH}{\|} \qquad \underset{NH}{/\!\!/} \; \underset{H}{|} \; \underset{NH}{\backslash\!\!\backslash}$$

## d. Carbodiimides

The hydrolysis of carbodiimides to urea is also catalyzed by alkali (similarity to acids):

$$RN\!\!=\!\!C\!\!=\!\!NR + H_2O \rightarrow RNHCONHR$$

but the polymerization of many carbodiimides is catalyzed preferentially. Thus, for example, di-$m$-nitrophenylcarbodiimide or di-$m$-cyanophenylcarbodiimide undergoes rapid polymerization in alkaline solution with scarcely any hydrolytic attack.

In contrast to acids, alkalies (such as sodium ethylate) catalyze alcoholysis of carbodiimides, but reaction rates are low. Other substances also catalyze alcoholysis, for example, $Cu_2Cl_2$.

Ammonia and primary amines react additively with the carbodiimide group to form substituted guanidines:

$$RN\!\!=\!\!C\!\!=\!\!NR + R'NH_2 \rightarrow RNHCNHR$$
$$\underset{NR'}{\|}$$

## 5. Miscellaneous Reactions

### a. Nitriles

A nitrile can be converted directly to an ester; in general, 1 mol of nitrile is refluxed with 1 mol of concentrated sulfuric acid and about 10 mol of an alcohol at as high a temperature as can be obtained, preferably in sealed tubes at about 150°C. Inasmuch as ammonia is released:

$$RCN + R'OH + H_2O \rightarrow RCOOR' + NH_3$$

and is retained by the acid, the process should be useful for quantitative analysis of nitriles. However, aromatic nitriles with substituents in the *ortho* position resist esterification.

Nitriles readily react with hydrogen sulfide to form thioamides:

$$RCN + H_2S \rightarrow R—\underset{\underset{S}{\|}}{C}—NH_2$$

Often, the reaction can be made to take place by passing hydrogen sulfide into an alcoholic solution of the nitrile containing ammonia or triethanolamine (205). Thioamides are readily hydrolyzed by heating with aqueous acids or alkalies:

$$RCSNH_2 + 2 H_2O + HCl \rightarrow RCOOH + H_2S + NH_4Cl$$

When thioamides are heated with benzophenone dichloride, a blue color is formed (probably thiobenzophenone); the color reaction may be used to test for thioamides as well as nitriles (after conversion to thioamides) (281).

### b. ISOCYANIDES

For the purposes of analytical chemistry, the reaction of isocyanides with sulfur at elevated temperatures is of interest (310):

$$RNC + S \rightarrow RNCS$$

Simple isocyanides add two halogen atoms (often violently) to form low-melting solids of the general formula $RNCX_2$ (167,199), but aliphatic isocyanides ($<C_{11}$) form dimeric or trimeric water-soluble products (166); the reaction with halides is of interest because the nitriles do not form addition products. Alkyl iodides also form addition products with isocyanides (but not with nitriles).

Isocyanides have the remarkable property of forming complexes with metal salts as well as metals, typically copper (161). For example, copper dissolves in liquid cyclohexyl isocyanide to form a soluble complex which shows absorption at $2180\ cm^{-1}$ ($4.59\ \mu$); cuprous chloride also forms complexes (290). So great is the stability of isocyanide complexes that carbon monoxide is displaced from metal carbonyls.

Isocyanides react with hydrogen sulfide to form thioformamides.

$$RNC + H_2S \rightarrow RN\underset{\underset{S}{\|}}{H}CH$$

Isocyanides also react with phenols and naphthols, for example, phenyl isocyanide and phenol (290):

$$C_6H_5NC + C_6H_5OH \longrightarrow$$

Silver cyanide dissolves in liquid isocyanides with evolution of heat, but not in nitriles; the reaction may be used as a convenient method for determining whether a sample is a nitrile or an isocyanide.

### c. CARBODIIMIDES

The reaction of certain carbodiimides with cyanide ion is of interest in analytical chemistry because it takes place essentially instantaneously and quantitatively; in general, the reaction yields $\alpha$-cyano-$N,N'$-disubstituted formamidines:

$$RN=C=NR + CN^- + H_2O \rightarrow RNHC\underset{\underset{CN}{|}}{C}=NR + OH^-$$

The quantitative addition of cyanide ion, however, takes place only with the monomeric aromatic carbodiimides, in which the nitrogen atoms are no less basic than those of diphenylcarbodiimide (139); moreover, although the reaction given above appears to be an addition of hydrogen cyanide, the free acid reacts very slowly and often not at all. In fact, the addition of hydrogen cyanide requires a substantial amount of free cyanide ion, such as can be provided by solutions above pH 7, but then the reaction with cyanide must compete with polymerization reactions as well as with hydrolytic reactions that are catalyzed greatly by alkaline conditions. The alkali-sensitive carbodiimides, which are hydrolyzed very slowly by acids, are dissolved in pure tetrahydrofuran, and the solution, first rendered acidic by HClO$_4$, is poured into an excess of an aqueous solution of alkali-metal cyanide mixed with methanol. Tetrahydrofuran solutions of other carbodiimides can be poured directly into an excess of a methanolic solution of alkali-metal cyanide (139) (see Section VI.A.7.e).

The reaction of carbon disulfide with carbodiimides is also of interest for analytical work because it provides a convenient method for cleavage into two isothiocyanates:

$$R'N=C=NR'' + CS_2 \rightarrow R'NCS + R''NCS$$

### C. GENERAL REFERENCES

The following list of references is included here for the analytical chemist who wishes more detailed descriptions of the chemical reactions which have been briefly summarized in the foregoing sections.

*Sidgwick's The Organic Chemistry of Nitrogen*, 3d ed., revised and rewritten by I. T. Millar and H. D. Springall, Clarendon Press, Oxford, 1966. Provides an authoritative summary of the theoretical aspects of the nitrile and isonitrile groups.

Mowry, David T., "The Preparation of Nitriles," *Chem. Rev.*, **42**, 189 (1948). A comprehensive review of syntheses and sundry information.

Astle, Melvin J., *Industrial Organic Nitrogen Compounds*, Reinhold, New York, 1961, Ch. 5. A general review of the syntheses and reactions of nitriles; copious references.

Kirk, R. W., and D. F., Othmer, *Encyclopedia of Chemical Technology*, Vol. 9, The Interscience Encyclopedia, New York, 1950, p. 352. A cursory review of all aspects of nitrile and isonitrile chemistry and applications.

Migrdichian, Vartkes, *The Chemistry of Organic Cyanogen Compounds*, (ACS Monograph Series 105), Reinhold, New York, 1947. A detailed treatment of the chemistry and preparation of nitriles with numerous references: cursory information on isonitriles.

Ugi, Ivar (Ed.), *Isonitrile Chemistry*, Academic Press, New York, 1971.

Rappoport, Z. (Ed.), *The Chemistry of the Cyano Group*, Wiley-Interscience, New York, 1970.

Dhar, D. N., "The Chemistry of Tetracyanoethylene," *Chem. Rev.*, **67**, 611 (1967).

Malatesta, L., and F. Bonati, *Isocyanide Complexes of Metals*, Wiley-Interscience, New York, 1969.

Kurzer, F., and K. Douraghi-Zadeh, *Chem. Rev.*, **67**, 107 (1967). An excellent summary of the chemistry of carbodiimides.

## IV. TOXICITY AND INDUSTRIAL HYGIENE

The use of nitriles as chemical intermediates for production of synthetic rubber, textiles, plastics, and other products has increased the probability that they will be found in significant quantities in the atmosphere or in surface waters. The nitriles most frequently used in large-scale syntheses are acrylonitrile, acetonitrile, propionitrile, cyanoacetic acid, cyanoacetamide, succinonitrile, adiponitrile, benzonitrile, phthalonitrile, and phenylacetonitrile. Small amounts of these compounds in surface waters are ordinarily destroyed by enzymatic action, with the formation of organic acids and ammonia (or associated materials) (177), but the time required for destruction varies considerably and is a function of the presence or the absence of an acclimated biota in the waters. Acrylonitrile and oxydipropionitrile require a long time for initial adaptation of organisms and for the adjustment from summer to winter temperatures. Lactonitrile appears to be especially toxic because it forms cyanide ion.

Although a large number of the nitriles of high molecular weight, unsubstituted, saturated fatty acids are essentially nontoxic, it is best to regard all nitriles as toxic unless accurate information is available which suggests relative safety (the presence of other substances such as by-products of synthesis may confer toxicity on a given sample of nitrile). Unsaturated nitriles such as acrylonitrile and fumaronitrile are toxic, and they may cause severe dermatitis; benzonitrile is also toxic. As can be anticipated, cyanohydrins, $\alpha$-aminonitriles, and tetra-cyanoethylene approach the toxicity of hydrogen cyanide; halogenated nitriles are lachrymators, are toxic, and usually show exceptional der-matitic activity. Isobutyronitrile may be absorbed directly through the skin. The toxic nitriles are insidious; repeated sublethal exposures may damage the central nervous system, leading to paralysis, speech impediments, vertigo, and anemia.

The isocyanide group is isoelectronic with carbon monoxide, and so it is often assumed that the isocyanides are toxic; indeed, these compounds should be regarded as potential hazards to mammals, although in actuality the danger is not as great as their disagreeable odor would lead one to believe. The toxicity of the isocyanides has not been studied in detail, but it has been found that 1,4-diisocyanobutane is surprisingly toxic ($LD_{50}$ for mice $< 10$ mg/kg); in contrast, for most isocyanides, oral and subcutaneous doses of 500–5000 mg/kg are tolerated by mice (286).

Cyanamide is a caustic toxic substance; on the other hand, dicyan-diamide is harmless to mice. Dry silver cyanamide explodes violently when heated; all metal salts of cyanamide should be regarded as explosive hazards. Most substituted cyanamides are extremely toxic and should be handled with exceptional care.

Although the carbodiimides are not regarded as toxic, each member must be handled with care because extreme and unpredictable reactivity may induce unforeseen reactions in the human body. It is reported that dicyclohexylcarbodiimide shows an $LD_{50}$ value for rats of the order of 2.6 grams/kg (9).

## V. QUALITATIVE DETERMINATIONS

### A. PREPARATION OF SAMPLE

Samples containing nitriles seldom need special treatment, inasmuch as nitriles are not overly sensitive materials. In contrast, isocyanides should be treated with caution largely because of stench and because of the possibility that they may be toxic; also, they are very reactive and often quite volatile. Moreover, since isocyanides are rather unstable and may deteriorate on standing, samples should be processed immediately. Polymers of isocyanides may form rapidly on glass, especially when in contact with ground or roughened surfaces. Also, since isocyanides react with metals, it is best to preserve samples in polytetrafluoro-ethylene vessels; storage at a low temperature is recommended. Compounds of cyanamide and the carbodiimides are frequently so reactive that they are best analyzed as soon as possible.

## B. DETECTION OF THE NITRILE GROUP

### 1. Chemical Methods

Basic, nitrogen-containing functional groups can be detected by their solubility in dilute hydrochloric acid (248) or by color changes with acid-indicator mixtures (70); the neutral nitrogen-containing groups such as nitro, nitroso, azoxy, azo alkyl nitrite, and alkyl nitrate can be reduced by ferrous hydroxide or zinc and ammonium chloride. Nitriles and amides can be detected but not differentiated by the hydroxamic acid test, provided that esters, acid anhydrides, acid chlorides, and trichloromethyl groups are absent (40).

There is no specific test for the nitrile group; however, the following tests can be used in combination with tests for other functional groups.

### a. AMIDOXIME TEST

Nitriles react with a buffered solution of hydroxylammonium chloride and hydroxylamine at the boiling point of propylene glycol to form amidoximes, which yield reddish colors when treated with ferric ions:

$$RCN + NH_2OH \rightleftharpoons RC(NH)NHOH \rightleftharpoons RC(NH_2)NOH$$
$$3\,RC(NH)NHOH + Fe^{3+} \rightleftharpoons (RC(NH)NHO)_3Fe + 3\,H^+$$

The following procedure is typical (257):

Mix 20–30 mg of the sample with 2 ml of a molar solution of hydroxylamine hydrochloride in propylene glycol, and then add 1 ml of a molar solution of potassium hydroxide in propylene glycol. Heat and boil the mixture for about 2 min; cool, and add about 1 ml of a 5% solution of ferric chloride ($6\,H_2O$) in ethanol. A red to violet color is a positive test; esters, acid chlorides, anhydrides, anilides, amides, and imides interfere. Dicyandiamide and diethylcyanamide do . not give a positive test. The time of reaction at elevated temperature should not be prolonged because amides slowly form hydroxamic acids, RCONHOH, which also give colors with ferric ions (23).

### b. SODA-LIME TEST FOR ALIPHATIC NITRILES

Most nitrogen-containing substances yield hydrogen cyanide on pyrolysis, and all do so in the presence of sodium or lime (Lassaigne; see Section II.B.1.a of "Detection of Nitrogen in Samples," Vol. 15); however, it has been found that aliphatic nitriles generate hydrogen cyanide when heated to 250°C, whereas aromatic nitriles and other nitrogen-containing substances do not (88). Since the cleavage of hydrogen cyanide is facilitated by the presence of calcium oxide and calcium carbonate, the reaction often may be made to occur at temperatures as low as 150°C.

The following procedure may be used to detect aliphatic nitriles such as cyanoacetic acid and substituted benzyl cyanides (88):

Place a few milligrams of the sample in the bottom of a small test tube. Cover with a few hundredths of a gram of a 1:1 mixture of calcium oxide and calcium carbonate. Place the tube in a glycerol bath which is at 250°C, and cover the mouth of the tube with a disk of filter paper moistened with copper(II) acetate–benzidine solution [equal parts of a solution of 2.86 grams of copper acetate/liter and a solution made by diluting 675 ml of water saturated with benzidine with 525 ml of water (250)]. A positive test is indicated by the formation of a blue stain on the paper disk after 3–5 min.

Inasmuch as benzidine is a carcinogen, 4,4'-tetramethyldiaminodiphenyl-methane may be substituted, and copper ethylacetoacetate complex is used in place of copper acetate; the test paper is prepared by dipping strips of paper in a solution containing about 50 mg of each reagent in 10 ml of chloroform (86).

### c. SODA-LIME TEST FOR ALL NITRILES

Pyrolysis of nitriles at elevated temperatures (i.e., above the 250°C limit imposed by the preceding test procedure) liberates ammonia. Trofimenko and Sease (280) found that simple amides ($CONH_2$) and amines also liberate ammonia; there is a possibility that $N$-substituted amides in which the substituent is a high molecular weight alkyl residue may decompose on heating to form a simple amide that can interfere. The test suggested by Trofimenko and Sease is predicated on the detection of ammonia in the soda-lime pyrolysate of the sample by a methanolic solution of copper sulfate; the behavior of methanolic copper sulfate with ammonia is readily distinguishable from its performance with amines.

Thoroughly mix 100 mg of the sample with enough low-moisture, 20-mesh soda-lime to pack a $13 \times 100$ mm test tube one-fourth full. Clamp the tube at a 45-deg angle; put a one-hole stopper in the mouth of the tube, and place an L-shaped glass delivery tube in the stopper so that the longer leg of the delivery tube (preferably drawn to a fine point) reaches to the bottom of a small test tube which contains 1 ml of methanolic copper sulfate solution. (Prepare the methanolic copper sulfate solution by dissolving 1 gram of $CuSO_4 \cdot 5 H_2O$ in 100 ml of anhydrous methanol; allow to stand for 24–48 hr and then filter.)

Heat the upper parts of the soda-lime layer gently at first, and then more vigorously, gradually extending the zone of heating until all volatile products are distilled through the soda-lime layer into the methanolic copper sulfate reagent; with the reagent, ammonia produces a white turbidity which quickly changes to a purple or blue precipitate that settles to the bottom and does not change color appreciably on standing. (Ethylenediamine and possibly other *vic*-diamines respond similarly.) Aliphatic amines give soluble blue or green complexes or pale green or greenish blue solutions from which a flocculent, pale green precipitate settles out; heavy, intensely colored precipitates are formed with primary aryl amines.

### d. SULFUR FUSION TEST FOR ALL NITRILES

Pyrolysis of an organic compound in the presence of sulfur usually yields hydrogen sulfide (even at temperatures as low as 250°C); when nitriles or isonitriles are heated with sulfur, thiocyanogen is formed, and the pyrohydrolysis products contain thiocyanic acid (88):

$$H_2S + (CNS)_2 \rightarrow 2\,HCNS$$

Thiocyanic acid is readily detected by formation of a characteristic red coloration with ferric ion.

Mix about 0.01 gram of powdered sulfur with about 1 mg of the sample in a micro test tube; put the tube in a hole in an asbestos sheet (to protect the opening of the tube from excessive heat), and cover the open end of the tube with a disk of paper moistened with acidified 1% ferric nitrate solution. Start heating the test tube just under the asbestos ring, and proceed gradually toward the closed end of the tube. A positive test is indicated by the appearance of a red stain on the paper; aliphatic nitriles react rapidly, but aromatic nitriles must be heated strongly.

Compounds that contain a CN group in an open or closed chain behave like nitriles on pyrolysis; thus purine derivatives, thiazoles, triazoles, oximes, and Schiff bases will release thiocyanic acid when pyrolyzed with sulfur.

### e. AMMONIA FROM REDUCTION OF NITRILES

Powerful reducing agents such as titanous ion in sodium hydroxide react with nitriles to form ammonia; Pasez et al. (214) recommend a silver-catalyzed manganese reagent for the detection of ammonia (formation of a black precipitate):

$$MnSO_4 + 2\,AgNO_3 + 4\,NH_4OH \rightarrow MnO_2 + 2\,Ag + (NH_4)_2SO_4 + 2\,NH_4NO_3 + 2\,H_2O$$

Volatile amines also react as ammonia; the test is performed as follows:

The sample (a few milligrams) is placed in a test tube or flask which can be stoppered with a polyethylene cap that holds a small piece of paper treated with silver-manganese reagent. The sample is moistened with 1 ml of water and then treated with 0.5 ml of a 1% solution of titanous chloride for 5 min at room temperature. Then, after adding 1 ml of a saturated solution of sodium hydroxide, the container is stoppered (with the test paper in place) and held at

70°C for 1 hr. Ammonia, which may be formed by reduction of nitriles, will diffuse and react with the manganese test paper, changing its color first to gray and finally to black.

Tet papers are prepared as follows. Dissolve 8.5 grams of silver nitrate and 8.5 grams of $MnSO_4 \cdot H_2O$ in 1 liter of water. Immerse high-quality filter paper in the solution, and dry in an atmosphere free of acidic or alkaline vapors.

The test can readily detect 60–100 $\mu$g of nitriles; unfortunately, ammonium salts, aliphatic nitro compounds, oximes, amines, and amides also give positive indications.

### f. Hydrolysis by Strong Base

Nitriles are hydrolyzed by alkalies to ammonia; ammonia can be readily detected by its odor, or by its action on moistened litmus paper or copper sulfate solution (280). However, the test is not specific, for amides also are hydrolyzed to ammonia, and ammonium salts obviously will interfere.

Put a boiling chip in the bottom of a $15 \times 150$ mm test tube and add enough $N$ sodium hydroxide solution to fill the tube one-fourth full; place 100–150 mg of the sample in the tube, and immediately close the mouth of the tube with a one-hole stopper which carries a delivery tube that extends to the bottom of a small tube containing 1 ml of methanolic copper sulfate solution (as in Section V.B.I.c). Heat the sodium hydroxide solution gently to boiling, and allow vapors to enter the copper sulfate solution for a few moments. Ammonia produces a white turbidity which quickly changes to a purple or blue precipitate.

Alternatively, the ammonia given off by hydrolysis can be detected at the mouth of the tube by moistened litmus paper, but other volatile basic materials (amines) will interfere.

The aqueous sodium hydroxide solution can be replaced with a normal solution of potassium hydroxide in 9–1 propylene glycol-water; in some instances, a 35% solution of alkali must be used.

### g. Hydrolysis by Strong Acids

Nitriles are hydrolyzed to amides and thence to ammonia by prolonged action of strong acids. Although many nitriles are readily hydrolyzed to ammonia by dilute solutions of mineral acids, the amides of others are so resistant that it is necessary to prolong hydrolytic attack to several hours and to use concentrated solutions. The following procedure is not specific, but it can be used freely on nearly any type of nitrile:

Prepare a solution of 50% (w/w) sulfuric acid in glacial acetic acid; place about 100 mg of the sample in the bottom of a $15 \times 150$ mm test tube, and add enough of the acid solution to cover the sample with a 5-mm layer. Reflux the mixture for 2–3 hr (use a small cold-finger inserted in the tube). With the aid of some water, transfer the hydrolysate to a 100-mm Kjeldahl flask fitted with a one-hole stopper and delivery tube (see the preceding test). Remove the stopper, and quickly add pellets of reagent-grade sodium hydroxide in rather large excess; replace the stopper, and test for ammonia as described previously.

### h. FLUORESCEIN CHLORIDE

On fusion with fluorescein chloride, nitriles form a symmetrical rhodamine which exhibits a yellow to yellow-green fluorescence (85). Amides and primary amines interfere.

## 2. Physical Methods

The nitrile functional group generally can be detected with greater certainty by infrared and mass spectrometric methods than by chemical methods. Whenever intensities of absorption are sufficient, the nitrile and isonitrile functional groups can be readily detected by ir spectrophotometry.

### a. INFRARED SPECTROPHOTOMETRY

The characteristic absorption of the nitrile functional group was noted as early as 1905 by Coblentz (54) and later by Bell (21), who reported that aryl nitriles absorb at $2222 \, cm^{-1}$ (4.5 $\mu$) and alkyl nitriles at $2273 \, cm^{-1}$ (4.4 $\mu$). An extensive study of about 70 different nitriles was made by Kitson and Griffith (160); in saturated mono- and dinitriles the -C≡N stretching frequency (split for some dinitriles) lies between 2260 and $2240 \, cm^{-1}$ (4.42 and 4.46 $\mu$). For unsaturated, aliphatic unconjugated nitriles, the stretching frequency lies within the same range; when the nitrile group is conjugated with a double bond, the frequency range is $2232–2218 \, cm^{-1}$ (4.48–4.51 $\mu$), but both bonds are often found in dinitriles that contain both types of groups. Aromatic compounds in which the nitrile group is attached to the ring absorb in the range 2240–$2221 \, cm^{-1}$ (4.46–4.50 $\mu$). Additional work by Besnainu et al. (26) has confirmed the frequency ranges found by Kitson and Griffith and has shown also that substitution of electron-attracting atoms such as oxygen or chlorine on the $\alpha$-carbon has little effect on the stretching frequency of the nitrile group (but the intensity of absorption is decreased); on the other hand, electronegative groups on the $\beta$-carbon appear to lower the frequency as much as $16 \, cm^{-1}$ from the central frequency of $2250 \, cm^{-1}$ (22). Moreover, Besnainu et al. (26), as well as Kitson and Griffith (160), have found that the intensity of absorption of the nitrile bond is relatively strong for nitriles containing unsaturated nonaromatic groups, and very strong for aromatic nitriles; however, when oxygenated groups are in the molecule, the absorption is quenched, and, indeed, in some compounds the nitrile absorption can scarcely be discerned (e.g., acetonecyanohydrin). Thus lack of an absorption band in the -C≡N region cannot serve as proof that the functional group is absent.

Following is a summary of the range of frequencies which are characteristic of nitrile group stretching vibrations (see also Table II):

| | |
|---|---|
| Saturated alkyl nitriles | $2260$–$2240$ cm$^{-1}$($4.42$–$4.46\ \mu$) |
| Aryl nitriles | $2240$–$2220$ cm$^{-1}$($4.46$–$4.50\ \mu$) |
| $\alpha,\beta$-Unsaturated nitriles | $2235$–$2215$ cm$^{-1}$($4.47$–$4.51\ \mu$) |

The lowest frequency given above for $\alpha,\beta$-unsaturated nitriles is shifted to still lower frequencies when enamine tautomers can be formed; thus Baldwin (15) found a shift of $17$–$50$ cm$^{-1}$ below $2215$ cm$^{-1}$ in a series of $\beta$-amino-$\alpha,\beta$-unsaturated nitriles for which the following enamine tautomers exist:

$$H_2N-\overset{|}{C}=\overset{|}{C}-C\equiv N \leftrightarrow H_2N-\overset{|}{\overset{+}{C}}-\overset{|}{C}=C=N^- \leftrightarrow H_2\overset{+}{N}=\overset{|}{C}-\overset{|}{\overset{-}{C}}-C\equiv N$$

Baldwin also found shifts of $5$–$46$ cm$^{-1}$ for $\beta$-hydroxy- and $\beta$-alkoxy-$\alpha,\beta$-unsaturated nitriles; shifts of $5$–$15$ cm$^{-1}$ were observed for nitroacetonitrile anions. Kazitsyna et al. (154) also found lowered (and sometimes absent) nitrile absorption frequencies ($2203$–$2215$ cm$^{-1}$) for dinitriles of the type

$$R_1-\underset{\underset{C\equiv N}{|}}{\overset{\overset{R_2}{|}}{C}}-NH-(CH_2)_n-NH-\underset{\underset{C\equiv N}{|}}{\overset{\overset{R_2}{|}}{C}}-R_1 \qquad (n = 2 \text{ or } 6)$$

Compounds of this type form stable hydrochlorides (not unstable iminochlorides) from which the original substances can be obtained by treatment with alkali; however, the ir spectra of the hydrochlorides do not show the characteristic nitrile functional group, and this lack is attributed to the decrease in polarity of the nitrile group because of the appearance of a positive charge on the amino nitrogen (salt formation). Other strong electron-acceptor substituents (e.g., an oxygen atom, halogen atom, sulfo group, or nitro group) near the nitrile group should also lead to a decrease in intensity of the nitrile absorption.

Nitriles react with a number of metal halides to form complexes; the lone pair of electrons on the nitrogen atom is involved (304), and studies of ir absorption indicate that palladium halide complexes have a *trans*-planar configuration, whereas platinum(II) halide complexes have a

Table II. Characteristic Infrared Absorption Frequencies for CN Stretching in the Functional Groups of Typical Nitrile, Isocyanide, N-Cyano, and Carbodiimide Compounds

| Compound | Formula | Medium | Frequency (cm⁻¹) | (μ) | Ref. |
|---|---|---|---|---|---|
| **Nitrile group compounds** | | | | | |
| Acetonitrile | $CH_3CN$ | $CHCl_3$ | 2257 | 4.43 | 90,319 |
| Iodoacetonitrile | $ICH_2CN$ | $CHCl_3$ | 2247 | 4.45 | 243 |
| Aminoacetonitrile | $H_2NCH_2CN$ | $CHCl_3$ | 2242 | 4.46 | 243 |
| Chloroacetonitrile | $ClCH_2CN$ | $CHCl_3$ | 2257 | 4.43 | 243 |
| Trichloroacetonitrile | $Cl_3CCN$ | $CHCl_3$ | 2249 | 4.45 | 146 |
| n-Butyronitrile | $CH_3(CH_2)_2CN$ | — | 2252 | 4.44 | 160 |
| Nonanonitrile | $CH_3(CH_2)_7CN$ | — | 2252 | 4.44 | 160 |
| Hexadecanonitrile | $CH_3(CH_2)_{14}CN$ | $CHCl_3$ | 2247 | 4.45 | 146 |
| Malononitrile | $NCCH_2CN$ | — | 2278 | 4.39 | 160 |
| Succinonitrile | $NC(CH_3)_2CN$ | — | 2261 | 4.42 | 160 |
| Adiponitrile | $NC(CH_2)_4CN$ | — | 2252 | 4.44 | 146 |
| Suberonitrile | $NC(CH_2)_6CN$ | — | 2247 | 4.45 | 160 |
| Acrylonitrile | $CH_2{=}CHCN$ | — | 2232 | 4.38 | 160 |
| Vinylacetonitrile | $CH_2{=}CHCH_2CN$ | — | 2252 | 4.44 | 160 |
| Tetracyanoethylene | $(NO)_2C{=}C(CN)_2$ | KBr | 2227 | 4.49 | 191 |
|  |  |  | 2260 | 4.42 |  |
| Fumaronitrile | $NCHC{=}CHCN$ (*trans*) | Solid | 2240 | 4.46 | 191 |
| Maleonitrile | $NCHC{=}CHCN$ (*cis*) | Liquid | 2260 (w) | 4.42 | 191 |
|  |  |  | 2230 | 4.48 |  |
|  |  |  | 2250 | 4.44 |  |
| Cinnamonitrile | $C_6H_5CH{=}CHCN$ | — | 2221 | 4.50 | 323 |
| Benzonitrile | $C_6H_5CN$ | $CHCl_3$ | 2231 | 4.48 | 243 |
| o-Hydroxybenzonitrile | $HOC_6H_4CN$ | $CHCl_3$ | 2227 | 4.49 | 243 |
| m-Hydroxybenzonitrile | $HOC_6H_4CN$ | $CHCl_3$ | 2234 | 4.48 | 243 |
| p-Hydroxybenzonitrile | $HOC_6H_4CN$ | $CHCl_3$ | 2227 | 4.49 | 243 |

| Compound | Formula | Solvent | cm⁻¹ | | Ref. |
|---|---|---|---|---|---|
| Phenylacetonitrile | $C_6H_5CH_2CN$ | — | 2252 | 4.44 | 160 |
| β-Phenylpropionitrile | $C_6H_5CH_2CH_2CN$ | — | 2247 | 4.45 | 160 |
| 1-Naphthonitrile | $C_{10}H_7CN$ | CCl₄ | 2227 | 4.49 | 91 |
| 3-Cyanopyrene | $C_{16}H_9CN$ | CCl₄ | 2221 | 4.50 | 91 |
| Isocyanide group compounds | | | | | |
| Methyl isocyanide | $CH_3NC$ | Gas | 2166 | 4.62 | 319 |
| Propyl isocyanide | $CH_3CH_2CH_2NC$ | CHCl₃ | 2148 | 4.66 | 110 |
| Ethyl isocyanide | $CH_3CH_2NC$ | — | 2151 | 4.65 | 48 |
| sec-Butyl isocyanide | $CH_3(C_2H_5)CHNC$ | CCl₄ | 2125 | 4.70 | 48 |
| Phenyl isocyanide | $C_6H_5NC$ | CHCl₃ | 2125 | 4.70 | 110 |
| | | | 2146 | 4.66 | 288 |
| p-Nitrophenyl isocyanide | $O_2NC_6H_4NC$ | — | 2116 | 4.73 | 288 |
| 2,4,6-Trimethyl phenyl isocyanide | $(CH_3)_3C_6H_2NC$ | — | 2109 | 4.74 | 288 |
| 2-Naphthyl isocyanide | $C_{10}H_7CN$ | — | 2122 | 4.71 | 288 |
| Derivatives of Cyanamides | | | | | |
| Dimethylcyanamide | $(CH_3)_2NCN$ | CCl₄ | 2221 | 4.50 | 146 |
| Diethylcyanamide | $(CH_3CH_2)_2NCN$ | CCl₄ | 2212 | 4.52 | 146 |
| Carbodiimides | | | | | |
| Dimethylcarbodiimide | $CH_3NCNCH_3$ | — | 2140 | 4.67 | 224 |
| Diethylcarbodiimide | $C_2H_5NCNC_2H_5$ | — | 2138 | 4.68 | 187 |
| Diisopropylcarbodiimide | $C_3H_7NCNC_3H_7$ | — | 2128 | 4.70 | 187 |
| Di-n-butylcarbodiimide | $C_4H_9NCNC_4H_9$ | — | 2138 | 4.68 | 187 |
| Di-sec-butylcarbodiimide | $C_4H_9NCNC_4H_9$ | — | 2128 | 4.70 | 187 |
| Dicyclohexylcarbodiimide | $C_6H_{11}NCNC_6H_{11}$ | — | 2130 | 4.69 | 198 |
| Dibenzylcarbodiimide | $C_6H_5CH_2NCNH_2C—C_6H_5$ | — | 2140 | 4.67 | 187 |
| Di-p-tolylcarbodiimide | $CH_3C_6H_4NCN—C_6H_4CH_3$ | — | 2120 | 4.72 | 187 |
| | | | 2145 | 4.66 | |
| Di-p-methoxyphenylcarbodiimide | $CH_3OC_6H_4NCN—C_6H_4OCH_3$ | — | 2120 | 4.72 | 187 |
| | | | 2148 | 4.65 | |
| Di-β-naphthylcarbodiimide | $C_{10}H_7NCNC_{10}H_7$ | — | 2100 | 4.76 | 187 |
| | | | 2152 | 4.65 | |

447

*cis*-planar configuration (305). The ir spectrum of bis(adiponitrilo)copper(I) nitrate has been examined in detail by Matsubara (181).

Studies of the Raman spectra of nitriles confirm ir observations in general; the absorption intensities of aromatic conjugated nitriles are abnormally great compared to those of aliphatic unconjugated nitriles (145,299).

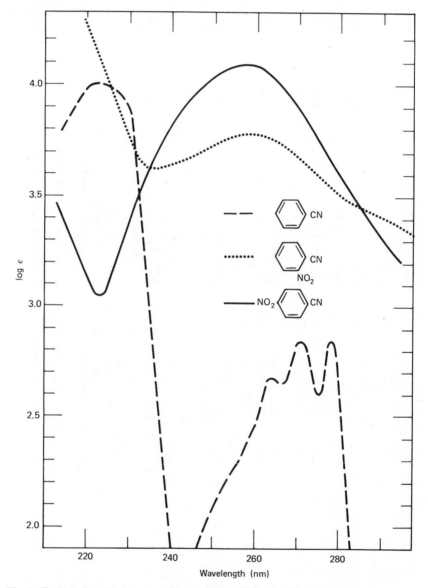

Fig. 1. Typical ultraviolet spectra of benzonitriles. [Data from Leandri and Spinelli (171).]

## b. ULTRAVIOLET SPECTROPHOTOMETRY

Ultraviolet spectral characteristics of the nitrile group are of little value for analytical work. In general, the group shows no definitive absorption bands; the aliphatic compounds that have been studied show nitrile absorption in the region 200–220 nm. In the far ultraviolet, Cutler (65) found a Rydberg series in the absorption spectrum of acetonitrile for which the limit corresponds to an ionizing potential of 11.96 eV. Grunfeld (113), Bielecke and Henri (28), and Schurz et al. (241) found weak characteristic absorptions in the uv spectrum of saturated aliphatic nitriles; the absorption is in the vicinity of 270 nm.

The uv spectra of unsubstituted, unsaturated aliphatic nitriles show absorption maxima at 203 nm. Substitution shifts the absorption maxima as follows: $\alpha$- or $\beta$-monosubstituted, $205 \pm 3$ nm; $\alpha,\beta$- or $\beta,\beta$-disubstituted, $210 \pm 3$ nm; and $\alpha,\beta,\beta$-trisubstituted, $216 \pm 2$ nm (127). Bathochromic shifts are observed with five- and six-membered ring nitriles: $\alpha$- or $\beta$-monosubstituted, 208 nm; $\alpha,\beta$- or $\beta,\beta$-disubstituted, 217 nm; $\alpha,\beta,\beta$-trisubstituted, 225 nm (312). The uv spectra of a short series of substituted benzonitriles have been studied by Leandri and Spinelli (171); benzonitrile shows two regions of strong absorption: 260–280 nm ($\log \epsilon \approx 2.70$) and 220–230 nm ($\log \epsilon \approx 4$). Nuclear substitution leads only to small shifts in the positions of the absorption bands; as a rule, two systems of bands are perceptible in the spectra of most substituted benzonitriles (see Fig. 1).

Existence of weak donor–acceptor complexes formed by interaction of aromatic hydrocarbons with propionitriles is suggested by the work of Weimer and Prausnitz (307); with $p$-xylene as a donor and $n$-heptane as acceptor-solvent, formation constants were computed at 229, 229.5, and 230 nm. Polycyano compounds show charge-transfer absorption bands at higher wavelengths. Complexes of pyrene with polycyano compounds have been studied by Wallenfels and Friederich (303); the following is a tabulation of typical data:

| Compound | Solvent | $\lambda_{max}$ (nm) |
|---|---|---|
| Tetracyanobenzoquinone | $CH_2Cl_2$ | 1129, 617, 485 |
| Tetracyanoethylene | $CH_2Cl_2$ | 724, 495 |
| 2,3-Dicyanobenzoquinone | $CH_2Cl_2$ | 718, 478 |
| Hexacyanobenzene | $CH_2Cl_2$ | 637, 450 |
| Pentacyanotoluene | $CH_2Cl_2$ | 533, 411 |
| 1,2,4,5-Tetracyanobenzene | $CHCl_3$ | 495 |
| 1,2,4,5-Tetracyanobenzene | $CH_3CN$ | 460 |
| Tetracyano-$m$-xylene | $CH_2Cl_2$ | 439 |

The energies of the longest wavelengths of maximum absorption of the complexes of some polycyano compounds with hexamethylbenzene or pyrene were found by Peover (211) to be linear with the polarographic half-wave potentials of the acceptor molecules.

## C. DETECTION OF THE ISOCYANIDE GROUP

There are no specific tests for the isocyanide group; however, the disagreeable and penetrating odors (may be toxic!) of the isocyanides may be sufficient identification for the more volatile compounds. The following tests, though not exclusive for the purpose, may aid in the detection of isocyanides.

### 1. Chemical Methods

#### a. SULFUR FUSION TEST FOR ISOCYANIDES

The sulfur fusion test described for the nitriles (Section V.B.I.d) will also respond positively in the presence of isocyanides; however, since the isocyanides are quite volatile, experimental conditions must be closely regulated to prevent escape of vapors from the zone of reaction.

#### b. REDUCTION BY SODIUM

Sodium in alcohol reduces isocyanides to a secondary amine of the type $RNHCH_3$; the amine can be detected by any convenient method. In the following procedure, volatile amines are detected by the color formed with copper sulfate in methanol.

Add a few drops of methanol to about 2 mg of the sample in a small test tube; after the sample has dissolved or dispersed, add tiny fragments of freshly cut sodium metal until about 10 mg of the metal has been added. Treat the residue in the test tube with soda lime as indicated in Section V.B.I.b. The volatile secondary aliphatic amine produced by reduction of the isocyanide will give a pale green or greenish blue solution with the methanolic copper sulfate reagent; a flocculent, pale green precipitate will settle out if enough amine has been volatilized.

#### c. REACTION WITH OXALIC ACID

Strong solutions of oxalic acid react with isocyanides in the cold, yielding carbon monoxide and carbon dioxide (118):

$$4\,RNC + 3\,H_2C_2O_4 + H_2O \rightarrow 3\,CO + 3\,CO_2 + 4\,RNHCHO$$

Release of the gaseous products may be considered a positive test, but it must be noted that the same gases are produced by spontaneous decomposition of oxalic acid or by reaction of oxalic acid with substances other than isocyanides (e.g., carbodiimides).

### d. COMPLEXATION OF COPPER

Pertusi and Gastaldi (213) have shown that the test for detection of cyanide ion by copper–benzidine mixture as described in Section V.B.I.b can be used directly for the detection of isocyanides. Crabtree et al. (59,219,296) have modified the test so that tiny amounts of isocyanides in air can be detected; air is passed over silica gel, and then the absorbed isocyanide is allowed to react with a solution containing $p,p'$-tetramethyldiaminodiphenylmethane, salicylic acid, cupric sulfate, and acetone. The strong copper-complexing action of isocyanides causes the reagent to turn blue.

### 2. Physical Methods

#### a. INFRARED SPECTROPHOTOMETRY

Relatively few isocyanides have been examined by infrared spectrophotometry, but it is fairly certain that the triple bond in compounds of this type absorb in the range of 2180–2110 cm$^{-1}$ (4.59–4.74 $\mu$). McBride and Beachell (182) and Gillis and Occolowitz (110) have compared the spectra of some alkyl and aryl nitriles and isocyanides; the alkyl isocyanides absorb between 2190 and 2140 cm$^{-1}$ (4.57 and 4.67 $\mu$), and the aryl isocyanides between 2125 and 2105 cm$^{-1}$ (4.70 and 4.75 $\mu$). (See Table II.) McBride and Beachell (182) also report that a band near 1592 cm$^{-1}$ (6.28 $\mu$) is observed in the spectra of isocyanides but not in the spectra of the corresponding nitriles; however, this band is absent in the spectrum of methyl isocyanide. As a general rule, the isocyanides absorb at lower frequencies (about 100 cm$^{-1}$) and with markedly greater intensities than the corresponding nitriles (see Section III.A.2 for polymeric isocyanides).

Results of ir spectral studies suggest that isocyanides are stronger Lewis bases than are nitriles. The absorption frequency of the isocyanide group increases with increasing solvent polarity, in contrast to other multiply bonded systems such as the carbonyl group; thus, for example, isocyanide absorption at 2131 cm$^{-1}$ in cyclohexane is shifted to 2138 cm$^{-1}$ in acetonitrile, and it appears that the polar solvent has enhanced the contribution of the polar triple-bonded structure to the ir spectrum and has increased the frequency of absorbance. Since a similar frequency shift is observed with $p$-tolyl isocyanide, resonance structures such as Ar$\overline{=}$N$\overset{+}{=}$C: must contribute negligibly to the ir spectrum (108,110,133,236,237). The frequency of the isocyanide fundamental stretching mode is greater and is slightly lower in intensity in chloroform and pentachloroethane than in carbon tetrachloride (110).

### b. Ultraviolet Spectrophotometry

The isocyanide group per se is not a chromophore; as a result, the ultraviolet spectra of aliphatic isocyanides show no absorption pattern worthy of comment. Figure 2 illustrates the general absorption characteristics of cyclohexyl isocyanide, which is a compound essentially aliphatic in nature and sufficiently stable to permit handling. However, the isocyanide group interacts strongly with structures capable of resonance. For example, when the isocyanide group is attached to a

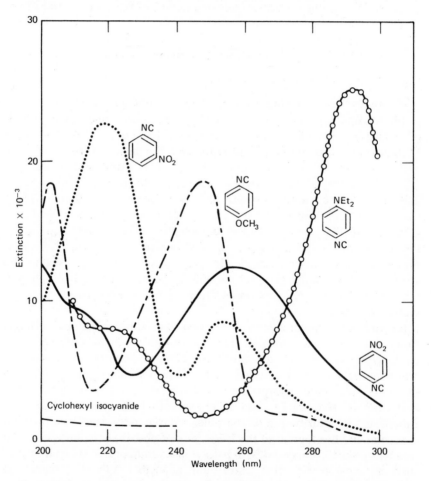

Fig. 2. Typical ultraviolet spectra of isocyanides. [Data from Ugi and Meyr (288).]

benzene ring, the intensities of the characteristic benzenoid absorption bands in the ultraviolet are strongly increased. Depending on the nature of other substituents and their positions on the benzene ring, more or less bathochromic shifts are obtained. The spectra of the aromatic isocyanide compounds included in Fig. 2 show the general characteristics of absorption in the uv region; identification of the isonitrile functional group from uv spectra is scarcely possible.

## D. DETECTION OF THE N-CYANO GROUP

Reactions of cyanamide and dicyandiamide useful as tests for their detection are reminiscent of those employed in qualitative inorganic analysis, for example, formation of colored silver salts, and determination of their solubilities in water, nitric acid, and ammonium hydroxide; unfortunately, there are very few reactions which can be used to detect the presence of the N-cyano group. Moreover, identification cannot be simple because many N-cyano compounds (such as cyanamide) are so reactive with water, small amounts of acids or bases, and other substances that they are seldom found in pure form. For example, in freshly prepared commercial calcium cyanamide, nitrogen is contained almost exclusively as the salt; during storage, reaction with carbon dioxide and moisture inevitably occurs, and the product then contains calcium hydroxide, calcium carbamate, dicyandiamide, urea, and a host of other nitrogen derivatives.

Detection of the N-cyano group by infrared spectroscopy is seldom definitive because the characteristic C—N absorption occurs essentially in the same region as it does when the group is attached to carbon.

### 1. Chemical Methods

There are no specific tests for detecting the presence of the N-cyano group; however, a number of tests for simple compounds such as cyanamide and dicyandiamide are available. The cleavage of N-cyano groups by metal-acid couples often allows enough hydrogen cyanide to escape for detection by sensitive tests.

#### a. SILVER SALTS

Cyanamide is identified readily by its bright, canary-yellow silver salt, which is quite insoluble in water and ammonium hydroxide, but soluble in nitric acid or in alkaline cyanides. In contrast, a silver nitrate double salt of dicyandiamide slowly separates as small white crystals (somewhat glistening) from nitric acid solutions, and it is soluble in ammonium hydroxide. The silver salts of cyanamide and dicyandiamide are insoluble in potassium hydroxide solution. Silver nitrate double salts of dicyandiamide having the composition $C_2N_4H_4AgNO_3$, $(C_2N_4H_4)_2AgNO_3$, and $(C_2N_4H_4)_3AgNO_3$ have been separated, but the first compound is converted to $C_2N_3H_4Ag$ by potassium hydroxide solution, while the second and third form silver oxide along with the same compound.

When mixtures of cyanamide and dicyandiamide are to be separated, a modification of the quantitative procedure of Caro (46) can be used: acetate is first added to a dilute ammoniacal solution of the sample until precipitation of yellow silver cyanamide is complete; after filtration, potassium hydroxide is added to precipitate dicyandiamide. Pure normal silver cyanamide is obtained only from very dilute solutions; the precipitate from stronger solutions contains an excess of silver.

The presence of sulfides or thiourea invalidates the silver tests indicated above because a brown or black precipitate forms. Cadmium hydroxide shaken with a sample solution that has been neutralized with sulfuric acid effectively removes thiourea and sulfides; the clear filtrate is then treated with a slightly ammoniacal solution of silver nitrate (29,163). Alternatively, a suspension of lead carbonate may be used to remove sulfides (39).

Dicyandiamide can be readily detected in the presence of thiourea by precipitation with silver nitrate in ammoniacal solution; after filtration, the remaining solution is acidified with nitric acid to allow a silver dicyandiamide precipitate to form. However, the silver dicyandiamide test can be performed only at high concentrations and low temperatures (163); a more sensitive test [in the absence of halides, sulfides, guanidine, and other compounds (140)] is the precipitation of a double compound of dicyandiamide by a solution of silver picrate (124,147) or by solutions of picric acid and silver nitrate (44).

### b. LEAD SALTS

Ammoniacal lead acetate forms a yellow precipitate with cyanamide; it is similar in appearance to silver cyanamide, and the reaction may be used to test for cyanamide in the presence of thiourea, after sulfides have been removed by lead carbonate (39).

Very small quantities of cyanamide in the presence of considerable thiourea can be detected by first shaking a solution of sample with cadmium hydroxide to remove sulfides and then treating the supernatant with ammonia and lead hydroxide; the surface of the white lead hydroxide precipitate gradually becomes yellow because of formation of lead cyanamide (29,163).

## c. Amidoxime Test

The amidoxime test described in Section V.B.I.a for the nitrile group also gives a positive indication with dicyandiamide, diethylcyanamide, and so on when a solution of hydroxylammonium chloride in propylene glycol is used (omit addition of potassium hydroxide).

## d. Evolution of Hydrogen Cyanide

Reduction of $N$-cyano compounds with a metal-acid couple such as zinc and hydrochloric acid releases enough hydrocyanic acid to permit detection by a sensitive test such as paper moistened with cupric acetate and benzidine solutions (test papers are described in Section V.B.I.b). The detection limits for cyanamide and dicyandiamide are reported to be about 8–10 $\mu$g (140).

## e. Reactions with Ammonia

When cyanamide is added to aqueous ammonia and the mixture evaporated to dryness (or when cyanamide is heated with ammonium salts), guanidine is formed; under similar conditions, dicyandiamide forms biguanidine. If the reaction products are heated to 150°C, guanidine is converted to biguanidine with release of ammonia:

$$(NH_2)_2C{=}NH \rightarrow [HN{=}C(NH_2)_2]NH + NH_3$$

Feigl et al. (89) used the reactions noted above as the basis for a series of spot tests for differentiating cyanamide from dicyandiamide. The test described in the following paragraph depends on the use of Nessler's reagent for detection of ammonia; urea interferes because it forms biuret at 150° with release of ammonia. The test can be used to detect calcium cyanamide in fertilizers; the limit of detection is of the order of 20 $\mu$g of cyanamide or about 50 $\mu$g in 2 mg of dicyandiamide.

Mix a tiny amount of solid sample or a drop of its solution with 1 drop of a concentrated aqueous solution of ammonia in the bottom of a micro test tube and evaporate to dryness; all ammonia must be removed. Position, over the top of the tube, a piece of filter paper moistened with Nessler's reagent, and dip the bottom of the tube into a bath of glycerol preheated to about 120°; raise the temperature of the bath to 150°C. A brown stain on the paper indicates the presence of cyanamide.

In view of the extreme sensitivity of the Nessler reagent and the difficulty of assuring *complete* removal of ammonia before application of the test, it is evident that the foregoing procedure leaves much to be desired. On the other hand, the following test (also described by Feigl and co-workers) can be used freely because it depends on the detection of the guanidine formed from reaction of ammonia and cyanamide:

Place 1 drop of a solution of the sample on a filter paper, and hold it over an open bottle of aqueous ammonia for 2–3 min; then let the paper stand in air until excess ammonia diffuses away. Apply 1 drop of a freshly prepared solution of 5% aqueous sodium 3,4-dihydro-3,4-dioxo-1-naphthalenesulfonate (206). A red stain appears if at least $25 \mu g$ of cyanamide is present; the detection limit is improved if the paper is heated in an oven.

Feigl and co-workers (89) have also used the test for guanidine to detect dicyandiamide; in this instance, the $N$-cyano group is cleaved by a metal-acid couple. Both cyanamide and dicyanamide form hydrogen cyanide when treated with zinc and hydrochloric acid, but only dicyandiamide forms guanidine. Accordingly, after a sample has been reduced, it is made alkaline with sodium hydroxide, and the resulting solution is tested for guanidine as indicated in the preceding paragraph. A red-violet color is indicative of dicyandiamide, but a blank test must be run in parallel because of the red color of the reagent.

## 2. Physical Methods

The infrared spectra of $N$-cyano compounds such as the substituted cyanamides exhibit no features which positively indicate the attachment of the nitrile group to nitrogen rather than to carbon. In comparison to nitriles, the intensity of the CN group vibration is much stronger in substituted cyanamides (146); there is a fall in frequency when cyanamides are compared with nitriles (see Table II), but there can be confusion because the CN group frequencies for the cyanamides, 2235–2200 cm$^{-1}$ (4.47–4.54 $\mu$), overlap the lower values in the range assigned to the nitrile group.

## E. DETECTION OF THE CARBODIIMIDE GROUP

Only a few of the reactions of the carbodiimide group are sufficiently selective to serve as a means for detecting it in the presence of similar nitrogen functional groups. For example, the reaction of oxalic acid with carbodiimides, which is distinctive in that it readily generates an equimolar mixture of carbon dioxide and carbon monoxide, also occurs with isocyanides.

Detection of the carbodiimide group by infrared spectroscopy is complicated by the fact that its absorption bands tend to coincide with the bands of the isocyanide group. However, the carbodiimide group does have a distinctive band in the ultraviolet, although in this region of the electromagnetic spectrum the same band may be common to some of the isocyanide group compounds.

Both chemical and physical tests are required to obtain positive evidence for the presence of the carbodiimide group.

## 1. Chemical Methods

No specific test has been reported for the detection of the carbodiimide group. However, there are a few reactions which can be used to amass evidence that the carbodiimide group may be at hand; positive detection is assured when chemical evidence and spectroscopic evidence are coincident.

### a. REACTION WITH OXALIC ACID

The reaction of oxalic acid with the carbodiimide group, which forms carbon monoxide and carbon dioxide, should take place in an anhydrous medium. A solution of oxalic acid in dry dioxane may be used.

For the test, small amounts of the sample may be added to the dioxane solution of oxalic acid; alternatively, the sample is also dissolved in dioxane and added to a solution of oxalic acid in dioxane. Effervescence may be presumed to be a positive test, but when small samples are involved, it is preferable to use as a reaction vessel a serum bottle that has been flushed free of air by a stream of helium and capped with a rubber septum through which hypodermic needles can be passed and reagents introduced. Of course, if the sample is to be handled dry, it is put into the bottle first and then an oxalic acid-dioxane solution is introduced via a syringe and needle. Immediately thereafter, the needle of a gas-sampling syringe is passed through the septum; gas pressure generated by reaction is relieved by the expansion volume provided by the syringe. After reaction has taken place, several samples of the ullage gas are analyzed for carbon monoxide and carbon dioxide by gas chromatography, mass spectroscopy, or other convenient technique. Formation of an equimolar mixture of carbon monoxide and carbon dioxide suggests that the sample contains the carbodiimide group (see Section V.C.1.c):

$$RN{=}C{=}NR + (COOH)_2 \rightarrow RNHCONHR + CO + CO_2$$

### b. FORMATION OF SUBSTITUTED UREA

The reaction of the carbodiimide group with oxalic acid described in the preceding section, as well as acidic hydrolysis of the functional group (see Section III.B.3.d), produces a compound of the type R'NHCONHR'', which can be isolated and then demonstrated to be a substituted urea. For the purpose of qualitative detection of the carbodiimide group, the colorimetric test described by Fearon (84) is usually adequate, provided that R'' in the structure given above is not an acyl radical (R' may be H or a simple alkyl group); however, other colorimetric tests (134) may need to be applied since no single test is applicable to all forms of substituted ureas. In many instances, it may be more appropriate to isolate the urea and use infrared absorption to detect the carbonyl band [about $1650\ cm^{-1}$ $(6.02\ \mu)$] and the amide II band [about $1590\ cm^{-1}$ $(6.29\ \mu)$] (may appear as a doublet).

### c. Cleavage by Carbon Disulfide

The carbodiimides are cleaved by carbon disulfide at 180–200°; ordinarily, the reaction is complete within a few hours (138) when performed in a sealed tube with a modest excess of carbon disulfide. The more reactive carbodiimides can be treated at lower temperatures; the reaction mixture must be thoroughly chilled before opening the tube. Symmetrically substituted carbodiimides form only one isothiocyanate, and the unsymmetrically substituted form two (see Section III.B.5.c). After driving off the excess carbon disulfide, the residue is tested for isothiocyanate by conversion to Prussian blue:

Mix a small amount (e.g., a few milligrams) of sample with 0.2–0.3 ml of butanol, a few drops of 10% sodium thiosulfate solution, and 0.2–0.3 ml of 2% sodium hydroxide solution. Evaporate the mixture to dryness, and add a few drops of 10% sodium thiosulfate solution; again evaporate to dryness. Now add 1 drop of 10% ferric chloride solution, and render the mixture acidic (about 3 drops of 15% sulfuric acid). A blue color is a positive test for organic thiocyanate (38).

### d. Formation of Substituted Thiourea

Substituted carbodiimides may be converted to the corresponding thioureas by hydrogen sulfide (see Section III.B.3.d), but the reaction is often not sufficiently well defined to produce a high yield of substituted thiourea. The thioureas so formed can be detected colorimetrically by the test described by Fearon (84), but since all types of thioureas may not respond to the test, it is best to use spectrophotometric procedures. For example, the shift in the ultraviolet absorption (starting material from about 212 to about 247 nm when converted to thiourea) may serve as an indication of the presence of a carbodiimide group in the starting material; however, further proof should be sought by examination of the infrared spectrum of the generated thiourea.

### 2. Physical Methods

#### a. Infrared Spectrophotometry

The infrared spectra of compounds containing the carbodiimide group have characteristic absorption bands in the region $2150–2100\ cm^{-1}$ (4.65–

4.76 $\mu$) (187). The alkyl carbodiimides usually show a single peak in the region 2140–2125 cm$^{-1}$ (4.67–4.71 $\mu$) owing to the antisymmetric stretching of the —N=C=N— system. The aromatic carbodiimides give rise to two bands, either because there is resonance coupling between the —N=C=N— fundamental and an aromatic overtone (in which instance the intensity of the normally weak aromatic band is enhanced) or because there is conjugation in the carbodiimide group system; other explanations may be more appropriate. Symmetric bending of the —N=C=N— group appears to give rise to absorption at about 1350 cm$^{-1}$ (7.41 $\mu$) (198). Characteristic absorption bands for the carbodiimide group in typical compounds are given in Table II.

### b. Ultraviolet Spectrophotometry

The ultraviolet spectra of a number of aliphatic and alicyclic disubstituted carbodiimides show a characteristic absorption band at 212–213 nm (18); in contrast, the corresponding thioureas have bands at 245–250 nm (18), and so it is possible to distinguish between substituted thiourea starting material and the carbodiimide formed from it (see Section II.D).

## F. IDENTIFICATION OF NITRILES

Identification of a specific nitrile compound by chemical methods usually involves hydrolysis to the corresponding acid (or amide) and identification of the acid via a derivative such as the p-bromophenacyl ester; in many instances, the acid or amide can be identified by infrared spectrophotometry or by X-ray diffraction (acids generally are first converted to a salt). Nitriles can be readily identified entirely by ir techniques or by a combination of chemical and instrumental methods.

Other derivatives suitable for identification of nitriles are amines (obtained by reduction), ketones (by a Grignard reagent), and aldehydes (by reduction with lithium aluminum hydride); derivatives such as α-iminoalkylmercaptoacetic acid hydrochlorides and alkyl(2,4,6-trihydroxyphenyl) ketones are not very useful because data for only a dozen or so of these compounds are available.

## 1. CHEMICAL METHODS

### A. Hydrolysis to Carboxylic Acids

#### (1) Alkaline Hydrolysis

Typical procedures for alkaline hydrolysis of nitriles involve rather concentrated aqueous solutions of caustic at reflux to ensure that refractory nitriles (and their amides) are completely converted to acids.

More general procedures specify the use of glycerol or diethylene glycol to raise the boiling point and increase the rate of hydrolysis (231). The following procedure may be used freely:

Reflux a mixture of 2 ml of glycerol (or diethylene glycol), about 1 gram of potassium hydroxide, and 0.1 gram of the nitrile for 1 hr. A convenient refluxing vessel can be fabricated from a 19/22 standard-taper joint by sealing and rounding the bottom of the tube attached to the joint so as to form a test tube; a suitable condenser can be slipped into the test tube joint.

Dilute with 0.5 ml of water, cool, and add 1–2 ml of ether. Mix well, and then allow immiscible layers to separate; carefully remove the ether from the viscous aqueous layer, and reject it. Cool the tube and render its contents acidic by cautious addition of 6 N hydrochloric acid. If a solid acid separates, filter it off and wash sparingly with acidulated water; alternatively, extract the acidic mixture with several 2-ml portions of ether and carefully evaporate the ether to obtain the acid. A derivative of the acid may be prepared (see the following procedure). Alternatively, the acid is recrystallized from methanol.

### (2)  Acid Hydrolysis

Nitriles are smoothly hydrolyzed in 75% sulfuric acid at 160–190°C. Often, phosphoric acid (51) is added to raise the boiling point of the sulfuric acid, and in some instances it is best to use a refluxing medium consisting of 75% sulfuric acid with sodium chloride or a mixture of sulfuric acid, phosphoric acid, and sodium chloride; hydrochloric acid increases the rate of hydrolysis and is much more effective for many nitriles than is sulfuric acid (159,165). Steam-volatile acids or acids that distil below 180–190°C without decomposition can be removed easily from the hot reaction mixture; extraction of the acid by ether is a more general procedure.

Prepare 75% sulfuric acid by carefully pouring 70 ml of concentrated sulfuric acid (98%) into 35 ml of water contained in a tared flask, cooling the resulting mixture, and then adding more water to make a total weight of 167 grams. Into the test tube apparatus described in the preceding procedure, place 2 ml of phosphoric acid (85%), 1 ml of 75% sulfuric acid, and 0.25 gram of the nitrile. (An alternative reflux mixture consists of 2.5 ml of 75% sulfuric acid and 0.1 gram of sodium chloride.)

Attach the reflux condenser and boil the mixture for 1 hr; if the acid is volatile, the condensed reflux will generally be cloudy because of separation of a solid or an oil. If the acid can be distilled, add 3–10 ml of water to the cool reaction mixture and arrange the apparatus so that the distillate can be collected in a convenient receiver (e.g., a test tube). If the acid cannot be distilled (or it is not convenient to distil it), dilute the reaction mixture with about 3 ml of water and neutralize most of the mineral acid with 6 N sodium hydroxide solution (at least

15 ml for the $H_2SO_4$–$H_3PO_4$ mixture). Extract the acid with three successive 5-ml portions of ether; carefully evaporate the ether to obtain the crude acid.

Acids collected as distillates (containing water) are neutralized with sodium hydroxide to a phenolphthalein end point; then 2–3 drops of dilute hydrochloric acid are added. The solution is refluxed for 2 hr with 100 mg of p-nitrobenzyl bromide and enough methanol to give a clear solution at reflux; the cold reaction mixture is filtered to remove the crude benzyl ester, and the ester is recrystallized from methanol.

Acids obtained from ether extractions are best converted to p-toluidides or anilides (51,248); however, they can be dissolved in water as the sodium salts (neutralized to phenolphthalein) and converted to benzyl esters as indicated above.

In many instances, the hydrolysis mixture can be neutralized and mixed (cold) with 3-ml increments of a cold alcoholic solution containing 400–500 mg of S-benzylisothiouronium chloride until a precipitate is formed; the derivative is filtered, washed with water, and recrystallized from the minimum amount of hot alcohol (add water until a turbid mixture is formed).

### b. HYDROLYSIS TO AMIDE

Partial hydrolysis of nitriles to amides can readily be brought about in concentrated phosphoric acid (72) or sulfuric acid:

$$RCN + H_2O \rightarrow RCONH_2$$

Characterization of nitriles by this procedure is especially attractive because amides are solids, and conversion of a number of aryl nitriles and hydroxy-substituted alkyl nitriles can be effected simply by treatment with a small amount of concentrated sulfuric acid at 20–80°C (see Section III.B.3.a). Wurz and Sharpless (321) have presented data on the X-ray diffraction patterns of the amides of saturated aliphatic acids.

Place about 100 mg of the nitrile in the bottom of a small test tube; add 5 drops of concentrated sulfuric acid, and mix well with a 3-mm stirring rod, spreading the mixture over the glass surface in the bottom area of the tube. If there is an immediate reaction, the mixture may set into a solid mass. Immerse the tube in a water bath at 60–70°C for 2–3 mins; stir the contents occasionally. Remove the tube from the bath, and cool it in running water; quickly add 1 ml of water, stir well, and cool the mixture. Add 2 ml of 10% sodium carbonate solution (the final solution should be just alkaline), mix well, and filter the amide; wash twice with water and then dry.

### c. REDUCTION TO AMINE

Nitriles can also be identified by reduction to the corresponding amines and preparation of suitable derivatives (aryl thiourea, 3,5-dinitrobenzoates, and aryl sulfonamides). Yields of amines obtained by reduction of nitriles are generally poor because side reactions also take place (see Section III.B.1); as a result, it may be necessary to start with rather large amounts of a nitrile in order to obtain sufficient amine.

Reduction by hydrogen (catalytic) is preferred because no interfering substances remain in the reaction mixture, but the experimental conditions and, especially, the catalyst system must be carefully selected. Consequently, reductions are customarily performed by sodium and alcohol or by lithium aluminum hydride.

### (1) Reduction by Sodium and Alcohol

Cutter and Taras (66) have found that sodium and absolute alcohol reduce many nitriles to the corresponding amines; usually amines are distilled out of the reaction products, but this cannot be done with the higher boiling amines or polyamines. The following procedure is a modification of the Cutter–Taras method:

Place 0.25–0.5 gram of the nitrile and 10 ml of absolute ethanol in a 25-ml flask equipped with a reflux condenser. Immerse the flask up to its neck in a water bath at $55 \pm 5°C$. Cautiously add about 0.75 gram of freshly cut thin slices of sodium through the top of the condenser at a rate which maintains a smooth reflux (about 15 min will be required). Cool the mixture to room temperature, and slowly add about 5 ml of concentrated hydrochloric acid through the top of the condenser; exercise care, inasmuch as spattering occurs when the acid falls into the strongly alkaline alcoholic mixture in the flask. Be certain that the mixture is acidic, and then arrange the equipment for distillation; distil more than 10 ml (mostly alcohol). Cool the residue in the flask, and cautiously add 7–8 ml of 50% sodium hydroxide solution in small increments to render the contents of the flask strongly alkaline; keep the mixture cool to avoid loss of amine by volatilization. Quickly connect the flask to its condenser, and heat with a small, soft flame until the contents are nearly dry; allow the condenser tube (or adapter) to dip into a mixture of 1 ml of water and 1 ml of $3 N$ hydrochloric acid. Neutralize the distillate (pH 8) with sodium hydroxide; then add 0.2–0.3 ml of phenylisothiocyanate and shake vigorously (cooling is helpful). Filter the derivative, wash sparingly with 1:1 alcohol–water, and recrystallize from hot, dilute alcohol.

The procedure gives excellent yields of readily volatile amines but cannot be applied to alkyl mononitriles above $C_8$. Aromatic nitriles give very low yields; moreover, Cutter and Taras (66) were not able to obtain satisfactory results with $\alpha$-naphthonitrile or with $m$-tolunitrile, and yields were only of the order of 10% with $o$- and $p$-tolunitrile. Amines can be readily extracted from alkaline reduction mixtures (after removal of alcohol), but some of the higher amines may be recovered more effectively from the reaction mixture by distillation under reduced pressure.

## (2) Reduction by Lithium Aluminum Hydride

Reduction by lithium aluminum hydride can be conveniently and rapidly performed at room temperature; the amine can be easily removed from the reaction medium by steam distillation or by extraction with a solvent:

$$2 \, RCN + LiAlH_4 \rightarrow (RCH_2N)_2LiAl$$

The intermediate complexes formed from lithium aluminum hydride and nitriles are stable in a nitrogen atmosphere, but are pyrophoric in contact with air (202). Inasmuch as the reduction of a nitrile may proceed through an intermediate aldimine, reduction to an amine may be slow or may require elevated temperature when only a small excess of lithium aluminum hydride is used; an excess of the order of 100% is prescribed in the procedure given below. Ordinarily, hydride reductions are carried out in diethyl ether, but tetrahydrofuran or di-$n$-butyl ether may also be used; obviously, the solvents must be anhydrous and be of high purity. The noisome flocculent aluminum precipitates obtained after hydrolysis of the lithium aluminum amine complex are avoided by conversion to aluminate or a tartrate complex. Following is the procedure of Nystrom and Brown (202):

Prepare a molar solution of lithium aluminum hydride by quickly transferring about 4 grams to 100 ml of anhydrous ether contained in a flask equipped with a reflux condenser and a desiccant guard tube. Gently reflux the mixture until the soluble fraction of the hydride has been brought into solution; usually the sediment in the solution is allowed to settle, but the solution can be used in any condition of turbidity.

Dissolve 0.01 mol of the sample in anhydrous ether; transfer at least 35 ml of molar lithium aluminum hydride solution to a dry flask equipped with a dry reflux condenser. Carefully add the sample solution in several aliquots to the hydride through the condenser; agitate the mixture after each addition. After the entire sample solution has been added, allow the mixture to reflux gently (steam bath) for not less than 15 mins and preferably for 2 hr; use a desiccant guard tube for prolonged reflux. Cool the reaction mixture; then cautiously add water (dropwise) until the hydride is decomposed. Dissolve precipitated aluminum hydroxide by adding a large excess of sodium hydroxide solution (6 $M$), and then extract the amine with ether or another solvent; alternatively, the amine can be extracted from the distillate obtained from the alkaline mixture by passing in steam, or (better) the amine can be distilled at a high temperature by supplying ethylene glycol to the distilling flask as water is removed. Aminosulfuric acids often cannot be distilled.

Add 1–2 ml of ethanol and 0.5 ml of phenylisothiocyanate to the residue obtained from evaporation of the solvent extract; reflux for a few minutes, and then add water to cause separation of the substituted phenylthiourea. Steam distillates are treated with phenylisothiocyanate, shaken vigorously, and cooled to obtain derivatives.

The above procedure is generally applicable and is based on a nitrile/hydride ratio of 1:3.5 (molar); however, since reduction of other structures (double bonds) and functional groups (e.g., nitro group) undoubtedly takes place, the resulting amine will not correspond to the original nitrile. Occasionally, reduction of double bonds can be avoided by limiting the amount of hydride and adopting an inverse order of addition at low temperature. Thus, in the foregoing procedure, the sample solution is placed in the flask and cooled to $-10°C$ in an ice bath; the solution of lithium aluminum hydride is added via the condenser at such a rate that the reaction mixture is not allowed to rise over 5–10°C, and the volume of lithium aluminum hydride is limited to not more than 12 ml (nitrile/hydride ratio of 1:2). LeMoal et al. (172) report that the nitrile group sometimes escapes when $\alpha,\beta$-unsaturated nitriles are reduced by lithium aluminum hydride or lithium borohydride.

Siggia and Stahl (253) studied the application of the lithium aluminum hydride reduction to the quantitative determination of amides; the resulting amines were steam-distilled into a volume of standard acid which was back-titrated with standard base (see Section VI.A.2.a). Reduction was found to be essentially quantitative for benzonitrile, butyronitrile, capronitrile, and p-chlorobenzonitrile. Markedly incomplete reductions (and steam distillations) were obtained for acetonitrile, acrylonitrile, succinonitrile, adiponitrile, phenylacetonitrile, 3-butenenitrile, γ-phenoxybutyronitrile, lactonitrile, m-nitrobenzonitrile, and 1-naphthonitrile. It is of interest to note that the procedure used by Siggia and Stahl (ratio 1 mol of nitrile to about 4 mol of LiAlH₄ for 15 min at room temperature) yielded quantitative amounts of amines from benzonitrile, butyronitrile, and capronitrile; in earlier work, Amundsen and Nelson (6) obtained the corresponding amines in yields of 72, 57, and 90%, respectively, at a nitrile/hydride ratio of 1:1; clearly, the lithium aluminum hydride reduction process is sharply dependent on the nitrile/hydride ratio. Moreover, when lithium aluminum hydride reduction procedures are applied to dinitriles, yields of the amines are often poor because the highly insoluble intermediate products that are formed cannot be reduced effectively; in contrast, monofunctional, soluble nitriles such as lauryl cyanide yield over 90% of the corresponding amine.

It is advisable to conduct the reduction of nitriles under purified nitrogen because the intermediate products are sensitive to oxygen (36,202).

### (3) Reduction by Catalytic Hydrogenation

Catalytic hydrogenation is the preferred route for conversion of nitriles to the corresponding amines, especially when less than 500 mg of the sample is available; however, the conditions of hydrogenation and the catalyst are critical, and results may not be entirely satisfactory. Huber (136) found that palladium oxide provides quantitative (98.5 + %) reduction of acetonitrile, benzonitrile, adiponitrile, and succinonitrile in glacial acetic acid; the hydrogenation takes place at atmospheric pressure. The palladium catalysts, such as palladium on charcoal, ordinarily used in micro- and semimicrohydrogenations often are ineffective or provide low yields of amines (64).

As was indicated in Section III.B.1.a, secondary amines are often formed by hydrogenation, but the tendency for reaction of aldimines with primary amines is minimized by the presence of ammonia:

$$RCH{=}NH + RCH_2NH_2 + H_2 \rightarrow (RCH_2)_2NH + NH_3$$

Consequently, ammonia must be present in the hydrogenation vessel in order to obtain high yields of primary amines. Freifelder (97) found that aliphatic nitriles are smoothly reduced to primary amines in the presence of ammonia and a rhodium-on-alumina catalyst (Baker and Co., Newark, N.J.); following is a typical procedure based on Freifelder's findings:

Dissolve about 1 mmol of the nitrile in about 30 ml of a 10% solution of ammonia in ethanol. Add 0.05–0.10 gram of rhodium-on-alumina catalyst, and hydrogenate at 2.5 atm (room temperature); monitor the hydrogen uptake. Ordinarily, hydrogenation will be complete within 2 hr (2 mmol of hydrogen). Filter the solution to remove the catalyst, and then carefully evaporate the solvent to leave a residue containing the amine.

Hydrogenations frequently can also be performed at ambient temperature and pressure with 5% palladium-carbon in 90% ethanolic solutions containing 10% of ammonium carbamate; typically, 100–500 mg of the sample is dissolved in 20–25 ml of a 10% ethanolic solution of ammonium carbamate, and 100–250 mg of 5% palladium-carbon is used (51), but poor conversions may be obtained (64).

A typical procedure for high-pressure hydrogenation with Raney nickel catalyst and ammonia is cited by Whitmore et al. (315).

### d. Formation of Ketones

Conversion of nitriles to ketones provides a useful method of identification, inasmuch as physical property data are widely available for ketones. A general method for forming ketones from nitriles consists of addition of a Grignard reagent followed by decomposition and hydrolysis; the ketones may be converted to semicarbazones or to other derivatives.

### (1) Ketones by Grignard Reagent

Shriner and Turner (249) used a Grignard reagent for conversion of nitriles to phenyl alkyl ketones; 4 mol of phenylmagnesium bromide/mol of nitrile was found to provide the best results for alkyl nitriles:

$$RCN + \phi MgBr \rightarrow R\phi C{=}NMgBr$$
$$R\phi C{=}NMgBr + 2\,HCl \rightarrow R\phi C{=}NH{\cdot}HCl + MgBrCl$$
$$R\phi C{=}NH{\cdot}HCl + H_2O \rightarrow RCO\phi + NH_4CL$$

The original reference should be consulted for details. Unfortunately, the Grignard procedure is inconvenient and limited in applicability for general work because the reagent combines with many substances, and the procedure can be applied only to pure anhydrous samples in anhydrous systems.

### (2) Trihydroxyphenyl Ketones

Ketones can be made from nitriles and phenols by the Hoesch reaction (condensation), and for this purpose Howells and Little (135) have recommended phloroglucinol as the phenol because it yields solid alkyl trihydroxyphenyl ketones:

$$2,4,6\text{-}(HO)_3\varphi + RCN + HCl \xrightarrow{\;ZnCl_2\;} (HO)_3\varphi\text{---}C(R)NH{\cdot}HCl \xrightarrow{\;H_2O\;}$$

$$+ NH_4Cl$$

The nitrile and phloroglucinol are dissolved in dry ether, and anhydrous zinc chloride is added; the mixture is treated with dry hydrogen chloride for about 30 min. The oil that is formed is taken up and treated with water; the aqueous solution is concentrated by evaporation until the ketone separates on cooling. Unfortunately, only a few melting points have been recorded (51).

### e. FORMATION OF ALDEHYDES

The controlled reduction of nitriles by lithium aluminum hydride provides a convenient procedure for preparation of the corresponding aldehydes (which can be identified readily when converted to suitable derivatives). Yields of aldehydes may be low. The following general procedure is based on the use of slightly more than 0.25 mol of lithium aluminum hydride/mol of nitrile, that is, a nitrile/hydride ratio of 4:1 (256):

Dissolve the nitrile (0.002 mol) in about 20 ml of dry ether in a reflux apparatus; boil the solution rather vigorously in order to provide adequate agitation. Over a period of 20 min, add 5 ml of an ethereal 0.1 M solution of lithium aluminum hydride. Continue to reflux the mixture for 1 hr; then cool, and add drops of water until the hydride is decomposed. Decant the ethereal solution, and evaporate it or use it to form a derivative of the aldehyde.

### f. ADDITION PRODUCT WITH MERCAPTOACETIC ACID

According to Condo et al. (57), nitriles are best identified by their addition products with mercaptoacetic acid:

$$RCN + HSCH_2COOH + HCl \rightarrow R\overset{\displaystyle NH \cdot HCl}{\underset{\displaystyle \|}{-C}}SCH_2COOH$$

The compound formed is the hydrochloride of an $\alpha$-iminoalkylmercaptoacetic acid (an iminothiol ester); the following procedure is attractive because it is simple and because it gives products of high purity.

Place 0.5 gram of the nitrile and 1 gram of mercaptoacetic acid in a 25-ml glass-stoppered flask, mix well, and add 12–15 ml of absolute ether which has been saturated with dry hydrogen chloride at 5–10°C. Generally, crystals will form within several hours with aliphatic nitriles, but aromatic nitriles may require as much as 24 hr. Filter the crystals that separate on standing in an ice bath, wash with ether, and dry in a vacuum desiccator.

The addition products are stable, but melting points are actually decomposition points. The products may be titrated as dibasic acids with thymol blue; the neutralization equivalent further assists in the identification of the original nitrile. Comparison of the infrared spectra or X-ray diffraction patterns with those of known derivatives may also help to establish the identities of unknowns. The following procedures (57) also may be used for preparing the adducts in better yields:

I. Mix 0.5 gram of the nitrile with 1 gram of mercaptoacetic acid in a test tube, and then add 7 ml of absolute ether and agitate until all is in solution; immerse the tube in an ice-salt mixture (−10°C), and saturate with dry hydrogen chloride. Stopper the tube, and keep in the ice-salt mixture until crystals separate.

II. Mix 0.5 gram of the nitrile with 1 gram of mercaptoacetic acid; cool the mixture in an ice bath, and then saturate with dry hydrogen chloride. Remove crystals to a filter with the aid of anhydrous ether; wash with ether, and dry in a vacuum desiccator.

### g. Miscellaneous Derivatives

### (1) N-(Diphenylmethyl)amides

Nitriles react readily with benzhydrol, $\phi(CHOH)\phi$, in acetic acid to form the corresponding N-(diphenylmethyl)amides. Prajsnar et al. (220) found that the derivatives have sharp melting points; unfortunately, very few derivatives of these types have been prepared.

In a small test tube, mix about 0.5 gram of the nitrile with an equal amount of benzhydrol; add 2 ml of glacial acetic acid and stir in 0.25 ml of concentrated sulfuric acid. Heat the mixture for 30 min at 60°C, and then let it stand for several hours at room temperature. Pour the mixture into 7–10 ml of ice water; filter the derivative, wash with water, and recrystallize from ethanol.

## 2. Physical Methods

The identification of a specific nitrile compound by instrumental methods is at present hampered by lack of sufficient reference data; fortunately, data are available for the simpler nitriles. Indirect indentification of many nitriles can be accomplished by conversion to acids or amides (or other derivatives) followed by X-ray diffractometry or infrared spectrophotometry; however, such procedures cannot be thoroughly endorsed at present because pertinent data are sparse and because derivatives of requisite purity are not easily obtained.

### a. Infrared Spectrophotometry

Direct identification of a specific nitrile by infrared methods is seldom possible because of the paucity of "fingerprint" data on any but the simplest of nitrile compounds. However, since rather extensive arrays of ir spectra are available for acids and amines, specific nitriles often can be identified after they have been converted to corresponding compounds.

### b. Ultraviolet Spectrophotometry

The spectral features of the nitriles in the visible and ultraviolet regions are insufficiently diagnostic to be of value for identification of specific nitriles; however, when closely similar compounds are under study, the relative positions of absorption bands may suggest structural differences.

### c. Mass Spectroscopy

Identification of a specific nitrile by mass spectroscopy appears to be quite straightforward; however, the molecular ion is formed in such small amounts that its low intensity may prove troublesome when samples are impure. In a study of 18 aliphatic nitriles, McLafferty (184) found the molecular ion to constitute at least 0.1% relative abundance in their mass spectra. The molecular ion is the most abundant in the mass spectrum of acetonitrile, but it is very small for isobutyronitrile; peculiarly, the abundance of the molecular ion seems to increase with the molecular weight of the nitrile (at least up to $C_{18}$).

A characteristic of the mass spectra of most nitriles, but not propionitrile, is a strong $m/e$ 41 peak, usually the base peak (up to $C_{10}$) formed by the cleavage of the bond beta to the nitrile group, and the appearance of rather significant abundances of $M-1$ ions ($M$ = mol. wt.); although peaks formed by loss of one hydrogen are not large, it is unusual that the $M-1$ ion in nitriles is more abundant than the molecular ion (especially of low molecular weight). Another characteristic of the spectra of the nitriles is the frequent appearance of an $M+1$ ion because of ion–molecule interaction; however, in many of the newer high-sensitivity instruments, formation of the $M+1$ ion is limited by low operating pressures in the ion source. The pressure-sensitive $M+1$ ion finds general analytical use, but it is an exceptional convenience in work with nitriles, inasmuch as it is stronger than the molecular-ion peak. The $M-41$ peak in nitriles also is the result of ion–molecule reactions (27) and serves to confirm the molecular weight of the nitrile.

In contrast to compounds that have electronegative groups such as halogen and carbonyl, alpha-bond cleavage is neither a dominant nor a significant characteristic. It is also of interest to note that the ions $m/e = 26$ ($CN^+$) and $m/e = 40$ ($CH_2CN^+$) are not significant (beta cleavage). The overall features of the mass spectra of aliphatic nitriles betray to the experienced eye the structural relationship of these compounds to the corresponding paraffins, especially for long chains. Fragments of the general formula $(CH_2)_n CN^+$ are found in the mass spectra of many nitriles, but the presence of nitrogen in the fragments can be confirmed only by measurements with high-resolution instruments. For example, the frequent occurrence of the $m/e$ 41 ion in the spectra of the nitriles suggests formation of an ion $(CH_3CN)^+$, possibly $(CH_2{=}C{-}NH)^+$, but the annoying likelihood that the ion at $m/e$ 41 is $(C_3H_5)^+$ can be disproved only by measurements at high resolution ($[CH_3CN]^+ = m/e$ 41.0397 and $[C_3H_5]^+ = m/e$ 41.0523).

The mass spectra of cyanogen (74), cyanoacetylenes (74), methyl and ethyl cyanoacetates (32), and halogenated acetonitriles (155) have been examined.

The mass spectra of the few aromatic nitriles that have been studied show a strong molecular ion; in contrast to the aliphatic nitriles, the strongest peak in the spectra of benzonitrile and the three isomeric monomethyl derivatives (50) is the molecular ion. Only one other strong peak appears in the spectrum of benzonitrile; it is formed by a one-step cleavage process (as shown by a metastable ion) by loss of HCN ($m/e = 76$). Fragments containing CN are very few, and the loss of the cyano group is scarcely visible. The spectra of the monomethylbenzonitriles are abundant in $M - 1$ ions and fragments formed by loss of HCN; in contrast to the parent compound, the methylated derivatives show a high tendency for formation of $H_2CN$ by elimination of HCN from the $M - 1$ fragment.

The molecular ion is also the strongest (base) peak in the spectrum of 1,4-dicyanobenzene, with the loss of HCN accounting for the next strongest peak; on the other hand, loss of HCN is not a prominent mode of fragmentation in the mass spectrum of 4-(*tert*-butyl)benzonitrile (186). The mass spectrum of phthalonitrile has been reported (130).

### d. GAS CHROMATOGRAPHY

Resolution of mixtures of nitriles (or mixtures of nitriles with other nitrogen-containing substances) by gas chromatography is somewhat difficult and often may require the use of two columns. One of the first procedures to be described for separation of nitriles made use of 30–60 mesh firebrick (C-22) impregnated with Carbowax-4000 monostearate operated at a temperature of 220–235°C with a helium flow of about 50 ml/min (175). Lysyj (178) noted that the order of elution of nitriles from a polar column may be different from the order obtained with a less polar column; thus a column containing Carbowax-400 (polar) does not resolve acrylonitrile and methacrylonitrile from a mixture of acetonitrile, acrylonitrile, methacrylonitrile, and propionitrile, and a nonpolar column containing Craig polyester succinate does not separate acetonitrile from acrylonitrile. However, by using both columns, it was found possible to obtain analytical values for the four components.

Taramasso and Guerra (274) considered that the differences in boiling points of the four nitriles used by Lysyj were sufficient to permit separation on a truly nonpolar column; indeed, good resolution was obtained with a waxy paraffin supported on Celite, but the peaks were asymmetric. Symmetrical peaks were obtained on Teflon; the best working conditions were obtained with a 5-m column (4-mm i.d.) packed with Haloport F (Hewlett–Packard) coated with 10% by weight of paraffin wax (mp, 56–58°C). The column was operated at 70°C, and hydrogen was used as a carrier at a flow of 1 ml/sec. The column separated acetonitrile, acrylonitrile, propionitrile, methacrylonitrile, and *trans*-crotononitrile, but allyl cyanide and *cis*-crotononitrile were not resolved. However, these two compounds were found to be resolved on a 60–80 mesh firebrick (C-22) impregnated with 30% Octoil S and operated at 100°C.

Other stationary phases useful for separation of nitriles are the following: 2–5% Carbowax-20M on polytetrafluoroethylene (75–225°C at 10°/min); 5% tetrahydroxyethylenediamine on polytetrafluoroethylene (or a silanized support) at 70–75°C; cyanosilicones on a silanized support at 150°C; and Apiezon L on kieselguhr (298). A mixture of sodium caproate and Apiezon L on Celite 22a is recommended for use with dinitriles (196); a support impregnated with 2% Versamid-900 appears to be effective with trisnitriles (308). Phosphoric acid (2%)-treated substrates coated with diethylene glycol polyester adipate are useful for temperature-programmed separations (70–200°C) (7,189). However, results reported by Gray et al. (111) suggest that such columns may create problems, for it was noted that oleamide was converted completely to a nitrile in a polyester-phosphoric acid column.

DiLorenzo and Russo (75) obtained good separation of nitriles, imines, and amines on a column containing Chromosorb P and 20% UCON LB-550-X treated with 20% KOH; a nitrogen flow of 75 ml/min and a temperature of 80°C were found adequate. Silicones on glass [Parimskii and Shelomov (208)] as well as on diatomaceous earth [Aminov and Terent'ev (5)] have proved useful for nitrile separations; Kourovtzeff (164) found that glass beads wet with oxyethyleneamines that have 10–20 oxyethylene residues provide adequate separation of nitriles and their corresponding amines.

Analysis of aliphatic nitriles in aqueous acidic solution is difficult because of excessive tailing of water. Ordinarily, a solvent extraction suffices to separate components from water, but Arad-Talmi et al. (7) found that extraction of nitriles was not complete unless large volumes of the solvent (o-dichlorobenzene) were used, thus making the extract too dilute for direct injection into a gas chromatograph. Anhydrous sodium sulfate could be used to take up the water contained in a sample that had been mixed with 5–6 vols. of the solvent; for the nitriles examined, the drying agent showed no preferential absorption. Generally, the total nitrile concentration in samples was about 15% (w/w) in aqueous hydrochloric acid saturated with potassium chloride; for a 10-ml sample mixed with 50–60 ml of o-dichlorobenzene, 50 grams of anhydrous $Na_2SO_4$ was added, and the mixture shaken for about 5 min. The mixture was filtered, and 10–15 ml injected into the chromatograph. A 2-m by 6-mm i.d. copper spiral column was used; it was packed with a 23% diethylene glycol adipate polyester phase (LAC-2-R-466, Cambridge, Ind.) with 2% phosphoric acid supported on

Chromosorb W (30–60 mesh). For monobasic nitriles, the column was operated at 70°C, and for dibasic nitriles at 220°C; for mixtures of mono- and dibasic nitriles, the column temperature was programmed at 10°/min, starting at 70° and ending at 220°C. Following is the order of elution: acrylonitrile, propionitrile, butyronitrile, o-dichlorobenzene, succinonitrile, and adiponitrile; no water peak is detected.

Murata and Takanishi (197) found that aqueous solutions of nitriles could be handled easily by column packings containing polyethylene glycols or glycerol.

A fusion hydrolysis reagent consisting of about 1% sodium acetate in potassium hydroxide has been found adequate for conversion of poly-acrylonitrile to ammonia (96). In essence, solid samples of 1–3 mg are weighed into platinum boats and covered with a layer of fusion reagent. The mixture is heated at a temperature of about 300°C for about 0.5 hr; gases are condensed in a liquid nitrogen trap and then released into a gas chromatograph for determination of released ammonia.

## G. IDENTIFICATION OF ISOCYANIDES

### 1. Chemical Methods

There is no general chemical method that can be used to identify individual isocyanides; however, isocyanides may be treated much as are nitriles, and thus it is quite commonplace to convert isocyanides with acid to corresponding primary amines for identification. The hydrolytic procedures described in Sections V.D.I.a and V.D.I.b may be used, but it must be recalled that formation of amines by the acid hydrolysis of isocyanides often takes place violently, and usually incompletely.

Isocyanides may be reduced to secondary amines ($RNC \rightarrow RNHCH_3$) which can be identified per se or be converted to recognizable appropriate derivatives; aryl isocyanides are best reduced by sodium and a higher alcohol such as butanol or pentanol, but aliphatic isocyanides are preferably reduced by hydrogen and nickel catalyst at an elevated temperature.

Conversion of isocyanides to isothiocyanates by heating with sulfur or phosphorus pentasulfide may provide a more tractable product for identification via a derivative, and often it is possible to convert isocyanides to nitriles by heating to about 250°C; however, these processes should be used only as a last resort because reactions at high temperatures quite frequently produce a variety of products, and often the analyst's fortune is such that the desired product is not in preponderance. The oxidation of isocyanides to isocyanates by mercuric oxide may provide an adequate route for the preparation of derivatives, but no formal studies of this method of identification have been reported.

Reactions of isocyanides with carboxylic acids are noted by Ugi (290); although such derivatives may in time be found useful for identification, they are of little value at present because physical data seldom are available.

## 2. Physical Methods

### a. SPECTROPHOTOMETRY

Absorptions of the isocyanides in the ultraviolet region are too broad and insufficiently characteristic to be used for identification of specific compounds. On the other hand, although infrared spectra of specific isocyanides are sufficiently different to permit identification, the required bank of data simply is not available. However, isocyanides can be converted into substances for which data exist, for example, to secondary amines which then can be identified by their ir spectra (or by the spectra of appropriate derivatives of the amines).

### b. MASS SPECTROSCOPY

Data on the mass spectral patterns of isocyanides are very scarce; Gillis and Occolowitz (109) have studied the mass spectra of methyl, ethyl, propyl, and butyl isocyanides and compared them with those of the corresponding nitriles. As with the nitriles, beta-bond cleavage is the principal mode of fragmentation; however, alpha cleavage, which is of minor importance in the fragmentation of aliphatic nitriles, is readily evident in the mass spectra of the aliphatic isocyanides, and this implies that the R-N bond of isocyanides is weaker than the R-C bond in nitriles. Concomitantly, the isocyanides show pronounced losses of $H_2CN$ and HCN. Appearance potential studies confirm the above observations of bond strength. Isocyanides, like nitriles, show a pressure-dependent $M + 1$ peak, but the disposition for alpha cleavage readily makes possible mass spectrometric differentiation of the aliphatic nitriles and isocyanides. Moreover, because the mass spectra of the aliphatic nitriles and isocyanides contain peaks corresponding to the hydrocarbon skeletons attached to the functional groups, an experienced mass spectroscopist generally will have little difficulty in deducing the structure of an unknown isocyanide, in spite of lack of reference spectra.

Zeeh (326) has studied a group of aromatic isocyanides; as with the aromatic nitriles, the main fragmentation mode for substituted phenyl and naphthyl isocyanides is loss of HCN.

### c. Gas Chromatography

Aliphatic isocyanides readily polymerize and often react with metals; consequently, gas chromatographic separations may be troublesome unless special precautions are taken to ensure that the column and its packing do not induce decomposition of the isocyanides and that operating conditions do not induce polymerization. Metal tubes coated with inert polytetrafluoroethylene (or with a silanizing agent) or glass tubes (156) are recommended, but conditioned metal columns operated at about 75° seem to be adequate for many purposes. Adequate separation of submicrogram quantities of alkyl isocyanides are obtained with glass columns packed with Chromosorb W-HMDS (a support treated with hexamethyldisilazane) operated at 90°C (156). Packings of firebrick or diatomaceous earth coated with polypropylene glycol or diisodecylphthalate have also been found adequate (75–90°C); it may be necessary to condition these packings by a few injections of any isonitrile before performing an analysis (48,49).

## H. IDENTIFICATION OF N-CYANO COMPOUNDS

Identification of a specific N-cyano compound by chemical methods usually involves one or more degradations to determine the nature of the structures attached to the N-cyano group. The simple N-cyano compounds are quite reactive and generally can be converted to substances that can be easily identified or characterized.

### 1. Chemical Methods

Hydrolytic methods offer a convenient route to derivatives of N-cyano compounds for purposes of identification.

### a. Hydrolysis to Amines

Disubstituted cyanamides, especially the dialkylcyanamides, can be hydrolyzed by acids or bases to secondary amines, which, in turn, are characterized by conversion to sulfonamides, thioureas, and so on:

$$R_2NCN + 2 H_2O \xrightarrow{H^+} NH_4^+ + R_2NH + CO_2$$

$$R_2NCN + H_2O \xrightarrow{2 OH^-} NH_3 + R_2NH + CO_3^{-2}$$

### b. Hydrolysis to Urea

Monosubstituted aryl cyanamides (e.g., benzylcyanamide), can be converted to aryl ureas by heating with hydrochloric acid:

$$C_6H_5NHCN + H_2O \xrightarrow{HCl} C_6H_5NH(CO)NH_2$$

The urea can be identified by its melting point, by conversion to a xanthydrol, or by hydrolysis and conversion of these products into identifiable derivatives.

Disubstituted carbodiimides, R—N=C=N—R, are also readily hydrolyzed to the corresponding disubstituted ureas:

$$\text{RNCNR} \xrightarrow{\text{H}_2\text{O}} \text{RNHCONHR}$$

### c. Formation of Guanidines

Amines such as aniline react with monosubstituted cyanamides to form $N$-substituted guanidines; for example, phenylcyanamide reacts with aniline to form diphenylguanidine (82):

$$\text{C}_6\text{H}_5\text{NHCN} + \text{C}_6\text{H}_5\text{NH}_2 \rightarrow (\text{C}_6\text{H}_5\text{NH})_2\text{CNH}$$

Disubstituted carbodiimides also form guanidines with amines:

$$\text{R—NCN—R} + \phi\text{—NH}_2 \rightarrow \text{RNHC}{=}\text{NR}$$
$$\underset{\text{NH—}\phi}{|}$$

### d. Alcoholysis

A number of the reactive $N$-substituted cyanamides can be converted to urethanes (carbamate esters) by alcohol:

$$\text{R}_2\text{NCN} \xrightarrow{\text{EtOH}} \text{R}_2\text{NCOOEt}$$

Similarly, alcohol converts disubstituted carbodiimides to $O$-substituted ureas:

$$\text{R—NCN—R} \xrightarrow{\text{EtOH}} \text{RNHC}{=}\text{NR}$$
$$\underset{\text{OEt}}{|}$$

## I. IDENTIFICATION OF CARBODIIMIDES

Chemical methods for the identification of carbodiimide group compounds usually involve formation of ureas because there is a sizable bank of data on the properties of substituted ureas; moreover, the substituted ureas are usually quite tractable and, if necessary, can be treated to obtain additional derivatives. It is fortunate that, since the carbodiimide group can readily undergo a wide variety of additive reactions, it is possible to prepare a large number of derivatives for corroboration of structural evidence should identification via a ureide not be possible.

Identification of carbodiimide group compounds by physical methods nearly always involves comparison of spectra of the carbodiimides themselves with references, or comparison of the spectra of the substituted ureas obtainable as hydrolytic products of the carbodiimides with the numerous urea spectra currently available. Highly purified liquid carbodiimides can possibly be identified via refractive index; molar refractions can be estimated fairly well with the available value of 10.62 for carbodiimide group refraction (240).

## 1. Chemical Methods

### a. FORMATION OF SUBSTITUTED UREA

The carbodiimides can readily be converted to corresponding ureas by hydrolysis in dilute acid, but it is best to avoid side reactions by using the general, quantitative gas-forming reaction of carbodiimides with oxalic acid (see Section V.E.I.a).

Dissolve about 0.1 gram of the carbodiimide in dry 1,4-dioxane or dry ether. Working under an atmosphere of dry, inert gas, add tiny drops of a saturated solution of oxalic acid in dioxane until gas evolution just ceases. Allow the dioxane to evaporate at a temperature less than about 90°C. The residue is nearly pure substituted urea, but it should be recrystallized at least once from alcohol.

The melting points of the substituted ureas are usually fairly sharp and well separated from each other. Hunig et al. (139) have cited melting points (or boiling points) for a group of substituted diphenylureas.

### b. CLEAVAGE TO ISOTHIOCYANATES

Carbodiimides are cleaved into corresponding isothiocyanates when heated in sealed tubes with carbon disulfide at temperatures of the order of 180–200°C. With symmetrically substituted carbodiimides, only one product is formed, but the unsymmetrically substituted form two products which may be somewhat difficult to separate for identification:

$$R'N{=}C{=}NR'' + CS_2 \rightarrow R'NCS + R''NCS$$

As an example, Huhn (138) treated phenyl-$p$-tolylcarbodiimide with carbon disulfide and, after removing the excess of carbon disulfide, separated the residual oil into two fractions by distillation, one boiling at 220–225°C and the other at 235–240°C. The higher boiling fraction was found to be $p$-tolylisothiocyanate (mp, 26°); it also formed a derivative with aniline which corresponded to phenyl-$p$-tolylthiourea (mp, 141°). The lower boiling fraction was readily identified as phenylisothiocyanate.

## 2. Physical Methods

### a. INFRARED SPECTROPHOTOMETRY

The infrared spectra of the various simple carbodiimide group compounds are sufficiently different to permit identification of these compounds by comparison with reference spectra. However, this simple approach does not provide confidence when more complex structures are to be identified. In these instances, it is best to compare the spectrum of a derivative of the carbodiimide (e.g., the ureas and thioureas noted above) with reference spectra.

### b. ULTRAVIOLET SPECTROPHOTOMETRY

Absorptions in the ultraviolet cannot be used for direct identification of the carbodiimides; however, uv spectrophotometry is a useful adjunct for establishing the effectiveness of steps used in the purification of a thiourea derivative. When data are available for comparison, the position of an absorption maximum of a pure derivative may be the decisive factor in establishing the identity of a carbodiimide.

## VI. QUANTITATIVE METHODS

General quantitative methods for the determination of nitriles are not available, largely because reactions of the functional group are seldom quantitative and because the reactivity of the group is governed in altogether too large a measure by the structures attached to it. A number of procedures have been developed, however, for analysis of specific nitrile compounds; some of these procedures can be extended to include a modest group of similar nitriles.

General methods for the determination of isocyanides have not been developed; however, Ugi et al. (287) suggest that all isocyanides may be determined titrimetrically by standardized sodium polysulfide solution, but procedural details and definitive studies of the applicability of the method are lacking.

General quantitative methods for $N$-cyano group compounds are not available; the few methods that have been developed for cyanamide and dicyandiamide are included in the following discussions.

### A. TITRIMETRIC PROCEDURES

Ordinarily, titrimetric procedures are used in preference to gravimetric ones whenever selection is possible, but analytical procedures for nitriles and isocyanides are so few that freedom of selection is not possible; unfortunately, the titrimetric methods which are available leave much to be desired.

### 1. Titrations with Reducing Agents

In sharp contrast with other nitrogen-containing functional groups, direct volumetric reductimetric methods are not the pivotal procedures upon which the analyst can depend for analysis of compounds containing nitrile and isocyanide functional groups. In fact, at present there are no direct volumetric reductimetric methods that can be recommended for the general analysis of nitriles or isocyanides.

### 2. Acidimetric Procedures

Titrimetric determination of the nitrile group by acidimetry depends upon conversion of the group to ammonia or to an amine; obviously, procedures are indirect and inexorably dependent upon the host of complex factors that govern formation of ammonia from the nitrile group. The indirect acidimetric procedures are best classified in accordance with the method used to convert the nitrile group into a substance determinable by acidimetry.

#### a. REDUCTION TO AMINES

Reduction of nitriles to their corresponding amines followed by determination of the amine group by acidimetry is one of the quantitative procedures that have been investigated sufficiently to establish limitations. In Section V.D.I.c(2) were described the results obtained by Siggia and Stahl (253) when lithium aluminum hydride is used as a reducing agent for the quantitative determination of amides and nitriles; the procedure is summarized below:

The sample (0.0006 mol) is treated for 15 min at room temperature with 5 ml of a 2% solution of lithium aluminum hydride in ether. Then the excess hydride is destroyed by water; 10 ml of 6 N sodium hydroxide is added (an excess, so as to dissolve precipitated aluminum), and the resulting solution is steam-distilled in a standard Kjeldahl apparatus to transfer the amine into 50.00 ml of 0.02 N sulfuric acid. The excess of sulfuric acid is titrated with 0.02 N sodium hydroxide to the green end point of methyl purple indicator.

In an alternative procedure, developed to force distillation of the fatty amines, which come over slowly, the sample is mixed with 10 ml of 2% hydride solution and refluxed on a steam bath for 0.5 hr. As before, the hydride reagent is decomposed with 10 ml of water and 5 ml of 6 N sodium hydroxide; 25 ml of ethylene glycol is added before the flask is attached to the distillation apparatus. The solution is distilled at a rapid rate nearly to dryness, and then ethylene glycol is added at such a rate that boiling does not stop until 100 ml of the glycol has been collected as distillate. The condenser is washed with about 50 ml of hot isopropyl alcohol; then the amine in the distillate is titrated with 0.02 N sulfuric acid (potentiometric end point).

Although the method may provide acceptable results for a small number of aliphatic and aromatic nitriles, it fails when applied to an important group of common nitriles (see the following tabulation):

### Reduction of Nitriles by Lithium Aluminum Hydride: Method of Siggia and Stahl (253)

| Quantitative | | Incomplete |
|---|---|---|
| Benzonitrile | Acetonitrile | Phenylacetonitrile |
| Butyronitrile | Acrylonitrile | 3-Butenenitrile |
| Capronitrile | Succinonitrile | γ-Phenoxybutyronitrile |
| p-Chlorobenzonitrile | Adiponitrile | Lactonitrile |
| | | m-Nitrobenzonitrile |
| | | 1-Naphthonitrile |

Moreover, as suggested in Section III.B.1.a(2), reaction of a large excess of lithium aluminum hydride with the nitrile group seldom provides quantitative yields of amines. Clearly, the method of Siggia and Stahl cannot be recommended for general use, especially since amides, imides, and aliphatic nitro compounds also interfere. In instances where the method can be shown to be applicable, periodic checks must be made of its performance under current laboratory conditions, for there is always a possibility that the presence of other substances in a sample may change the rates of hydrolysis of the nitriles (251) or interfere in indefinable ways.

Catalytic hydrogenation may often provide a more effective route for conversion of nitriles to amines. Thus, as noted in Section III.B.1a(3), platinum oxide catalyst appears to perform efficiently in glacial acetic acid. Huber (136) treated platinum oxide with dilute nitric acid to remove basic contaminants and then hydrogenated nitriles dissolved in glacial acetic acid at a pressure of 1.5 atm and at temperatures less than 80°C. The amines formed by hydrogenation in the glacial acetic acid solution were titrated directly by 0.2 N perchloric acid (dioxane) in the usual way. Unfortunately, Huber's work included only four nitriles; other reports are too sparse to permit the drawing of conclusions as to the general utility of the method, but it can be anticipated that catalytic hydrogenation will not always provide quantitative conversion of nitriles to amines.

### b. Conversion to Ammonia

Nitriles may be converted to ammonia by strongly alkaline reducing agents such as silver-catalyzed ferrous hydroxide; it is not clear whether the formation of ammonia proceeds by a hydrolytic attack (via amides) which is aided by a reducing environment or whether the nitrile group is reductively cleaved. The application of alkaline ferrous hydroxide for conversion of nitriles to ammonia was first described by Ludzack et al. (177) for analysis of a group of nitriles. The strong reducing action of the nearly white ferrous hydroxide which has been prepared in the absence of oxygen was first studied by Sandonnini and Bezzi (233); in solutions of low alkalinity, ferrous hydroxide slowly reduces nitrate to ammonia with formation of a brown hydrated ferric oxide:

$$8 \, Fe(OH)_2 + NaNO_3 \rightarrow 4 \, Fe_2O_3 + NH_3 + NaOH + 6 \, H_2O$$

but in high alkalinity a black, mixed oxide is formed:

$$12 \, Fe(OH)_2 + NaNO_3 \rightarrow 4 \, Fe_3O_4 + NH_3 + NaOH + 10 \, H_2O$$

The reduction obviously can be extended to include the nitrite ion, and it is observed that the reaction is very much faster than with the nitrate ion. Szabo and Bartha (267) found that the rate of reduction of nitrate ion could be markedly accelerated by colloidal silver formed *in situ* by homogeneous precipitation; large crystallites of silver are not catalytic. Also, cations of lead, mercury, bismuth, antimony, and chromium(III) or anions such as cyanide, thiocyanate, ferri- and ferrocyanide, and chromate appear to cause formation of colloidal ferric oxide, which acts as a poison to colloidal silver (probably because of formation of silver ferrite). Naturally, silver precipitants such as iodide, sulfide, and thiosulfate ions also interfere, but a surprisingly large number of other cations and anions can be tolerated. Copper is also an effective catalyst for the reduction of nitrate ion by ferrous hydroxide (268).

Varner et al. (297) used catalyzed and uncatalyzed ferrous hydroxide to determine nitrates and nitrites, respectively, in mixtures containing ammonia, amide, nitrite, and nitrate. Ammonia was stripped at 50°C from pH 10 solution (tetraborate), made alkaline and boiled to hydrolyze amides and remove ammonia, and then treated with ferrous solution and boiled to remove ammonia formed from nitrites; finally, silver was added to force reduction of nitrate to ammonia. A ferrous reduction and distillation of ammonia required less than 5 min. Ludzack et al. (177) modified Varner's method, mostly by lengthening the reduction time to 2 hr at 65°C; a mixture of equal parts of silver sulfate, nickel chloride, copper chloride, and mercury chloride was used as a catalyst. [Note that

Szabo and Bartha (267) report mercury to be a catalyst poison.] With the modified procedure, acetonitrile, acrylonitrile, adiponitrile, and benzonitrile provided ammonia corresponding to conversions between 90 and 100% at nitrile compound concentration levels of 5–50 mg/liter. Lactonitrile gave disappointingly low values (less than 20% recovery), probably because of formation of cyanide ion. Additional studies of alkaline ferrous hydroxide reductions seem warranted.

### c. Ammonia via Alkaline Peroxide

Reaction of nitriles with alkaline peroxide has been described in some detail in Section III.B.2, where it was noted that conversion to amide occurs much faster than in alkaline hydrolyses. Although the reaction was first noted by Radziszewski (222) about 1885, its use in quantitative procedures (185,309,314) has not been completely exploited. Petersen and Radke (215), as well as Whitehurst and Johnson (314), used the reaction as the basis for the determination of low concentrations of nitriles in water or air, and also for the assay of acetonitrile, propionitrile, butyronitrile, and succinonitrile. The assay method, given below in brief, depends on detemination of the hydroxide ion consumed in the hydrolysis of a nitrile.

Into each of two flasks are pipetted 50 ml of $N$ potassium hydroxide and 100 ml of 3% hydrogen peroxide solution. A sample containing 6–10 mequiv. of nitrile is put into one flask; the other flask serves as a blank. The flasks are allowed to stand for 5 min at room temperature (occasional swirling). Then a few glass beads are put into each flask, and each flask is attached to its condenser. Distillation is continued until the residue in each flask has a volume of about 10 ml. After cooling and adding 100 ml of water, 50 ml of 0.5 $N$ sulfuric acid is pipetted into each flask and then titrated with 0.5 $N$ sulfuric acid until the pink color of phenolphthalein is discharged. The difference between the blank and sample titrations is the measure of nitrile.

The reactions involved are as follows:

$$RCN + 2 H_2O_2 \rightarrow RCONH_2 + O_2 + H_2O$$
$$RCONH_2 + NaOH \rightarrow RCOONa + NH_3$$

The hydrolysis of amides is slow, even at elevated temperature in strong caustic; nitriles are completely converted to amides in about 5 min at room temperature, however, and only small amounts of the amides are hydrolyzed to the carboxylic acid salts. In fact, West (309) was able to prepare benzamide from benzonitrile in quantitative yields, and quantitative conversion of $\alpha$-naphthonitrile was found to be possible at slightly higher temperatures (40°); in these instances, the remarkable resistance of the amides to hydrolysis provides a convenient method for their synthesis.

Whitehurst and Johnson (314) found it necessary to ensure complete saponification of amides by evaporating the reaction mixture so that the initial alkali concentration was increased to about 2 N. In view of the great possibility for cleavage of hydrolysates or formation of acidic substances by alkali attack, and the small losses of alkali by spatter and carryover during distillation, it is remarkable that assay values were found to be $100 \pm 0.5\%$.

The effect of hydrogen peroxide concentration on the rate of conversion of nitriles appears to be an important parameter; Whitehurst and Johnson note that, if the volume of peroxide prescribed in their procedure is reduced by one half, the assay for butyronitrile is 3% low even though the assays for the other nitriles are unaffected.

Results as high as 120% were obtained for assays of benzonitrile, acrylonitrile, ethylenecyanohydrin, and 3-methoxypropionitrile, possibly because of side reactions or oxidation at points of unsaturation. Approximately 2 mol of base was required for each mole of lactonitrile; however, quantitative results could not be obtained. Alkyl nitriles generally do not give satisfactory results, and aromatic nitriles with *ortho* substituents usually resist hydrolysis under the conditions imposed by the procedure.

All compounds that are oxidized by alkaline peroxide to an acid will interfere; methanol, ethanol, and 2-propanol interfere only slightly and can be tolerated in small amounts. Obviously, a correction can be applied for free acid initially in the sample; amines that can be steam distilled will not interfere. Most esters and amides in samples will also react quantitatively with peroxide, but they can be determined independently.

The alkaline peroxide method was applied as follows to permit determination of low concentrations of nitriles in water:

Exactly 25 ml of 0.2 N potassium hydroxide solution is pipetted into each two flasks, and then 20 ml of 30% hydrogen peroxide is added. To one flask is added 200 ml of water sample, and 200 ml of distilled water is added to the other (blank). The flasks are allowed to stand at room temperature for 5 min; then the water content of each flask is distilled away until the residual volume is at least 2 ml. The residue (after dilution with water) is titrated with standard 0.1 N acid to the disappearance of phenolphthalein red.

Concentrations of acetonitrile between 1000 and 5 ppm in water were determinable by the alkaline peroxide method; propionitrile, butyronitrile, and succinonitrile were also readily determinable at the 5-ppm level.

## d. Direct Hydrolysis to Ammonia

The quantitative hydrolysis of nitriles, discussed in Section III.B.3.a and III.B.4.a, usually requires a high concentration of hydroxyl or hydrogen ion; amides are formed first, but (as a rule) the overall rate at which a sample of nitrile can be converted to ammonia is governed largely by the rate of hydrolysis of the amide. It is important to recognize that, although hydrolytic procedures for analysis of nitriles are widely used, they are not specific (30).

The following procedures may be used freely for nitriles that can be quantitatively converted to ammonia:

### Acid Hydrolysis

Transfer about 3 mmol of the nitrile to a 250-ml round-bottom boiling flask which has a neck made from the outer member of a standard-taper ground-glass joint, and add 50 ml of 5.5 N hydrochloric acid. Connect the flask to a mating reflux condenser, and gently reflux the mixture for at least 10 hr in a hood. Allow to cool to room temperature, and transfer to a 500-ml Kjeldahl flask with the aid of about 100 ml of water. Add a few boiling stones and about 5 grams of sodium hydroxide pellets, and quickly connect the flask to a Kjeldahl trap and a distillation condenser which has been arranged to collect ammonia in 100 ml of 1.5% boric acid neutralized to methyl red. Distil ammonia, and then titrate as in the usual Kjeldahl method.

### Alkaline Hydrolysis

Place about 50 ml of glycerol or diethylene glycol in a 500-ml Kjeldahl flask, and add about 5 grams of potassium hydroxide. Dilute with about 50 ml of water, and then transfer about 3 mmol of the nitrile to the flask; quickly connect the flask to a trap and condenser, and very slowly distil the water in the reaction mixture into 100 ml of 1.5% boric acid, as indicated in the above procedure. If the hydrolysis takes place too slowly and cannot be completed before all water is distilled from the Kjeldahl flask, it is best to maintain the reaction mixture at a temperature of about 100°C and to strip ammonia by a gentle stream of nitrogen saturated with water vapor at 100°C. Siggia et al. (252) used as a reaction mixture 100 ml of water, 20 ml of tetrahydrofuran, and 10 ml of 50% aqueous sodium hydroxide solution; ammonia was stripped by a stream of dry nitrogen.

Although polymeric amides formed from polymeric nitriles with functional groups on the polymer backbone are exceptionally resistant to alkaline hydrolysis, there are exceptions, because Stafford and Toren (260) have found acidic hydrolysis to be as effective as sodium hydroxide for determination of acrylonitrile in acrylonitrile–methylvinylpyridine copolymers. For this purpose, 0.2–0.4 gram samples are refluxed for 2 hr in 50 ml of 54% sulfuric acid, and then ammonia is distilled in a Kjeldahl apparatus in the usual manner; acid concentrations greater than 54% cause charring. Alkaline hydrolysis of acrylonitrile-methylvinylpyridine copolymers did not provide quantitative release of ammonia unless distillation in a Kjeldahl apparatus was carried out to the point where the sodium hydroxide in the distilland had been concentrated to 90–95%, where glass-etching is severe. Moreover, in such concentrated alkaline solutions, decomposition of methylvinyl-pyridine releases pyridine, which interferes in the subsequent titration of ammonia. In contrast, Gibson and Heidner (107) found that alkaline hydrolysis with 50% aqueous potassium hydroxide was satisfactory for acrylonitrile–vinyl acetate copolymers.

Inasmuch as hydrolysis of a nitrile group to an amide generally can be made to take place readily, hydrolytic conditions in an analytical procedure are controlled to provide a maximum rate of hydrolysis for the amide. On the other hand, since the rates of hydrolysis of various nitriles are different and, in turn, the rates of hydrolysis of amides are also different from each other, as well as from those of the nitriles, it is often possible to resolve mixtures of nitriles, of amides, or of nitriles and amides. In fact, Livengood and Johnson (176) and Ranny et al. (223) have noted that there is a considerable difference in the rates of saponification of mono- and diethanolamides. Siggia et al. (252) have demonstrated that a differential-reaction-rate technique is applicable to quantitative resolution of binary mixtures of primary amides and of nitriles and amides, as well as binary mixtures of nitriles. For this purpose, the progress of an alkaline hydrolysis is followed by distilling and then titrating the ammonia formed. In the presence of a large excess of water and alkali, the hydrolysis follows first-order kinetics fairly well. Plots of the logarithm of the total concentration of amide and nitrile or of total nitriles remaining against time show a straight line for a single amide or nitrile; for a mixture of two compounds hydrolyzing at different rates, data plots usually suggest the existence of two lines, for which slopes can be evaluated by graphical methods or by mathematical procedures (104,225). Quantitative hydrolysis of amides and nitriles to ammonia seldom takes place, but the reactions are usually sufficiently complete to enable determination of hydrolysis rates. However, the total original concentration of nitrogen must be obtained by a Kjeldahl or Dumas determination.

### 3. Iodometric Procedures

Nitriles may be reduced by hydriodic acid or a mixture of an alkali metal iodide and sulfuric acid; iodine is released. The reaction may take place in a closed tube in accordance with a procedure outlined by Aldrovandi and De Lorenzi (4), and released iodine may be titrated by thiosulfate; however, the procedure has not been studied sufficiently to establish its limitations, and the consumption or release of iodine by other reactive functional groups may prove to be an insurmountable interference.

### 4. Direct Titration of Cyanide Ion

As a rule, the nitrile group is attached so firmly that hydrolysis or reaction with water does not form hydrogen cyanide; however, there are some exceptions, notably the cyanohydrins, the acyl cyanides, and compounds such as tetracyanoethylene (see Section III.B). In instances where the cyanide ion is rapidly formed, titration with standard silver nitrate solution provides a direct means for determining these types of nitriles.

Cyanohydrins are readily analyzed because they release hydrogen cyanide very rapidly; thus Berther et al. (25) caused cyanohydrins to react with nickel ammonia ions to form nickel tetracyanide complex, and then titrated uncomplexed nickel ions with 0.1 M EDTA (murexide indicator).

### 5. Aquametric Procedures

A quantitative procedure for determination of the water taken up by the hydrolysis of nitriles was developed by Mitchell and Hawkins (193). The nitrile is hydrolyzed by a slight excess of water in glacial acetic acid; boron trifluoride is used as a catalyst:

$$RCN + H_2O \rightarrow RCONH_2$$

The amount of water in excess of that needed to hydrolyze the nitrile is determined by Karl Fischer reagent.

Accurately weigh a sample of the nitrile (less than 10 mmol) into a 250-ml glass-stoppered volumetric flask (use another flask for a blank). Add 20 ml of a solution of 300 grams of boron trifluoride and 6.5 ml of water in 500 ml of glacial acetic acid. Stopper the flask, and place in a water bath at 80°C; after 2 hr, remove the flask to an ice bath and add 15 ml of anhydrous pyridine to the cold reaction mixture. Titrate the solution with Karl Fischer reagent in the usual manner; determine the water content of the sample by direct titration, and the water content of the hydrolysis reagent by titration of the blank. The net water consumed by hydrolysis is equivalent to the nitrile content of the sample.

The procedure outlined above was found to provide quantitative values for assay of lower aliphatic mono- and dinitriles such as acetonitrile, propionitrile, butyronitrile, valeronitrile, adiponitrile, and sebaconitrile. Several aromatic and mixed aliphatic-aromatic nitriles were also found to react quantitatively (phenylacetonitrile, $p$-chlorobenzonitrile, and $\beta$-naphthonitrile), but although $m$- and $p$-toluonitrile react quantitatively, methyleneaminoacetonitrile, cyanoacetonitrile, and cyanoacetic acid react incompletely; little or no interference is observed with amides. Obviously, alcohols and other substances that interfere with the Karl Fischer titration must be absent.

## 6. Nonaqueous Titrations

### a. NITRILES

Nitriles with activated $\alpha$-hydrogens often can be titrated as acids in nonaqueous media. Thus Fritz and Yamamura (101) obtained satisfactory end-point breaks and quantitative results when malononitrile, cyanoacetamide, and ethyl cyanoacetate were titrated in acetone solution with $0.1 N$ triethyl-$n$-butylammonium hydroxide (glass-calomel electrodes). The compounds are structures of the type A—CH$_2$—B, where A and B are the following electron-withdrawing groups:

$$-CN \qquad -C{\overset{\displaystyle O}{\underset{\displaystyle NH_2}{}}} \qquad -C{\overset{\displaystyle O}{\underset{\displaystyle OC_2H_5}{}}}$$

### b. CYANAMIDE

In a study of nonaqueous titrations of basic compounds with acetic anhydride as a solvent and perchloric acid in acetic acid or dioxane as a titrant, Wimer (320) found that cyanamide itself exhibited no measurable basic properties toward perchloric acid–acetic anhydride mixtures. However, $N,N$-dimethylcyanamide responded as a base, and it could be titrated as rapidly as most amides. A glass electrode was found suitable after soaking in acetic anhydride for 12 hr; the aqueous salt bridge normally used with the reference calomel cell was replaced with a $0.1 M$ solution of lithium perchlorate in acetic anhydride.

### c. CARBODIIMIDES

The rapid quantitative reaction of carbodiimides with oxalic acid was discussed in Section V.E.1.a and V.I.1.a; in anhydrous solvents such as dioxane, an equimolar mixture of carbon dioxide and carbon monoxide is released and a substituted urea is formed. Zarembo and Watt (325) added a known volume of a standard solution of oxalic acid in dry dioxane to a solution of a sample of carbodiimide in dioxane and then titrated the excess of oxalic acid by a solution of sodium methoxide, using thymol blue as indicator. The entire determination is performed in the absence of air and water.

## 7. Miscellaneous Titrimetric Procedures

Some of the most important nitriles are also unsaturated compounds for example, acrylonitrile, which is of great commercial importance principally because it polymerizes easily. Accordingly, a great deal of effort has been expended on the development of specific methods for the industrially important unsaturated nitriles; although most of the methods presented below are specific reactions of C-C double-bond unsaturation, they are included in this section because they are widely used in laboratories specializing in the analysis of nitrile compounds.

### a. ALKYL THIOL REACTION WITH $\alpha,\beta$-UNSATURATED NITRILES

Acrylonitrile and some $\alpha,\beta$-unsaturated carbonyl compounds can be determined by reaction with an excess of a mercaptan:

$$R—CH{=}CH—C{\equiv}N + R'SH \to RCH(SR')—CH_2C{\equiv}N$$

The unconsumed mercaptan is determined iodimetrically or by an amperometric titration with silver nitrate (17,143). The method is useful because it permits assay of acrylonitrile. More importantly, however, it can be used to determine small amounts of the substance in gas streams or in air (8,126); for this purpose, the gas is bubbled through a cold solution of the mercaptan in isopropyl alcohol, and after an iodine solution equivalent to the initial amount of mercaptan has been added, the residual iodine is determined spectrophotometrically.

The determination of acrylonitrile can also be performed in essentially nonaqueous media, for example, in ethanol; mercaptoacetic acid is added, and the excess thiol is titrated with iodine in ethanol.

### b. REACTION OF SODIUM SULFITE WITH $\alpha,\beta$-UNSATURATED NITRILES

Critchfield and Johnson (61) have developed an acidimetric method for determination of $\alpha,\beta$-unsaturates such as acrylonitrile and other common industrial compounds, for instance, acrylic acid, maleic acid, and methyl acrylate. In brief, sodium sulfite solution is acidulated under an atmosphere of nitrogen with a measured amount of standard sulfuric acid, isopropanol is added as a cosolvent to minimize side reactions, and then the sample is added. After standing for a length of time at room temperature or at an elevated temperature (in pressure bottles), the remaining acid is titrated with standard alkali to the disappearance of the green color of Alizarin Yellow (C.I. 14030)—Xylene Cyanol (C.I. 43535) mixed indicator. A substituted sulfonate is formed:

$$NaHSO_3 + CH_2{=}CH{-}CN \rightarrow NaO_3S{-}CH_2CH_2{-}CN$$

with corresponding decrease in acidity; strong electron-attracting groups other than nitrile will also cause reaction to take place, for example, —COOH, —COOR, and —CONH$_2$. Unfortunately, other types of compounds—typically, reactive structures such as epoxides, vinyl ethers, and oxidizing agents—will react more or less with sodium bisulfite. Moreover, the pH of the reaction medium usually determines the rate of reaction, and spurious reactions can sometimes take place. Thus application of the method must be approached with deliberation, and its successful use can be assured only after thorough study with samples of defined composition. The original work should be consulted for the experimental conditions that were found satisfactory for analysis of a number of industrially important compounds.

Sodium sulfite itself reacts with $\alpha,\beta$-unsaturates; since the reaction releases alkali, it can be a source of error in the method discussed above:

$$CH_2{=}CHCN + Na_2SO_3 + H_2O \rightarrow NaSO_3CH_2CH_2CN + NaOH$$

On the other hand, the reaction can be made the basis of a method for analysis of $\alpha,\beta$-unsaturates; the alkali that is formed can be titrated with standard acid. Terent'ev et al. (275) developed a method using the reaction as a basis for the semimicrodetermination of acrylonitrile; the alkali was titrated with 0.05 $N$ sulfuric acid, using a mixture of Alizarin Yellow (C.I. 14030) and thymolphthalein as indicator.

### c. REACTION OF MORPHOLINE WITH $\alpha,\beta$-UNSATURATED NITRILES

The reaction of primary and secondary amines with $\alpha,\beta$-unsaturated compounds has been applied by Critchfield et al. (60) to the analysis of industrially important nitriles such as acrylonitrile and methacrylonitrile; the method uses an excess of morpholine as the secondary amine and acetic acid as catalyst:

$$O(CH_2)_4NH + R{-}\overset{H}{C}{=}\overset{H}{C}{-}CN \rightarrow O(CH_2)_4N{-}\underset{R}{\overset{H}{C}}{-}CH_2CN$$

Inasmuch as the reaction is quite general, compounds with strong electron-attracting groups other than nitrile will also react as shown. After reaction is complete, the excess of morpholine is acetylated with acetic anhydride in an acetonitrile medium to form the corresponding amide and acetic acid; because the amide and acetic acid are neutral in the medium, the tertiary amine formed in the reaction can be titrated with standard alcoholic hydrochloric acid, using Methyl Orange (C.I. 13025)—Xylene Cyanol (C.I. 43535) mixed indicator.

Acrylonitrile (or acrylamide) reacts completely with morpholine in less than 1 hr at room temperature; however, allyl cyanide (3-buteneni-trile) must be kept in contact with morpholine at 98° for more than 1 hr, and methacrylonitrile will require treatment for 2–4 hr at 98°C.

### d. Oxidative Titration of Acrylonitrile

The terminal methylene group in compounds where the C-C double bond is conjugated with a strongly polar group such as carbalkoxyl, cyano, or aryl is readily attacked by strong oxidizing combinations such as periodic acid and potassium permanganate. Dal Nogare et al. (68) developed a method for the determination of residual methylmethacryl-ate in polymerized masses by permanganate titration in an aqueous solution containing periodic and sulfuric acids; it was found that acry-lonitrile can also be determined. However, since a variety of reactions are involved, and the nature of the compound being titrated determines the reaction rates, a calibration plot must be prepared for each com-pound; typically, linear plots are observed for monomer masses up to 4 or 5 mg, but the slopes are different for each monomer type.

### e. Argentimetric Procedures for N-Cyano Compounds

The solubility differences between the silver salts of cyanamide and dicyandiamide form the basis for titrimetric determination of cyanamide in dicyandiamide.

When dicyandiamide is also to be determined in the presence of cyanamide, the insolubility of the silver salts of dicyandiamide in caustic alkali serves as the basis of most analytical procedures; however, since the silver content of silver dicyandiamide is not stoichiometric (see Section V.D) argentimetric procedures cannot be used.

Perotti (212) was among the first to use a procedure for quantitative analysis of cyanamide in the presence of dicyandiamide. An aqueous solution of the sample was slowly run into a measured volume of standard silver nitrate solution that had been made slightly ammoniacal and was sufficient for complete precipitation of cyanamide. After being warmed to coagulate the precipitate, the solution was filtered. The excess silver in the filtrate was determined by titration with standard thiocyanate solution; the cyanamide content was computed from the amount of silver consumed.

A minor modification of the Perotti procedure was recommended by Kappen (151); the cyanamide sample solution, first made acidic with nitric acid and then slightly ammoniacal, was mixed with an excess of silver nitrate solution. Brioux (33) also precipitated silver cyanamide from ammoniacal solution, but found it more convenient to determine the silver content of the precipitate; more recently, Capitani and Gambelli (44) employed a potentiometric titration of the excess silver remaining after a precipitation of cyanamide.

Grube and Kruger (112) made a critical study of the argentimetric determination of cyanamide and reported that ammonium hydroxide in the presence of ammonium salts markedly increased the solubility of silver cyanamide. Hager and Kern (120) found that some silver dicyandiamide also may be occluded by the silver cyanamide precipitate, and Marqueyrol and Desvergnes (180) also questioned the accuracy of argentimetric procedures. Pinck (216) stressed the need for reprecipitation of the first-isolated silver cyanamide when large amounts of dicyandiamide are present.

More recent studies by Takei and Kato (271) and Takei (270) have demonstrated that there is no need to precipitate silver cyanamide in an analysis of calcium cyanamide because cyanamide can be titrated directly in ammoniacal acetone solution with silver nitrate. Typically, 8 grams of calcium cyanamide is dissolved and diluted to 1 liter; a 25-ml aliquot is acidified with nitric acid and transferred to a titration vessel. Then 20 ml of acetone and 1 ml of 0.25 $M$ aqueous ammonium hydroxide solution are added, and the mixture is titrated with 0.1 $N$ silver nitrate solution to a Malachite Green (C.I. 4200) end point. Dicyandiamide, guanidine, and urea do not interfere. Thus Sato et al. (234) extracted a sample with water, treated an aliquot with oxalate to remove calcium, and titrated with standard silver nitrate (sulfide removed by cadmium acetate).

The separation of dicyandiamide as a double compound with silver picrate also can be used as the basis for argentimetric procedures. Thus, for a sample containing both cyanamide and dicyandiamide, Capitani and Gambelli (44) precipitated dicyandiamide by picric acid in a solution containing a measured amount of silver nitrate, and then determined the remaining silver in the filtrate by titration. In a slightly different approach, Johnson (147) used an excess of a standard solution of silver picrate instead of silver nitrate and determined the excess silver as usual. Halides and sulfides interfere with the picrate precipitation; these ions must be determined separately, and suitable corrections made. Unfortunately, melamine, guanidine, and many other substances are precipitated as picrates, but Inaba (140) claims that interference is avoided by a volumetric procedure in which excess silver is back-titrated.

## f. CYANOMETRIC PROCEDURE FOR CARBODIIMIDES

The reaction of cyanide ion with monomeric carbodiimides, which have nitrogen atoms that are at least as basic as those of diphenyl-carbodiimide, was described in Section III.B.5.c. A brief study of some of the analytical aspects of the reaction was made by Hunig et al. (139). In general, it was found that, although there is formed an $\alpha$-cyano-$N,N'$-disubstituted formamidine which appears to involve simply the addition of hydrogen cyanide to a carbodiimide, the reaction takes place only at pH values greater than 7, where cyanide ion is prevalent. Accordingly, the reaction may be written as follows:

$$\text{ArN=C=NAr} + \text{CN}^- + \text{H}_2\text{O} \rightarrow \underset{\underset{\text{CN}}{|}}{\text{ArNHC=NR}} + \text{OH}^-$$

Unfortunately, most carbodiimides are exceptionally sensitive to the presence of alkali, forming polymers and substituted ureas (by hydrolysis). For the alkali-sensitive carbodiimides, a procedure was developed in which an acidic solution of a sample was run into an excess of alcoholic cyanide solution and the excess of cyanide back-titrated with mercury(II) nitrate, using *sym*-diphenylcarbazide as indicator. Following are brief descriptions of three procedures developed by Hunig et al. (139):

Prepare a standard $N/10$ solution of mercuric nitrate by dissolving 4.3 grams of the reagent-grade salt in a little water containing about 3 ml of nitric acid and diluting with water to about 250 ml. Standardize with exactly 0.1000 $N$ KCl; use a 0.1% methanolic solution of *sym*-diphenylcarbazide as indicator in a medium which is about 0.1 $N$ in nitric acid (end point is a violet color). Prepare a solution of 0.1 $N$ KCN, and determine its titer with the mercuric nitrate solution as follows. Pipet 15 ml of the KCN solution into 40 ml of pure methanol. Add 5 ml of the purest tetrahydrofuran, 1 ml of *sym*-diphenylcarbazide indicator, and 0.1–0.2 ml of glacial acetic acid; titrate with standardized mercury(II) solution until the indicator turns blue.

**Procedure A** [for alkali-sensitive compounds such as di-*p*-tolyl-, dianisyl-, and bis(dimethylaminophenyl)carbodiimide]

Dissolve up to 0.5 mmol of the carbodiimide in 5 ml of tetrahydrofuran, and add a mixture of 15 ml of 0.1 $N$ KCN in 40 ml of methanol. Add 1 ml of indicator, and titrate the excess cyanide with standardized mercury(II) solution until the indicator turns blue.

## Procedure B (for alkali-sensitive compounds)

Dissolve 0.5 mmol of sample in 3.75 ml of tetrahydrofuran, and mix quickly with 1.25 ml of 0.4 $N$ perchloric acid. Immediately run this solution into a mixture of 15 ml of 0.1 $N$ KCN and 40 ml of methanol. Back-titrate the excess of KCN as described above.

## Procedure C (for iodo- and nitrophenylcarbodiimides)

Dissolve the sample as in Procedure B; quickly run this solution into a mixture of 30 ml of 0.1 $N$ KCN and 80 ml of methanol. Determine the excess of KCN.

Owing to the competitive reactions catalyzed by hydroxyl ions, results for carbodiimide contents usually are about 5% low (absolute); moreover, the manner in which solutions are mixed plays an important part because the pH environment in the transition zone between a mass of acidified carbodiimide solution and the alkaline cyanide solution during mixing determines which of the possible reactions will predominate. Clearly, the analytical procedure leaves much to be desired, but it is useful since it can be simply and rapidly performed and results can be obtained that are marginally acceptable.

## B. GRAVIMETRIC PROCEDURES

### 1. Nitriles

As was noted in Section VI.A.4, certain nitriles readily release cyanide ion on hydrolysis. Berinzaghi (24) utilized a gravimetric procedure for the determination of nitrile groups in acylated nitriles of aldonic acids by allowing 15–25 mg of a sample (dissolved in 2 ml of ethanol) to remain in contact with 50 mg of silver nitrate dissolved in 2 ml of concentrated ammonium hydroxide for 24 hrs. After acidulation with nitric acid, silver cyanide was filtered off, washed, dried at 100°C, and weighed. An average error of 3.6–8.2% is reported.

### 2. Cyanamide and Dicyandiamide

An indirect method for the gravimetric determination of cyanamide has been suggested by Fosse et al. (94); it is based on determination of the urea formed by hydrolysis of cyanamide in dilute nitric acid. Xanthydrol in the presence of acetic acid is used to precipitate dixanthylurea from the hydrolysate.

Inasmuch as urea can interfere in certain dicyandiamide determinations, von Dafert and Miklauz (67) hydrolyzed dicyandiamide with dilute nitric acid to guanylurea and then precipitated a nickel guanylurea compound which was ignited to nickel oxide. Ammonium salts, guanidine salts, and urea do not interfere; Garby (103) established that melamine, amidodicyanic acid, and sodium nitrate also do not interfere, but guanylurea and biguanidine must be removed by preliminary precipitation as their nickel salts (for dry samples, dicyandiamide can be extracted by absolute acetone, in which the salts of guanylurea and biguanidine are insoluble).

Dicyandiamide forms an insoluble double compound with silver picrate directly or when its solutions containing silver nitrate are treated with picric acid. Harger found that the compound of dicyandiamide with silver picrate can be precipitated quantitatively, filtered, dried, and weighed (124) (see also Section VI.A.7e).

## C. SPECTROPHOTOMETRIC PROCEDURES

There are only a few spectrophotometric procedures which can be applied for determination of nitriles and isocyanides. In general, infrared procedures are more valuable, but in some instances, such as for quality control, ultraviolet spectrophotometry can be used whenever specific nitrile or isocyanide compounds are involved. Color reactions for nitrile and isocyanide groups are essentially nonexistent, but colorimetric procedures for specific compounds are often based on structural peculiarities. Cyanamide and dicyandiamide have sufficiently different absorption maxima in the ultraviolet to make possible independent determination of each constituent.

### 1. Infrared

Analyses based on absorption of the nitrile group in the infrared are quite frequently performed in chloroform solutions. Typical absorption regions for the nitriles and isonitriles were given in Section V.B.2.a and in Table II. A special procedure has been described by Dinsmore and Smith (76) in an extensive and comprehensive article on the ir analysis of elastomers; the ratio of the absorbance at about $2237 \, cm^{-1}$ $(4.47 \, \mu)$ from the nitrile group to the absorbance at about $2857 \, cm^{-1}$ $(3.50 \, \mu)$ from C—H groups was plotted against known concentrations of acrylonitrile to obtain a working curve, which was then used to determine the acrylonitrile content of Buna-N rubbers. A similar procedure was employed by Nelson et al. (200) for the determination of nitrile residuals in the high molecular weight fatty amines which are prepared from the corresponding nitriles. An isopropyl alcohol solution of the amine is passed through a column of sulfuric acid ion-exchange resin in its acid form; the amines are retained on the column. After evaporation of the solvent, the residue is dissolved in chloroform and its absorbance is determined at $2247\,\text{cm}^{-1}$ ($4.45\,\mu$) in a 1-mm cell. The concentration of nitrile is obtained by computation, using a chloroform solution of octadecylnitrile as calibrant; it is worthy of note that the calibrant nitrile must be very similar to the one being determined.

The potassium bromide pellet technique obviously has important advantages insofar as ease of sampling is concerned. For the determination of the nitrile group in polyacrylonitriles, Doerffel (77) found the technique satisfactory; similarly, Putiev and Tashpulatov (221a) report that excellent results were obtained for cyanoethylcellulose.

Cross and Rolfe (63) have cited data on molar extinction coefficients, and Skinner and Thompson (254) have given integrated band intensities for some alkyl cyanides and a few substituted benzonitriles; Flett (93) recommends that the total area under the absorption envelope be used in quantitative measurements.

A procedure for determining dicyandiamide in melamine (265) by ir spectrophotometric methods has been described; potassium ferrocyanide is used as an internal standard, and measurements are made between 1800 and $2400\,\text{cm}^{-1}$ (5.56 and $4.17\,\mu$) in paraffin oil. The determination of dicyandiamide in technical urea may also be made with the aid of the potassium bromide disk technique with potassium ferrocyanide as internal standard, using $2020\,\text{cm}^{-1}$ ($4.95\,\mu$) for the standard and about $2170\,\text{cm}^{-1}$ ($4.61\,\mu$) for the sample (242).

## 2. Ultraviolet

The ultraviolet absorption spectra of nitrile, isocyanide, and $N$-cyano compounds are not sufficiently distinctive for general quantitative work; however, in many instances, strong characteristic absorption in a given region of the ultraviolet can be used as an index of purity or as a quantitative indication of a nitrile in a mixture. The complexity of the analytical problem becomes evident when it is recognized that a large number of functional groups, ring structures, amides, and carboxylic acids exhibit absorption in the very same region as the nitrile, isonitrile, and $N$-cyano groups. The important regions of absorption of uv energy are summarized in Section V.B.2.b.

As an example of a specific instance where uv spectrophotometry has been found useful, Pawlik (210) recommends determination of dicyandiamide in aqueous extracts of soil at 215 nm because the spectrum is sufficiently different from the spectra of its decomposition products (guanidine, urea, and nitrate ion). In later work, Finkel'shtein et al. (92) have extended Pawlik's findings; cyanamide exhibits maximum absorption at 225 nm in solutions at pH 13, and dicyandiamide at pH 7.

Thus, by measuring absorption of a solution at 225 nm at pH 13 and pH 7, it is possible to determine cyanamide to better than ±3% and dicyandiamide to better than ±8%.

### 3. Visible

As noted in preceding sections, there are no colorimetric procedures for nitriles and isocyanides which can be recommended for general use. However, the amidoxime test described in Section V.B.I.a can be utilized for quantitative purposes; Bergmann (23) has demonstrated that the same type of procedure can also serve for the colorimetric determination of amides, provided that experimental conditions are appropriately selected. Robertson et al. (227) have also made use of the amidoxime reaction for determination of amides and nitriles. A suitable procedure is given in detail in Section VII.D.

A spectrophotometric procedure for determination of cyanamide in solutions or extracts of complex mixtures (such as blood, urine, and soil) has been developed by Buyske and Downing (43); it is based on the development of a red color when cyanamide and pentacyanoammineferroate ion interact at pH 10.5 in carbonate buffer (83). The intensity of the red color is determined at 530 nm, the maximum point of absorption in the visible region of the spectrum. A calibration curve is linear up to at least 50 $\mu$g of cyanamide. The lower limit of detection is about 2 $\mu$g/ml of sample, but an ethyl acetate extraction of the sample solution at pH 2 and subsequent evaporation of the solvent proved effective for concentrating cyanamide or removing it from highly colored solutions which cannot be used directly; recoveries of at least 93% were obtained. Compounds often associated with cyanamide as a result of degradation and polymerization (or because they are deliberately introduced) do not interfere at levels of from 5 to 10 times the amount of cyanamide in the test solution; typical compounds include urea, ammonia, guanidine, dicyandiamide, melamine, cyanide, and citric acid.

Sodium nitroprusside in alkaline solution reacts with cyanamide to produce a red color which has an absorption maximum at 530 nm; moreover, even ferrocyanide will form a red complex when it is activated by light, and so the colored complex ion formed from cyanamide in each instance may be $[Fe(CN)_5H_2NCN]^{3-}$. However, different complexes may be involved because equilibrium conditions are not the same, and the color intensity developed with nitroprusside is only one-fourth as intense as that obtained with pentacyanoammineferroate (43).

A combination of nitroprusside and ferricyanide in alkaline solution has been used as the color-forming reagent in a spectrophotometric procedure for determination of cyanamide and its derivatives (272); however, the reagent does not distinguish cyanamide from dicyandiamide, for a magenta color is formed with both compounds. Moreover, as reported by Milks and Janes (190), who used the reagent for spotting paper chromatograms, red to orange colors are formed with potassium cyanoureate, biguanidine, guanidine carbonate, and so on, and a transitory blue to green color is formed with thiourea. A mixture of alkali, hydrogen peroxide, and sodium nitroprusside also develops colors with cyanamide and related compounds (132).

## 4. Specific Compounds

A variety of spectrophotometric procedures have been developed for specific compounds, usually nitriles of economic importance. Brief descriptions of a representative group of such procedures are given below; some of the procedures can possibly be used for analysis of other nitriles and isonitriles.

### a. TETRACYANOETHYLENES

Primary amines and some secondary amines react with tetracyanoethylene and its derivatives to give N-tricyanovinylamines (183).

A weighed 10-mg sample (containing tetracyanoethylene) is dissolved in 100 ml of N,N-dimethylaniline by shaking the mixture at room temperature for a few minutes; then the solution is allowed to stand for 15–20 hr at room temperature to form 4-tricyanovinyl-N,N-dimethylaniline. A 5-ml portion is diluted to 100 ml with acetone. An absorption maximum near 515 nm indicates the presence of tetracyanoethylene; calibration plots can be prepared to convert the absorbance to concentration of the nitrile.

Tetracyanoethylene and many of its derivatives form colored complexes with a large number of aromatic hydrocarbons and olefins. The color complexes are of a Lewis acid-base type with partial transfer of a $\pi$ electron from the aromatic hydrocarbon to the tetracyanoethylene (the acid). (See Section V.B.2.b.) The naphthalene–tetracyanoethylene complex is red, but it is not as tightly bound as an anthracene–tetracyanoethylene complex. Thus it is possible to estimate anthracene colorimetrically by a reaction based on the destruction of the red color of the naphthalene–tetracyanoethylene complex after anthracene and tetracyanoethylene have combined via a Diels-Alder reaction (235). (See Section VI.C.4.e for a possible application of this color reaction.)

### b. α-AMINONITRILES

Compounds such as aminoacetonitrile, α-aminopropionitrile, α-aminobutyronitrile, phenylaminoacetonitrile, methyleneaminoacetonitrile, and iminodiacetonitrile can be determined by the following procedure (269):

Mix 0.5 ml of 0.001 M α-aminonitrile with 0.5 ml of 10% aqueous trichloroacetic acid. Add 0.25 ml of a saturated solution of bromine water, and allow to stand for 1 hr at 37°C. Cool, and add 0.25 ml of a solution of 2 grams of arsenic trioxide in 100 ml of 0.1 N sodium hydroxide; add 3.5 ml of benzidine–pyridine reagent, and allow to stand for 15 min before determining absorbance at 530 nm. Calibration standards are prepared from 0.0001 N potassium cyanide. Aminonitriles can be determined in the presence of cyanide ion because they react with aqueous bromine in 10 sec, whereas the nitriles need about 1 hr. (Also see the following section on α-hydroxynitriles.)

The benzidine–pyridine reagent is prepared by mixing 1 vol of a solution containing 1 gram of benzidine dissolved in 13 ml of N hydrochloric acid and diluted to 30 ml (H$_2$O) with 3 vols. of a solution made by pouring a mixture of 80 ml of water and 20 ml of concentrated hydrochloric acid into 120 ml of pyridine.

### c. α-HYDROXYNITRILES

The procedure given above for α-aminonitriles is also applicable to α-hydroxynitriles (cyanohydrins) such as glycolonitrile and lactonitrile. Both of the α-substituted nitriles liberate cyanide ions under alkaline conditions; in principle, therefore, any sensitive colorimetric procedure for quantitative determination of cyanide ion can be used as an indicator of small amounts of these nitriles (3,80,115,116,121). Cyanide (or thiocyanate ion) can be readily converted to cyanogen bromide by reaction with bromine water in acidic or neutral solution; for purposes of analysis, cyanogen bromide is allowed to react with pyridine to form glutaconic aldehyde and then with benzidine to form a red condensation product (3). If 3-methyl-1-phenyl-2-pyrazolin-5-one is used in place of benzidine, a blue dye is formed (80). The above procedures, with minor modifications, can be used to determine nitriles in blood, plasma, serum, and other biological materials (37,137,150).

### d. ACRYLONITRILE

Kanai and Hashimoto (150) have found that ultraviolet light causes bromine in acidic solutions to react with acrylonitrile quantitatively to form cyanogen bromide, which can then be allowed to react with benzidine-pyridine reagent to form a pink color in procedures similar to those given above for α-aminonitriles and α-hydroxynitriles. Methods based on this finding have been developed for determining the acrylonitrile contents of air, blood, urine, and other samples of biological origin (150). The following procedure is applicable to samples containing 1–5 $\mu$g of acrylonitrile/ml.

Acidify 1 ml of the sample with 1 ml of 10% trichloroacetic acid solution, and then add 0.5 ml of bromine water (1 ml of saturated bromine water diluted with 9 ml of water). Allow the mixture to stand for 20 min under a 20-W fluorescent lamp. Remove excess bromine by adding 0.5 ml of a 2% aqueous solution of arsenic trioxide, and aspirate air over the solution. Add 2.5 ml of benzidine-pyridine reagent (see the procedure for $\alpha$-aminonitriles), and allow to stand for 20 min. Determine the absorbance of the pink color at 532 nm; the absorbance will be linear over the range of 1–5 $\mu$g of acrylonitrile/ml of sample. Since the determination is heavily influenced by the concentration of bromine, the same solution of dilute bromine water should be used to establish a standard curve for every set of measurements.

e. VINYLIDENE CYANIDE

The reaction of vinylidene cyanide with anthracene is quantitative under proper conditions:

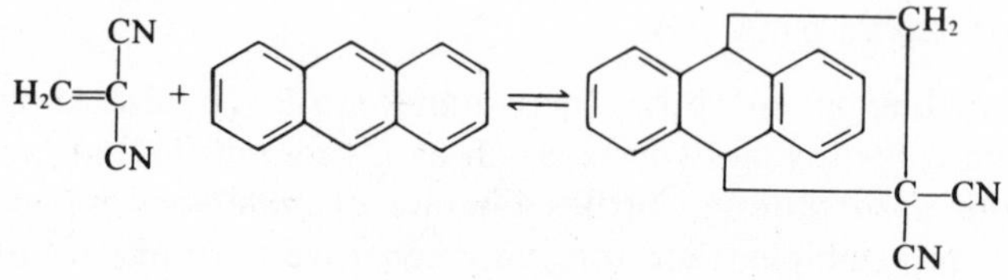

Vinylidene cyanide functions as a dienophile in Diels-Alder reactions; accordingly, it is anticipated that the above reaction can be applied to other dienophiles which react quantitatively with anthracene or with other dienes. Procedures for determination of vinylidene cyanide are based on photometric methods for determination of unreacted anthracene at 360 nm (283). Another method for determining anthracene was noted in Section VI.C.4.a.

## D.  POLAROGRAPHIC METHODS

Available data suggest that nitrile compounds can be reduced electrochemically if proper conditions can be found, that is, the right combination of solvent and electrode system. Nevertheless, the nitrile group is seldom reduced, and so acetonitrile is a well-established solvent and supporting electrolyte for polarography and for electrochemical studies at the most negative potentials that are made available by the use of tetrasubstituted ammonium salts; but in spite of its remarkable resistance, however, acetonitrile can be reduced to ethylamine in aqueous media with electrodes other than mercury (e.g., catalytic hydrogenation).

As a rule, polarographic analytical procedures for aliphatic nitriles take advantage of the presence of other functional groups; for example, acrylonitrile is polarographically active because of conjugation of the C-C double bond with the nitrile group. Also, when combinations of other functional groups sensitize the nitrile compound or the nitrile group, polarographic reduction may take place.

### 1. Aromatic Nitriles

a. AQUEOUS MEDIA

Aromatic nitriles generally are not polarographically active, but there are exceptions; for example, 2- and 4-cyanopyridine and $o$-phthalonitrile are reducible, presumably because the nitrile group is strongly conjugated with nuclear double-bond systems. The aliphatic benzonitrile and phenylacetonitrile are not reduced.

The polarographic behavior of the aromatic nitriles has only recently been clarified sufficiently to permit the drawing of valid conclusions. For example, the work of Manousek and Zuman (179) has established that $p$-acylbenzonitriles ($p$-RCOC$_6$H$_4$CN, where R is H or CH$_3$) form the corresponding benzylamines when reduced at mercury cathodes in 0.1 $M$ sulfuric acid; reduction occurs at $-0.70$ V versus SCE in one well-defined wave. Protonation first occurs at the nitrogen atom of the nitrile group:

$$\overset{\displaystyle O}{\underset{\displaystyle \|}{R-C}}-\phi CN + H^+ \rightarrow \overset{\displaystyle O}{\underset{\displaystyle \|}{R-C}}-\phi \overset{+}{C}NH \xrightarrow[4H^+]{4e} RC-\phi CH_2NH_2$$

Accordingly, the wave height is a function of the acidity of the suspension medium, and at pH values greater than about 1.5 there is observed a decrease in wave height. In contrast, with type XC$_6$H$_4$CN, where X is $p$-COO$^-$, $p$-SO$_2$NH$_2$, or $m$-SO$_2$NH$_2$, free cyanide ion is formed in neutral solutions during controlled-potential electrolysis:

$$XC_6H_4CN + 2\,e + H^+ \rightarrow XC_6H_5 + CN^-$$

Laviron (170) found that 2- and 4-cyanopyridines in acidic suspension media are reduced irreversibly at the dropping mercury electrode in reactions involving four electrons; irreversible reductions in basic media involve two electrons. Similar observations were made by Volke et al. (301), and in later work Volke and Holubek (300) confirmed the formation of cyanide ion from 4-cyanopyridine in a two-electron reduction process at pH 10–11.

A typical procedure for the polarographic analysis of an aromatic nitrile has been described by Jarvie et al. (144). Aliquots of $0.01\ M$ solutions of pure 2-cyanopyridine in ethanol were added to 50 ml of aqueous $KH_2PO_4$ and diluted to 100 ml with ethanol; the pH of the solution was adjusted to values between 5.65 and 7.4 by additions of concentrated sodium hydroxide solution. Polarograms were recorded in the usual fashion at 25°C versus SCE. Typically, the polarograms showed a single, fairly well-defined wave with a half-wave potential and wave height dependent on the pH of the solution (more negative half-wave potential versus SCE and lower wave height at higher pH values). At pH 5.65, the half-wave potential was found to be $-1.35\ V$ versus SCE, and at pH 7.4, $-1.42\ V$; the optimum pH for analysis is stated to lie between 5 and 6 at concentration levels between 2 and $12 \times 10^{-4}\ M$, but accurate assays of 2-cyanopyridine were made at pH 5.65 as well as at 7.40 (±5 parts per thousand).

b. NONAQUEOUS MEDIA

The polarographic behavior of aromatic nitriles in nonaqueous suspension media is markedly different from that in aqueous media. For example, benzonitrile in a suspension medium composed of N,N-dimethylformamide 0.05 M in tetrabutylammonium iodide is reduced at the dropping mercury electrode in a one-electron step which is independent of small amounts of proton donor (water). No dimers or polymers were found to be formed by reduction, but since 2,3-dihydrobenzonitrile was isolated, the following reduction scheme is suggested (246):

$$2C_6H_5CN + 2e \longrightarrow 2[C_6H_5CN]^- \overset{2H^+}{\rightleftharpoons} C_6H_5CN + \langle\!\!\!\bigcirc\!\!\!\rangle\!\!-CN$$

An extensive study of the polarographic behavior of aromatic nitriles in N,N-dimethylformamide was made by Rieger et al. (226). Direct-current polarographic, cyclic voltammetric, and esr techniques were used in a particular effort to examine the parameters for electrolytic generation of nitrile radicals. Benzonitrile was found to be reversibly reduced in a first wave to the radical anion (red-orange solution), but cyclic voltammetry did not indicate that the process was complicated by chemical kinetic factors. When a second polarographic wave is observed, it is considered to represent further reduction of the radical anion to the dianion. The following equations represent the reduction of phthalonitrile:

$$C_6H_4(CN)_2 + 1e^- \rightleftharpoons [C_6H_4(CN)_2]^- \qquad E_{1/2} = -2.12 \text{ V}$$

$$[C_6H_4(CN)_2]^- + 1e^- \rightleftharpoons [C_6H_4(CN)_2]^{2-} \qquad E_{1/2} = -2.76 \text{ V}$$

$$C_6H_5CN + S^- + CN^- \xleftarrow{\overset{\displaystyle |\text{SH}}{\text{(solvent)}}}$$

$$C_6H_5CN + 1e^- \rightleftharpoons [C_6H_5]^- \qquad E_{1/2} = -2.74 \text{ V}$$

The polarographic reduction of 4-aminobenzonitrile is assumed to proceed as follows:

$$H_2NC_6H_4CN + e^- \longrightarrow [H_2NC_6H_4CN]^- \longrightarrow [C_6H_4CN] + NH_2^-$$

$$2[C_6H_4CN] \longrightarrow NC-C_6H_4-C_6H_4-CN$$

$$NC-C_6H_4-C_6H_4-CN + e^- \longrightarrow [NC-\underset{}{\bigcirc}-\underset{}{\bigcirc}-CN]^-$$

Table III.

Polarographic Data for Reduction of Nitriles in Nonaqueous Media [$N,N$-Dimethylformamide; half-wave potentials vs. Ag-AgClO$_4$] Data taken from Rieger et al. (226).

| Compound | First wave | | Second wave | |
|---|---|---|---|---|
| | $-E_{1/2}$ | Est. $n$ | $-E_{1/2}$ | Est. $n$ |
| Tetracyanoethylene | 0.17[a,b] | — | 1.17[a,b] | — |
| Tetracyanoquinodimethane | 0.19[a,b] | — | 0.75[a,b] | — |
| 3,5-Dinitrobenzonitrile | 0.96 | 1 | 1.5[c,d] | 1 |
| Pyromellitonitrile | 1.02 | 1 | 2.07 | 1 |
| 4-Nitrobenzonitrile | 1.25 | 1 | 1.9[c] | 2 |
| Benzoyl cyanide | 1.45[a,c] | — | — | — |
| 4-Cyanobenzoic acid | 1.91 | 1 | 2.53 | 1 |
| Terephthalonitrile | 1.97 | 1 | 2.64 | 1[e] |
| 4-Cyanopyridine | 2.03 | 1 | 2.87 | 1 |
| Benzoylacetonitrile | 2.09[a] | — | — | — |
| Phthalonitrile | 2.12 | 1 | 2.76 | 2 |
| Isophthalonitrile | 2.17 | 1 | — | — |
| 4-Chlorobenzonitrile | 2.4[a,c] | — | — | — |
| 4-Fluorobenzonitrile | 2.69 | 1.5 | — | — |
| Benzonitrile | 2.74 | 1 | — | — |
| 4-Toluonitrile | 2.75 | 1 | — | — |
| 4-Anisonitrile | 2.95[a] | — | — | — |
| 4-Aminobenzonitrile | 3.12 | 1.5 | — | — |

[a] Possibly inconsistent (reference electrode drift).
[b] Run in acetonitrile.
[c] Wave obscured by maximum.
[d] Third one-electron wave at −2.75 V.
[e] Two waves poorly separated.

Most of the compounds studied (see Table III) provided evidence of the formation of a radical ion, and compounds showing a second reduction wave displayed a complex behavior, depending on the fate of the dianion.

### 2. Aliphatic Nitriles

#### a. AQUEOUS MEDIA

Aliphatic nitriles without conjugation are polarographically inactive. Unsaturated aliphatic nitriles such as acrylonitrile are polarographically active; in nearly every instance, a double bond is reduced rather than the nitrile group. Bobrova and Matveeva (31) have studied the polarographic reduction of a group of unsaturated aliphatic nitriles and have confirmed that reduction always occurs at the double bond. In fact, the polarographic reduction of 2,2′-azobisisobutyronitrile occurs because of the presence of the azo group.

Bird and Hale (30) were among the first to study the polarographic behavior of acrylonitrile and to devise methods for its assay, as well as its determination in water, butadiene, or air. A half-wave potential of about −2.05 V (vs. SCE) was observed in an aqueous supporting electrolyte 0.02 M in tetramethylammonium iodide. Platnova (217) also has reported that the half-wave potential in ethanol-water mixtures is −2.05 V and that the irreversible reduction process involves two electrons. Over a concentration range of 0.003–0.13 $M$, the relationship between the diffusion current and the concentration of acrylonitrile was shown to be a straight line. Propionitrile, $\beta$-chloropropionitrile, cyanoacetamide, and ethylene cyanohydrin were not reducible, but lactonitrile showed a reduction wave at the same potential as acrylonitrile. Oxygen does not interfere.

Spillane (259) described a similar method for analysis of methacrylonitrile in aqueous 0.1 $M$ tetramethylammonium bromide as a suspension medium. The half-wave potential was found to be −2.07 V versus SCE; the diffusion current was proportional to concentration, but it showed some time dependence (after a few hours). Isobutyronitrile, oxygen, and $\alpha,\beta$-unsaturated aldehydes do not interfere; however, acrylonitrile cannot be differentiated from methacrylonitrile.

A minor variation of the above methods was developed by Strause and Dyer (264) for analysis of acrylonitrile in mixtures that also contain potassium persulfate, cyanides, formaldehyde, and other products from the persulfate-initiated attack of oxygen on acrylonitrile. Successful application of the method is made possible by addition of hydroquinone and a special calibration technique which compensates for reduction of potassium. Daues and Hammer (69) applied the general polarographic methods discussed above to the determination of as little as 0.1 ppm of acrylonitrile in industrial effluents; the acrylonitrile is concentrated and separated from other components in the effluent by an azeotropic distillation with methanol.

b. Nonaqueous Media

A nonaqueous supporting electrolyte is used in a method for the polarographic determination of residual acrylonitrile in styrene–acrylonitrile copolymers as developed by Claver and Murphy (53). The supporting electrolyte, 0.1 M tetrabutylammonium iodide in $N,N$-dimethylformamide, is an excellent solvent for the copolymer and thus eliminates the need for extraction of acrylonitrile from an insoluble polymer matrix; moreover, since the solution tolerates small amounts of water, electrical conductivity can be increased to a convenient level. In a supporting electrolyte containing 5% water, acrylonitrile is reduced in a single, well-defined wave at a half-wave potential of $-1.63$ V versus SCE. No maxima are observed, but some distortion occurs at acrylonitrile concentrations greater than 500 ppm. The diffusion current is proportional to concentration up to 200 ppm of acrylonitrile (in the cell) in spite of the increased viscosity of the supporting medium (because of polymer). The relative accuracy of the method is stated to be $\pm 3.6\%$, and the lower limit of detection is about 30 ppm. Crompton and Buckley (62) modified the procedure so as to make possible detection of as little as 2 ppm of acrylonitrile.

A summary of other work on the polarography of aliphatic nitriles in nonaqueous media is given in Section III.B.1.a(4).

### 3. Isocyanides

The behavior of isocyanides at the dropping mercury electrode has not been studied in sufficient detail. Cappellina and Lorenzelli (45) have performed some experiments with isocyanides, but their results are inconclusive.

### E. DUMAS AND KJELDAHL METHODS

The nitrogen content of a nitrile or isocyanide is best determined by the Dumas method. Very volatile compounds and ring-nitrogen compounds will prove to be more troublesome, but an experienced analyst has at his or her disposal techniques that can overcome such problems. The cobalto-cobaltic oxide packing described in Section VII.E of "Nitrate and Nitrite Ester Groups" is recommended for analysis of nitriles and isocyanides by the micro Dumas combustion procedure. For nitriles that are difficult to burn, Chumachenko (52) and Abramyan et al. (1) recommend that the Dumas-Pregl combustion tube be packed with nickel oxide (or a mixture of nickel oxide and copper oxides), and operated at 900–950°C; samples are loaded into a quartz tube together with NiO or a mixture of NiO and CuO. Nickel packing is especially useful for fluorine compounds because nickel fluoride is very stable.

The nitrogen contents of substituted cyanamides and related compounds also may be determined by a micro Dumas combustion procedure, but Kjeldahl methods often are adequate.

## 1.  Nitriles and Isocyanides

The nitrogen contents of nitriles are very frequently determined by Kjeldahl methods, but results are not always satisfactory, especially for complex and unsaturated aliphatic nitriles; presumably, the nitriles are converted to resistant nitrogen-containing ring structures. Of course, sealed tube digestions of the type described by White and Long (313), Baker (14), Guillemard (119), or Belcher et al. (19) are satisfactory for nitriles, but they are inconvenient.

Rose and Zilioto (228) found that prehydrolysis with 80% sulfuric acid followed by a Kjeldahl digestion gave quantitative yields with saturated, but not with unsaturated, aliphatic nitriles. However, all types of aliphatic as well as aromatic nitriles yielded quantitative results when treatment with hydriodic acid in the digestion flask was adopted as a preliminary step to a standard Kjeldahl-Gunning-Arnold nitrogen procedure, which includes selenium and copper sulfate. [The strong reducing action of hydriodic acid in sealed tubes has been observed by Friedrich et al. (100) and by Aldrovandi and De Lorenzi (4).] A detailed procedure for Kjeldahl determination of nitrogen in nitriles including hydriodic acid pretreatment, is given in Section VII.A.

The standard salicyclic acid–thiosulfate Kjeldahl procedure often provides accurate results for the nitrogen contents of nitriles; the procedure given in Section VII.B of "Azo Group," Vol. 15, also may be used for analysis of nitriles.

Burleigh et al. (42) modified the semimicromethod of Cole and Parks (56) so that it could be used with large samples of elastomeric polymers such as butadiene–vinylpyridine copolymers, butadiene–vinylpyridine–carbon black master batches, and vulcanizates of these polymers. Following is a brief description of the procedure:

Transfer a sample corresponding to about 0.5 mequiv. of nitrogen into a 500-ml Kjeldahl flask. Add 5 grams of a 30 : 1 : 2 mixture (by weight) of anhydrous potassium sulfate, selenium, and mercuric oxide; then add 25 ml of concentrated sulfuric acid, and digest at full heat of a 550-W electric heater for 6 hrs. Cool the flask and its contents; cautiously add 250 ml of distilled water, and again cool the contents of the flask. Carefully add 65 ml of 48% sodium hydroxide so that the heavier alkali solution forms a layer on the bottom of the flask. Add 20 ml of 44% sodium thiosulfate pentahydrate solution in such a way that it too does not mix with the acidic supernatant solution. Add a piece of mossy zinc to prevent bumping; distil about half of the solution in the Kjeldahl flask into 50 ml of 4% boric acid solution. Complete the determination in the usual fashion.

## 2. Cyanamide and Dicyandiamide

Kjeldahl methods are often used subsequent to silver precipitation of the salts of cyanamide and dicyandiamide. For example, Caro (46) applied Kjeldahl procedures to the determination of cyanamide and dicyanamide; silver cyanamide was separated first (see Section VI.A.7.e) by adding silver acetate to an ammoniacal solution of the sample and then separating silver dicyandiamide from the filtrate by adding potassium hydroxide, expelling ammonia at the boiling point, and filtering. A Kjeldahl determination for nitrogen was performed on both precipitates. Marqueyrol and Desvergnes (180) considered Caro's method to be accurate; however, Grube and Kruger (112) found that dicyandiamide coprecipitated with silver cyanamide, and they recommended that Caro's procedure be modified so that ammoniacal silver acetate is added to a sample solution acidified with acetic acid. Hager and Kern (120) also noted the problems caused by coprecipitation; in addition they reported that, when silver dicyandiamide precipitate is boiled in alkaline solution, there is loss of nitrogen as ammonia. More importantly, Kappen (152) found that the presence of urea in samples causes errors as high as 50% in the dicyandiamide content. However, Hene and Haaren (128) noted that initial precipitation of the silver salts of cyanamide and dicyandiamide from potassium hydroxide solution eliminated most of the urea; the combined precipitates were dissolved in nitric acid, and the Kjeldahl nitrogen determinations on precipitated silver cyanamide (from ammoniacal media) and silver dicyandiamide (from caustic alkali solution) were found to be more representative of the sample composition. Pinck (216) also claimed that, since silver cyanamide occludes dicyandiamide, reprecipitation is necessary if accurate results are to be obtained when large amounts of dicyandiamide are present (see Section VI.A.7.e).

Brioux (33) precipitated both cyanamide and dicyandiamide as their silver salts from a caustic alkali solution and determined total nitrogen; the cyanamide content of the sample was determined as described in Section VI.A.7.e on a separate sample.

## F. GASOMETRIC PROCEDURES

### 1. Nitriles and Isocyanides

Reaction of methylmagnesium iodide with nitriles or isocyanides has been recommended by Niederl and Niederl (201) as the basis of an indirect gasometric method. In accordance with Soltys (258) and others, a sample is treated with a known amount of the Grignard reagent; after reaction with the functional group has occurred and evolution of methane from reaction of the Grignard reagent with active hydrogen is complete, the unused Grignard reagent is treated with an excess of aniline and the incremental volume of methane is measured. Thus, by difference, groups which react with methylmagnesium iodide *without* evolution of methane can be determined by the amount of Grignard reagent consumed.

The following functional groups yield 1 or more mol of methane: —OH, —SH, —COOH, —SO$_2$OH, —CONH$_2$, —CONHR, —SO$_2$NH$_2$, —SO$_2$NHR, —NH$_2$, =NH, and —NO$_2$; 1 mol of water yields 2 mol of methane. On the other hand, 1 mol of CH$_3$MgI is consumed by 1 mol of each of the following, without evolution of methane: —CHO, =CO, —CH$_2$X, nitrile, and isocyanide. Two moles of CH$_3$MgI is consumed by —COX and —COOR without methane being evolved.

In view of the wide variety of functional groups that can react with the Grignard reagent, a sample of nitrile or isonitrile must be dry and quite pure before it can be effectively analyzed by the Grignard gasometric method; it is necessary to exclude oxygen during the determination.

The attack of the isocyanide group by hypobromite ion results in release of nitrogen and carbon dioxide:

$$2\,RNC + 5\,NaOBr + H_2O \rightarrow 2\,ROH + 5\,NaBr + 2\,CO_2 + N_2$$

Guillemard (117) reported that the reaction is satisfactory for determination of isocyanides; the nitrogen is measured in an azotometer charged with potassium hydroxide solution.

Isocyanides also react with cold, concentrated solutions of oxalic acid to release an equimolar mixture of carbon dioxide and carbon monoxide (118) (see Section V.C.1.c).

### 2. Cyanamide

Monnier (194) hydrolyzed cyanamide to urea by evaporating a sample to dryness with formic acid; the urea was determined gasometrically from the nitrogen evolved when the residue was treated with hypobromite.

### 3. Carbodiimides

The reaction of oxalic acid with carbodiimides (release of equimolar amounts of carbon dioxide and carbon monoxide) was known for some time before Zetsche and Fredrich (327) first demonstrated its application for the quantitative determination of carbodiimides. Following is a modification of the procedure of Zetsche and Fredrich:

The reaction vessel consists of a 100-ml round-bottom flask which is equipped with a small dropping funnel (about 25-ml capacity), a gas inlet tube passing through a rubber stopper at the top of the dropping funnel, and a gas exit tube. The gas exit tube is connected by a three-way stopcock to a 50-ml azotometer charged with 50% potassium hydroxide solution. Arrangements are made so that a stream of carbon dioxide at a low pressure can be passed through the flask via the inlet tube. Initially, the flask and the gas exit tube are flushed free of air. Then a solution of 0.2 gram (1 mmol) of the carbodiimide in 10 ml of pure anhydrous dioxane is introduced into the flask, and air again removed by a stream of carbon dioxide (preferably, the exit gas stream should bypass the azotometer, save for an occasional test for nitrogen content). When all air has been removed, the stopcock of the dropping funnel is closed and a warm, freshly prepared solution of 1 gram of anhydrous oxalic acid in about 10 ml of dioxane is transferred to the funnel; air in the funnel is displaced by a gentle stream of carbon dioxide which is allowed to pass over the dioxane solution and leave through a crevice formed by dislodging the stopper at the top of the funnel. When all is in readiness, the three-way stopper is turned so that carbon dioxide can pass into the azotometer, and the oxalic acid solution is run into the carbodiimide solution over a period of a few minutes. Since the generation of gas may be quite rapid, it is necessary to add the oxalic acid solution in small increments and at a rate which ensures that carbon dioxide is absorbed in the azotometer; usually, difficultly soluble ureas precipitate. After the release of gas has subsided, carbon dioxide is allowed to enter the reaction flask and continued until most of the generated gas has been swept into the azotometer. To remove gases trapped by precipitated ureas the flask is heated in a bath at 80–90° and shaken occasionally. Passage of carbon dioxide is continued (about 1 hr) until all carbon monoxide has been transferred to the azotometer. The purity of the carbodiimide is computed from the volume of carbon monoxide collected:

$$RN{=}C{=}NR + (COOH)_2 \rightarrow RNHCONHR + CO_2 + CO$$

The reaction of oxalic acid with monomeric carbodiimides ordinarily is quite rapid, but since the polymeric forms react very slowly, the procedure described above will seldom provide quantitative conversion of these carbodiimide groups; prolonged action (e.g., days) may yield complete conversion, but resinous carbodiimides simply appear to resist attack by oxalic acid.

The carbon dioxide released by carbodiimides on reaction with oxalic acid can be weighed after absorption; the above procedure is suitable for this purpose, provided that an inert gas or air free of carbon dioxide is used to sweep out the reaction vessel. For the gravimetric determination, from 2–3 mmol of carbodiimide should be taken; it is necessary to use a system for collecting carbon dioxide without interference from the dioxane vapors which are also swept out of the reaction flask by the stream of inert gas.

Zarembo and Watt (325) have modified the procedure so that the analysis of the released gases is performed by gas chromatography (helium carrier; 0.25% Apiezon L on 80–100 mesh glass).

## G. COULOMETRIC PROCEDURES

An indirect microcoulometric procedure for the determination of cyanamide has been described by Yamada and Sakai (322); interfering substances are removed by paper chromatography. The ascending technique is used to separate 50–1000 $\mu$g of cyanamide in a volume of 10–30 $\mu$l of test solution; a dilute aqueous solution of ammonia is employed as developer, and then the cyanamide is converted to its silver salt by a spray of ammoniacal silver nitrate. Finally, the silver salt zone is removed and rechromatographed; the new zone is dissolved in nitric acid, and the silver content determined coulometrically. A silver cathode, a platinum anode, and a capillary microcoulometer (sens., $10^{-4}$ C) are used; electrolysis is complete in 0.5–2 hr, and error is less than ±3%. The silver in the silver salt zone may be determined more conveniently by atomic absorption.

## VII. LABORATORY PROCEDURES

The analysis of any variety of nitriles, isocyanides, or cyanamides is usually accomplished by determination of total nitrogen using the Dumas method. However, a large number of compounds perform satisfactorily in modified Kjeldahl determinations, and so this method of analysis is used whenever possible for quality control and in situations where a large number of analyses must be performed. The first procedure given below (Section VII.A) is a modified Kjeldahl method developed specifically for nitriles.

Nearly every compound with a functional group containing the —CN structure shows an absorption maximum in the region 2273–2198 cm$^{-1}$ (4.40–4.85 $\mu$); the procedure described in Section VII.B is a rapid, accurate method for determination of —CN compounds by infrared spectrophotometry.

The third method (Section VII.C) is a widely used procedure for the assay of acrylonitrile (also other $\alpha,\beta$-unsaturated compounds). The method outlined in Section VII.D is a general procedure which can be used for the colorimetric determination of nearly any nitrile, but it is subject to interference from a wide variety of other functional groups and classes of compounds.

## A. DETERMINATION OF NITRILES BY A KJELDAHL PROCEDURE

The nitrogen in the nitrile group of many compounds cannot readily be determined by ordinary Kjeldahl methods. As was noted in Section VI.E.1, a preliminary step is required to reduce the nitrile group to an amine so that the nitrogen content can be converted easily to ammonia. The procedure described below is a modification of the method of Rose and Ziliotto (228): it is based on the preliminary reduction of the nitrile group by hydriodic acid.

### Reagents

*Sulfuric acid, conc.* Use reagent-grade acid (sp. gr., 1.84).
*Potassium sulfate.* Use reagent-grade crystals.
*Copper sulfate, anhydrous.* Use reagent-grade material.
*Selenium.* Use high-quality, finely powdered metal.
*Potassium iodide.* Use reagent-grade material.
*Zinc metal.* Use reagent-grade 20-mesh material.
*Sodium hydroxide solution, 50%.* Dissolve sodium hydroxide in an equal weight of water; use the clear supernatant.
*Standard sulfuric acid, 0.1 N.* Prepare and standardize by a convenient method.
*Boric acid solution, 1.5%.* Dissolve 15 grams of reagent-grade boric acid in 1 liter of warm water; cool.
*Methyl red indicator solution.* Dissolve 0.1 gram of the sodium salt of methyl red in 100 ml of water.

### Procedure

Transfer a weighed sample estimated to contain 40–60 mg of nitrogen to a 500-ml Kjeldahl digestion flask; add about 1.5 grams of potassium iodide and 30 ml of concentrated sulfuric acid. Heat on a steam bath for about 45 min, occasionally swirling the contents to prevent stratification of the digestion mixture. Add 10 grams of potassium sulfate, 0.1 gram of selenium, 0.3 gram of anhydrous copper sulfate, and a few ceramic boiling chips or glass beads. (Selenized chips may be used.) Heat the mixture gently at first; then bring it to a boil, gradually increasing the rate of heating until boiling is brisk. When the digestion mixture becomes clear green, continue a brisk boil for at least 1 hr; adjust the rate of heating so as to maintain a volume of not less than 20 ml (add sulfuric acid if necessary). If iodine has condensed in the neck of the flask, drive it off by gentle application of a flame.

Cool the digestion mixture, cautiously dilute it with 250 ml of water, and cool again. Carefully pour 90–100 ml of 50% sodium hydroxide solution down the side of the neck of the digestion flask so that most of the base forms a layer at the bottom of the flask; then, in rapid succession, add a few granules of zinc metal (to prevent bumping), and connect the flask via a Kjeldahl spray trap to a condenser. With the usual precautions, distil off ammonia into 100 ml of 1.5% boric acid that has been made exactly neutral to methyl red. After 125–150 ml of distillate has been collected, titrate the ammonia in the boric acid solution with standardized 0.1 N sulfuric acid (methyl red end point).

**Calculations**

Compute the nitrogen content by the following formula:

$$\% \, N = \frac{\text{ml of acid} \times N \times 1.401}{\text{grams of sample}}$$

## B. DETERMINATION OF A NITRILE COMPOUND BY INFRARED SPECTROPHOTOMETRY

It was indicated in Section V.B.2.a that the nitrile group and other functional groups that include —CN have characteristic absorption maxima in the region 2273–2198 cm$^{-1}$ (4.40–4.55 $\mu$). Infrared analytical procedures are particularly useful for following the hydrolysis of nitriles, or for determining residual nitrile in hydrolysates; the following general procedure can be modified to suit the peculiarities of a sample or to increase the sensitivity of detection of the nitrile group by use of a thicker cell. The method cannot be used with samples that have absorption peaks which interfere with the nitrile peak.

**Reagents**

*Carbon tetrachloride or chloroform.* Reagent-grade material is usually satisfactory, but specially purified solvents (commercially available) provide assured and reproducible performance.

*Pure nitrile.* The nitrile compound used for calibration should be of the highest purity available.

## Special Apparatus

An infrared spectrophotometer that has a resolution of the order of 5–10 cm$^{-1}$ (0.01 $\mu$) at about 2250 cm$^{-1}$ (4.44 $\mu$) is required. A number of matched pairs of sealed liquid cells (sodium chloride) about 0.100 mm in thickness should be available.

## Calibration Solutions

Prepare a standard solution of the nitrile by transferring 20.00 grams to a 100-ml, glass-stoppered volumetric flask, dissolving, and then diluting to the mark with the same solvent as will be used for the sample (preferably carbon tetrachloride). Transfer the solution to a 10-ml buret.

Obtain eight dry, 10-ml, glass-stoppered volumetric flasks, label each as indicated below, and then deliver the requisite volume of the nitrile standard solution from the 10-ml buret; finally dilute each flask to the mark with solvent and mix well.

| Nitrile (%) | Standard Nitrile Solution (ml) |
|---|---|
| 0 | 0.00 |
| 0.5 | 0.05 |
| 1 | 0.10 |
| 5 | 0.50 |
| 10 | 1.00 |
| 20 | 2.00 |
| 40 | 4.00 |
| 60 | 6.00 |
| 80 | 8.00 |

## Procedure

Fill one of a pair of matched sealed liquid cells with pure solvent; this cell becomes the "reference" cell. In consecutive order, load the other cell with the calibrant solutions and obtain for each a partial infrared spectrum over the interval between approximately 2400 and 2000 cm$^{-1}$ (4.17 and 5.00 $\mu$). Draw a baseline between fixed points of minimum absorption on both sides of the characteristic nitrile absorption, for example, near 2325 and 2175 cm$^{-1}$ (4.3 and 4.6 $\mu$). From the baseline, measure the net maximum absorbance of the nitrile peak. Plot net maximum absorbance for each of the calibrant solutions versus percent nitrile to obtain a smoothed calibration curve.

Prepare a solution of the sample by dissolving 2.000 grams in the selected solvent and diluting to exactly 10 ml. Fill the sample cell with this solution, and obtain the nitrile absorbance over the baseline; read the percent nitrile directly from the calibration curve.

### C. TITRIMETRIC ASSAY OF ACRYLONITRILE BY DODECANETHIOL

The reaction of acrylonitrile with a mercaptan was discussed in Section VI.7.a, where it was noted that some $\alpha,\beta$-unsaturated nitriles, as well as similarly unsaturated compounds, can be determined by reaction with a mercaptan. Acrylonitrile reacts rapidly (about 2 min) and quantitatively in basic solutions containing a moderate excess of the mercaptan (about 50%), but methacrylates, crotonates, and aldehydes require as much as a 150% excess of mercaptan for reaction times of 10–15 min. Ketones in about 300% excess of mercaptan react somewhat incompletely and slowly (20–30 min). The method given below is a modification of the procedure described by Beesing et al. (17).

### Reagents

*Alcoholic dodecanethiol solution.* Dissolve approximately 50 grams of 1-dodecanethiol (laurylmercaptan) in about 1 liter of ethanol or isopropanol (mercaptan at 98% purity is sufficient). **Caution:** stench! The solution of mercaptan is best stored and dispensed from a self-filling buret which is permanently attached to a reservoir.

*Basic catalyst solution.* Dissolve 5 grams of potassium hydroxide in 100 ml of ethanol; alternatively, use commercially available 40% methanolic solutions of *N*-benzyltrimethylammonium hydroxide.

*Standard iodine solution,* 0.2 N. Place about 80 grams of potassium iodide, 25.4 grams of iodine, and 10 ml of water in a glass mortar; triturate with a pestle until solution is complete. Filter the solution through a fine sintered-glass frit, and dilute to 1 liter; store in a glass-stoppered bottle. Standardize by any convenient procedure.

*Nitrogen gas.* Use a purified grade so as to ensure low amounts of oxygen.

*Glacial acetic acid.* Use reagent-grade material.

*Alcohol.* Use reagent-grade 2-propanol or Formula 2-B ethyl alcohol.

## Procedure

Weigh 0.3 gram ($\pm 1$ mg) of the acrylonitrile sample into a 250-ml glass-stoppered flask; use another flask for the blank. Flush air out of each flask with a brisk stream of nitrogen gas. Transfer exactly 40 ml of alcoholic dodecanethiol solution to the sample flask, and mix well by swirling the flask. Add 2 ml of alcoholic potassium hydroxide solution (or about 10 drops of $N$-benzyltrimethylammonium hydroxide solution); immediately stopper the flask and swirl the contents. Allow to stand for 3 min.

Unstopper the flask, and add 4 ml of glacial acetic acid and 150 ml of alcohol. Titrate with 0.2 $N$ standard iodine solution to a faint yellow end point, which persists for 30 sec while the solution is gently agitated by swirling of the flask.

Run a reagent blank, using the same amount of dodecanethiol solution and basic catalyst solution; allow the stoppered flask to stand for 3 min before adding glacial acetic acid and diluting with alcohol (residual air in the flask slowly oxidizes alkaline mercaptan).

## Calculation

Calculate the percent acrylonitrile by the following formula:

$$\% \text{ Acrylonitrile} = \frac{5.306(B - A)N}{W}$$

Where  $A$ = iodine solution required for sample (ml),
       $B$ = iodine solution required to titrate the "blank" (ml)
       $N$ = normality of iodine solution,
       $W$ = weight of sample (grams).

## D. COLORIMETRIC DETERMINATION OF NITRILES AS AMIDOXIMATES

The reaction of the nitrile group with a buffered solution consisting of a mixture of hydroxylammonium chloride and hydroxylamine in propylene glycol was discussed in Section V.B.1.a. The reaction is apparently catalyzed by alkali, but it proceeds slowly (hours) in aqueous media at room temperature. Soloway and Lipschitz (257) found that the reaction rates of many nitriles were increased dramatically in boiling propylene glycol; moreover, ferric hydroxide is not precipitated in hydroxylammonium chloride–hydroxylamine buffered solutions. Unfortunately, interferences are manifold because the reaction medium and experimental conditions used in the test are propitious for the formation of hydroxamic acids from a wide variety of substances, for example, all types of amides, guanidines, ureas, hydrazides, amidines, imides, O-acetyl and N-acetyl aromatics, esters, anhydrides, and acid chlorides; the hydroxamic acids form complexes with ferric iron that are similar in color to those of the amidoximes.

The procedure given below may be used with a wide variety of nitriles, but it does not prescribe a rigidly fixed set of experimental conditions inasmuch as the rates of conversion of nitriles to their amidoximes, and the intensities of coloration of their ferric complexes, vary widely. Accordingly, it is necessary to establish a calibration plot with the particular nitrile being sought and to determine optimum reaction times, as well as appropriate dilutions for photometry.

### Reagents

*Hydroxylammonium chloride solution*, 1 M. Dissolve 70 grams of reagent-grade hydroxylamine hydrochloride in 1 liter of propylene glycol.

*Potassium hydroxide solution*, 1 M. Dissolve 56 grams of reagent-grade potassium hydroxide in 1 liter of propylene glycol; use the clear, supernatant solution.

*Ferric chloride solution*, 5%. Dissolve 5 grams of iron(III) chloride hexahydrate in 100 ml of ethanol.

*Propylene glycol*. Use purified 1,2-propanediol (bp, 186–187°) or U.S.P. grade.

### Procedure

Prepare a calibrant solution of the nitrile to be determined by accurate weighing and dilution with propylene glycol to a fixed volume. If the nitrile is not soluble in propylene glycol, use a readily volatile solvent (such as chloroform or carbon tetrachloride) which can be removed after aliquots have been dispensed into reaction flasks. The calibrant solution preferably should be about 0.05 M (i.e., 50 $\mu$mol/ml) for initial trials.

Transfer to 50-ml Erlenmeyer flasks aliquots of the calibrant corresponding to 0, 20, 40, 60, 80, and 100 $\mu$mol of the nitrile compound. If a solvent other than propylene glycol was used to prepare the calibrant, carefully remove nearly all of the solvent at a low temperature. Add to each flask 4 ml of hydroxylamine hydrochloride solution and 2 ml of potassium hydroxide solution. Rapidly heat the solutions in the flasks to the boiling point, and then maintain just at the boiling point for not less than 2 min. Set aside to cool.

Add 1.0 ml of alcoholic ferric chloride solution to each flask, and swirl to mix; transfer to a 50-ml volumetric flask with propylene glycol, and bring to the mark with propylene glycol. (Centrifuge the solution if it is turbid because of precipitation of potassium chloride or ferric hydroxide.) Using the blank as a reference, determine the absorbances of the calibrant solutions with a filter photometer (500–700 nm) or with a spectrophotometer (1-cm cells) at the wavelength of maximum absorption in the range of 500–570 nm. Plot absorbance versus concentration.

It is necessary to ensure that the reaction time at the boiling point is adequate; some nitriles may require prolonged boiling (maintain volume). Also, since the color intensities of the ferric amidoximate complexes of various nitriles may differ by as much as tenfold, it may be necessary to use more or less concentrated calibrant or to vary the volume of the final dilution. It is also necessary to establish the stability of the colored complex under the selected experimental conditions; ordinarily, the color fades slowly over a period of hours. The addition of a fixed quantity of water in the final dilution often leads to an increased intensity of coloration of the ferric complex; also, the position of the absorption maximum is shifted toward the violet.

For analysis of a sample, a solution of known concentration is prepared and treated according to the optimized procedure by which the calibration plot was established.

## REFERENCES

1. Abramyan, A. A., A. A. Kocharyan, and A. G. Karapetyan, *Izv. Akad. Nauk. Arm. SSR, Khim., Nauki*, **15**, 225 (1962); through *Chem. Abstr.*, **58**, 6192 (1963).

2. Ahrens, F. B., *Z. Elektrochem.*, **5**, 99 (1896).

3. Aldridge, W. N., *Analyst*, **69**, 262 (1944); **70**, 474 (1945).

4. Aldrovandi, R., and F. DeLorenzi, *Ann. Chim. (Rome)*, **42**, 298 (1952); through *Chem. Abstr.*, **46**, 10051 (1952).

5. Aminov, S. N., and A. B. Terent'ev, *Uzb. Khim. Zh.*, **1968**, 35; through *Chem. Abstr.*, **69**, 37312 (1968).

6. Amundsen, L. H., and L. S. Nelson, *J. Am. Chem. Soc.*, **73**, 242 (1951).

7. Arad-Talmi, Y., M. Levy, and D. Vofsi, *J. Chromatogr.*, **10**, 417 (1963).

8. Ascik, K., *Polimery (Poland)*, 8(2), 66 (1963); through *Chem. Abstr.*, **59**, 12076 (1963).

9. Aumuller, W., *Angew. Chem.*, **75**, 857 (1963).

10. Baizer, M. M., *J. Electrochem. Soc.*, **111**, 215 (1964).

11. Baizer, M. M., *Tetrahedron Lett.*, **15**, 973 (1963).

12. Baizer, M. M., and J. D. Anderson, *J. Electrochem. Soc.*, **111**, 223 (1964).

13. Baizer, M. M., and J. D. Anderson, *J. Electrochem. Soc.*, **111**, 226 (1964).

14. Baker, P. R. W., *Analyst*, **80**, 481 (1955).

15. Baldwin, S., *J. Org. Chem.*, **26**, 3288 (1961).

16. Bargain, M., *Compt. Rend. Congr. Natl. Soc. Savantes, Sect. Sci.*, **88**(1), 139 (1963); through *Chem. Abstr.*, **64**, 17041 (1966).

17. Beesing, D. W., W. P. Tyler, D. M. Kurtz, and S. A. Harrison, *Anal. Chem.*, **21**, 1073 (1949).

18. Behringer, H., and H. Meier, *Annalen.*, **607**, 67 (1957).

19. Belcher, R., T. S. West, and M. Williams, *J. Chem. Soc.*, **1957**, 4323.

20. Belgian Pat. 647,423 (Oct. 30, 1964), W. T. Somerville and E. O. Shuster; through *Chem. Abstr.*, **63**, 12968 (1965).

21. Bell, F. K., *J. Am. Chem. Soc.*, **57**, 1023 (1935).

22. Bellamy, L. J., *The Infrared Spectra of Complex Molecules*, 2nd ed., Methuen, New York, 1958, p. 263.

23. Bergmann, F., *Anal. Chem.*, **24**, 1367 (1952).

24. Berinzaghi, B., *Anal. Assoc. Quim. Argent.*, **44**, 120 (1956); through *Chem. Abstr.*, **51**, 1777 (1957).

25. Berther, C., K. Kreis, and O. Bochmann, *Z. Anal. Chem.*, **169**, 184 (1959).

26. Besnainu, S., B. Thomas, and S. Bratoz, *J. Mol. Spectrosc.*, **21**, 113 (1966).

27. Beynon, J. H., *Mass Spectrometry and Its Applications to Organic Chemistry*, Elsevier, New York, 1960, p. 382.

28. Bielecke, J., and V. Henri, *Compt. Rend.*, **156**, 1860 (1913).

29. Bieling, H., and W. Laabs, *Z. Anal. Chem.*, **215**, 35 (1965).

30. Bird, W. L., and C. H. Hale, *Anal. Chem.*, **24**, 586 (1952).

31. Bobrova, M. I., and A. N. Matveeva, *J. Gen. Chem. USSR* (Eng. trans.), **27**, 1219 (1957).

32. Bowie, J. H., R. Grigg, S.-O. Lawson, P. Madsen, G. Schroll, and D. H. Williams, *J. Am. Chem. Soc.*, **88**(8), 1699 (1966).

33. Brioux, C., *Ann. Chim. Anal.*, **15**, 341 (1910); through *Chem. Abstr.*, **4**, 2025 (1910).

34. British Pat. 855,027 (Nov. 30, 1960), D. Szabo; through *Chem. Abstr.*, **55**, 17501 (1961).

35. British Pat. 951,770 (Mar. 11, 1964); through *Chem. Abstr.*, **63**, 6923 (1965).

36. Brown, W. E., in R. Adams, Ed., *Organic Reactions*, Vol. VI, John Wiley, New York, 1951, Ch. 10.

37. Bruce, R. B., J. W. Howard, and R. F. Hanzal, *Anal. Chem.*, **27**, 1346 (1955).

38. Brückner, H., F. Friedrich, D. Göckeritz, M. Schüssler, and R. Pobloudek-Fabini, *Pharm. Zentralhalle*, **100**, 452 (1961); through *Anal. Abstr.*, **9**, 1545 (1962).

39. Buchanan, G. H., *Ind. Eng. Chem.*, **15**, 637 (1923).

40. Buckles, R. E., and C. J. Thelen, *Anal. Chem.*, **22**, 676 (1950).

41. Burckhalter, J. H., E. M. Jones, W. F. Holcomb, and L. A. Sweet, *J. Am. Chem. Soc.*, **65**, 2012 (1943).

42. Burleigh, J. E., O. F. McKinney, and M. G. Barker, *Anal. Chem.*, **31**, 1684 (1959).

43. Buyske, D. A., and V. Downing, *Anal. Chem.*, **32**, 1798 (1960).

44. Capitani, C., and G. Gambelli, *Chim. Ind. (Milan)*, **35**, 890 (1954); through *Chem. Abstr.*, **48**, 7495 (1954).

45. Cappellina, F., and I. Lorenzelli, *Ann. Chim. (Rome)*, **48**, 855 (1958).

46. Caro, N., *Z. Angew. Chem.*, **23**, 2405 (1910).

47. Carothers, W. H., and G. A. Jones, *J. Am. Chem. Soc.*, **47**, 3051 (1925).

48. Casanova, J., R. E. Schuster, and N. D. Werner, *J. Chem. Soc.*, **1963**, 4280.

49. Casanova, J., N. D. Werner, and R. E. Schuster, *J. Org. Chem.*, **31**, 3473 (1966).

50. *Catalog of Mass Spectral Data*, American Petroleum Institute Research Project 44, Carnegie Institute of Technology, Pittsburgh, Pa.

51. Cheronis, N. D., J. B. Entrikin, and E. M. Hodnett, *Semimicro Qualitative Organic Analysis*, 3rd ed., Wiley-Interscience, New York, 1965.

52. Chumachenko, M. N., *Izv. Akad. Nauk SSR, Ser. Khim.*, **1963**, 1893; through *Chem. Abstr.*, **60**, 7462 (1964).

53. Claver, G. C., and M. E. Murphy, *Anal. Chem.*, **31**, 1682 (1959).

54. Coblentz, W. W., *Investigations of Infrared Spectra*, Carnegie Inst. Publ. 35, 1905; abstracted in *Phys. Rev.*, **20**, 273 (1905).

55. Coetzee, J. F., G. P. Cunningham, D. K. McGuire, and G. R. Padmanabhan, *Anal. Chem.*, **34**, 1139 (1962).

56. Cole, J. O., and C. R. Parks, *Ind. Eng. Chem., Anal. Ed.*, **18**, 61 (1946).

57. Condo, F. E., E. T. Hinkel, A. Fassero, and R. L. Shriner, *J. Am. Chem. Soc.*, **59**, 230 (1937).

58. Cook, P. O., *J. Org. Chem.*, **27**, 3873 (1962).

59. Crabtree, E. V., E. J. Poziomek, and D. J. Hoy, *Talanta*, **14**, 857 (1967).

60. Critchfield, F. E., G. L. Funk, and J. B. Johnson, *Anal. Chem.*, **28**, 76 (1956).

61. Critchfield, F. E., and J. B. Johnson, *Anal. Chem.*, **28**, 73 (1956).

62. Crompton, T. R., and D. Buckley, *Analyst*, **90**, 76 (1965).

63. Cross, L. H., and A. C. Rolfe, *Trans. Faraday. Soc.*, **47**, 354 (1951).

64. Curran, D. J., and S. Siggia, in Z. Rappoport, Ed., *The Chemistry of the Cyano Group*, Wiley-Interscience, New York, 1970, Ch. 4, pp. 167–207.

65. Cutler, J. A., *J. Chem. Phys.*, **16**, 136 (1948).

66. Cutter, H. B., and M. Taras, *Ind. Eng. Chem., Anal. Ed.*, **13**, 830 (1941).

67. von Dafert, F. W., and R. Miklauz, *Z. Landw. Vers.-Wesen Deutschoesterr.*, **22**, 1 (1919); through *Chem. Abstr.*, **14**, 586 (1920).

68. Dal Nogare, S., L. R. Perkins, and A. H. Hale, *Anal. Chem.*, **24**, 512 (1952).

69. Daues, G. W., and W. F. Hamner, *Anal. Chem.*, **29**, 1035 (1957).

70. Davidson, D., *J. Chem. Educ.*, **19**, 154, 221, 532 (1942).

71. Davidson, D., and H. Skovronek, *J. Am. Chem.*, **80**, 376 (1958).

72. Dehn, W. M., and K. E. Jackson, *J. Am. Chem. Soc.*, **55**, 4285 (1933).

73. Dhar, D. N., *Chem. Rev.*, **67**, 611 (1967).

74. Dibeler, V. H., R. M. Reese, and J. L. Franklin, *J. Am. Chem. Soc.*, **83**, 1813 (1961).

75. DiLorenzo, A., and G. Russo, *J. Gas Chromatogr.*, **6**, 509 (1968).

76. Dinsmore, H. L., and D. C. Smith, *Anal. Chem.*, **20**, 11 (1948).

77. Doerffel, K., *Wiss. Z. Tech. Hochsch. Chem. Leuna-Merseberg*, **7**, **71**, (1965); through *Chem. Abstr.*, **64**, 814 (1966).

78. Eiichi, I., and O. Keijiro, *Yuki Gosei Kagaku Kyokoi Shi*, **19**, 716 (1961); through *Chem. Abstr.*, **56**, 1337 (1962).

79. Eisner, H. E., T. Eisner, and J. J. Hurst, *Chem. Ind.*, **1963**, 24.

80. Epstein, J., *Anal. Chem.*, **19**, 272 (1947).

81. van Es, T., and B. Staskun, *J. Chem. Soc.*, **1965**, 5775.

82. Evans, T. W., and W. M. Dehn, *J. Am. Chem. Soc.*, **52**, 3647 (1930).

83. Fearon, W. R., *Analyst*, **71**, 562 (1946).

84. Fearon, W. R., *Biochem. J.*, **33**, 902 (1939).

85. Feigl, F., *Spot Tests*, Vol. 2, Elsevier, New York, 1954, p. 197.

86. Feigl, F., and V. Anger, *Analyst*, **91**, 282 (1966).

87. Feigl, F., and V. Gentil, *Mikrochim. Acta*, **1**, 44 (1959).

88. Feigl, F., V. Gentil, and E. Jungreis, *Mikrochim. Acta*, **1**, 47 (1959).

89. Feigl, F., D. Goldstein, and E. Libergott, *Chemist-Analyst*, **53**, 37 (1964).

90. Felton, D. G. I., and S. F. D. Orr, *J. Chem. Soc.*, **1955**, 2170.

91. Figeys-Fauconnier, M., H. Figeys, G. Geuskens, and J. Nasielski, *Spectrochim. Acta*, **18**, 689 (1962).

92. Finkel'shtein, A. I., B. I. Sukhorukov, T. M. Kornienko, and Y. I. Mushkin, *Mater. Tret'ego Ural. Soveshch. Spektrosk., Inst. Sverdl.*, 168 (1960); through *Chem. Abstr.*, **58**, 9633 (1963).

93. Flett, M. St. C., *Spectrochim. Acta*, **18**, 1537 (1962).

94. Fosse, R., P. Hagene, and R. Dubois, *Compt. Rend.*, **179**, 214 (1924).

95. Franklin, T. C., and R. D. Sothern, *J. Phys. Chem.*, **58**, 951 (1954).

96. Frankoski, S. P., and S. Siggia, *Anal. Chem.*, **44**, 2078 (1972).

97. Freifelder, M., *J. Am. Chem. Soc.*, **82**, 2386 (1960).

98. French Pat. 1,384,209 (Jan. 4, 1965), L. Ugi et al.; through *Chem. Abstr.*, **63**, 6924 (1965).

99. French Pat. 1,397,573 (Apr. 30, 1965); through *Chem. Abstr.*, **63**, 6923 (1965).

100. Friederich, A., E. Kuhaas, and R. Schurch, *Z. Physiol. Chem.*, **216**, 68 (1933).

101. Fritz, J. S., and S. S. Yamamura, *Anal. Chem.*, **29**, 1079 (1957).

102. Galat, A., *J. Am. Chem. Soc.*, **70**, 3945 (1948).

103. Garby, C. D., *Ind. Eng. Chem.*, **17**, 266 (1925).

104. Garman, R. G., and C. N. Reilly, *Anal. Chem.*, **34**, 600 (1962).

105. Gautier, A., *Ann. Chim. Phys.*, **17**(4), 203 (1869).

106. German Pat. 877,302 (May 21, 1953), P. Kurtz, H. Bretschneider, and H. Möller; through *Chem. Abstr.*, **50**, 8708 (1956).

107. Gibson, M. E., Jr., and H. Heidner, *Anal. Chem.*, **33**, 1825 (1961).

108. Gillis, R. G., *J. Org. Chem.*, **27**, 4103 (1962).

109. Gillis, R. G., and J. L. Occolowitz, *J. Org. Chem.*, **28**, 2924 (1963).

110. Gillis, R. G., and J. L. Occolowitz, *Spectrochim. Acta*, **19**, 873 (1963).

111. Gray, L. S., L. D. Metcalfe, and S. W. Leslie, *Gas Chromatogr., Int. Symp.*, **4**, 203 (1963); through *Chem. Abstr.*, **60**, 15116 (1964).

112. Grube, G., and J. Kruger, *Z. Angew. Chem.*, **27**, 326 (1914).

113. Grunfeld, M., *Ann. Chim.*, **20**, 304 (1933).

114. Guccione, E., *Chem. Eng.*, **72**(5), 150 (1965).

115. Guilbault, G. C., and D. N. Kramer, *Anal. Chem.*, **38**, 834 (1966).

116. Guilbault, G. C., and R. J. McQueen, *Anal. Chim. Acta*, **40**, 251 (1968).

117. Guillemard, H., *Ann. Chim. Phys.*, **14**, 327 (1908).

118. Guillemard, H., *Ann. Chim. Phys.*, **14**, 330 (1908).

119. Guillemard, H., *Bull. Soc. Chim.*, **1**(4), 196 (1907).

120. Hager, G., and J. Kern, *Z. Angew. Chem.*, **29**, 309 (1916).

121. Hanker, J. S., R. M. Gamson, and H. Klapper, *Anal. Chem.*, **29**, 879 (1957).

122. Hantzsch, A., *Berichte*, **64**, 661 (1931).

123. Hantzsch, A., *Berichte*, **64**, 674 (1931).

124. Harger, R. N., *Ind. Eng. Chem.*, **12**, 1107 (1920).

125. Hartung, W. H., *J. Am. Chem. Soc.*, **50**, 3370 (1928).

126. Haslam, J., and G. Newlands, *Analyst*, **80**, 50 (1955).

127. Heilmann, R., J. M. Bonnier, and G. de Gaudemais', *Compt. Rend.*, **244**, 1787 (1957).

128. Hene, E., and A. Haaren, *Z. Angew. Chem.*, **31**, 129 (1918).

129. Hertler, W. R., and E. J. Corey, *J. Org. Chem.*, **23**, 1221 (1958).

130. Hill, H. C., and R. I. Reed, *Tetrahedron*, **20**, 1359 (1964).

131. Hofmann, A. W., *Annalen*, **144**, (1867).

132. Hofmann, E., and A. Wuensch, *Naturwissenschaften*, **45**, 338 (1958).

133. Horrocks, W. D., and R. H. Mann, *Spectrochim. Acta*, **19**, 1375 (1963).

134. Hoseney, R. C., and K. F. Finney, *Anal. Chem.*, **36**, 2145 (1964).

135. Howells, H. P., and J. G. Little, *J. Am. Chem. Soc.*, **54**, 2451 (1932).

136. Huber, W., *Z. Anal. Chem.*, **197**, 236 (1963).

137. Hudson, J. R., and J. R. A. Pollock, *Analyst*, **83**, 374 (1957).

138. Huhn, A., *Chem. Ber.*, **19**, 2404 (1886).

139. Hunig, S., H. Lehmann, and G. Grimmer, *Annalen*, **579**, 77 (1953).

140. Inaba, H., *Jap. Anal.* (Eng. trans.), **3**, 107 (1954); through *Chem. Abstr.*, **48**, 9866 (1954).

141. Janardhan, P. B., *J. Sci. Ind. Res.* (*India*), **12B**, 183 (1953); through *Chem. Abstr.*, **47**, 10378 (1953).

142. Janz, G. J., and S. S. Danyluk, *J. Am. Chem. Soc.*, **81**, 3846, 3850 (1959).

143. Janz, G. J., and N. E. Duncan, *Anal. Chem.*, **25**, 1410 (1953).

144. Jarvie, J. M. S., R. A. Osteryoung, and G. J. Janz, *Anal. Chem.*, **28**, 264 (1956).

145. Jesson, J. P., and H. W. Thompson, *Proc. Roy. Soc.*, **268A**, 68 (1962).

146. Jesson, J. P., and H. W. Thompson, *Spectrochim. Acta*, **13**, 217 (1958).

147. Johnson, E., *Ind. Eng. Chem.*, **13**, 533 (1921).

148. Johnson, H. W., Jr., and P. H. Daughhetes, Jr., *J. Org. Chem.*, **29**, 246 (1964).

149. Kaabak, L. V., A. P. Tomilov, and S. O. Varshavskii, *J. Gen. Chem. USSR* (Eng. trans.), **34**, 2121 (1964).

150. Kanai, R., and K. Hashimoto, *Ind. Health* (*Jap.*), **3**, 47 (1965).

151. Kappen, H., *Landw. Vers.-Sta.*, **70**, 445 (1909); through *Chem. Abstr.*, **3**, 2483 (1909).

152. Kappen, H., *Z. Angew. Chem.*, **31**, 31 (1918).

153. Kawamura, F., and S. Suzuki, *J. Chem. Soc. Jap., Ind. Chem. Sect.*, **55**, 476 (1952); through *Chem. Abstr.*, **48**, 3167 (1954).

154. Kazitsyna, L. A., B. V. Lokshin, and O. A. Glushkova, *Zh. Obschch. Khim.*, **32**, 1391 (1962).

155. Kelen, G. P. van der, and P. J. DeBievre, *Colloquium Spectroscopicum Internationale, Luzern, Sept. 14–19,* 1959, pp. 272–277.

156. Kelso, A. G., and A. B. Lacey, *J. Chromatogr.*, **28**, 156 (1965).

157. Kent, R. E., and S. M. McIlvain, *Organic Syntheses,* Vol. 25, John Wiley, New York, 1945, p. 61.

158. Kenyon, R. L., D. V. Stingley, and H. P. Young, *Ind. Eng. Chem.*, **42**, 202 (1950).

159. Kilpatrick, M. L., *J. Am. Chem. Soc.*, **69**, 40 (1947).

160. Kitson, R. E., and N. E. Griffith, *Anal. Chem.*, **24**, 334 (1952).

161. Klages, F., and K. Mönkemeyer, *Chem. Ber.*, **85**, 126 (1952).

162. Kolbe, H., *Liebig's Ann.*, **98**, 344 (1856).

163. Korinfskii, A. A., *Zavod. Lab.*, **11**, 816 (1945); through *Chem. Abstr.*, **40**, 7071 (1946).

164. Kourovtzeff, K., *Rev. Fr. Corps Gras*, **13**, 271 (1966); through *Chem. Abstr.*, **65**, 2480 (1966).

165. Krieble, V. K., and C. I. Noll, *J. Am. Chem. Soc.*, **61**, 560 (1939).

166. Kühle, E., B. Anders, E. Klauke, H. Tarnow, and G. Zumach, *Angew. Chem., Int. Ed. Eng.*, **8**, 20 (1969).

167. Kühle, E., B. Anders, and G. Zumach, *Angew. Chem., Int. Ed. Eng.*, **1**, 649 (1962).

168. Kuroki, N., and K. Konishi, *Kogyo Kagaku Zasshi*, **59**, 619 (1956); through *Chem. Abstr.*, **52**, 4988 (1958).

169. Lapworth, A., and R. H. F. Manske, *J. Chem. Soc.*, **1930**, 1976.

170. Laviron, E., *Compt. Rend.*, **250**, 3671 (1960).

171. Leandri, G., and D. Spinelli, *Bull. Sci. Fac. Chim. Ind. (Bologna)*, **15**(3), 90 (1957).

172. LeMoal, H., R. Carrie, and M. Bargain, *Compt. Rend.*, **251**, 2541 (1960).

173. Lewis, R. N., and P. V. Susi, *J. Am. Chem. Soc.*, **74**, 840 (1952).

174. Liler, M., and D. Kosanovic, *J. Chem. Soc.*, **1958**, 1084.

175. Line, W. E., H. M. Hickman, and R. A. Morrissette, *J. Am. Oil Chem. Soc.*, **36**, 20 (1959).

176. Livengood, S. M., and C. A. Johnson, "Symposium on Analytical Methods for Surfactants," *Chem. Spec. Mfrs. Assoc. Proc.*, December 1957, p. 123.

177. Ludzack, F. J., R. B. Schaffer, R. N. Bloomhoff, and M. B. Ettinger, *J. Sewage Ind. Wastes*, **31**, 33 (1959).

178. Lysyj, I., *Anal. Chem.*, **32**, 771 (1960).

179. Manousek, O., and P. Zuman, *Chem. Commun.*, **1965**, 158.

180. Marqueyrol, P. O., and L. Desvergnes, *Ann. Chim. Anal.*, **2**, 164 (1920).

181. Matsubara, I., *Bull. Chem. Soc. Jap.*, **35**, 27 (1962).

182. McBride, J. J., Jr., and H. C. Beachell, *J. Am. Chem. Soc.*, **74**, 5247 (1952).

183. McKusick, B. C., *J. Am. Chem. Soc.*, **80**, 2806 (1958).

184. McLafferty, F. W., *Anal. Chem.*, **34**, 26 (1962).

185. McMaster, L., and F. B. Langreck, *J. Am. Chem. Soc.*, **39**, 103 (1917).

186. Mekhtiev, S. D., R. G. Pisaev, R. Kasimov, and Y. G. Kamarov, *Azer. Khim. Zh.*, **4**, 70 (1965).

187. Meakin, G. D., and R. J. Moss, *J. Chem. Soc.*, **1957**, 993.

188. Mendius, O., *Liebig's Ann.*, **121**, 129 (1862); through *Chem. Centralbl.*, **33**, 415 (1862).

189. Metcalfe, L. D., *J. Gas Chromatogr.*, **1**, 7 (1963).

190. Milks, J. E., and R. H. Janes, *Anal. Chem.*, **28**, 846 (1956).

191. Miller, F. A., O. Sala, P. Devlin, J. Overend, E. Lippert, W. Luder, H. Moser, and J. Varchim, *Spectrochim. Acta*, **20**, 1233 (1964).

192. Millich, F., *Chem. Rev.*, **72**, 101 (1972).

193. Mitchell, J., Jr., and W. Hawkins, *J. Am. Chem. Soc.*, **67**, 777 (1945).

194. Monnier, R., *Chem. Ztg.*, **35**, 601 (1911); through *Chem. Abstr.*, **5**, 2793 (1911).

195. Mosettig, E., in R. Adams, Ed., *Organic Reactions*, Vol. VIII, John Wiley, New York, 1951, Ch. 5.

196. Mugnaini, E., and G. Gambelli, *Chim. Ind. (Milan)*, **45**, 44 (1963); through *Chem. Abstr.*, **60**, 15146 (1964).

197. Murata, Y., and T. Takanishi, *Kogyo Kagaku Zasshi*, **64**, 676 (1961); through *Chem. Abstr.*, **57**, 2824 (1962).

198. Nakanishi, K., *Infrared Absorption Spectroscopy—Practical*, Holden-Day, San Francisco, 1962.

199. Nef, I. U., *Ann. Chem.*, **270**, 267 (1892).

200. Nelson, J. P., L. E. Peterson, and A. J. Milun, *Anal. Chem.*, **33**, 1882 (1961).

201. Niederl, J. B., and V. Niederl, *Organic Quantitative Microanalysis*, 2nd ed., John Wiley, New York, 1942.

202. Nystrom, R. F., and W. G. Brown, *J. Am. Chem. Soc.*, **70**, 3738 (1948).

203. Ogura, K., *Mem. Coll. Sci. Kyoto Imp. Univ.*, **12A**, 339 (1929); through *Chem. Abstr.*, **24**, 2060 (1930).

204. Ohta, M., *Bull. Chem. Soc. Jap.*, **17**, 485 (1942).

205. Olin, J. F., and T. B. Johnson, *Rec. Trav. Chim.*, **50**, 72 (1931).

206. Otsuka, S., K. Mori, and K. Yamagami, *J. Org. Chem.*, **31**, 4170 (1966).

207. Papa, D., E. Schwenk, and B. Whitman, *J. Org. Chem.*, **7**, 587 (1942).

208. Parimskii, A. I., and I. K. Shelomov, *Malob. Zhir. Prom.*, **30**, 28 (1964); through *Chem. Abstr.*, **61**, 11339 (1964).

209. Passerini, M., *Gazz. Chim. Ital.*, **54**, 529 (1924).

210. Pawlik, A., *Z. Pflanzenernaehr. Dueng.*, **89**, 181 (1960); through *Chem. Abstr.*, **8**, 2184 (1961).

211. Peover, M. E., *Trans. Faraday. Soc.*, **58**, 2370 (1962).

212. Perotti, R., *Gazz. Chim. Ital.*, **35**, 228 (1905).

213. Pertusi, C., and E. Gastaldi, *Chem. Ztg.*, **37**, 609 (1913).

214. Pesez, M., J. Bartos, and J-C. Lampetaz, *Bull. Soc. Chim. Biol.*, **1962**, 719.

215. Petersen, G. W., and H. H. Radke, *Ind. Eng. Chem., Anal. Ed.*, **16**, 63 (1944).

216. Pinck, L. A., *Ind. Eng. Chem.*, **17**, 459 (1925).

217. Platnova, M. N., *Zh. Anal. Khim.*, **3**, 310 (1956).

218. Popp, F. D., and H. P. Schultz, *Chem. Rev.*, **62**, 19 (1962).

219. Poziomek, E. J., and E. J. Crabtree, *Anal. Lett.*, **1**, 929 (1968).

220. Prajsnar, B., A. Maslankiewicz, and Z. Najzarek, *Chem. Anal. (Warsaw)*, **10**, 1221 (1965); through *Chem. Abstr.*, **65**, 2990 (1966).

221. Pyryalova, P. S., and E. N. Zil'berman, *Izv. Vyssh. Uchebn. Zaved., Khim. Khim. Tekhnol.*, **8**, 82 (1965); through *Chem. Abstr.*, **63**, 6913 (1965).

221a. Putiev, Yu. P., and Yu. T. Tashpulatov, *Uzbeksk. Khim. Zh.*, **10**, 41 (1966); through *Chem. Abstr.*, **66**, 20069 (1967).

222. Radziszewski, B., *Berichte*, **17**, 1289 (1884).

223. Ranny, M., J. Prachor, and J. Novak, *Prum. Potravin.*, **14**, 211 (1963).

224. Rapi, G., and G. Sbrana, *Chem. Commun.*, **1968**, 128.

225. Reilley, C. N., and L. J. Poppa, *Anal. Chem.*, **34**, 801 (1962).

226. Rieger, P. H., I. Bernal, W. H. Reinmuth, and G. K. Fraenkel, *J. Am. Chem. Soc.*, **85**, 683 (1963).

227. Robertson, E. B., B. D. Sykes, and H. B. Dunford, *Anal. Biochem.*, **9**, 158 (1964).

228. Rose, E. L., and H. Ziliotto, *Anal. Chem.*, **17**, 211 (1945).

229. Rosenthaler, L., *Pharm. Acta Helv.*, **4**, 196 (1929).

230. Rothe, W., *Pharmazie*, **5**, 190 (1950).

231. Rovira, S., and L. Palfray, *Compt. Rend.*, **211**, 396 (1940).

232. Sandmeyer, T., *Berichte*, **17**, 1633, 2650 (1884).

233. Sandonnini, C., and S. Bezzi, *Gazz. Chim. Ital.*, **60**, 693 (1930).

234. Sato, M., M. Sato, T. Fujisawa, and S. Sato, *J. Electrochem. Soc. Jap.*, **23**, 238 (1955); through *Anal. Abstr.*, **3**, 585 (1956).

235. Schenk, G. H., and M. Ozolins, *Talanta*, **8**, 109 (1961).

236. Schleyer, P. von R., and A. Allerhand, *J. Am. Chem. Soc.*, **84**, 1322 (1962).

237. Schleyer, P. von R., and A. Allerhand, *J. Am. Chem. Soc.*, **85**, 866 (1963).

238. Schmidt, E., and W. Striewsky, *Chem. Ber.*, **75**, 286 (1940).

239. Schmidt, E., F. Hitzler, E. Lahda, R. Herbeck, and M. Pezzati, *Berichte*, **71B**, 1933 (1938).

240. Schmidt, E., D. Ross, J. Kittl, H. H. vôn Dusel, and K. Wamsler, *Annalen*, **612**, 11 (1957).

241. Schurz, J., H. Sah, and A. Ullrich, *Z. Phys. Chem. (Frankfrut)*, **21**, 185 (1959).

242. Seiffarth, V. K., W. Wohlrabe, and H. W. Ardelt, *Z. Anal. Chem.*, **217**, 345 (1966).

243. Sensi, P., and G. G. Gallo, *Gazz. Chim. Ital.*, **85**, 224 (1955).

244. Sevast'yonova, I. G., and A. P. Tomilov, *J. Gen. Chem. USSR*, (Eng. trans.), **33**, 2741 (1963).

245. Sevast'yonova, I. G., and A. P. Tomilov, *Sov. Electrochem.*, **2**, 1026 (1966).

246. Sevast'yonova, I. G., and A. P. Tomilov, *Sov. Electrochem.*, **3**, 563 (1967).

247. Sherk, K. W., M. V. Augur, and M. D. Soffer, *J. Am. Chem. Soc.*, **67**, 2239 (1945).

248. Shriner, R. L., R. C. Fuson, and D. Y. Curtin, *Systematic Identification of Organic Compounds*, 4th ed., John Wiley, New York, 1956.

249. Shriner, R. L., and T. A. Turner, *J. Am. Chem. Soc.*, **52**, 1267 (1930).

250. Sieverts, A., and A. Hermsdorf, *Z. Angew Chem.*, **34**, 3 (1921).; through *Chem. Abstr.*, **15**, 1269 (1921).

251. Siggia, S., and J. G. Hanna, *Anal. Chem.*, **36**, 228 (1964).

252. Siggia, S., J. G. Hanna, and N. M. Serencha, *Anal. Chem.*, **36**, 227 (1964).

253. Siggia, S., and C. R. Stahl, *Anal. Chem.*, **27**, 550 (1955).

254. Skinner, M. W., and H. W. Thompson, *J. Chem. Soc.*, **1955**, 487.

255. Smirnov, Yu, D., S. K. Smirnov, and A. P. Tomilov, *Zh. Org. Khim.*, **4**, 216 (1968).

256. Smith, L. I., and E. R. Rogier, *J. Am. Chem. Soc.*, **73**, 4047 (1951).

257. Soloway, S., and A. Lipschitz, *Anal. Chem.*, **24**, 898 (1952).

258. Soltys, A., *Mikrochemie*, **20**, 107 (1936).

259. Spillane, L. J., *Anal. Chem.*, **24**, 586 (1952).

260. Stafford, C., Jr., and P. E. Toren, *Anal. Chem.*, **31**, 1687 (1959).

261. Staskun, B., and T. van Es, *J. Chem. Soc. (C)*, **1956**, 4695.

262. Stephen, H., *J. Chem. Soc.*, **1925**, 1874.

263. Stephen, H., and T. Stephen, *J. Chem. Soc.*, **1956**, 4695.

264. Strause, S. F., and E. Dyer, *Anal. Chem.*, **27**, 1906 (1955).

265. Sukhorukov, B. I., and A. I. Finkel'shtein, *Izv. Akad. Nauk. SSSR, Ser. Fiz.* (Eng. trans.), **23**, 1230 (1959).

266. Sullivan, M. X., and W. C. Hess, *J. Am. Chem. Soc.*, **58**, 47 (1936).

267. Szabo, Z. G., and L. Bartha, *Acta. Chim. Hung.*, **1**, 116 (1951); through *Chem. Abstr.*, **45**, 10123 (1951).

268. Szabo, Z. G., and L. Bartha, *Magyar Kim. Foly.*, **57**, 84 (1951); through *Chem. Abstr.*, **45**, 10133 (1951).

269. Szewczuk, A., *Chem. Anal. (Warsaw)*, **4**, 971 (1959); through *Chem. Abstr.*, **54**, 16255 (1960).

270. Takei, S., *Jap. Analyst* (Eng. trans.), **3**, 243 (1954); through *Anal. Abstr.*, **3**, 584 (1956).

271. Takei, S., and T. Kato, *Technol. Rep. Tohoku Univ.*, **18**, 159 (1954); through *Chem. Abstr.*, **56**, 6639 (1962).

272. Takimoto, M., and K. Koeda, *Kogyo Kagaku Zasshi*, **63**, 799 (1960); through *Chem. Abstr.*, **56**, 6639 (1962).

273. Tamele, M., C. J. Ott, K. E. Marple, and G. Hearne, *Ind. Eng. Chem.*, **33**, 115 (1941).

274. Taramasso, M., and A. Guerra, *J. Gas. Chromatogr.*, **3**, 138 (1965).

275. Terent'ev, A. P., S. I. Obtemperanskaya, and M. M. Buzlanova, *Vestn. Moskov Univ.*, **11**(1), 187 (1956); through *Chem. Abstr.*, **51**, 11935 (1957).

276. Thesing, J., D. Witzel, and A. Brehm, *Angew. Chem.*, **68**, 425 (1956).

277. Tomilov, A. P., L. V. Kaabak, and S. L. Varshavskii, *J. Gen. Chem. USSR*, **33**, 2737 (1963).

278. Tomilov, A. P., L. V. Kaabak, and S. L. Varshavskii, *Khim. Prom.*, **1962**, 562; through *Chem. Abstr.*, **59**, 1282 (1963).

279. Travagli, G., *Gazz. Chim. Ital.*, **87**, 673 (1957).

280. Trofimenko, S., and J. W. Sease, *Anal. Chem.*, **30**, 1433 (1958).

281. Tschugaeff, L. A., *Berichte*, **35**, 2482 (1902).

282. Turner, L., *J. Chem. Soc.*, **1956**, 1686.

283. Tyler, W. P., D. W. Beesing, and S. J. Averill, *Anal. Chem.*, **26**, 674 (1954).

284. Ugi, I., W. Betz, U. Fetzer, and K. Offermann, *Chem. Berichte*, **94**, 2814 (1961).

285. Ugi, I., and F. Bodesheim, *Chem. Ber.*, **94**, 1157 (1961).

286. Ugi, I., U. Fetzer, U. Eholzer, H. Knupfer, and H. Offermann, *Angew. Chem.*, **77**, 492 (1965).

287. Ugi, I., R. Meyr, U. Fetzer, and C. Steinbrückner, *Angew. Chem.*, **71**, 386 (1959).

288. Ugi, I., and R. Meyr, *Chem. Ber.*, **93**, 239 (1960).

289. Deleted.

290. Ugi, I. Ed., *Isonitrile Chemistry*, Academic Press, New York, 1971.

291. U.S. Pat. 2,426,014 (Aug. 19, 1947), W. F. Gresham; through *Chem. Abstr.*, **41**, 7409 (1947).

292. U.S. Pat. 2,715,138 (Aug. 9, 1955), G. B. Crane; through *Chem. Abstr.*, **50**, 7126 (1956).

293. U.S. Pat. 2,732,397 (Jan. 24, 1956), D. C. Hull; through *Chem. Abstr.*, **50**, 12097 (1956).

294. U.S. Pat. 2,794,043 (May 28, 1957), J. E. Jansen and M. E. Roha; through *Chem. Abstr.*, **51**, 16514 (1957).

295. U.S. Pat. 2,806,872 (Sept. 17, 1957), N. J. Kartinos and J. B. C. Normington; through *Chem. Abstr.*, **52**, 3354 (1958).

296. U.S. Pat. 3,567,382 (Mar. 2, 1971), E. V. Crabtree, E. J. Poziomek, and D. J. Hoy; through *Chem. Abstr.*, **74**, 119902 (1971).

297. Varner, J. E., W. A. Bulen, S. Vanecko, and R. C. Burrell, *Anal. Chem.*, **25**, 1528 (1953).

298. Vasilescu, V., *Fette Seifen Anstrichm.*, **63**, 132 (1961); through *Chem. Abstr.*, **55**, 14941 (1961).

299. Venkateswarlu, K., and C. Balasubramanian, *Proc. Indian Acad. Sci.* (*A*), **51**, 151 (1960).

300. Volke, J., and J. Holubek, *Collect. Czech. Chem. Commun.*, **28**, 1597 (1963).

301. Volke, J., R. Kubicek, and F. Santory, *Collect. Czech. Chem. Commun.*, **25**, 1510 (1960).

302. Voronkov, M. G., *Latv. PSR Zinatnu Akad. Vestis Kim. Ser.*, **1961**, 25; through *Chem. Abstr.*, **57**, 9324 (1963).

303. Wallenfels, K., and D. Friederich, *Tetrahedron Lett.*, **1963**, 1223.

304. Walton, R. A., *Quart. Rev.*, **19**, 126 (1965).

305. Walton, R. A., *Spectrochim. Acta*, **21**, 1795 (1965).

306. Webb, J. L., and A. H. Corwin, *J. Am. Chem. Soc.*, **66**, 1456 (1944).

307. Weimer, R. F., and J. M. Prausnitz, *Spectrochim. Acta*, **22**, 77 (1966).

308. Weissert, N. H., and R. A. Coelho, *J. Gas Chromatogr.*, **5**, 160 (1967).

309. West, B. L., *J. Am. Chem. Soc.*, **42**, 1656 (1920).

310. Weith, W., *Chem. Ber.*, **6**, 210 (1873).

311. Weith, W., *Chem. Ber.*, **7**, 10 (1874).

312. Wheeler, O. H., *J. Org. Chem.*, **26**, 4755 (1961).

313. White, L. M., and M. C. Long, *Anal. Chem.*, **23**, 363 (1951).

314. Whitehurst, D. H., and J. B. Johnson, *Anal. Chem.*, **30**, 1332 (1958).

315. Whitmore, F. C., H. S. Mosher, R. R. Adams, R. B. Taylor, E. C. Chapin, C. Weisel, and W. Yanko, *J. Am. Chem. Soc.*, **66**, 725 (1944).

316. Wiberg, K. B., *J. Am. Chem. Soc.*, **75**, 3961 (1953).

317. Wiberg, K. B., *J. Am. Chem. Soc.*, **77**, 2519 (1955).

318. Wiemann, J., and M. L. Bouguerra, *Compt. Rend.*, **265**, 751 (1967).

319. Williams, R. L., *J. Chem. Phys.*, **25**, 656 (1956).

320. Wimer, D. C., *Anal. Chem.*, **30**, 77 (1958).

321. Wurz, D. H., and N. E. Sharpless, *Anal. Chem.*, **21**, 1446 (1949).

322. Yamada, T., and Y. Sakai, *Denki Kagaku*, **29**, 852 (1961); through *Chem. Abstr.*, **62**, 7110 (1965).

323. Yukhnovski, I., B. Iordanov, and M. Agova, *Izv. Inst. Org. Khim. Bulgar, Akad. Nauk*, **2**, 13 (1965); through *Chem. Abstr.*, **64**, 13563 (1966).

324. Zakharkin, L. I., D. N. Maslin, and V. V. Gavrilenko, *Izv. Akad. Nauk. SSSR, Ser. Khim.*, **1964**(8), 1511; through *Chem. Abstr.*, **64**, 17643 (1966).

325. Zarembo, J. E., and M. M. Watt, *Microchem. J., Symp. Ser.*, **2**, 591 (1962).

326. Zeeh, B., *Org. Mass Spectrom.*, **1**, 315 (1968).

327. Zetsche, F., and A. Fredrich, *Chem. Berichte*, **72**, 363 (1939).

328. Zil'berman, E. N., *Usp. Khim. (Russian Chem. Rev.)* (in Eng.), **29**, 331 (1960).

# INDEX